Restoring the Nation's Marine Environment

Restoring the Nation's Marine Environment

Edited by

Gordon W. Thayer
National Oceanic and
Atmospheric Administration

A Maryland Sea Grant Book
College Park, Maryland

Published by the Maryland Sea Grant College.

Publication of this book is the result of a collaborative effort between the University of Maryland Sea Grant College Program and the U.S. Department of Commerce's National Oceanic and Atmospheric Administration/ National Marine Fisheries Service, pursuant to NOAA Award No. NA16RG0458. The views expressed herein are those of the authors and do not necessarily reflect the view of NOAA or any of its subagencies.

The maps that appear on pages 145 and 148-156 are reproduced from *Coral Reefs of the World*, Volumes 1 and 3, with the permission of the United Nations Environment Programme.

Book design by Sandy Harpe
Typography by Charles Fletcher

Maryland Sea Grant College
Publication UM-SG-TS-92-06
Library of Congress Card Catalog Number: 92-060721
ISBN 0-943676-57-6

For information on Maryland Sea Grant books, contact:

Maryland Sea Grant College
Skinner Building
University of Maryland
College Park, Maryland 20742

Printed on recycled paper

Contents

Introduction, *Gordon W. Thayer*   1

Chapter 1. Restoring Cordgrass Marshes in Southern California   7
Joy B. Zedler
    Intertidal Salt Marshes   8
    Impacts and Responses of Southern California Marshes   16
    Restoration and Creation   23
    Future Needs and Directions   41

Chapter 2. Restoring Tidal Marshes in North Carolina
and France   53
Ernest D. Seneca and Stephen W. Broome
    Intertidal Marsh Habitat Type   54
    Impacts and Responses   55
    Restoration and Creation   58
    Future Needs and Directions   73

Chapter 3. Restoring Seagrass Systems in the United States   79
Mark S. Fonseca
    Overview of Habitat Type   80
    Impacts and Responses   83
    Restoration and Creation   86
    Future Research Directions and Needs   104

Chapter 4. Large-Scale Restoration of Seagrass Meadows   111
Hugh Kirkman
    Habitat   112
    Impacts and Responses   118
    Restoration and Creation   122

Protocols   133
Future Needs and Directions   135

Chapter 5. Restoring Coral Reefs with Emphasis
on Pacific Reefs   141
James E. Maragos
    Overview of Habitat Type   142
    Human Impacts on and Responses by Coral Reefs   164
    Response of Reef Habitat and Communities to Impact   173
    Restoration and Creation of Coral Reefs   174
    Additional Techniques that Could Be Pursued
      to Restore Coral Reefs   192
    Discussion of Current or Recommended
      Experimental Approaches That Should Be Pursued   199
    Establishing Priorities for Reef Restoration   205
    Future Needs and Directions   208

Chapter 6. Restoring Mangrove Systems   223
Gilberto Cintron-Molero
    Introduction   224
    Ecosystem Responses to Disturbance   230
    Major Causes of Habitat Loss and Alterations
      in the Region   239
    Restoration and Creation   243
    Future Needs and Direction   266

Chapter 7. Restoring Kelp Forests   279
David R. Schiel and Michael S. Foster
    Introduction   280
    Overview of Kelp Forests   282
    Impacts and Responses   288
    Restoration and Creation   301
    Restoration Techniques   312
    Guidelines for Future Restoration and Creation   319
    The Science of Kelp Forest Restoration   326

Chapter 8. Restoring Stream Habitats Affected
by Logging Activities   343
K.V. Koski
   Introduction   344
   Characteristics of Streams   349
   Habitat Requirements of Anadromous Salmonids   356
   Effects of Logging on Anadromous Salmonid Habitat   364
   Concept of Habitat Restoration   371
   Habitat Restoration Procedures   377
   Needs and Considerations   389

Chapter 9. The Columbia River: Fish Habitat Restoration
Following Hydroelectric Dam Construction   405
John G. Williams and Merritt E. Tuttle
   Introduction   406
   Past Habitat Degradation   409
   Habitat Restoration   412
   Legislation   416
   The Federal Investment   419
   Future Needs for Anadromous Fish Migration Habitat   419
   Satisfying These Needs Will Benefit
      the Northwest and the Nation   420

Chapter 10. Restoring Wetland Habitats in Urbanized
Pacific Northwest Estuaries   423
Charles A. Simenstad and Ronald M. Thom
   Introduction   424
   Estuarine Wetlands in the Pacific Northwest
      and Their Status   425
   Historic Habitat Loss   426
   Resource and Ecosystem Responses   429
   History and Experience of Estuarine
      Habitat Restoration   433
   Functional Approaches to Restoration "Ecotechnology"   460
   Future Needs and Directions   463
   Summary and Conclusions   466

Chapter 11. Restoring and Managing Disused Docks in
Inner City Areas   473
S.J. Hawkins, J.R. Allen, G. Russell, K.N. White,
K. Conlan, K. Hendry and H.D. Jones
    Introduction   475
    General Problems: A Nationwide Appraisal   477
    Case Study 1—Liverpool   486
    Case Study 2—Preston Dock   506
    General Solutions for Improvement of Disused Docks   515
    Value of Restored Dock Complexes   521

Chapter 12. Developing Prefabricated Reefs: An Ecological
and Engineering Approach   543
Daniel J. Sheehy and Susan F. Vik
    Introduction   544
    Restoration Requirements   545
    Systems Approach   548
    Ecological Design and Engineering
       for Restoration Applications   551
    Application of Designed Reef Modules   559
    Recommendations   571
    Conclusions   576

Chapter 13. The *Torrey Canyon* Oil Spill: Recovery of
Rocky Shore Communities   583
S.J. Hawkins and A.J. Southward
    Introduction   584
    Background Information on Rocky Shores   585
    Impacts on Rocky Shores   588
    The *Torrey Canyon*   591
    Lessons from the *Torrey Canyon* Oil Spill   612
    Wider Perspective   614
    Conclusions   619

Panel Discussion 1: What Is the United States Doing to
Restore Habitats?   633
    Panel Participants   635
    Question-and-Answer Session   661

Panel Discussion 2: What Have We Learned,
Where Are We Headed?   669
    Panel Participants   671
    Question-and-Answer Session   680

Index   697

The Science of Restoration: Status and Directions

Gordon W. Thayer
NOAA, National Marine Fisheries Service
Beaufort, North Carolina

Recent workshops and symposia on coastal fish habitat have indicated that the increasing loss of fish habitat due to unwise development, pollution and other human activities is the single largest long-term threat to the future viability of marine fisheries in the United States. For example, the National Oceanic and Atmospheric Administration annually evaluates between 8,000 and 10,000 proposals for development in estuarine and coastal habitats that have the potential to affect upwards of one million acres. Habitat loss and degradation, along with overharvesting and climate change, is one of the principal reasons for the decline in a number of living marine resources. These declines in both habitat and living marine resources have been documented in current peer-reviewed and symposium literature and the NOAA Symposium on Habitat Restoration, held in Washington, D.C., September 25-26, 1990, from which this book derives. As a direct consequence of the recognized importance of coastal habitats and the losses that are occurring in both living marine resources and their habitats, restoration, mitigation and habitat creation are receiving increasingly greater attention.

1

NOAA is the Federal trustee for coastal resources and their habitats, including wetlands. Under the Magnuson Act, Marine Mammal Protection Act, the Endangered Species Act, Superfund and Oil Pollution Act, NOAA is required to represent resource interests in habitat decisions, maintain and enhance fisheries, and restore habitats of living marine resources. Under Superfund, as amended, and the Oil Pollution Act of 1990, natural resource trustees, frequently NOAA, are authorized to act on behalf of the public to file claims for damages for injury to, destruction of, or loss of natural resources caused by releases of hazardous substances or discharges of oil, and the damages recovered must be used to restore, rehabilitate, or acquire the equivalent of such natural resources. These responsibilities are consistent with the philosophy of no net habitat loss.

Unfortunately, as noted earlier, coastal marine and estuarine habitats are being lost through natural and man-induced causes, the latter frequently the result of inconsistencies in the application of existing legislation. The goal of no net loss cannot be attained without active programs of habitat conservation and of restoration and creation that are based on sound scientific approaches. Technologies may exist to restore some habitats which, if done properly, have a chance of succeeding. However, creation, enhancement and restoration of coastal habitats involves more than just cultivating the vegetation, breaching dikes, transplanting corals or boulders, or nourishing beaches. Even where there have been documented successes (and these are few), there are problems that need to be addressed regarding our ability to restore functional attributes of habitats to the level of natural habitats. In the case of restored or created wetland habitats, the simple fact that they are green does not necessarily mean that they have developed into functional habitat providing services to living marine resources and to man. In fact, there is a growing body of literature to the contrary.

Wetland habitats provide water quality, hydrologic, and life support functions. We must strive for restoration of all of these functional values and not just pieces of habitats that have been previously altered or degraded. If functional values are not adequately replaced, we will fail in our efforts to enhance the productivity of living marine resources. It is essential that a comprehensive evaluation of restoration, mitigation and enhancement processes be initiated on a national scale. Research may demonstrate that the design criteria for projects need only be improved to approach functional levels of natural habitats. On the other hand, research may show that we cannot emulate nature as easily as has been assumed. If this latter point proves to be the case, then policies toward the mitigation process must become more conservative, and the application of restoration technologies in permitted developmental projects, hazardous waste and oil spill-related claims cases must take into consideration the need for research and/or pilot studies, and scientific monitoring to facilitate the potential need for mid-course corrections, and for the assessment of replacement of functional values or services. If these aspects are not incorporated into restoration plans it is highly probable that lost services will not be adequately restored and the system and public will receive less than desired by the developers of the restoration plan.

NOAA, with its stewardship for living marine resources, has both the responsibility and capability to conduct such an evaluation. NOAA has recognized the need to address the status of the science of restoration across a wide range of habitat types in different geographic regions for which it is trustee, particularly because NOAA has begun to pursue legal action against those accused of substantial damage to the natural resources of coastal areas. Need for this information provided impetus to the organization of the NOAA Symposium on Habitat Restoration. Its goal was to provide technical information to improve government leadership on restoration practices, i.e., to provide an information base from which decisions can be made on how best to approach restoration of NOAA trust habitats, including those restoration activities that may result from monies received for compensation of losses of natural resources under legal actions. We should not limit our application of this information, however, to cases in the legal arena. We should push appropriate coastal restoration and assessment practices on a national scale because this can be a major conservation measure that can enhance the productivity of living marine resources.

A panel of international experts from federal, university and private sector research and management programs was assembled for the technical session. They were asked to provide an overview of the state-of-the-art for restoration of a wide breadth of NOAA trustee habitat types. Each panel member was asked to identify the impacts to which the habitats have been subjected; describe the steps that have been made to restore the habitats; identify those areas where we need to gain further information; and, perhaps equally important, provide recommendations on how we could get this information accepted and incorporated into restoration planning efforts, i.e., how do we develop a restoration ethic. While not all authors are in agreement as to the success of restoration activities or even the need for active human involvement in the recovery process, there was a general consensus that restoration as a science is relatively young, remains experimental for many habitat types, and that, although in some instances the feasibility of restoring or creating habitats has been demonstrated, their functional value, stability, and resiliency are largely unknown. In addressing restoration projects, clear goals must be established and assessment of progress is necessary, frequently over long time periods. It was also evident that it is mandatory that those individuals responsible for developing habitat clean-up and remedial action plans must work in concert with those responsible for developing the restoration plans, because the former will dictate the feasibility of restoration and the technologies that can be applied.

Each of the papers has been refereed by at least two reviewers, and I want to thank both the referees and the authors for their contribution to the value of the Symposium on Habitat Restoration to future decision-making processes. Their willingness to address the issues, maintain time schedules, and provide the following documentation made my job that much easier. I am particularly grateful to five other individuals who contributed extensively to this book and to the symposium as a whole: Ann Manooch of the NMFS Beaufort Laboratory who worked closely with the authors and with me in the review and editing process; Commander Ted Lillestolen of the NMFS Protected Resources Office who worked with every facet of the symposium from start to finish; Victor Omelczenko of NOAA's National Sea Grant Office who was instrumental in the public relations aspect of the symposium and this publication; and Merrill Leffler and Sandy Harpe of the Maryland Sea Grant College who guided the book from manuscript through production.

Since this NOAA Restoration Symposium, the Under Secretary of Commerce has established a Damage Assessment and Restoration Program. This program is comprised of two professionally-staffed interrelated Centers. One is the Damage Assessment Center within the National Ocean Service that works with the Office of General Counsel to conduct damage assessments and bring claims against potentially responsible parties. The second is the Restoration Center within the National Marine Fisheries Service. This Center provides the institutional focus within NOAA required to identify and evaluate restoration methodologies for specific cases during the damage assessment process; uses recovered damages to restore the injured resources; and addresses research and development priorities within NMFS necessary for successful resource habitat restoration. This Damage Assessment and Restoration Program already has benefitted from the symposium and will continue to do so with the publication of this volume. The Symposium on Habitat Restoration of 1990 was the first in a series of symposia and workshops NOAA intends to convene that will deal with the restoration and damage assessment processes.

CHAPTER 1

Restoring Cordgrass Marshes in Southern California

Joy B. Zedler

Pacific Estuarine Research Laboratory (PERL)
San Diego State University, San Diego, California

Abstract

The salt marshes of southern California have largely been lost to urbanization. The remnants of natural habitat are continually disturbed, yet they still support rare and endangered species. Efforts to restore intertidal marshes have focused on revegetation with cordgrass (Spartina foliosa), but full ecosystem functioning has not yet been demonstrated. In San Diego Bay, an area of high-intertidal marsh that was excavated to create lower-intertidal marsh lacked functional equivalency with natural wetland remnants at 4-5 years of age. Several comparisons indicated less than 60% functional equivalency with an adjacent reference wetland. Understanding what causes functional inequivalency has led to new methods for improving restoration projects. Research is now underway to test the effectiveness of adding organic matter and nutrients to areas planted with cordgrass, and preliminary results indicate accelerated plant growth rates. Two kinds of research programs need to be funded: long-term studies of coastal restoration sites and development of new ecotechnological methods to accelerate the development of functional equivalency.

Intertidal Salt Marshes

Like all wetlands, the salt marsh ecosystem is a function of its hydrology. The two main hydrologic controls are tidal flushing and streamflow, which influence both the inundation and salinity regimes of the intertidal marsh soils. Tidal action is the environmental moderator—in the lower intertidal zone, it stabilizes soil moisture and salinity by daily or twice-daily inundation with seawater. Mudflats and sandflats occur at extreme low water, where exposure to the air is brief; salt marsh vegetation develops where soils are more often exposed than inundated—usually above mean sea level. Cordgrass (*Spartina* spp.) usually dominates the lower-marsh. Invertebrates (e.g., crabs, bivalves, ghost shrimp) and fishes (gobies) live in burrows in the channel banks and bottoms. The cordgrass marsh and intertidal flats are linked by tidal flows on a daily basis. Mobile animals (e.g., crabs, fishes) move into the tidal marsh to feed at high tide; detritus and algae float into the channel waters and may be consumed there; marsh birds move onto the flats to feed at low tide; nutrients move into the marsh for uptake by the algae and halophytes. Each habitat is important to the other, and a highly productive, interlinked food web results.

Tidal cordgrass marshes are highly valued coastal wetlands, because the associated estuaries support commercially valuable fishes and shellfishes. However, the salt marshes of the Pacific Coast are most valued for their support of habitat-dependent species and maintenance of biodiversity. Because so much of coastal marsh habitat has been lost to development, several species are now endangered with extinction.

The estimates of habitat loss of coastal wetlands in California range from 70 to over 90 percent. The most recent calculation of wetland loss—including both coastal and inland habitats—shows California leading the 50 states with 91% loss since the 1780s (Dahl 1990). E. O. Wilson, world authority on the relationship between habitat area and biodiversity on islands, predicts that a 90% loss in habitat area would produce a 50% decline in numbers of species (Wilson, public address, San Diego Natural History Museum, October 1990). Whether such a high rate of species loss has occurred in California's coastal wetlands may never be known. The State's current list of threatened and endangered species (California Dept. of Fish and Game 1989) includes four salt marsh plants, one salt marsh mammal, and nine coastal birds. Intertidal cordgrass is the preferred home of one of these—the light-footed clapper rail (*Rallus longirostris levipes*). Additional plants and several insects of coastal wetlands are regarded as threatened with extinction. Smaller plants (e.g., algae) and animals (e.g., mites, worms) have barely been catalogued, let alone evaluated for their rarity or declining populations. Many may have gone extinct without notice.

Geographic Setting of California's Cordgrass Marshes (including California, U.S.A., and Baja California, Mexico)

The cordgrass marshes of California are distinctive in their physical and biological characteristics. The climate is Mediterranean in nature, with long dry summers. The cool winters have low annual precipitation and highly variable rainfall events. High evaporation rates, especially during summer, lead to hypersaline soils, even though the region's semi-diurnal mixed tides inundate the marsh twice daily. In the San Diego area, evapotranspiration averages around 2 meters a year, while rainfall averages 0.3 m.

Because the coastal topography is rugged, many coastal wetlands have small watersheds. Except for San Francisco Bay, which drains 40% of the State of California (Conomos 1979), the bays and estuaries have limited freshwater inflow. Most North American estuaries develop a salinity gradient from the coastal inlet toward the inland extremes. In Mediterranean-type regions, however, there are few rainfall events during the year, and streamflows are more variable. As a result, channels are marine in character during most of the year. Occasional floods briefly lower the water salinity of channels from seawater (34 ppt) to brackish, making the system intermittently estuarine.

The California Coast has mixed, semi-diurnal tides.Each day there are two high tides of different magnitude and two low tides of different magnitude. Spring tide amplitudes are highest in January and June, with the lowest low tides during daytime in winter and during nighttime in June. The cordgrass marsh is usually wetted twice a day, with the higher tide producing a longer period of inundation. Estuaries with continuous tidal flushing, i.e., those that rarely if ever have been closed to tidal flushing, are in the minority. They include Bahía de San Quintín and Estero de Punta Banda in Mexico and Tijuana Estuary, San Diego Bay, Mission Bay, Upper Newport Bay, and Anaheim Bay in southern California.

Structure of the Cordgrass Marsh

Pacific cordgrass (*Spartina foliosa*) dominates the lower-marsh of southern California's continuously tidal wetlands. It occurs in only 7 of the region's 26 coastal wetlands (Zedler 1986). Cordgrass forms a canopy of about 1 m in height by late summer, with the most robust stems near tidal creeks. Just beyond the creek edges, where elevations are 10-20 cm higher, it co-occurs with pickleweed (*Salicornia virginica*). According to field and mesocosm experiments (Griswold 1988), pickleweed competes with cordgrass and reduces its density. Part of our understanding of cordgrass derives from comparative studies of wetlands that lack the species but support pickleweed. Pickleweed has a much broader ecological range, both within and among marshes—it occupies a broader elevation range, has a wider salinity tolerance, and can persist long after tidal flushing has been disrupted. Pickleweed occurs in all the region's coastal salt marshes. Cordgrass is clearly dependent on tidal flushing; it has been eliminated from at least one site (Los Peñasquitos Lagoon) that shifted from being continuously tidal to infrequently tidal sometime after 1939 (Zedler 1986).

In Baja California, two studies provide quantitative data on the cordgrass marsh. Ibarra-Obando and Escofet (1987) followed the decline of cordgrass marsh after tidal exclusion (by a dike), in comparison with tidal areas. Neuenschwander et al. (1979) described the marsh and transitional vegetation at Bahía de San Quintin. Elsewhere, the cordgrass marshes of San Francisco Bay have been described by Josselyn (1983). The most detailed work on pickleweed marsh dynamics is that of Onuf (1987).

A diverse animal community inhabits the lower intertidal marsh. The cordgrass canopy supports several arthropods that are not particularly aquatic in nature. Spiders and adults of several insect species can probably survive occasional inundation by high tides, but they seem to thrive where their host plants are tall enough to remain above water (K. Williams, San Diego State University, pers. comm.). The epibenthic community is supported by a productive algal mat that is dominated by several filamentous blue-green algae and a wide variety of diatoms (Zedler 1980, 1982). Within and among the mats of algae, microbes (including nitrogen-fixing bacteria) and small invertebrates make their living. The California horn snail, *Cerithidea californica*, is the most widespread and conspicuous invertebrate living on the marsh surface. Other species are much harder to locate. In a recent study of the epibenthic community, Rutherford (1989) trapped 45 species of invertebrates using litterbags anchored on the soil of a San Diego Bay cordgrass marsh. A tiny dipteran, *Pericoma* sp., was the most abundant organism, with polychaetes of the genus *Capitella*, the isopod, *Orchestia traskiana*, and the gastropod snail, *Assiminea californica*, less common. Animals living in the fine marsh muds include the fiddler crab, *Uca crenulata*, the yellow shore crab, *Hemigrapsus oregonensis*, and the longjaw mudsucker, *Gillichthys mirabilis*. Areas of sandier substrate would also have abundant bivalve clams, ghost shrimp and associated gobies (e.g., arrow goby, *Clevelandia ios*).

Habitat-Dependent Species

Several insects live on and in the cordgrass and may be restricted to *Spartina foliosa* marshes. Two tiny flies, *Incertella* sp.and *Cricotopus* sp., and a scale insect, *Haliaspis spartina*, live on the leaves. *Spartina foliosa* is also the host for the plant-hopper, *Prokelisia dolus*. The predacious beetle, *Coleomegilla fuscilabris*, in turn feeds on some of these herbivorous insects. A spider, *Tetragnatha laboriosa*, is well camouflaged against the cordgrass leaf, suggesting a tight association with the host plant.

The distribution of the light-footed clapper rail (*Rallus longirostris levipes*) coincides almost entirely with that of cordgrass. This large bird nests in cordgrass, using dead stems to construct the nest platform and live stems to weave a canopy over the nest, conferring protection from aerial predators. According to Jorgensen (1975), the rail nest is anchored to cordgrass stems, allowing it to float with rising tides but not be washed away. A tall grass canopy is no doubt very important for both predator avoidance and nest stability, especially during higher tides. The bird's foods include crabs, snails, and insects of the lower marsh and intertidal flats. While it has nested in marshes that lack cordgrass (e.g., Carpinteria Marsh) or have very little cordgrass (e.g., Mugu Lagoon), its largest populations are in wetlands with large areas of cordgrass (e.g., Upper Newport Bay). It is especially abundant in Baja California's extensive cordgrass marshes.

Several characteristic invertebrates and fishes are most abundant in tidal channel habitats; their association with cordgrass marshes is based on the consistently high water quality of fully tidal channels, compared to non-tidal lagoons. Where water salinities shift suddenly from marine to brackish, or where stagnant waters become anoxic during summer, such species are quickly killed (Nordby and Zedler, In press).

Functioning of the Cordgrass Marsh Ecosystem

Cordgrass marshes support vascular plant and algal growth all year in southern California, although growth rates are lower during winter. Studies have shown that the marsh is quite productive, especially of algae. Annual above ground productivity of cordgrass was measured at about 1 kg m^{-2}, which is similar to the August standing crop; this is approximately 220 gC m^{-2}yr^{-1} (Winfield 1980). Epibenthic algae added another 185-304 gC m^{-2}yr^{-1} (Zedler 1980). Algal productivity is high, relative to Atlantic and Gulf of Mexico cordgrass marshes, a consequence of hypersaline soils, lower cordgrass production, and more open canopies (ibid.). The main variables controlling seasonal growth of algae are temperature and light, which must also set the seasonal pattern of vascular plant growth. Interannual variations in rainfall and streamflow produce year-to-year variability in ecosystem functioning. Years with above-average rainfall have increased streamflows, which reduce water salinity and introduce nutrients. Vascular plant growth was stimulated by freshwater flooding and nutrient inflows in 1980, as indicated by the August standing crop of 1.4 kg m^{-2} (Table 1).

Covin's experiments (Covin and Zedler 1988) demonstrated that the cordgrass marsh is nitrogen limited, and that in areas of mixed pickleweed and cordgrass, pickleweed is the superior competitor for nitrogen. Phosphorus levels in leaf tissues do not indicate P-limitation (Langis et al. 1991). Zalejko (1989) found low levels of nitrogen fixation in one cordgrass marsh in San Diego Bay; if those results are typical for California wetlands, this helps to explain why nitrogen is limiting. Since seawater is low in nitrogen, supplies must come from the watershed. Results obtained by Winfield (1980), Covin and Zedler (1988), Fong and Rudnicki (data published as Figure 50 in Zedler and Nordby 1986), and Langis et al. (1991) all agree that levels of inorganic nitrogen in the channels and marsh soils are low, except during sewage spills or other inflow events. The system seems to receive its nitrogen in infrequent pulses, primarily from the watershed. Over time, nutrients and organic matter accumulate in the soil and plant biomass, but we do not know whether the primary mode of nutrient accumulation is the slow process of trapping inorganics from tidal waters or through pulsed events of flooding by nutrient-rich freshwater and rapid (though rare) uptake and storage.

The fate of materials produced in the salt marsh is less clear. Winfield (1980) measured decomposition rates and export of materials from the cordgrass marsh; he found rapid decomposition of cordgrass leaves in the wetter sites. The marsh showed a net export (105 gC m^{-2}yr^{-1}) in the form of dissolved, rather than particulate, organic carbon. Unfortunately, these detailed studies of productivity and transport have not been repeated in other cordgrass marshes of California. Dead plants quickly lose their soluble organic material to the water and soil. The dissolved organic carbon feeds microbes and perhaps invertebrate larvae. Bacteria and fungi attack the dead leaves and stems and transform them to detritus.

The salt marsh food web includes both detritus- and grazer-based food chains, with a variety of insects making use of the vascular plants, and numerous insects and benthic invertebrates feeding on the productive algal mats. Smaller grazers, detritivores, and omnivores are fed upon by larger omnivores (e.g., crabs) and carnivores, such as resident and migratory shorebirds and fishes. Birds that feed on the crabs and larger invertebrates include the willet (*Catoptrophorus semipalmatus*), marbled godwit (*Limosa fedoa*), long-billed curlew (*Numenius phaeopus*), and at least one migratory duck, the pintail (*Anas acuta*). Shorebird feeding is probably restricted to the mudflats where probing is easier (J. Boland, Pacific Estuarine Research Laboratory, pers. comm.). Birds of the mid- and upper-marsh habitats may well move into the cordgrass during low tide. The endangered Belding's Savannah sparrow, a bird of the mid-marsh pickleweed habitat, may feed on insects produced within the cordgrass.

Table 1. August biomass in the cordgrass marsh of Tijuana Estuary. Modified from Zedler and Nordby 1986.

Year	g m^{-2}	s.e.
1976	0.9	0.38
1978	1.0	0.38
1980	1.4	1.13

At the top of the food web are a few species that prey on bird eggs and chicks, e.g., raptors and snakes. Some of the organic material and smaller consumers move into the channels and support aquatic food chains. While there is some export of marsh foods to the nearshore environment, it is unlikely that the small, isolated wetlands of southern California have a major influence on coastal fisheries. The California halibut (*Paralichthys californicus*) is an exception. Kramer (1990) has shown that this species uses shoreline habitats (eelgrass beds) of the region's tidal wetlands as nursery habitat.

Consumption of plant foods by benthic invertebrates, insects, mammals, or birds has never been quantified. One recent study (C. Nordby, Pacific Estuarine Research Laboratory, unpubl. report) failed to catch fish on the salt marsh, despite repeated seasonal sampling during high tide. Since the flumes and nets were designed after those used in Atlantic coastal marshes, where thousands of fishes can be seined from the marshes after high tide, we conclude that the absence of fish use is a functional difference between tidal marshes of the two coasts.

Impacts and Responses of Southern California Marshes

Losses to Filling and Dredging

The loss of southern California's coastal wetland habitat is largely due to filling and dredging for urban developments. In the Los Angeles area, only a few tiny remnants of coastal wetland remain. One is Ballona Wetland, where most of the deeper water areas have become Marina del Rey, an affluent housing and boating development, but about 250 acres of salt marsh persist despite greatly reduced tidal flushing. Filling of salt marshes has slowed, but nearby coastal habitats are still in jeopardy. The Ports of Los Angeles and Long Beach propose to fill 2,400 acres of subtidal habitat within San Pedro Bay by the year 2020 at a cost of $4.8 billion (1987 dollars; San Pedro Bay Ports, Operations, Facilities and Infrastructure Requirements Study, April 1988). In San Diego Bay, plans to expand marina, residential, commercial, and recreational facilities abound, including a proposal to dredge a second entrance to the bay.

In Baja California, losses and disturbances have been much less severe, although the threats of development are accelerating in number and extent (S. Ibarra-Obando, CICESE, Ensenada; Juan Guzman, University of Baja California Sur, La Paz, pers. comm.). Estero de Punta Banda is close to San Diego and supports important reserves of one plant and several animal species that are endangered in the United States. In the past five years, a major hotel and dozens of houses have been built on the narrow sand dune that separates this Estero from the Pacific Ocean. The adjacent salt marsh has been filled and dredging of a boat channel has been proposed.

Altered Hydrology

Two basic types of hydrologic alterations have devastating effects on the intertidal salt marsh. These are decreased tidal influence and augmented freshwater inflows. In some locations, e.g., Buena Vista

Lagoon and San Elijo Lagoon, both alterations have occurred, and the salt marsh has been replaced by brackish cattail and bulrush marshes.

Reduced tidal flows have been a problem since the turn of the century, when roads were constructed across the coastal inlets. Constructing roads and bridges confines the inlet to a narrow location, and the tidal channel can no longer migrate across the floodplain. When fill is placed in wetlands, the tidal prism and flushing ability is reduced, allowing more sediment to accumulate in the wetland and even greater reductions in tidal prism. With less scouring action at the ocean inlet, sand accumulates and closure becomes likely. Tidal flushing diminishes or ceases until the next flood blasts through the sand berm. In some locations, the beaches have been starved of sand by upstream dams, and underlying cobbles are exposed. When storms move these cobbles into the inlet, tidal closure becomes more permanent, as the floods required to displace cobbles occur only rarely.

Tidal flows have also been deliberately excluded to use the marshlands as salt evaporation ponds (e.g., south San Diego Bay, south San Francisco Bay), as duck-attracting ponds for hunters (e.g., Bolsa Chica), and as pastures (e.g., along San Francisco Bay and Elkhorn Slough). In oil-producing areas, dikes were built to eliminate tidal action around oil wells (e.g., Bolsa Chica). In Estero de Punta Banda, which is about 120 km south of San Diego, a 0.45-km^2 area was diked for construction of an oil-platform factory, which was never built; enclosed water and soils have become extremely hypersaline, and the once luxuriant cordgrass marsh has been almost entirely eliminated (Ibarra-Obando and Escofet 1987).

Contamination

Wastewater Inflows. The impacts from wastewater inflow are greatest at Tijuana Estuary, where 12-14 million gallons a day of raw sewage flows north from the metropolis of Tijuana down the Tijuana River (Seamans 1988; City of San Diego, pers. comm.). Additional spills and northward flows occur when there are breaks in the sewer pipe that collects raw sewage from western Tijuana. The continuous flows of sewage have played a role in eliminating fish and invertebrate populations from the estuarine channels. We believe that the unusual flows of non-saline water are more to blame than the contaminants carried in the sewage (Nordby and Zedler, In press). Without sewage inflows, the estuarine waters would be as saline as seawater (except after rain events), whereas now the river channel is measurably diluted all the way to the ocean inlet (Zedler et al. 1990). Low salinity appears to be intolerable for many of the marine organisms.

The impact of sewage on the cordgrass marsh is less clear. Slight reductions in soil salinity, along with nutrient inflows, can increase cordgrass growth, at least in the short term (Zedler 1986). To date we have not seen invasions of brackish marsh species into the lower marsh, primarily because tidal effects maintain high soil salinities. In contrast, sewage spills significantly affect the estuary's pickleweed marsh vegetation, which is used by the state-endangered Belding's Savannah sparrow. An "invasion plume" of exotic weeds has developed in areas of reduced tidal flow, where soil salts are readily diluted by freshwater inflows. The cover of pickleweed has declined by 50% where exotic weeds are most dense (Zedler et al. 1990).

Sewage inflows no doubt bring contaminants into the estuary. Cadmium concentrations are high in soil samples from the upper salt marsh along the main channels of Tijuana Estuary (Zedler et al. 1990). Gersberg et al. (1989) also found cadmium contamination in sediments of Tijuana Estuary, especially in areas downstream of Goat Canyon sewage flows. Other toxic materials have been documented in river water upstream (Southern California Coastal Water Project, unpubl. data), but no samples have been taken at Tijuana Estuary. Nutrients are brought downstream, but by the time the sewage reaches Tijuana Estuary, concentrations of nitrogen and phosphorus are at levels comparable to secondary treated wastewater. The 3-mile-long riparian and marsh system along the Tijuana River apparently filters some of the nutrients from the sewage.

Inflows from Street Drains. In Mediterranean-type climates, freshwater inflows are usually restricted to winter rainstorms. In urban areas, however, there is runoff all year from irrigation, car-washing, and other activities. At Tijuana Estuary, we have looked for evidence of contamination from street runoff and exotic species invasions from artificially lowered soil salinities. High levels of six heavy metals (especially copper) were found in sediment samples near six street drains adjacent to the cordgrass marsh (Zedler et al. 1990).

Invasions of exotic species occur where soil salinities are reduced, with the extent related to the amount (volume and flow frequency) of freshwater inflow (Zedler et al. 1990). These effects do not extend to the cordgrass marsh, because of the dissipating force of daily tidal action.

Oil Spills. Coastal wetlands are likely candidates for oil spill impacts, as evidenced by the event at Huntington Beach, California, in February 1990 (cf. Smith 1990). The Talbert Marsh had been restored one year earlier by excavating channels and reopening the system to tidal flushing. The 394,000-gallon oil spill occurred just offshore and presented a month-long threat to the wetland. This event demonstrates the need for on-site, rapid-response adaptive management of restoration sites.

The Talbert Marsh could not have been protected without the Huntington Beach Wetlands Conservatory, a group of local observers and innovators who used several preventative and containment techniques in their attempt to keep the oil from entering the tidal inlet. Most of the safeguards failed, but the experience shows how important an on-site, rapid-response management group is for minimizing the impacts of oil spill disasters. About a dozen containment booms were placed across the ocean inlet to stop flows and absorb oil. Because these were known to be ineffective under storm conditions, a berm was built to block tidal flow into the marsh. Unfortunately, this had to be breached to drain stagnant (low oxygen) water and prevent a fish kill. Immediate closure was then delayed by a bulldozer breakdown, and some oil entered the channel. On day 10, heavy rains and runoff threatened to inundate the Pacific Highway, so the inlet had to be breached again. The outflowing water eroded the entire dike, and oil was again brought in by the flood tides. This time, crude oil contaminated the marsh. The dike was rebuilt on day 15, but storm swells washed it away on day 18. The dike was rebuilt again, this time further up the inlet channel. It held until the threat of oil contamination subsided a month later and tidal flows could be readmitted to the restored marsh.

Despite all of these incredibly rapid actions, oil entered the marsh, causing two kinds of problems—direct contamination and disturbance from clean-up crews. Still, damages would have been substantially greater had there been no local protection capability. The lesson is clear—coastal wetlands and restoration sites need to have on-site, rapid-response, adaptive management in order to reduce the problems caused by oil spills.

Human Disturbances

Any wetland near a metropolis will experience a multitude of daily human disturbances, in various combinations, including noise, trampling, night lighting, flyovers, and freeways. Not much is know about their direct impacts, because so few undisturbed systems are available for comparison. It is clear that bird species differ in their sensitivity to the presence of humans (Josselyn et al. 1989;

Pacific Estuarine Research Laboratory, unpubl. data). There are also temporal differences in sensitivity. Belding's Savannah sparrows are more readily flushed during their non-territorial phase than during nesting time (White 1986).

Trampling has both direct and indirect impacts on the marshes. At Tijuana Estuary, there is direct damage by Mexicans who travel north on foot—dozens of undocumented aliens move through the estuary every day. Vehicular and foot traffic has had indirect effects on the salt marsh because of the denudation of dune and strand vegetation. Trampled dunes are no longer stable, and sea storms wash sand into estuary channels. Most southern California dunes have lost their native plant cover, and shorelines have moved inland, eliminating salt marsh in the process. At Tijuana Estuary, the seaward edge of the barrier beach retreated more than 300 feet between 1852 and 1986 (Williams and Swanson 1987). A secondary impact of dune washovers is the reduced tidal prism, which increases the chance of inlet closure. The 1984 closure of Tijuana Estuary, which caused drastic reductions in cordgrass cover and the clapper rail population, was caused by dune washovers and tidal prism reduction during the 1983 winter storm (Zedler and Nordby 1986).

Responses to Impacts

There are clearly major impacts due to filling, dredging, trampling and reduced tidal flushing, all of which damage habitat. The cumulative response to all these disturbances is that several species have become threatened and endangered with extinction. A reduction in biodiversity has most certainly occurred, since the great majority of coastal salt marsh habitat has been destroyed. The cordgrass marsh was shown to be especially susceptible, when an 8-month period without tidal flushing nearly decimated the light-footed clapper rail population, its invertebrate foods, and its protective cordgrass canopy at Tijuana Estuary (Zedler and Nordby 1986). Frequent and prolonged closure of Los Peñasquitos Lagoon appears responsible for its loss of cordgrass. Permanent closure to tidal flushing, by diking a large segment of Estero de Punta Banda, eliminated the enclosed cordgrass marsh as a functional ecosystem (Ibarra-Obando and Escofet 1987).

Reduced tidal flows and augmented freshwater inflows act together to shift habitat type from tidal creeks to shallow-water brackish lagoons. Sedimentation fills former creeks and salts are diluted in both the water and soils. The impact of reduced salinity is that marsh habitats are invaded by brackish marsh plants. Major inflows following reservoir discharges to San Diego River allowed a pickleweed-cordgrass marsh to be invaded and dominated by cattails (*Typha domingensis*) within six months. Experimental removal of cattails in replicate plots demonstrated competitive interactions, with cattails reducing the density and biomass of cordgrass transplants (Table 2). Marine invertebrates and fishes are eliminated when coastal wetlands are cut off from tidal flows and dominated by freshwater inflows (C. Nordby, Pacific Estuarine Research Laboratory, unpubl. data on San Elijo Lagoon).

Table 2. The influence of cattails (*Typha domingensis*) on cordgrass at the San Diego River salt marsh, from experimental removal experiments in 1980 (modified from Zedler 1984). Data are number of cordgrass transplants present on sequential sampling dates.

	No. Planted	Number Alive at Later Dates		
Cordgrass	on 4/30/80	5/20/80	12/29/80	8/18/81
With cattails present	35.0	35.0	71.0	58.0
Without cattails	35.0	35.0	258.0	608.0
Ratio, without/with	1.0	1.0	3.6	10.5

Restoration and Creation

Review of Salt Marsh Restoration Activities

The ultimate goal of wetland restoration is to provide self-sustaining ecosystems that closely resemble the natural systems in both structure and function. Restorationists are far from achieving this goal, although attempts to restore wetlands to pre-disturbance conditions are ongoing, using a variety of construction measures (e.g., breaching of dikes to restore tidal flows, excavation of fill to restore original topography). The status of the science of wetland restoration was recently reviewed by the Environmental Protection Agency, with the help of over 30 authors from around the nation (Kusler and Kentula 1989). Several problems with constructed (i.e., created or man-made) wetlands were noted, including inadequate food chain support, inappropriate species composition, likelihood of inadequate genetic diversity, invasions by exotic species, impaired nutrient functions, low sediment organic matter content, and lack of resilience or self-sustaining capacity. Attempts to restore southern California salt marshes also have problems—plans are often faulty; numerous errors are made during implementation; unforeseen difficulties arise; and long-term adaptive management is lacking (Zedler 1988a,b).

There is no guarantee that constructed wetlands will persist in perpetuity. We have yet to learn if they can withstand events such as catastrophic floods and sea storms, invasions of exotic species, outbreaks of herbivorous insects, and rising sea level. Mechanisms that confer resilience are poorly understood. Seed banks and genetic diversity are important for long-term maintenance of plant communities. If transplants are derived from only a few clones, they will probably have little genetic diversity and may not include the genotypes necessary for population persistence. Sites that appear successful in the short term may fail in the long run.

The many problems of restored marshes can be grouped as hydrologic, chemical, and biological problems. Of these, hydrologic problems are the most often documented, in part because they are easier to see, and in part because the impacts on the entire system are dramatic.

Hydrologic and Topographic Problems. These include overexcavation, underexcavation, unexpected sediment accumulation, and shifting of tidal access during site construction with resultant erosion (undercutting transplants) and deposition (smothering transplants). Either improper planning or mistakes in implementation of restoration plans can result in hydrologic errors. A common problem concerns diked wetlands that have been removed from tidal flushing for many years, during which time the substrate subsides and elevations are lowered. Simply breaching the dikes to such wetlands does not restore them to their pre-dike condition. The salt marsh of Oregon's Salmon River Estuary is such an example—due to subsidence, the restored site is mostly open water habitat that may take a century or more to accrete enough sediments to bring the topography back to an intertidal marsh (Frenkel and Morian 1990).

On the other hand, too much accretion causes problems where tidal channels are cut into semi- or non-tidal areas. Such excavated areas often sediment in, especially for sites dredged deeper than the downstream channel that feeds them. Corrective measures for areas of subsidence (e.g., filling) or accretion (e.g., continued dredging) can be developed, but implementation would prolong the period of ecosystem disruption and increase the likelihood of invasion by exotic species (see below).

Providing both the proper tidal flushing and topography is critical to wetland restoration. Marsh plants are extremely sensitive to the degree of stagnation (anoxia) in the soil and to the salinity. Even a 10-cm mistake in elevation can prevent the desired marsh plants from growing well. Areas that receive too little tidal current may become stagnant, thus reducing plant growth; areas with too strong a current will erode, taking marsh transplants with them.

Perhaps the most obvious hydrologic/topographic error is that constructed marshes rarely "look like" natural marshes. The complex networks of small tidal creeks are always lacking, and the broad, flat marsh plains are compressed or absent. In their place, bulldozers often create a collage of habitat types that would rarely abut one another—steep-sided islands and deep, straight channels. Isolated habitats can pose serious problems for the biota that need transitions between habitats (discussed below).

"Chemical" Problems. The quality of the substrate and water often differ for natural and constructed wetlands. The substrate of excavated sites may become acid upon exposure to air, as accumulated sulfides are oxidized to sulfuric acid. Excavation can inadvertently uncover toxic materials if the former wetland had been used as a landfill. Restoration sites may also be downstream of contaminated waters. In urban areas, there is the additional problem that marshes can accumulate trash, which smothers vegetation. Such locations will not support high-quality wetlands.

Organic sediments are a basic feature of natural wetlands. It has been suggested that organic matter influences nearly every aspect of wetland ecosystem functioning, by changing sediment porosity and water-holding capacity, by influencing nutrient dynamics, by altering the growth rates and nutrient content of plants, and by influencing the species composition and abundance of invertebrates associated with the sediments. The microbes, plants and animals, in turn, affect the rate of organic matter accumulation in wetland sediments. For most wetlands, the development of organic sediments occurs over time periods that are closer to centuries or millennia, than to years or decades.

In three regions of the United States—North Carolina (Craft et al. 1986, 1988), Texas (Lindau and Hossner 1981) and southern California (Langis et al. 1991)—comparisons of constructed and restored wetlands indicate that sediment organic matter and nitrogen levels are lower than in the natural reference wetlands (Table 3).

Table 3. Organic carbon (%) and total nitrogen (mg kg^{-1} dry weight) in natural and constructed marshes of Texas (Lindau and Hossner 1981), North Carolina (Craft et al. 1988), and southern California (Langis et al. 1991). "Marsh Age" = years between marsh construction and sampling. Table compiled by R. Langis (PERL).

Location	Marsh Age	Organic Carbon		Total Nitrogen	
		Natural	Constructed	Natural	Constructed
Texas	1	0.3-1.12	0.13	227-588	95
North Carolina	10-15	0.6-8.6	0.6-1.8	364-1680	322-924
California	4	2.0-2.5	0.1-1.1	1740-2270	870-960

Soils of all of the constructed marshes had less than half the organic matter content of natural marshes. Low organic levels may impair microbial activities (nitrogen fixation, nutrient recycling, sulfate reduction). They may also influence faunal density and composition, not only through trophic linkages but also by altering sediment compactness and the ability of burrowing fauna to colonize the sediment. Because it is uncertain how long a constructed marsh will take to match the organic levels and nutrient status of natural wetlands, predictions of the time required for functional equivalency are, at best, highly speculative.

Biological Problems. Pacific cordgrass is not a ready invader of newly created sites. It has limited establishment ability and must be transplanted with great care (Zedler 1984, 1986). Several kinds of biological problems have developed in the attempts to restore Pacific cordgrass marshes.

1. Transplantation Problems. Because there are few natural marshes, there are few sources of transplants for restoration sites. Removal of plants from the region's wetland reserves is not advisable. Hence, there is a need to salvage plants whenever vegetation is being destroyed, so that it can be transplanted. It may be necessary to build interim nurseries to hold plants until restoration sites are available. In one San Diego Bay project, an intertidal nursery was built in 1983, planted with salvaged cordgrass, and allowed to grow for about four years. Mere construction of nurseries is not enough, however. The cordgrass nursery was accidently destroyed in 1990, when the adjacent topography was being bulldozed to expand habitat for cordgrass marsh. Another salvage effort failed when blocks of high-marsh vegetation were stockpiled and irrigated with fresh water for several years. Upland weeds invaded as soil salinity declined, reducing the quality of the sod for restoration efforts. For successful salvaging of coastal vegetation and soil, there must be a long-term plan and maintenance program for the nurseries.

Perhaps the most common problem with transplantation in southern California is the hypersaline soils. Lower marsh soils of natural marshes average over 40 ppt, and only a flood event or major reservoir discharge can lower the salinities to seawater or brackish levels. Marshes that have been diked become even more saline, and dredge spoils develop some of the highest salinities. At San Diego Bay, newly excavated dredge spoils averaged 80 ppt upon initial tidal influence; values dropped to 55 ppt in the first two months despite lack of rainfall and runoff. However, levels did not decline significantly over the next four months. Transplantation of cordgrass may need to be delayed until tidal flushing lowers salinities to at least 50-60 ppt.

2. Disconnected Habitats. When habitats are constructed as islands, the lack of terrestrial corridors can reduce or eliminate natural movements of animals—even flying insects. Predatory beetles were lacking at San Diego Bay's dredge spoil island—and their absence seems to be responsible for the extensive scale insect attacks on cordgrass transplants. Where beetles abound, i.e., on the mainland surrounding the Bay, scales are kept in check.

Islands of high-marsh habitat at another San Diego Bay restoration site (Figure 1) provided suitable habitat for reintroduction of an endangered plant, the salt marsh bird's beak (*Cordylanthus maritimus* ssp. *maritimus*). However, only 20% of the plants were pollinated (B. Fink, Pacific Estuarine Research Laboratory, pers. comm.). Where there is no seed bank and insufficient pollination to produce new seeds, annual species cannot persist. Bee pollinators need marsh-upland transition habitat for their ground nests and for alternative nectar sources that feed the bees when bird's beak is not in flower (ibid.).

3. Exotic Species. There is too often an abundance of exotic species—both animal and plant—that establish in constructed marshes. Besides taking up space and reducing the potential for native species to develop we have little information on how these aliens change food webs and wetland functioning. Rutherford (1989) found a Japanese mussel (*Musculista senhousia*) invading the constructed marsh at San Diego Bay in large numbers, but only rare occurrences in the natural reference wetland. Because many Pacific Coast estuaries already have large numbers of introduced invertebrates, there is concern that moving transplant cores from one wetland to another (e.g., San Francisco Bay to Elkhorn Slough) will bring unwanted species along with the desirable plants (M. Silberstein, Elkhorn Slough National Estuarine Research Reserve, pers. comm.).

4. Food Chain Support Functions. The few data available on invertebrates of constructed marshes have demonstrated considerably lower abundances or vastly different species complements than present in reference wetlands. Rutherford (1989) examined the epibenthos of San Diego Bay cordgrass marshes and found similar species but greatly reduced densities in the 4-year-old constructed site. Cammen (1976a, b) reported significant differences in the infauna of constructed wetlands along coastal North Carolina. At a 1-year-old marsh, the dominant taxa were insect larvae, whereas polychaetes dominated the natural wetland. Total densities and calculated secondary production also were markedly lower than in the natural system. Similar observations were made at a 2-year-old site. Sacco et al. (1987) revisited the latter site when it was approximately 15 years old and reported a 10-fold increase in densities and high similarity with the infauna of natural marshes. Moy (1989) and Sacco (1989) found that the fauna of an upland site that was graded to support *Spartina alterniflora* had fewer nematodes, ostracods, harpacticoids and oligochaetes than natural marshes.

Species composition does not always become more similar through time. Moy (1989) demonstrated a convergence of species lists with one adjacent marsh but a divergence in composition over a 2-year period in another marsh receiving the same source water as the created site. Sacco (1989) evaluated the composition, density and trophic structure of infauna in coastal North Carolina marshes 1-17 years after construction. Although variable, the data showed that the six constructed marshes had similar faunal components and trophic groupings (deposit feeders, suspension feeders, and carnivores) but that densities were uniformly lower than in natural marshes.

(a)

(b)

Figure 1. Wetland mitigation site at Sweetwater Marsh National Wildlife Refuge on San Diego Bay. The site has an urban setting, which includes a major freeway, a new interchange under construction, a flood control channel under construction, an abandoned railway and associated fill, and adjacent

(c)

commercial development. The low-tide view (a) shows one of eight islands that were constructed by removing fill, excavating channels, contouring the topography and planting cordgrass. This island is representative of problems at the 12-acre site. The island tops are unvegetated, and the cordgrass is sparse and short. The islands and the cordgrass canopy are fully inundated during the highest tides (b), unlike the natural reference wetland, which has a dense canopy of cordgrass emerging above the water line (c, Paradise Creek). The emergent canopy provides cover, retains debris for bird roosting, and harbors non-aquatic insects and spiders; all are important functions for the endangered light-footed clapper rail, the target species for this mitigation project. The constructed marsh was built to the proper elevation for cordgrass, but the plants do not grow well because of substrate (nutrients, organic matter) limitations. Photos were taken by P. H. Zedler on Jan. 1, 1991. Photos (b) and (c) were taken during the extreme high (+7.2-ft MLLW) tide in the morning, photo (a) during low tide in that afternoon.

New Experimental Approaches—"Ecotechnology"

Restoration sites can be modified to accelerate ecosystem development or provide functions that managers and decision-makers desire. Such efforts comprise ecotechnology, a young area within restoration science. To date, only a few ideas have been tested in cordgrass marshes.

Test Plantings. Because little is known of the requirements of individual plant species, and almost nothing of the interactions among factors that reduce plant growth, an initial investment in experimental transplantation can identify suitable soils and elevations for individual sites. Where plant species are known to have special requirements, e.g., the endangered salt marsh bird's beak (*Cordylanthus maritimus maritimus*), which is a hemiparasitic annual plant, we recommend a combination of greenhouse and field experiments to identify habitat preferences.

Soil Augmentation. The comparison of soils in constructed and natural marshes (cf. Table 3) indicated that restoration sites in three states had lower nitrogen, soil organic matter, and plant biomass. Elsewhere, experiments in natural cordgrass marshes have shown that nitrogen additions increase foliar nitrogen and plant biomass (e.g., Covin and Zedler 1988). Also, Valiela et al. (1984) have shown that nutrient additions affect a wide range of ecosystem functions, e.g., a 24% increase in decomposition of vascular plant material following enrichment. While restoration sites have previously included fertilizer applications, studies of nutrient dynamics following augmentation of soils with organic matter and nitrogen have not been carried out. Thus, NOAA's Coastal Ocean Program began to support such experiments in new cordgrass marshes of North Carolina (near Beaufort) and southern California (San Diego Bay).

Habitat Heterogeneity. Tidal creeks have long been recognized as important to the functioning of salt marshes. Chalmers (1982) presented a conceptual model to explain how several attributes of creekbanks (high nutrient influx, greater aeration and higher sediment redox, reduced sulfide concentrations), cause the tallest and most productive cordgrass. With more plant biomass, there should be more detritus production and greater productivity of consumers. Creeks are known to be productive of fishes and invertebrates, especially benthic molluscs and crustaceans. The benthic and creekside habitats are closely linked due to interchanges of organisms and materials with daily tidal flooding. These "edges" are also suitable foraging sites for birds, such as rails, that find both food and cover in close proximity.

To provide the appropriate hydrologic setting, careful attention must be paid to the grading of restoration sites. While many natural tidal wetlands are dissected by a network of tidal channels and creeks, most constructed marshes begin with smoothly graded terrain. Observations of such wetlands in southern California suggest that functioning is impaired by topographic homogeneity. Recent experiments in Texas (reported by Tom Minello and Roger Zimmerman, NMFS Galveston Laboratory, at the 1989 Estuarine Research Federation Conference) support this suggestion. After cutting channels through transplanted marshes, densities of brown shrimp, blue crabs, grass shrimp, and small forage fish increased in the interior of the marsh, along with densities of benthic infauna and epifauna. Increasing the "edge" and improving access seems to increase the secondary productivity of salt marshes.

Some of the shortcomings of constructed marshes may be correctable by excavating tidal creeks. New sites would be easiest to modify. The problem is that the benefits of adding topographic complexity to constructed marshes have not been documented, so the need is not clear and the increased construction costs are not easily justified. A total ecosystem experiment is needed to quantify the effects of adding a network of tidal creeks to restored wetlands and to devise recommendations for this new management tool. It is hypothesized that areas with a tidal creek network would have greater plant and animal productivity than areas with topographic homogeneity. The sediment should develop a higher root biomass and more rapid accumulation of organic matter.

Recommendations for Improved Restoration Planning and Integration of Research and Restoration

Because of the many problems with salt marsh restoration (Zedler 1988a, b), especially past failures to match natural ecosystem functioning (PERL 1990), it is important to use research findings to guide restoration projects. At the same time, major restoration sites are needed by scientists to test new methods of accelerating ecosystem development. The restoration process may take several years, and agency personnel should meet frequently with scientists who are evaluating restoration progress and field experiments. In this way, the necessary ecotechnology can develop and be incorporated as restoration efforts proceed.

One model of interactive research and planning is the Tidal Restoration Project for Tijuana Estuary. This 1,000-ha wetland has experienced an 80% decline in its tidal prism over the past 140 years, with the result that continuous tidal flushing is no longer assured. Closure of the ocean inlet for eight months in 1984 nearly decimated the clapper rail population, and some plant species (e.g., the annual, *Salicornia bigelovii*) will never recover unless replanted. While the planning process has been prolonged by the interactive approach, the result is a greatly improved plan. Support for the work has come from the State Coastal Conservancy, beginning in 1986.

The first step was to photograph and map the topography, measure the existing tidal prism, estimate historic tidal prisms, identify causes of tidal prism reductions (sedimentation events), select a restoration target (the 1852 tidal prism), and plan the necessary dredging. Hydrologists (Williams and Swanson 1987) carried out the initial planning, which included a map of proposed excavations plus two "river training berms" designed to protect newly dredged areas from future sedimentation events.

The second step was to identify sensitive habitats that might be damaged by increased tidal flushing and to identify necessary experimental work needed to construct new wetlands as mitigation for any unavoidable impacts. A vegetation map was prepared from the aerial photos and from field sampling, and surveys of birds, mammals, reptiles, amphibians, insects and spiders, and channel benthos were undertaken. A "constraints map" indicating sensitive species was thus developed. A Geographic Information System was used to enter all maps (1-foot elevation contours, channels, vegetation, and constraints) on the computer for printing at various scales and combinations. Overlays of proposed dredging and constraints indicated areas where alternatives should be sought or where mitigation would be needed if dredging plans could not be changed.

The third step was for the hydrologists to modify the initial dredging plan to accommodate the concerns of ecologists and managers. Simultaneously, ecologists identified information needs that dictated a phase-one experimental program to precede implementation of a long-term, large-scale (200-ha) tidal restoration project. An 8-ha experimental salt marsh will be constructed initially to monitor channel evolution (erosion, sedimentation) and to test the importance of incorporating a network of tidal creeks into the marsh plain prior to transplantation. Within the 8-ha, six subunits will be assigned to treatment with and without tidal creeks. Scientists will follow nutrient dynamics, plant growth, invertebrate colonization, and fish and bird use, testing the general hypothesis that cordgrass ecosystem development will be accelerated by providing a more heterogeneous topography at the onset of restoration. There were also several questions about how to mitigate unavoidable disturbance to pickleweed habitat. A second experimental area employing 24 tidal mesocosms was thus planned. This smaller experimental site is currently being developed with funds from NOAA as part of the National Estuarine Research Reserve program. Studies have been proposed to test methods of restoring pickleweed marsh under three tidal flushing regimes (full tidal, wetter semi-tidal, and drier semi-tidal), with and without freshwater inflows (six treatments in all, four replicate mesocosms per treatment).

The fourth step—evaluating the environmental impacts of the resulting "integrated" restoration plan—is in progress as of October 1990. Upon approval of phase one, the State Coastal Conservancy will proceed with excavation, transplantation, and monitoring of the 8-ha experimental wetland. Further work will be delayed until funds, permits, and results from the experimental work are available.

Ongoing Research

Long-Term Comparisons of Constructed and Natural Salt Marshes. Studies of the Connector Marsh at San Diego Bay began four years after construction and cordgrass transplantation. Four theses (Swift 1988; Cantilli 1989; Rutherford 1989; and Zalejko 1989) and a research paper (Langis et al. 1991) report the detailed findings. The Connector Marsh was found to be less than 60% functionally equivalent to the adjacent reference marsh, based on comparisons of 11 soil, plant, and invertebrate characteristics (Table 4). Sediments contained half the nitrogen; soil water had less than 20% as much inorganic nitrogen; cordgrass biomass was less than half as great, plants were significantly shorter and had lower nitrogen content, and epibenthic invertebrates were about a third as abundant. These ecosystem attributes are interrelated, and differences can be traced to the coarser soil texture and lower organic

Table 4. Functional equivalency of the constructed and natural cordgrass marshes comparing soils, nutrients, plants and epibenthos. Soil and plant data are from Langis et al. (1991) except for plant heights (unpubl. PERL data). Epibenthic invertebrate data are from Rutherford (1989). Further discussion of the comparison can be found in PERL (1990) and in Zedler and Langis (unpubl. data).

Data Set	Percent
Organic matter content	51
Sediment nitrogen (inorganic N)	45
Sediment nitrogen (TKN)	52
Pore-water nitrogen (inorganic N)	17
Nitrogen fixation (surface cm)	51
N fixation (rhizosphere)	110
Biomass of vascular plants	42
Foliar nitrogen concentration	84
Height of vascular plants	65
Epibenthic invertebrate numbers	36
Epibenthic invertebrate species lists	78
Average of comparisons =	~57%

matter content. Invertebrates are influenced by the quantity and quality of their plant food supplies, which are in turn influenced by nitrogen, which is released by microbes, which are limited by energy supplies from soil organic matter. The short plant canopy is fully inundated by high tides, making it unsuitable for use by non-aquatic arthropods, such as spiders and beetles, some of which are fed upon by birds. Finally, with less food and a shorter plant canopy, the site is probably not yet suitable for residency by top carnivores, such as the endangered light-footed clapper rail.

Several questions require continued research: Is it likely that the Connector Marsh will ever match the functions of the natural marsh? How long will it take to achieve equivalency, or at least 85% similarity in functions? How does nitrogen accumulate in natural marshes—gradually or as occasional events (e.g., uptake following flooding)? Are there ways to accelerate the development of ecosystem functions? What are the best indicators of functional equivalency?

Soil Augmentation to Accelerate Cordgrass Growth. Our experiment at San Diego Bay followed field studies of the Connector Marsh at ages 4-6 years. When a large excavation project was nearing completion, we proposed to augment the sandy substrate with different types of organic matter (OM) and nitrogen (N), in order to test methods of enhancing nutrient-supply and food-chain-support functions.

Excavation of the 17-acre intertidal wetland was completed on February 27, 1990. In early March, just before it was opened to tidal flushing, we set up 7 soil treatments: no treatment; rototill only; straw (low-N organic matter); alfalfa (high-N organic matter); inorganic N fertilizer (ammonia sulfate); straw + N fertilizer; and alfalfa + N fertilizer. Treatments were replicated within 4 blocks, for a total of 28 plots. My graduate ecology class planted cordgrass (10 pots/treatment plot) on March 21, 1990, and soil salinities, plant numbers, and plant heights have been censused monthly.

Soil ammonia concentrations in July 1990 averaged 1.2 $\mu g\ g^{-1}$ sediment dry wt where neither N fertilizer nor N-rich organic matter were added and 4.9 $\mu g\ g^{-1}$ where either alfalfa, N fertilizer, or both were added (Zedler and Langis, unpubl. report to California Sea Grant). Cordgrass responded strongly to treatments with both nitrogen and organic matter added, and it grew slowly where neither was added. After 6 months of growth, the highest total stem length (with alfalfa plus N fertilizer added) was 3.4 times the lowest (with rototilling but no amendments). Stem densities increased 4-fold over the first 6 months of study, and the large number of new shoots indicates substantial belowground growth of roots and rhizomes. The differences in canopy density are readily visible in the field, and the potential for preventing the kinds of problems that persist at the 6-year-old marsh is good.

While initial results are very promising, the long-term effects of augmenting soils must still be determined, with careful study of the processes that are responsible for ecosystem development. While biomass has increased substantially with soil augmentation, assessment of height responses must wait for year two. Most of the plants are still very short, reflecting their recent vegetative propagation. It is not clear how long the effects of initial soil augmentation will continue. If the nitrogen is held on site and recycled, then growth may continue to increase until N becomes limiting. The added organic matter may be able to fuel N-fixation and gradually increase the nutrient status of the site. Algal mats may develop and affect cordgrass growth either favorably (through N-fixation or trapping N from tidal waters) or negatively (by competing for N). It is also possible that changes will develop in the consumer communities. Scale insects are a problem in two other constructed wetlands of San Diego Bay, and such herbivores may be stimulated by nutrient-rich vegetation. Whether or not fertilized plots will be damaged by herbivores may not be known for several years. Finally, whether any of the effects will be sufficient to allow the target species, the light-footed clapper rail, to occupy the site remain unknown. Funding for continuing studies has been sought from NOAA's Coastal Ocean Program.

Research Needs

A review of information and research needs has just been completed by the California Department of Transportation (Conners et al. 1990). The authors list research needs for restoring and creating salt marshes in seven areas: (1) hydrodynamic processes, to include studies of optimal marsh surface and channel characteristics, studies of flows and sediment movement, development of new measurement techniques for gathering tidal data and sampling suspended solids, hydrodynamic modeling; (2) soil development, to include new methods of improving plant growth, determinations of the time needed for new marsh soils to mimic natural soils, and to study pollutant impacts; (3) vegetation establishment, to include planting guidelines for more species, studies of herbivory and competition, new propagation techniques, and exotic plant control; (4) monitoring, to include protocols for what and when to sample; (5) spatial requirements, to include studies of optimal habitat size and mixtures, plus buffer requirements; (6) construction practices, new equipment, and effects of equipment; and (7) stormwater treatment, especially problems of pollutants.

Wetland scientists would certainly add to the above recommendations. There is a critical need for information on how to attract or transplant native animal species to restored wetlands. And we need to know how to maintain native animal populations, especially in regions where extreme events (floods, droughts) occur, and where persistence of a population may require frequent reestablishment from refugia. We need to understand what types of terrestrial, riparian, coastal "corridors" are most effective in facilitating dispersal and movements between habitat remnants.

We also need to know existing levels of genetic diversity in our remnant wetlands and to estimate the genetic diversity that is needed for transplanted or seeded populations. At present, it is occasionally recommended or required that transplants come from local stock (M. Guinon, Sycamore Associates, pers. comm.), however, there is never a requirement that the full range of local genotypes be included in seed or propagules destined for restoration sites. For restored wetlands to persist in perpetuity, it is important to build in the mechanisms that will allow resilience. Transplants may not include the appropriate levels of genetic diversity, because many of the species used are vegetatively propagated both in nature and in culture, with the latter perhaps produced from a few initial clones. Furthermore, the natural systems available for collecting seeds or transplants already have greatly reduced areas, and the gene pool from which to build new habitats has been greatly diminished. Small, isolated populations with reduced genetic diversity are more susceptible to extreme events (S. Williams, Pacific Estuarine Research Laboratory, pers. comm.). Populations with lower heterozygosity may be less vigorous. Inbreeding can further reduce genetic diversity, as well as the viability and fecundity of offspring (ibid.).

Future Needs and Directions

Assessment Protocols

There is a need to standardize criteria for judging restoration, construction, and enhancement activities. The no-net-loss policy should underscore the need for mitigation sites to be functionally equivalent to natural habitats. Detailed, functional criteria must be used to judge restoration success, especially in southern California, where several large projects are being planned and where endangered species populations are at risk. Three proposed projects illustrate the amount of habitat and money that will be involved in restoring southern California salt marshes. The first three projects are located in San Diego County, which has an estimated 1870 acres (757 ha) of saline marsh (Table 5 based on U.S. Fish and Wildlife Service draft maps from the National Wetland Inventory.)

Table 5. San Diego County wetland habitats, determined from US FWS draft maps of the National Wetland Inventory by S. Lockhart. Data are hectares (1 ha = 2.471 ac). Br/Fr = Brackish/Freshwater. Taken from PERL (1990).

Location	Bay	Channel	Saline Marsh	Br/Fr Marsh	Im-pounded Waters	Other
Tijuana Estuary	0.0	223.6	171.1	88.1	0.0	1.4
San Diego Bay	4483.2	0.2	11.0	23.3	430.9	0.0
Sweetwater	0.0	1.9	116.1	18.6	0.0	1.0
Famosa Slough	0.0	2.6	1.5	0.0	0.0	0.0
San Diego River	62.6	0.0	47.1	19.8	0.0	1.1
Mission Bay	620.1	0.0	35.3	39.9	0.0	0.0
Los Peñasquitos	0.0	3.1	130.7	50.0	0.0	0.0
San Dieguito	30.0	0.0	28.7	36.7	0.0	0.8
San Elijo	0.0	31.6	87.6	74.2	0.0	0.0
Batiquitos	42.7	99.9	8.6	52.7	0.0	0.0
Aqua Hedionda	105.9	0.1	31.4	26.9	0.0	0.0
Buena Vista	0.0	0.0	0.0	15.6	64.6	0.0
Loma Alta	0.0	0.0	0.0	0.7	0.0	0.0
San Luis Rey	0.0	0.7	0.0	82.6	0.0	2.0
Oceanside Harbor	85.3	0.0	0.2	0.0	0.0	0.0
Santa Margarita	30.4	0.0	87.3	79.0	0.0	0.0
Cockleburr Cyn	0.0	0.0	0.0	2.0	0.0	0.0
French Canyon	0.0	0.0	0.0	6.3	0.0	0.0
Aliso Creek	0.0	0.0	0.0	10.2	0.0	0.0
Unnamed Canyon	0.0	0.0	0.0	3.8	0.0	0.0
Las Pulgas Cyn	0.0	0.0	0.0	46.8	0.0	0.6
San Onofre Creek	0.0	0.0	0.0	8.8	0.0	0.0
San Mateo Creek	0.0	0.0	0.0	56.3	0.0	32.2

Tijuana Estuary. To prevent inlet closures, plans are being made to restore much of the historic tidal prism of this estuary. In the preferred configuration, about 24 acres of existing salt marsh would be damaged by dredging and grading activities, and 276 acres of new salt marsh would be created from disturbed upland areas. The cost of the entire project is estimated at $40 million, with no funding source yet identified. Much of the estuary is a U.S. Fish and Wildlife Service Refuge for Endangered Species (the light-footed clapper rail, the California least tern, and salt marsh bird's beak). The Belding's Savannah sparrow, a state-listed endangered bird, is abundant on the site.

Marisma de Nación. The 17-acre mitigation marsh that was excavated from dredge spoil adjacent to San Diego Bay will be planted to cordgrass in January 1991. Because the U.S. FWS has laid out very strict requirements for successful mitigation of lost habitat, the site must have a detailed monitoring program (Pacific Estuarine Research Laboratory 1990). At issue are three endangered species that must have appropriate habitat on site (the light-footed clapper rail, the California least tern, and salt marsh bird's beak). In addition, the marsh is expected to be used by the Belding's Savannah sparrow.

Batiquitos Lagoon. A plan to dredge this non-tidal lagoon has just been completed (City of Carlsbad, California, and U.S. Army Corps of Engineers 1990). The preferred alternative would provide a fully-tidal system, with 7 of the 123 acres of existing salt marsh damaged during construction. The Port of Los Angeles would provide over $20 million as mitigation for lost marine fish and bird habitat at San Pedro Bay (Batiquitos Lagoon Enhancement Project Newsletter, July 1990).

Ballona Wetland. About 250 acres of salt marsh will be restored to tidal flushing (a $10 million project that may involve some mitigation credits for port expansion in San Pedro Bay). A declining population of the Belding's Savannah sparrow currently uses the site.

Scientific Information Needs

Restoration projects need to be based on the best scientific information available and to incorporate experimentation that will answer any remaining questions about the site's ability to support necessary wetland functions. The resulting "ecotechnology" will gradually accumulate for a range of case studies, so that what is now a series of trials and efforts will become a restoration science.

The basic need is to become predictable. We need to be able to predict what we will achieve and when we will achieve it. To determine what we can achieve requires comparison of restored and natural wetlands for periods of 10 years or longer. Long-term research goes beyond systematic sampling by following promising leads and discarding hypotheses that do not stand up to testing. Gradually, an understanding of the complexity and patterns emerge.

To accelerate the rate at which ecosystems functions develop, we need new ecotechnological methods, such as soil augmentation procedures, means of adding topographic heterogeneity to construction sites, ways of introducing predators that control herbivorous insects, methods of controlling the invasions of exotic animal and plant species, improved use of wastewater wetlands to treat inflowing waters, bioremediation measures for decontaminating soils that contain organic toxins (e.g., petroleum derivatives, chlorinated hydrocarbons), and procedures that will ensure ecosystem persistence, even if catastrophic floods occur or sea levels rise more rapidly than anticipated. Research will be needed to determine how well such new methods work.

Where corrective measures are not workable, we need to know the tolerance limits of species for various limiting factors. Contaminated sites pose special problems. For soils that are contaminated with heavy metals, it is critical to know what hydrologic conditions would mobilize (make available to the food chain) or immobilize the toxic substances. We need to know which plant species are most likely to take up metals from wetland soils, and which have low potential for making such materials available to food chains. In any contaminated soil, it is essential to know what levels of contaminant can occur without harm to organisms, especially those at the top of the food chain.

The Importance of Research Funding Agencies

Research funds are essential to bring scientific predictability to restoration. A close working relationship between funding agencies and project proponents can have major benefits. First, restoration research at the ecosystem scale is best carried out in large areas that are part of restoration projects. Access to such sites and some control over what goes on at the experimental part of the site is needed, and it cannot happen without good rapport between the builders and the researchers. Second, because the marsh builders need to comply with certain criteria, any research funding for assessment will help the proponent, and monitoring monies provided by the latter will help researchers. Project proponents may have a shorter-term interest in the project, once the site has met the construction standards; funding agencies have their most important role in allowing long-term research, e.g., studies to identify latent problems, to follow ecosystem maturation, and to understand why functional equivalency does or does not develop.

A successful model of agency cooperation was the San Diego Bay study, which was initially funded by the California Sea Grant Program and the California Department of Transportation (Caltrans). Sea Grant funded the study of more basic-science issues (e.g., functional equivalency of paired sites) than Caltrans was willing to support, but Caltrans paid for routine monitoring of several reference wetlands and ecosystem components (including fishes, benthos and birds). The collective efforts were synergistic, and what emerged was a comprehensive comparison of constructed and natural cordgrass marshes. Longer-term studies and development of new ecotechnological methods, however, await future funding.

NOAA's National Estuarine Research Reserves (NERRs) combine several important attributes that will certainly foster restoration of Tijuana Estuary, the most disturbed of the 19 NERRs. Most importantly, several estuarine sites have been set aside expressly for use in research. In addition, the program provides funds for selected monitoring procedures that provide long-term data on ecosystem structure and function. Finally, the program supports experimental research at the designated sites. This combination of activities makes Tijuana Estuary a good candidate for successful restoration.

Information Transfer

NOAA's Habitat Restoration Symposium (September 1990) was an important opportunity for decision-makers to obtain an overview of the problems encountered while carrying out restoration projects. No one who listened to the talks could conclude that habitats are easy or quick to restore. The take-home message was clearly that it is better not to disturb or destroy in the first place. Only the areas with the least disturbance recover to the point where differences from natural wetlands are undetectable, and then only slowly. This type of symposium should be available to decision-makers throughout the nation and at intervals when sufficient new information is available.

A second type of information transfer is needed for planners, resource managers, restoration ecologists, and consultants. As new techniques and monitoring recommendations become available, it will be important to transmit the findings to persons who are planning, reviewing or monitoring restoration projects. While much of the information developed by scientists will be presented at scientific meetings, these audiences may not include the majority of information users. Local workshops are needed to highlight new findings, along with publication of regional guidelines for accelerating restoration and protocols for assessing restored sites.

Acknowledgments

The information and ideas presented grew out of discussions with many resource agency staff and researchers, and especially my colleagues at the Pacific Estuarine Research Laboratory: R. Langis, S. Williams, R. Gersberg, J. Boland, B. Fink, T. Griswold and C. Nordby. The paper is an outgrowth of research supported by NOAA, National Sea Grant College, Department of Commerce, under grant number NA85AA-D-SG140, project number R/CZ-82, through the California Sea Grant College, and by the State Resources Agency and the California Department of Transportation. Comments from B. Nyden, G. Thayer and an anonymous reviewer improved the manuscript.

Literature Cited

California Department of Fish and Game. 1989. 1988 annual report on the status of California's State listed threatened and endangered plants and animals. The Resources Agency, Sacramento. 129 p.

Cammen, L. M. 1976a. Abundance and production of macroinvertebrates from natural and artificially established salt marshes in North Carolina. Am. Midl. Natur. 96:244-253.

Cammen, L. M. 1976b. Macroinvertebrate colonization of *Spartina* marshes artificially established on dredge spoil. Estuarine Coastal Mar. Sci. 4:357-372.

Cantilli, J. 1989. Sulfide phytotoxicity in tidal salt marshes. M. S. Thesis, San Diego State University, San Diego, California. 115 p.

Chalmers, A. 1982. Soil dynamics and the productivity of *Spartina alterniflora*, p. 231-242. *In* V.S. Kennedy (ed.), Estuarine comparisons. Academic Press, New York.

City of Carlsbad, California, and U.S. Army Corps of Engineers. 1990. Batiquitos Lagoon Enhancement Project Final EIR/EIS, Vol. I-III. Carlsbad, California.

Conners, D. H., F. Riesenberg IV, R. D. Charney, M. A. McEwen, R. B. Krone and G. Tchobanoglous. 1990. Research needs: Salt marsh restoration, rehabilitation, and creation techniques for Caltrans construction projects. Dept. of Civil Engineering, University of California, Davis. 61 p.

Conomos, T. J. 1979. Properties and circulation of San Francisco Bay waters, p. 47-84. *In* T.J. Conomos (ed.), San Francisco Bay: the urbanized estuary. Pacific Division, AAAS, San Francisco.

Covin, J. D. and J. B. Zedler. 1988. Nitrogen effects on *Spartina foliosa* and *Salicornia virginica* in the salt marsh at Tijuana Estuary, California. Wetlands 8:51-65.

Craft, C.B., S.W. Broome and E.D. Seneca. 1986. Carbon, nitrogen and phosphorus accumulation in man-initiated marsh soils, p. 117-131. *In* A. Amoozegar (ed.). Proceedings of the 29th annual meeting of the Soil Science Society of North Carolina, Raleigh, North Carolina.

Craft, C.B., S.W. Broome and E.D. Seneca. 1988. Nitrogen, phosphorus and organic carbon pools in natural and transplanted marsh soil. Estuaries 11: 272-280.

Dahl, T. E. 1990. Wetland losses in the United States 1780s to 1980s. U.S. Dept. of the Interior, Fish and Wildlife Service, Washington, D.C. 21 p.

Frenkel, R. E. and J. C. Morian. 1990. Restoration of the Salmon River salt marshes: Retrospect and prospect. Final Report to the U.S. Environmental Protection Agency, Seattle. 142 p.

Gersberg, R., F. Trindade and C. Nordby. 1989. Heavy metals in sediments and fish of the Tijuana Estuary. Border Health 5:5-15.

Griswold, T. 1988. Physical factors and competitive interactions affecting salt marsh vegetation. M. S. Thesis, San Diego State University, San Diego, California. 84 p.

Ibarra-Obando, S. E. and A. Escofet. 1987. Industrial development effects on the ecology of a Pacific Mexican estuary. Environ. Conserv. 14:135-141.

Jorgensen, P. 1975. Habitat preferences of the light-footed clapper rail in Tijuana Estuary Marsh, California. M.S. Thesis, San Diego State University, San Diego, California. 115 p.

Josselyn, M. 1983. The ecology of San Francisco Bay tidal marshes: A community profile. U.S. Fish and Wildlife Service, Division of Biological Services, FWS/OBS-83/23. 102 p.

Josselyn, M., M. Martindale and J. Duffield. 1989. Public access and wetlands: Impacts of recreational use. Romberg Tiburon Center Tech. Rep. 9. 56 p.

Kramer, S. H. 1990. Habitat specificity and ontogenetic movements of juvenile California halibut, *Paralicthys californicus*, and other flatfishes in shallow waters of southern California. National Marine Fisheries Service, Southwest Fisheries Science Center Administrative Report LJ-90-22. La Jolla, California.. 157 p.

Kusler, J. and M. Kentula (eds.). 1989. Wetland creation and restoration: the status of the science. U.S. Environmental Protection Agency, EPA 600/3-89/038a,b. 2 vols.

Langis, R., M. Zalejko and J. B. Zedler. 1991. Nitrogen assessments in a constructed and a natural salt marsh of San Diego Bay, California. Ecol. Applications 1:40-51.

Lindau, C.W. and L.R. Hossner. 1981. Substrate characterization of an experimental marsh and three natural marshes. Soil Sci. Soc. Amer. J. 45: 1171-1176.

Moy, L. D. 1989. Are *Spartina* marshes renewable resources? A faunal comparison of a man-made marsh and two adjacent natural marshes. M.S. Thesis, North Carolina State University, Raleigh.

Neuenschwander, L., T. H. Thorsted, Jr. and R. Vogl. 1979. The salt marsh and transitional vegetation of Bahía de San Quintin. Bull. S. Calif. Acad. Sci. 78:163-182.

Nordby, C. S. and J. B. Zedler. In press. Responses of fishes and benthos to hydrologic disturbances in Tijuana Estuary and Los Peñasquitos Lagoon, California. Estuaries.

Onuf, C. 1987. The ecology of Mugu Lagoon, California: An estuarine profile. U.S. Fish. Wildl. Serv. Biol. Rep. 85(7.15). 122 p.

PERL (Pacific Estuarine Research Laboratory). 1990. A manual for assessing restored and natural coastal wetlands with examples from southern California. Calif. Sea Grant Rep. T-CSGCP-021.

Rutherford, S. 1989. Detritus production and epibenthic communities of natural versus constructed salt marshes. M.S. Thesis, San Diego State University, San Diego, California. 79 p.

Sacco, J. N. 1989. Infaunal community development of artificially established salt marshes in North Carolina. M.S. Thesis, North Carolina State University, Raleigh.

Sacco, J. N., F. L. Booker and E. D. Seneca. 1987. Comparison of the macrofaunal communities of a human-initiated salt marsh at two and fifteen years of age, p. 282-285. *In* J. Zelazny and S. Feierabend (eds.), Increasing our wetland resources. National Wildlife Federation, Washington, D.C.

Seamans, P. 1988. Wastewater creates a border problem. J. Water Pollut. Control Fed. 60:1797-1804.

Smith, Gordon W. 1990. The fight for a restored wetland in Huntington Beach. Coast and Ocean 6(2):42-45.

Swift, K. L. 1988. Salt marsh restoration: Assessing a southern California example. M.S. Thesis, San Diego State University, San Diego, California. 84p.

Valiela, I., J. M. Teal, C. Cogswell, J. Hartman, S. Allen, R. Van Etten and D. Goehringer. 1984. Some long-term consequences of sewage contamination in salt marsh ecosystems, p. 301-316. *In* P. J. Godfrey, E. R. Kaynor, S. Pelczarski and J. Benforado (eds.), Ecological considerations in wetlands treatment of municipal wastewaters. Van Nostrand Reinhold, New York.

Williams, P. and M. Swanson. 1987. Tijuana Estuary enhancement hydrologic analysis. California State Coastal Conservancy, Oakland. 49 p.

Winfield, T. P., Jr. 1980. Dynamics of carbon and nitrogen in a southern California salt marsh. Ph.D. Dissertation. University of California, Riverside, and San Diego State University, San Diego, California. 76 p.

White, A. 1986. Effects of habitat type and human disturbance on an endangered wetland bird: Belding's Savannah sparrow. M.S. Thesis, San Diego State University, San Diego, California. 73 p.

Zalejko, M. 1989. Nitrogen fixation in a natural and constructed salt marsh. M.S. Thesis, San Diego State University, San Diego, California. 71 p.

Zedler, J. B. 1980. Algal mat productivity: comparisons in a salt marsh. Estuaries 3:122-131.

Zedler, J. B. 1982. Salt marsh algal mat composition: Spatial and temporal comparisons. Bull. S. Calif. Acad. Sci. 81:41-50.

Zedler, J. B. 1984. Salt marsh restoration: A guidebook for southern California. Calif. Sea Grant Rep. T-CSGCP-009. 46 p.

Zedler, J. B. 1986. Catastrophic flooding and distributional patterns of Pacific cordgrass (*Spartina foliosa Trin.*). Bull. S. Calif. Acad. Sci. 85:74-86.

Zedler, J. B. 1988a. Salt marsh restoration: Lessons from California, p. 123-138. *In* J. Cairns (ed.), Rehabilitating damaged ecosystems. CRC Press, Boca Raton, Florida.

Zedler, J. B. 1988b. Restoring diversity in salt marshes: Can we do it?, p. 317-325. *In* E. O. Wilson (ed.), Biodiversity. National Academy Press, Washington, D.C.

Zedler, J. B. and C. S. Nordby. 1986. The ecology of Tijuana Estuary: An estuarine profile. U.S. Fish Wildl. Serv. Biol. Rep. 85 (7.5). 104 p.

Zedler, J. B., C. S. Nordby and T.Griswold. 1990. Linkages: Among estuarine habitats and with the watershed. Tech. Memo., National Oceanic and Atmospheric Administration, National Ocean Service, Office of Coastal Resource Management. Washington, D.C. 71p.

CHAPTER 2

Restoring Tidal Marshes in North Carolina and France

Ernest D. Seneca and Stephen W. Broome

Department of Botany and Soil Science
North Carolina State University, Raleigh, North Carolina

Abstract

Tidal marshes are important ecosystems that provide life support, water quality and hydrologic functions. These marshes are often severely impacted by dredge and fill, subsidence, tidal restriction, oil spills, pipelines, drilling, defense operations, highway construction and other development activities. When such actions occur, it is important to use appropriate techniques to rehabilitate or restore the affected systems or establish equivalent new marshes. Critical aspects of tidal marsh restoration or creation that should be considered are careful site selection, elevation, slope, and tidal range, wave climate, salinity, soil physical and chemical properties, cultural practices, sedimentation, wildlife predation and traffic. The feasibility of restoring or creating tidal marshes has been adequately demonstrated, but functional equivalency is difficult to document and requires extensive sampling over long periods of time. Requirements for assessment of functional equivalency are likely to remain controversial; however, wetland creation and restoration technology should continue to be developed and used particularly when unavoidable wetland losses occur.

Intertidal Marsh Habitat Type

Estuarine tidal marshes occupy that portion of the intertidal zone from about mean sea level (m.s.l.) to the upper level of spring tides along many temperate coastlines of the world. At high latitudes (ca. 50°) on the North American and European coasts, the tide range is often 5-8 m, but marshes only occupy the upper 1 m or so of this intertidal zone. Along the Gulf coast of the United States, marshes often occupy half of the <0.5 m tide range. The relationship between tide range and marsh occurrence becomes especially important in restoration undertakings. This paper deals primarily with the vegetation component of tidal marshes, but we recognize that the faunal component should be understood and monitored in initial restoration projects within a geographic location.

Although the plant species change from one coastline of the world to another, the emergent macrophytes which characterize these marshes fill similar niches, have similar adaptations, and perform similar functions. Together with epiphytic, benthic, and planktonic algae and submerged aquatic vegetation, they form the base of the estuarine food web, and contribute especially to the detrital component. Further, these macrophytes provide valuable habitat for other organisms, including those actually attached to the plants like barnacles, epiphytic algae, and salt marsh periwinkles, plus those that move about more freely like insects and spiders. The vegetation also provides foraging habitat for birds and mammals and protective habitat for smaller finfish as they seek shelter from predators.

An important function of emergent macrophytes and one absolutely essential to perpetuation of the marsh itself along coastlines undergoing subsidence or sea level rise is that of sediment removal or trapping of sediments. When nutrients such as phosphorus are attached to sediment particles, trapping of sediments by vegetation retains an important earthbound nutrient in the estuarine system rather than it being carried onto the continental shelf and deposited or being lost to the deep ocean depths where recycling is dependent on longer term biogeochemical events. Emergent vegetation traps sediments, takes up, removes, and recycles certain nutrients that might otherwise cause eutrophication. These plants increase the residence time of certain nutrients in the system and in selected cases may remove toxic materials and give the system more time to assimilate them by transformation and decomposition processes.

Emergent macrophytes not only trap sediments, they hold them and by so doing function to stabilize shorelines. The leaves and stems of these plants serve as natural groins to trap sediments and the roots and rhizome system bind the sediment particles together to provide stability. Estuarine shoreline erosion is not as severe in areas with extensive marshes, which function as buffers to storm waves and stabilize sediments.

Collectively, these functional values of estuarine tidal marshes make them some of the most vital systems on earth. Their protection is ecologically justified and their timely rehabilitation and restoration following damage or destruction should be a high priority.

Impacts and Responses

With the established value and accepted importance of these vital systems, logic would dictate that humans would never violate them. What then, are the threats or what represents potential impacts of these systems? First, it must be acknowledged that scientists, policy makers, managers, and the public have or do not always realize the values of these systems. Because of the lack of a good data base, indifference, and other priorities, it is estimated that from 30-50% of our nation's wetlands had been lost by the 1950s (U.S. OTA 1984), and another 10% of our coastal wetlands were lost from the mid-1950s to the mid-1970s, 8% of which were tidal marshes (Mitsch and Gosselink 1986). Much of this loss was due to filling in activities which occurred from the turn of the present century until dredge and fill legislation in the 1970s. We are still continuing to lose marshes in a slow, piecemeal, cancerous-like manner by development projects which seek to convert them to uplands or dredge them out for marinas and boat basins.

With sea level rise and subsidence of some coastlines, marshes must accrete, migrate landward/upstream, or disappear. Along the Louisiana Gulf coast of the United States, accretion in streamside marshes is keeping up with apparent sea level rise, 90% of which is due to coastal subsidence (Delaune et al. 1983). The interior marshes, removed from the stream channels or more inland, are not accreting enough to keep pace with apparent sea level rise and are disappearing.

Dredging for navigation channels is a necessity for intra- and interstate commerce, national security, commercial fishing, and recreation vessels. Management of dredged material disposal to include consideration of upland sites and estuarine habitat creation/enhancement sites is appropriate. Conscious, informed decisions must guide the dredge and disposal activities, no longer purely cost-benefit to the receiver of services (users).

Along the New England coast of the United States, many marshes have been altered by tidal restriction (Roman et al. 1984). It has been estimated that 10% of Connecticut's salt marshes have been influenced by tidal restriction. A specific site, the Pine Creek marsh in the Town of Fairfield, experienced a reduction of 96% from 1931 (175 ha of undisturbed marsh) to 1974 (7 ha of undisturbed marsh) (Steinke 1974). Current unpublished work by Salvatore Bongiorno, Fairfield University, has subsequently documented partial recovery of vegetation and invertebrates following removal of a tidal restricting dike on a portion of this marsh. *Phragmites*-dominated marsh is giving way to *Spartina*-dominated marsh and fiddler crabs, salt marsh periwinkles, and ribbed mussels are returning.

In recent decades, petroleum and other toxic materials spills have become more prevalent. The *Torrey Canyon* (Hawkins and Southward, this volume) and *Amoco Cadiz* oil spills off the Brittany coast of France and more recently the *Exxon Valdez* along the Alaskan coast have not only influenced sensitive coastal ecosystems but have incited public awareness of the severe effects of these types of impacts. The deliberate spill in the Persian Gulf also may have far-reaching ecological effects. Not only are ecological systems, their biotic and abiotic components damaged, stressed, and in more severe cases destroyed, human livelihoods including tourism, mariculture, and fishery resources are also influenced detrimentally.

The examples given of impacts to these valuable systems, plus others to include pipelines and drilling operations for offshore petroleum resources, military defense operations, and willful alteration by a variety of human activities such as highway construction, emphasize the importance of developing appropriate techniques and procedures to establish, rehabilitate, and restore tidal marshes as conditions dictate. We must consider humans and their activities in the grand scheme of coastal ecosystems/human interactions. Aside from filling, and given sufficient time, marshes can recover from most human perturbations. The time required, however, is difficult to predict. Marsh establishment and restoration technology can lend some increased probability of successful recovery in much shorter spans of time, often cutting recovery time in half or less under appropriate management.

An important consideration of impact to and response of marsh systems to perturbation, especially human-initiated, is whether the impact is permanent or temporary. In the case of fill activities, which are designed to convert marshes to other uses, the perturbation usually results in a permanent loss of habitat. However, where illegally-placed or accidentally-placed fill material has been on the marsh only a short period of time, a few weeks during the growing season or perhaps longer during the colder part of the year, the marsh can recover naturally following removal. The marsh may in fact recover with robust growth due presumably to the addition of nutrients from the fill material. In the case of a toxic materials spill, naturally occurring microbes and physical/chemical processes often enable the system to recover naturally within some time frame also. The loss in this case is temporary in the grand scheme of things, but the loss of certain critical functions during this time inflicts a cost to the ecosystem and to society. Tidal restriction, likewise may be temporary because current studies suggest that it can be reversed. In this case, it is a management decision and usually an alteration from one type of marsh to another, possibly high marsh to low marsh. Permanent changes of these systems should be avoided where at all possible and mitigation should be required where permanent alteration is unavoidable.

Restoration and Creation

Case Histories

An example of successful restoration of marsh vegetation on a portion of the marshes damaged and/or destroyed by the *Amoco Cadiz* oil spill and associated cleanup operations was accomplished at Ile Grande, France (Figure 1). At this site we adapted techniques and procedures developed along the North Carolina coast, U.S.A., to the plants and sediments at Ile Grande (Seneca and Broome 1982, Broome et al. 1988). Although several native marsh plants were tested, the most successful transplants were sprigs (roots without attached substrate) of *Halimione portulacoides* and plugs (roots with a core of substrate) of *Puccinellia maritima* planted on a 0.5-m spacing. After we selected the best plant species, determined the best type of transplant, fertilizer amendment, and season to plant, complete cover could be achieved within two years.

In addition to these determinations, we established nursery areas for each of the species and compared transplants from the nurseries

(a)

(b)

Figure 1. *(a)* *Transplanting* Puccinellia maritima *and* Halimione portulacoides *in May, 1981, at Ile Grande, France where marsh vegetation was destroyed by the* Amoco Cadiz *Oil Spill; (b) the same location in May, 1987.*

with those taken from natural stands in the vicinity. Those transplants from the nurseries survived and grew as well as those taken from natural stands. A comparison of planted areas with areas of the marsh not planted indicated that natural processes were very slow to develop and plants were slow to recolonize these sites, especially perennial macrophytes (Levasseur and Jory 1982). It is estimated that our transplanting and restoration efforts produced vegetative cover in two years that would have taken natural processes 5-10 years to accomplish.

A second example of successful restoration of marsh vegetation was accomplished at Pine Knoll Shores, Bogue Banks, North Carolina. In the late 1960s, dredged material from a small boat channel about 100 m offshore was deposited on existing marsh. The marsh vegetation was smothered and shortly thereafter the shoreline began to erode. Installation of an asbestos sheet-piling type bulkhead failed to solve the erosion problem and was in danger of being undermined throughout much of its length. At the request of a homeowners' association, we established 12-m wide experimental plantings of *Spartina alterniflora* along the shoreline adjacent to the bulkhead.

Transplants taken from a nearby dredged material disposal site performed equally well with greenhouse-grown seedlings. Spacings of 45- and 60-cm were more desirable than a 90-cm spacing at this site. By the end of the second growing season or about 18 months after planting, there was a good stand over much of the experimental area except in the immediate vicinity at the base of the bulkhead. By the end of the third growing season, rhizomes had reclaimed this critical area also and there was a continuous stand of vegetation over most of the experimental area. Details of this reclamation effort and growth of the marsh vegetation were documented over a 10-year period (Broome et al. 1986).

Experimental Approaches

Restoration or creation of tidal salt marshes may be accomplished in a number of ways depending on the environment and circumstances at a particular location (Lewis 1982; Daiber 1986; Broome 1989). For some altered tidal marshes, simply removing obstructions like dikes and culverts that prevent or restrict tidal exchange may be sufficient to cause reversion of the flora and fauna to the original species composition. Under many conditions, establishment of marshes can be greatly accelerated by seeding or transplanting the dominant vegetation. For example, shorelines exposed to moderate wave energy nearly always require transplanting to establish marsh. Also, establishment of vegetation on freshly deposited dredged material can be greatly accelerated by seeding or transplanting (Figure 2). This is especially true where dredged material is exposed to wave action or isolated from a source of seed.

Availability of plant propagules is often a limiting factor in transplanting or seeding tidal salt marsh vegetation. Seed or transplants should come from near the planting site to avoid introducing ecotypes that are not adapted to local conditions. Ecotypic variation is prevalent in populations of plants that have wide geographic ranges (Kadlec and Wentz 1974) including certain dominant marsh macrophytes (Seneca 1974).

Propagation techniques for marsh vegetation include direct seeding, transplanting sprigs or plugs dug from natural stands, and growing seedlings in pots or flats. The description of propagation methods that follows is from our experience in North Carolina and the Brittany Coast in France. The plant species that were successfully propagated in North Carolina were *Spartina alterniflora, S. patens, S. cynosuroides* and *Juncus roemerianus*. Those established in France were *Puccinellia maritima* and *Halimione portulacoides*.

Seed production of *Spartina* species is most prolific in recently colonized open stands or along marsh edges such as creek banks. Seed should be harvested when mature, just prior to shattering.

(a)

(b)

Figure 2. (a) Transplanting Spartina *on a dredged material disposal island in Core Sound, North Carolina in March, 1987; (b) the same island six months later in September, 1987. This plant growth illustrates how development of new marsh is accelerated by transplanting vegetation.*

This occurs generally from late September to mid-October, with *S. cynosuroides* seed maturing about two weeks earlier than those of *S. patens* and *S.alterniflora*. Maturity dates vary from North to South, year to year, and even within stands. Seed heads can be clipped with knives, clippers or any suitable mechanical harvester. Storing the seedheads under refrigeration (2-4°C) for 3-4 weeks before threshing results in easier separation of the seeds and stems. Threshing may be done by hand or with threshers used for small grain.

The threshed seed of *S. alterniflora* and *S. cynosuroides* must be stored (refrigerated) in containers filled with seawater or artificial seawater so that the seeds are submerged. The seed are not viable if allowed to dessicate. The salinity of the water used should be about 35 ppt for *S. alterniflora* seed and about 10 ppt for those of *S. cynosuroides. Spartina patens* seed may be stored dry under refrigeration.

Spartina alterniflora can be established by direct seeding on sites protected from waves, but successful germination and growth is limited to the upper half of the intertidal zone. After preparing the seed bed, seed should be evenly distributed at the rate of about 100 viable seed per square meter and incorporated to a depth of 2-3 cm using any type of tillage equipment (optimum seeding dates are April 1-May 30). Direct seeding is economical but there are disadvantages. It is not usually successful on sites exposed to wave energy, and the large quantities of seed required are usually difficult to obtain. Growing potted seedlings is a more efficient way of utilizing limited quantities of seed.

Seedlings can be grown in pots or flats in a greenhouse or outdoors in warm climates. Plastic tray pack liners with 36 or 42 individual compartments are good containers for plant production. Plastic liners are better than peat pots because roots and rhizomes are confined to individual planting units. Roots and rhizomes grow through peat pots creating a solid mat that must be cut apart. Growth media may be a commercial potting mix or a mixture of equal parts sand, peat and sterilized topsoil. Immediately before planting, the seed should be treated with a fungicide. An effective treatment method is soaking seed in a solution of 25% household bleach for 10-15 minutes and rinsing with tap water. Plant several seeds in each pot or compartment to depths of 1-2 cm and keep the potting media wet with daily watering. Flooding is not necessary. The seedlings should be fertilized as needed with Hoagland's solution or commercial fertilizers according to rates recommended on the package. Diseases, such as damping off, are often a problem and must be treated with a fungicide.

Some advantages of using seedlings grown in tray packs or pots are that survival when planted in the field is virtually 100% since the root system remains intact; disturbance of natural stands is avoided; pot-grown seedlings provide a source of plants when suitable digging sites are not available; and potted seedlings can be held longer than bare-root plants in case of delays in construction or site preparation.

Disadvantages of growing seedlings include the high cost (40-60 cents per plant); the necessity of advance planning to collect seed and allowing enough time to grow the seedlings; the difficulty of transporting bulky pots or tray-packs; and potting media that does not contain the marsh soil flora and fauna associated with the roots of field-dug transplants.

Transplants can be obtained from recently established stands growing on sandy substrates such as dredged material disposal sites, around inlets where sand is accumulating or along marsh edges. Old marshes have dense root mats making the plants difficult to obtain and usually of poor quality. Plants from new stands are generally more vigorous, have larger stems and are easier to dig and separate. Good transplants should have a well developed root system and have small shoots and rhizomes attached. They can be dug by loosening the soil with a shovel and separating individual plants at the rate of 200-500 plants per man hour. Damage to young stands on sandy soils is minimal since shoots and rhizomes that remain quickly revegetate the disturbed area. After plants are dug, the roots must be kept moist by packing in moist sand or peat moss until transplanted. Plants also may be kept for a few weeks by placing the roots in a trench dug at an intertidal site.

Plugs are planting units that include a soil core, root mass and associated shoots. Because more labor is required to harvest, transport and transplant plugs they are usually less desirable than single stem transplants for the *Spartina* species; however, if the planting stock must be obtained from old marshes growing on peat, clay or silty soils, plugs may be the best alternative. We found that plugs were the best propagule for transplanting *Puccinellia maritima*, a fine-stemmed grass.

Transplanting techniques are similar for field-dug and pot-grown plants and can be done manually or mechanically. Mechanical transplanters, such as used for tobacco or vegetable plants, are efficient on large, accessible sites with soil material that will support equipment and is relatively free of debris. Mechanical planters produce a uniform planting and are much faster and less laborious. Manual planting, however, is the method most often used for intertidal vegetation because of conditions of the substrate and availability of equipment. Tree-planting dibbles or spades work well to open planting holes to a depth of about 15 cm. Both manual and machine planting should be done at low tide when there is no water on the surface. Portable power augers may be advantageous on compact soils (Garbisch et al. 1975). Transplanting plugs of *P. maritima* at Ile Grande, France was accomplished manually using a 6.5-cm diameter soil auger.

Transplant spacing is an important consideration since it affects the number of plants required, how rapidly cover is achieved and the probability of success on sites exposed to wave action. Rate of spread differs among species and should be taken into account. On favorable protected sites, a 1-m plant spacing for *Spartina alterniflora* and *S. patens* is adequate to achieve complete cover in one year.

A 0.5- to 0.6-m spacing should be used for *S. cynosuroides, J. roemerianus, P. maritima* and *H. portulacoides* on protected sites and for *S. patens* and *S. alterniflora* where wave action is a factor.

Reintroduction of Fauna

The type of faunal community that develops in restored or created tidal marsh is likely to be site dependent. If a physical environment with hydrology, vegetation and soils similar to natural marshes in the area is created, it is logical to predict that fauna, which utilize such habitat, will colonize the new marsh. Tidal flooding facilitates transport of immature or less mobile stages of organisms into a restored or created marsh, while birds, mammals, reptiles, invertebrates, fish and other mobile fauna can migrate to the new habitat. If the assumption of natural recolonization by fauna is accepted, an important question to ask is the length of time required for this to occur. The time required is difficult to predict and is affected by a number of environmental factors. Studies of faunal components of planted marshes have been reported and several are summarized below.

Cammen (1976a, b) sampled macroinvertebrates in a two-year old planted marsh and found the average density to be much lower and species composition quite different from that of natural marshes. The same planted marsh was sampled after 15 years and the average density of macrofauna had increased 10-fold (Sacco et al. 1988). Species composition also changed over time. At the two-year sampling date, the dominant macrofauna were amphipods, mysids and dipteran larvae, but after 15 years, oligochaetes and polychaetes accounted for 94% of the density. The density and species composition after 15 years was similar to that of natural marshes.

Newling and Landin (1985) reported that plant communities at six U.S. Army Corps of Engineers marsh creation sites observed over an eight-year period were equal to or more productive than those of natural reference marshes. This study also determined that wildlife use was greater in the created marshes than in the natural marshes. A study in Texas found consistently lower densities of brown shrimp, grass shrimp, pinfish and gobies (but equal densities of juvenile blue crab) in planted marshes five to six years in age as compared to natural reference marshes (Minello et al. 1986).

Sacco (1989) compared species composition, density and trophic structure of transplanted marshes (ranging in age from 1-17 years) with natural marshes at six sites along the North Carolina coast. The two marsh types were found to have similar organisms and community trophic structure, but total density was lower in the planted marshes. Sacco postulated that the lower densities may be related to lower organic matter content in the substrate of the planted marshes. The study indicated that recreating the infaunal community of a natural marsh is possible, but the time required is variable and related to site-specific factors.

West (1990a) compared growth and survival of the benthic feeding fish *Leiostomus xanthurus* (spot) in enclosures in creeks in natural and created brackish water marshes that were planted two years earlier. After being held for two weeks, growth and survival of the fish were similar in the two marsh creeks.

The benthic invertebrate communities in these creeks were also evaluated over a period of two to four years after the planted marsh was established (West 1990b). Results indicated differences in invertebrate community structure among the creeks, but the results suggested that creeks in the created marsh area are capable of supporting a benthic invertebrate community with density and diversity similar to that of the natural areas. The species diversity of the benthos in the natural marshes is low because of the stress of the salinity extremes, and the species adapted to that environment are apparently able to adapt quickly to a wide range of conditions in the created marsh creeks.

Current Recommended State-of-the-Art

Project Goals. The objective of tidal marsh restoration or creation is to initiate development of an ecosystem that has the structural characteristics and provides the functional values of hydrology, water quality improvement and food chain support of natural marshes in the area. Typical goals of these projects are dredged material stabilization, creation of habitat using dredged material, shoreline erosion control, and replacement of marshes lost to adverse impacts such as filling or diking.

The feasibility of creating tidal marshes has been adequately demonstrated, but functional equivalency is difficult to document and often requires a long period of time (Kusler and Kentula 1989). Many wetland scientists are concerned that restoration and habitat construction are not producing marshes that carry out the broad range of wetland functions and that transplant survival has become the measure of success (Pacific Estuarine Research Laboratory 1990).

When setting project goals, it is important to determine what constitutes success, whether functional assessment is to be required, the parameters that should be measured, and the length of time required to complete the evaluation. There is a need for strict assessment criteria; however, meaningful criteria must be developed that are also economical and practical to measure.

Preconstruction Considerations. Careful planning and site selection increases the probability of success of wetland creation or restoration. There are a number of physical, chemical and biotic conditions that must be considered in determining the feasibility of marsh establishment. "Hydrology is probably the single most important determinant for the establishment and maintenance of specific types of wetlands and wetland processes" (Mitsch and Gosselink 1986). Attempts to create or restore tidal marshes are certain to fail without providing the proper elevation, tidal flooding, and drainage. Other factors that must be considered are wave climate, salinity, soil physical and chemical properties, sedimentation, plant species, wildlife and human activity and cultural practices such as plant material and fertilization.

Critical Aspects to Ensure Success. Projects should be planned to allow time for obtaining permits and completing construction well in advance of optimum planting dates. Permitting and construction delays that result in postponement of planting jeopardize plant survival and growth, and expose the site to erosion. If natural seeding is relied upon, opening a site to tidal flow or grading to the correct elevation should occur just before seed drop by the desired plant species.

Elevation, slope and tidal range interact to determine the areal extent of the intertidal zone and the depth and duration of submergence. These factors control the elevation range and zonation of plant species within the marsh. Elevation requirements of marsh vegetation at a given site is best determined by observing the upper and lower limits of the dominant plant species in a nearby natural marsh. A surveyors level can be used to relate elevation limits of the reference marsh to the planting site. The elevation limits are especially critical in areas with a small tidal amplitude that results in a narrow elevation range for marsh establishment.

Marsh vegetation grows on a wide range of slopes, but they should be as gentle as possible (1-3%) while still providing good surface drainage. Gentle slopes provide more intertidal area and dissipate wave energy, reducing the erosion potential. It is generally easier to establish marsh where there is a wide, regular tidal range than in an area with irregular wind-driven tides. This is especially true along shorelines exposed to waves, because wider marsh fringes are more effective in damping waves and resisting erosion.

Channels or creeks should be provided for large marsh systems to insure good drainage and tidal exchange. The tide is in effect, an energy subsidy facilitating the exchange of sediments, nutrients and organic matter between the marsh and the estuary. Tidal channels also promote greater use by fishes, benthos, and shorebirds (Newling and Landin 1985). Closing of the tidal inlet may be a problem at some created marsh sites because of sediments being deposited by waves or long-shore drift.

Initial establishment and long-term stability of marshes is affected by exposure to wave energy. Shoreline characteristics that are useful indicators of severity of wave climate are average fetch, longest fetch, shore configuration and grain size of sediments. Knutson et al. (1981) developed a site evaluation form using these indicators. Other factors to consider are boat traffic and depth of water offshore. Along channels, boat and ship traffic create significant waves, while shallow offshore water reduces severity of waves reaching the shore (Knutson and Innskeep 1982).

Salinity of the tidal and interstitial water determine the plant species that should be planted and the flora and fauna that will eventually colonize a site. Excess salinity can be a problem in marsh plant establishment on some sites. Concentrated soil solutions may result from evaporation of water standing in surface depressions at low tide or held by restrictive clay layers. Freshwater seepage from adjacent uplands benefits plant growth by lowering salinity and providing nutrients.

Physical and chemical properties of the soil are important aspects of marsh creation and restoration. Mechanical operations such as grading, shaping and planting are generally easier to accomplish on sandy soils than on silt, clay or organic matter. A disadvantage of sandy material is its low nutrient supplying capacity, but this may not be a problem where tidal waters deposit nutrient rich sediments.

We have found nitrogen to be a limiting factor in plant growth on most planting sites, and phosphorus became a limiting factor on some sites when adequate nitrogen was applied (Broome et al. 1983). The growth response to fertilization depends on the inherent fertility of the soil and the amount of nutrients supplied by tidal inputs, seepage, runoff, precipitation and nitrogen fixation. Success of some marsh creation or restoration projects depends on application of nitrogen and phosphorus fertilizers at planting and several years thereafter to restore soil nutrient pools. Soil tests should be used to determine the rate of phosphorus to apply. Nitrogen applied at the rate of 112 kg ha^{-1} has been effective in increasing growth. The fertilizer should be incorporated into the substrate or a slow-release fertilizer material may be applied in the planting hole. On most sites an application of fertilizer before or at planting was sufficient to supply nutrients until the natural nutrient cycling processes were restored. Plant growth on poor sites was enhanced by topdressing the second growing season with soluble materials. Ammonium sulfate was the most effective source of nitrogen and triple superphosphate was the best phosphorus material.

Where grading is necessary to establish proper elevations, stockpiling and replacing the topsoil should be considered. At most locations, topsoil has better chemical and physical properties and higher organic content than the underlying material. A favorable substrate material is likely to facilitate functional development of the marsh and colonization by infauna.

Cultural practices are the techniques required for fostering plant growth. These include selecting plant species adapted to environmental conditions at the site, seed collection, seed storage, seedling production, site preparation, using vigorous transplants that are handled and planted with care, proper fertilization and plant spacing. Attention to such details, which might seem inconsequential, can be the difference in success and failure of establishment of marsh vegetation.

Other factors that may affect success at some sites are sedimentation, traffic, wildlife predation, contaminated sediments and availability of sunlight. A moderate amount of sedimentation stimulates growth by supplying nutrients and providing a mechanism for the marsh surface to keep pace with sea level rise; however, excessive accumulation can damage plants and increase elevations above the range of intertidal vegetation. Blowing sand is a common problem on dredged material that may result in burial of marsh vegetation. Sand fencing and/or vegetation should be used to trap sand to protect marsh vegetation. Excessive foot or vehicular traffic will damage marsh vegetation and must be excluded from new plantings. Providing wildlife habitat is a key function of tidal marshes, but excessive use and feeding during the establishment period can destroy a planting. New plantings are vulnerable to feeding by Canada and snow geese and muskrat. The possibility of toxic contaminants in dredged material from industrial or heavily populated areas should also be considered. Plant uptake of toxic elements may introduce these materials to the food web. Lastly, insufficient sunlight due to shading by trees was found to be a limiting factor to marsh establishment along some creeks in Virginia (Hardaway et al. 1984).

Follow-Up Needs

Ideally, a restored or created marsh should be self-sustaining and free of maintenance; however, maintenance during the first few years is necessary in many cases to insure success. Along shorelines exposed to wave action, one of the first maintenance requirements is replacement of transplants that wash out. Litter along drift lines should be removed if there is a danger of smothering plants. Topdressing with fertilizer in the second or third growing season may be necessary on infertile sites. Another potential problem on brackish and freshwater sites is invasion by undesirable plant species such as *Phragmites australis*. These weeds may have to be removed to develop the type of marsh desired. One means of discouraging invasion by unwanted plants on restoration or creation sites is to promote rapid cover of the transplanted vegetation by close spacing and other good management practices. Some more drastic methods for controlling undesirable plants are cutting, draining, saltwater flushing and herbicides. It should be remembered that marsh restoration or creation is not necessarily a one-time effort. Problems that arise should be addressed to improve the chances of success.

Many scientists feel that there is a need for strict assessment criteria and long-term monitoring to determine if marsh restoration and creation projects carry out the broad range of functions of natural systems (Pacific Estuarine Research Laboratory 1990). The Pacific Estuarine Research Laboratory (1990) recommends that the following attributes be considered in assessing how well constructed wetlands replace the function of natural wetlands: hydrology, topography, soils, nutrient dynamics, algae, vascular plants, and consumers. A sampling program that provides meaningful data to prove or disprove functional equivalency is not easy. The degree of monitoring and the length of time required for an accurate assessment is likely to remain controversial. Detailed sampling programs are impractical for many small projects, but perhaps through research, indicator criteria can be developed that are both meaningful and practical to measure.

Future Needs and Directions

The primary goal of those of us involved in marsh restoration is to assist the altered system to revert to its original structural composition and functional capacity in a timely manner. It is not reasonable or possible to re-establish all organisms (plant, animals, microbes) that constituted the original system. Our objective initially is to restore the dominant emergent vegetation or some part of it with the assumption that once this component is established, the other organisms will follow as conditions favor their establishment. Although this assumption is not valid in every case, establishing the dominant vegetation is a key step in restoration. Emergent macrophytes serve as substrate for certain other organisms, produce live and dead organic matter both above and below ground for the food web, provide habitat for other organisms, cycle nutrients, trap and stabilize sediments, and serve as a buffer for storm waves. Ultimately, we are concerned with how favorable the human-established/human-assisted system compares structurally and functionally to a natural system counterpart in the geographic vicinity. In addition to the density and production (above and below ground) of the macrophytes, it is important to determine use of the site by other organisms, to include birds (Bruton 1988), infauna (Sacco 1989), and larger invertebrates such as the salt marsh periwinkle and fiddler crabs, and also finfish and shellfish that use these habitats when they are flooded.

Because in restoration operations we are putting back a system that has already been damaged or destroyed, long term detailed sampling/monitoring is only justified from a scientific perspective to serve as a data base. It is a different game plan than setting out to replace a system that, for some reason, is to be destroyed (e.g., by development) and must be mitigated for by replacement in kind. In this latter case, it is not unreasonable to require monitoring of the site for selected criteria (primary production, infaunal species composition and density, avian use, etc.) for a specified time period. Monitoring the vegetation composition and production, rate of organic matter incorporation in the sediments to determine the carbon base (Craft et al. 1988, 1989), and faunal utilization are important from a knowledge perspective and to determine recovery time. A reasonable monitoring period might be five years to make a valid assessment of success and to make reasonable predictions of further development.

All this knowledge of marsh restoration is of little use unless it is incorporated into contingency plans and management decisions to ensure utilization. Communication among those who develop restoration technology, those who ensure its use, and those who use it is paramount to successful restoration of our valuable estuarine tidal marshes. There must be a common level of understanding and expertise as to what can be achieved within a given time frame in these undertakings. As natural forces and human populations interact with these estuarine systems, restoration technology will become increasingly more important and necessary as we strive to maintain biodiversity and ecosystem productivity, and a healthy, productive and aesthetically pleasing environment.

Acknowledgments

This paper is based on research supported by the North Carolina Agricultural Research Service, NOAA/CNEXO Joint Scientific Commission, and the Coastal Engineering Research Center, U.S. Army Corps of Engineers. Special thanks for technical assistance go to Messrs. Carlton L. Campbell and Larry L. Hobbs.

Literature Cited

Broome, S.W. 1989. Creation and restoration of tidal wetlands of the Southeastern United States, p. 37-66. *In* J.A. Kusler, and M.E. Kentula (eds.), Wetland creation and restoration: the status of the science. U.S. Environmental Protection Agency, EPA600/3-89-038a.

Broome, S.W., E.D. Seneca and W.W. Woodhouse, Jr. 1983. The effects of source, rate and placement of N and P fertilizers on growth of *Spartina alterniflora* transplants in North Carolina. Estuaries 6:212-226.

Broome, S.W., E.D. Seneca and W.W. Woodhouse, Jr. 1986. Long-term growth and development of transplants of the salt-marsh grass *Spartina alterniflora*. Estuaries 9:63-74.

Broome, S.W., E.D. Seneca and W.W. Woodhouse, Jr. 1988. Tidal salt marsh restoration. Aquat. Bot. 32:1-22.

Bruton, J.G. 1988. Avian response to man-initiated brackish water marshes. M.S. Thesis, North Carolina State University, Raleigh. 75p.

Cammen, L.M. 1976a. Abundance and production of macroinvertebrates from natural and artificially established salt marshes in North Carolina. Am. Midl. Nat. 96:487-493.

Cammen, L.M. 1976b. Macroinvertebrate colonization of *Spartina* marshes artificially established on dredge spoil. Estuarine Coastal Mar. Sci. 4:357-372.

Craft, C.B, S.W. Broome, E.D. Seneca and W.J. Showers. 1988. Estimating sources of soil organic matter in natural and transplanted marshes using stable isotopes of carbon and nitrogen. Estuarine Coastal Shelf Sci. 26:633-641.

Craft, C.B., S.W. Broome and E.D. Seneca. 1989. Nitrogen, phosphorus and organic carbon pools in natural and transplanted marsh soils. Estuaries 11:272-280.

Daiber, F.C. 1986. Conservation of tidal marshes. Van Nostrand Reinhold, New York. 341p.

Delaune, R.D., R.H. Baumann and J.G. Gosselink. 1983. Relationships among vertical accretion, coastal submergence, and erosion in a Louisiana Gulf Coast marsh. J. Sediment. Petrol. 53:328-334.

Garbisch, E.W., Jr., P.B. Woller and R.J. McCallum. 1975. Saltmarsh establishment and development. U.S. Army Corps of Engineers, Tech. Memo. 52. 110p.

Hardaway, C.S., G.R. Thomas, A.W. Zacherle and B.K. Fowler. 1984. Vegetative erosion control project: Final report. Virginia Institute of Marine Science, Gloucester Point. 275p.

Kadlec, J.A. and W.A. Wentz. 1974. State-of-the-art survey and evaluation of marsh plant establishment techniques: induced and natural. Vol. 1. U.S. Army Corps of Engineers, Waterways Exp. Stn., Vicksburg, Mississippi, Contract Rep. D-74-9.

Knutson, P.L., J.C. Ford and M.R. Inskeep. 1981. National survey of planted salt marshes (vegetative stabilization and wave stress). Wetlands 1:129-157.

Knutson, P.L. and M.R. Inskeep. 1982. Shore erosion control with salt marsh vegetation. U.S. Army Coastal Engineering Research Center, Fort Belvoir, Virginia, Technical Aid 8203. 23p.

Kusler, J.A. and M.E. Kentula (eds.). 1989. Wetland creation and restoration: the status of the science. U.S. Environmental Protection Agency, Environmental Research Lab., Corvallis, EPA 600/3-89/038A. 473p.

Levasseur, J.R. and M-L. Jory. 1982. Reestablishment naturel d'une vegetation de marais maritimes alteree par les hydrocarbures de l'Amoco-Cadiz: modalites et tendances, p.329-362. *In* Ecological study of the

Amoco Cadiz oil spill, report of the NOAA-CNEXO Joint Scientific Commission. U.S. Dept. Commerce, Washington, DC.

Lewis, R.R. II (ed.). 1982. Creation and restoration of coastal plant communities. CRC Press, Boca Raton, Florida. 219p.

Minello, T.J., R.J. Zimmerman and E.J. Klima. 1986. Creation of fishery habitat in estuaries, p. 106-120. *In* M.C. Landin and H.K. Smith (eds.), Beneficial uses of dredged material. U.S. Army Corps of Engineers, Waterways Exp. Stn., Tech. Rep. D-87-1.

Mitsch, W.J. and J.G. Gosselink. 1986. Wetlands. Van Nostrand Reinhold, New York. 539p.

Newling, C.J. and M.C. Landin. 1985. Long-term monitoring of habitat development at upland and wetland dredged material disposal sites, 1974-1982. U.S. Army Corps of Engineers, Waterways Exp. Stn., Tech. Rep. D-85-5. 224p.

Pacific Estuarine Research Laboratory. 1990. A manual for assessing restored and natural coastal wetlands with examples from southern California. Calif. Sea Grant Rep. T-CSGCP-021. 105p.

Roman, C.T., W.A. Niering and R.S. Warren. 1984. Salt marsh vegetation change in response to tidal restriction. Environ. Manage. 8:141-149.

Sacco, J.N., F.L. Booker and E.D. Seneca. 1988. Comparison of the macrofaunal communities of a human-initiated salt marsh of two and fifteen years of age, p. 282-285. *In* J. Zelazny and J.S. Feierabend (eds.), Increasing our wetland resources. National Wildlife Federation, Washington, D.C.

Sacco, J.N. 1989. Infaunal community development of artificially established salt marshes in North Carolina. M.S. Thesis, North Carolina State University, Raleigh. 41p.

Seneca, E.D. 1974. Germination and seedling response of Atlantic and Gulf coasts populations of *Spartina alterniflora.* Am. J.Bot. 61:947-956.

Seneca, E.D. and S.W. Broome. 1982. Restoration of marsh vegetation impacted by the *Amoco Cadiz* oil spill and subsequent clean-up operations at Ile Grande, France, p. 363-419. *In* Ecological study of the *Amoco Cadiz* oil spill, report of the NOAA-CNEXO Joint Scientific Commission. U.S. Dept. Commerce, Washington, D.C.

Steinke, T.J. 1974. A proposal concerning the restoration of the Pine Creek Estuary, Fairfield, Connecticut. Fairfield Conservation Commission, Fairfield, Connecticut. 78p.

U.S. Office of Technology Assessment (OTA). 1984. Wetlands: their use and regulation. U.S. Congress, Washington, D.C. OTA-0-206. 208p.

West, T.L. 1990a. Growth and survival of *Leiostomus xanthurus* (spot) in man-made and natural wetlands. Report to Texasgulf, Inc., Aurora, North Carolina. 30p.

West, T.L. 1990b. Benthic invertebrate utilization of man-made and natural wetlands. Report to Texasgulf, Inc., Aurora, North Carolina. 100p.

CHAPTER 3

Restoring Seagrass Systems in the United States

Mark S. Fonseca

NOAA, National Marine Fisheries Service
Beaufort, North Carolina

Abstract

Seagrass restoration has been viewed as a proven technology, capable of producing a product upon specification, when in fact it has really been the product of a series of disjunct experiments. Rather than focusing on ways to improve development actions and ecological resilience, protection of seagrass beds (and most other natural resources) has been concerned with defensive or remedial action. Depletion of seagrass resources and the services they provide are not perceived in policy-making circles as serious problems because of an undeserved faith in the technology of restoration.

Project goals have historically been inappropriate such that seagrass restoration has never prevented a net loss in habitat. Suggested goals for future projects should: develop persistent cover, generate equivalent acreage, increase acreage, replace the same seagrass species, and restore secondary (faunal) production. These goals are differentiated from sole measures of density and percent survival. Monitoring for cover and persistence should continue for three years. Restoration should only be performed on areas previously supporting seagrass, unless uplands currently zoned for development are available for conversion. Otherwise, substitution of one habitat type for another may be anticipated. Intrinsic population growth rate provides guidance as to the spe-

cies used for restoration. Site selection remains a complex problem which requires inclusion of specific conditions in the permit to help determine project location and success.

Research needs include: (1) a definition of functional restoration, (2) a compilation of population growth and coverage rates, (3) the resource role of mixed species plantings, (4) the impact of substituting pioneer for climax species on faunal composition and abundance, (5) culture techniques for propagule development, (6) transplant-optimization techniques such as the use of fertilizer, (7) the importance of maintaining genetic diversity, and (8) the implementation of a consistent policy on seagrass restoration and management among resource agencies wherein the role of seagrass restoration as mitigation is clearly defined and restoration technique, monitoring, performance and compliance guidelines are standardized. A system which does not merely document but tracks permits and evaluates them as to performance and compliance must be developed. Most important is the willingness of management to not only collect performance and compliance data but to strictly enforce compliance and, if necessary, prosecute violators regardless of acreage involved, a process that can only succeed through extensive intra- and interagency coordination.

Overview of Habitat Type [1]

Geographic Range and Characteristics

There are approximately 60 species of seagrass in the world, of which at least 12 occur in the United States (Table 1). Similar to other flowering plants, seagrasses have vascularized leaf, root, and rhizome (tiller or runner) systems as opposed to nonvascular algae or seaweeds (especially rhizophytic algae). Seagrasses differ from most other wetland plants in that they lead an almost exclusively subtidal existence, carry on both sexual and asexual reproduction

---

[1] This paper is adapted largely from Fonseca (1989) which contains more detailed treatment of some sections presented here as well as case histories from each coast of the United States.

in the water, reside for the most part in marine salinities, and utilize the water column for support.

Recent (but rather incomplete) surveys have shown that there may be as much as 6.2 million acres of seagrass habitat in the area from New Jersey through Texas (Iverson and Bittaker 1986; Orth and van Montfrans 1990; USEPA unpubl. data; R. Ferguson, NMFS/NOAA, Beaufort, North Carolina, pers. comm.). This acreage is not included in current estimates of wetland abundance by any agency, despite the classification of seagrass as a wetland under the Cowardin system (Cowardin et al. 1979).

Seagrasses occur across a wide vertical range, from rocky intertidal (e.g., *Phyllospadix* sp.) to 40 meter depths (*Halophila* sp., pers. obs.), and, for some species, broad latitudinal ranges. For example, eelgrass (*Zostera marina*) extends from near the Arctic circle on both coasts of the United States to North Carolina on the east

Table 1. List of seagrass species by family, genus and species, and common names (if given) that are found in the United States and adjacent waters.

Family and Species	Common Name
Hydrocharitaceae	
Enhalus acoroides Royle	
Halophila decipiens Ostenfeld	paddle grass
Halophila engelmanni Ascherson	star grass
Halophila johnsonii Eiseman	Johnson's seagrass
Thalassia testudinum Konig	turtlegrass
Potamogetonaceae	
Halodule wrightii Ascherson	shoalgrass
Phyllospadix scouleri Hook	surf grass
Phyllospadix torreyi S. Watson	surf grass
Ruppia maritima L.	widgeongrass
Syringodium filiforme Kutz	manatee grass
Zostera japonica Aschers. et Graebner	eelgrass
Zostera marina L.	eelgrass

coast and to the Gulf of California on the west coast. Across this extraordinary latitudinal range, this species exhibits either annual or perennial growth, with growth peaks either in the summer or in the fall and spring (Thayer et al. 1984; Phillips 1984). Because variation in growth season and life history within a seagrass species is often great, species and region-specific planting times and projected coverage rates are considered in subsequent sections in terms of determining performance and compliance of seagrass restoration/ mitigation projects.

Functions of Seagrass Beds

Wood et al. (1969) and Thayer et al. (1984) described ten basic functional roles for seagrasses: (1) a high rate of leaf growth, (2) the support of large numbers of epiphytic organisms, (3) the leaf production of large quantities of organic material, (4) the utilization of seagrass-derived detritus supports a complex food web (although few organisms graze directly on the living seagrasses), (5) the damping of waves and slowing currents enhance sediment stability and increase the accumulation of organic and inorganic material, (6) the binding by roots of sediments, thereby reducing erosion and preserving sediment microflora, (7) the mediation by the plants along with their associated detritus production of nutrient cycling between sediments and overlying waters, (8) the decomposition of roots and rhizomes providing a significant and long-term source of nutrients for sediment microheterotrophs, (9) the role of roots and leaves in providing horizontal and vertical complexity which, coupled with abundant and varied food resources, leads to densities of fauna generally exceeding those in unvegetated habitats, and (10) the movement of water and fauna which transports living and dead organic matter (particulate and dissolved) out of seagrass systems to adjacent habitats.

Impacts and Responses

Types of Impacts and Severity

Natural Mortality. Seagrass beds are subject to natural disturbances such as bioturbation (rays, crabs, etc.) and storm-associated scour (Continental Shelf Associates, Inc. and Martel Laboratories, Inc. 1985). These disturbances are probably the mechanisms which provide the spatial and temporal basis for the mosaic of seagrass beds which can easily be observed in nature. The relative ability of different seagrasses to respond to disturbance (e.g., pioneering vs. climax) determines, in part, how quickly a species can colonize disturbance areas and how long they can reside there in the face of potential competition from other seagrass species.

Intrinsic mortality rates particularly come into play in high water motion environments. In these environments, small hummocks of seagrass form a semi-permanent feature of the bottom, existing in a temporary equilibrium with the physical forces. As the patch of seagrass spreads, the older shoots in the middle of the patch die first, destabilizing the center and facilitating the disintegration of the patch. The fringe patch fragments which survive either merge with other fragments or themselves become the nuclei of a new patch, thus forming a dynamic, yet persistent seagrass community.

Oil and Chemicals. Subtidal seagrasses have historically suffered little damage from oil spills. Impacts on intertidal beds have been significant and impacts on rocky intertidal species are unreported. However, impacts on the seagrass-associated fauna has been great during oil spills. Zieman et al. (1984) and Zieman and Zieman (1989) summarized impacts by petroleum products. They point out that beside acute physical impacts, such as smothering, exposure to these products will lower stress tolerance, reduce market values, incorporate carcinogenic and mutagenic substances into the food chain, all of which may induce mortalities which lower the diversity and functional value of the seagrass system.

The influence of other chemicals on seagrass growth and survival have not been well-studied. Zieman (1982), Thayer et al. (1984) and Zieman and Zieman (1989) reviewed the available literature. They pointed out that heavy metal accumulation by seagrasses is well documented, again suggesting potential biomagnification problems. Many of the impacts by chemicals in nature have been difficult to discern because they often occur in concert with other impacts, such as increased turbidity, which in and of itself can reduce or kill seagrasses. Little funding has been directed at elucidating chronic and sublethal affects of chemicals, especially those associated with non-point source runoff.

Dredge and Fill. Dredging of seagrass beds often destroys the existing plants because most root systems are concentrated in the top 20 cm of the sediment, and thus are easily dislodged, leaving no sedimentary record of their prior existence. Also, dredged channels typically create depths greater than the compensation depth of seagrass, inhibiting natural recolonization. Channel side slopes are often unstable and the use of the channel imparts physical disturbance (propeller turbulence and boat wake waves) previously inexperienced in the vicinity and beyond the operational ability of most restoration techniques and natural colonization. Filling is equally devastating. Seagrasses cannot tolerate much more than a quarter of their leaf area to be occluded (pers. obs.) and still maintain an ability to redirect their rhizome growth to regain the sediment surface. Also, seagrasses are almost never buried in an upright position. They bend over with the typically rapid addition of sediments found in dredge operations and become completely buried even with the addition of a few centimeters of sediment.

Light Reduction. Seagrasses are completely dependent on water clarity to allow adequate light for photosynthesis and thus, survival. Water clarity (low turbidity) is critical for the restoration and management of this ecosystem as has been seen in Chesapeake Bay (Orth and Moore 1982a), San Francisco Bay (pers. obs.), and Tampa Bay (Lewis et al. 1985). Without adequate water clarity at a restoration site, or the preservation of water clarity over existing seagrass meadows, seagrass restoration is not feasible. Numerous disparate studies on the light requirements of different species of seagrass provide remarkably consistent results: seagrass beds require a minimum of 20% of incident solar radiation (at the water surface) to survive (Kenworthy and Haunert 1991). Standing crop and production were shown in this paper to be linearly related to insolation, indicating that for every incremental increase/decrease in water clarity there is a corresponding increase/decrease in seagrass biomass and production. Significant enhancement of seagrass beds can be expected from improved water clarity, providing a potential means of reclaiming acreage lost to past increases in turbidity.

Impact Response in the Context of Attempting Restoration
Although all seagrasses have high rates of leaf production, (Zieman and Wetzel 1980), not all seagrasses add new shoots (via vegetative reproduction) at rates similar to each other within a geographic area. Many seagrasses do not add new shoots for weeks at a time. Other species of seagrass add new whole shoots daily. Those species that add new shoots rapidly have a concomitantly high bottom coverage rate, a factor which enhances their suitability in seagrass planting projects. Those species that are slow in their addition of shoots are too slow in coverage of the bottom for effective habitat restoration on their own, a phenomena noted not only in this country (Fonseca et al. 1987c, 1988) but in Australia as well (Kirkman, this volume).

In order to quickly stabilize the sediment at a site, the fastest-covering species should be chosen (Fonseca et al. 1987c). If the species that was lost was one of the slow-covering species, plantings of this seagrass may be interspersed among the faster-growing species to compress (in time) the successional process (e.g., planting *Halodule* interspersed with *Thalassia*) (Derrenbacker and Lewis 1982).

Slower-covering genera (e.g., *Thalassia*) present a particularly difficult problem in restoration attempts. As development proceeds in the coastal zone, the older, climax meadows are more and more often the ones being damaged. Since they are slow-growing, those meadows may represent decades of development. Even artificially-assisted restoration of these systems, such as transplanting, will take many years. For example, to attain shoot densities equivalent to natural beds via transplanting takes an average 3.38 years in Tampa Bay (author's unpubl. data).

Restoration and Creation

Historical Review

Seagrass transplanting has been a subject of experimentation for nearly half a century (Addy 1947). Phillips (1960) initiated the first serious effort to apply seagrass transplanting to restoration. Most of the early work, however, utilized transplanting as a means to assess phenotypic and genotypic aspects of seagrasses with tangential application toward management problems. With the advent of the mitigation concept in the 1970s, the ability to move seagrass plants and establish them elsewhere was often broadly interpreted as a "technology" by which damage to seagrass beds could be repaired. Unfortunately, the pressure to make management decisions grew disproportionately to the increase regarding seagrass habitat restoration—as a consequence, there has never been a seagrass restoration project which restored more acreage than what was lost (Fonseca et al. 1987c). Most of the documented seagrass transplanting and restoration efforts have been reviewed by Fonseca et al. (1987b, 1988), Fonseca (1989) and Thom (1990).

Seagrass restoration has been viewed as a proven technology, capable of producing a product upon specification, when in fact it has really been the product of a series of disjunct experiments (Zieman and Zieman 1989). Rather than focusing on ways to improve development actions and ecological resilience, protection of seagrass beds (and most other natural resources) has been concerned with defensive or remedial action. Depletion of seagrass resources and the services they provide are not perceived in policy-making circles as serious problems, because of an undeserved faith in the technology of restoration (Colby 1989). In turn, this stance is founded by the apparent belief that ecological costs are external to costs to the economy. The paradox here is that the value of seagrass beds, not being realized for the services which it has provided free of charge, is now being increased due to the fact that in economics, value is inversely proportional to abundance (Colby 1989).

As mentioned earlier, seagrass restoration often has been unsuccessful. This is because seagrass transplanting only addresses a narrow field of the impacts which destroy seagrass beds; typically dredge and fill. If environmental conditions have otherwise deteriorated, due for instance to eutrophication, to the point where indigenous seagrass beds are destroyed, then transplanted material will die as well. Once this deterioration has occurred, it is very difficult to reverse. This is because the sediment stabilization and water column filtration benefits of the seagrass cover has been lost. Sediments are therefore easily resuspended, adding to the turbidity of the water column and the decreased likelihood of effective restoration.

Project Goals: Success and Failure

In retrospect, there have been few quantitative goals set for seagrass restoration projects. Goals have been of a highly qualitative nature, such as "successful restoration of the seagrass community." Although one usually has a visual image of a "successful" seagrass restoration that evokes images of thick, lush meadows, parameters are evaluated as success criteria that typically fall wide of defining the status of seagrass meadow development as it relates to functional habitat recovery. With this lack of guidance, goals provided by the private sector have, for the most part, been biologically unsound.

Unfortunately, the cost of determining functional recovery in anywhere near the complete meaning of the term (faunal abundance, composition, productivity, biomass, energy flow, nutrient cycling) is not affordable on a project scale. Therefore, a parallel track of research and management is being pursued wherein the efficacy of simple, diagnostic parameters, such as coverage and shoot density, may be coupled with a quantifiable expectation of the recovery of specific resources (Fonseca et al. 1990). As research expands our understanding of these systems, those data can be incorporated into a more holistic determination of functional roles.

Data have often been collected on percent survival, blade density, and percent success. Taken individually, these parameters do not provide useful, operational definitions of success. Measurements of planting unit survival may do little to convey how the meadow is developing. Blade and shoot density and biomass are products of local gradients of environmental conditions, a complex relationship that is not a readily predictable or controllable process. This is a poor basis on which to solely define success. A measurement of percent survival, blade density, biomass, or percent survival are too frequently adopted as a single project goal.

Goals for seagrass restorations vary widely, depending on the perspective of the participants. Developers often find themselves dealing with a subject about which they know little (seagrass restoration), and for which there is no set path to approach the issue. This is largely due to the fact there is no consistent policy on the subject between, or indeed, even within local, state and federal resource agencies.

Setting Project Goals
Suggested goals are:

1. Development of persistent vegetative cover (cover being defined as the area where rhizomes overlap).

2. Equivalent acreage of vegetative cover gained for cover acreage lost (where with naturally patchy distributions, such as in wave-exposed areas (Figure 1), it may require setting aside more planting area to accommodate the sum total of vegetative cover).

3. Increase in acreage where possible.

4. Eventual replacement of same seagrass species as was lost.

5. Development of faunal population structure and abundance in the new bed equivalent to natural, reference beds.

These goals need to be applied with the realization that no truly successful seagrass restoration project has yet occurred in a holistic sense [equal or greater persistent acreage being generated, although Thom (1990) has classified successful projects as having at least some persistent growth], even though successful establishment of some seagrass coverage has occurred. Failures have not been detected due to improper success criteria, site selection, technique, and monitoring. As a result, mitigation of seagrass losses is sometimes viewed in the same manner as salt marsh creation, which is technically less difficult.

Maintaining Genetic Diversity
The existence of seagrass races was identified by Backman (1984, 1985) and is currently a topic of several studies (S.L. Williams, San Diego State University, pers. comm.). Maintenance of genetic diversity is important so as to retain perturbation resiliency,

Figure 1. Aerial photograph of mixed Halodule wrightii *and* Zostera marina *seagrass bed in Back Sound, Carteret County, North Carolina. Bar scale ⁓=30 m. Note patchy beds on outer edges of shoal area which begin to coalesce farther onto the shoal.*

for example, how large a disturbance can a habitat sustain before losing population viability? Knowledge of genetic structuring of seagrass beds is in an embryonic stage at this point and has not been demonstrated to pose a problem in restoration and mitigation projects, especially given the relatively small size of most applications. However, given that races of seagrasses have been shown to exist, those involved in restoration of seagrass beds (especially on broad scales) should keep the concept of genetic diversity in mind. Until better guidance on this subject is established, donor material from a number of different locations should be utilized in an attempt to maintain genetic diversity.

Restoration vs. Mitigation

One typical problem is that resource managers tend to equate restoration with mitigation. These two terms more accurately describe two very different scenarios of habitat establishment, as noted in the introduction to this book. Restoration implies that the physical setting previously supported the habitat type in question and, under proper conditions, might do so again. Use of the term mitigation has a much broader application. Most important, mitigation is typically associated with a specific permit-related activity, but may apply to both natural and anthropogenically-induced losses of a habitat. Restoring a habitat also implies a returning to previous conditions, whereas the term mitigation does not clearly indicate the kind or scope of activity that may be instituted. Resource managers must employ clear and consistent terminology to avoid intra- and interagency confusion and conflict. Without enforcement of ecologically-sound and biologically-relevant success criteria, seagrass restoration will not become a predictable management tool. Enforcement of these success criteria have been achieved through strictly written permit conditions (Thayer et al. 1986).

Methods/Techniques

This section contains several components which are listed here to facilitate reference by readers:

- Overview
- Site selection
- Timing of construction
- Environmental conditions
- Fertilizer
- Monitoring
- Duration of monitoring
- Interpreting monitoring results

Overview. There is a large body of information on revegetation and transplanting projects in general. Reviews by Phillips (1982) and Fonseca et al. (1987a, b) provide specifics regarding the subject. By way of a general summary, seagrass revegetation projects are conducted largely with wild-harvested, vegetative shoots. A timely paper by Lewis (1987) describes the hazards of damaging existing natural beds and the options available in the State of Florida for alternative planting stock.

Previous work by Riner (1976) and more recently by Roberts et al. (1984) have pioneered research into the use of seeds for temperate species. Turtlegrass has been easily transplanted by seed for many years (Thorhaug 1974). Some work is being conducted on cultivation of turtlegrass seeds, but this process still is only a grow-out procedure, dependent on the harvest of wild seed. Impressive medium-scale eelgrass bed creation has been accomplished by seeding (author, pers. obs.; R.J. Orth, Virginia Institute of Marine Science, pers. comm.) in the Chesapeake Bay. This method has promise in reducing large-scale planting costs.

Seagrass transplants are typically fragile and must be handled with care. The plants are extremely susceptible to desiccation and must be kept soaked, or preferably, in ambient temperature water during the whole planting process. The technique of transplanting is well-documented and involves using bare-root shoots stapled into the bottom (Figure 2), plugs, or in some instances (as mentioned earlier), sowing of seed. While the technology is well-developed, its application is not.

Of the 12 species that occur in the United States, only 7 have been reported as being transplanted (*Halophila decipiens*, *T. testudinum*, *H. wrightii*, *Ruppia maritima*, *S. filiforme*, *Zostera japonica*, and *Z. marina*). Of these, *T. testudinum*, *H. wrightii*, *S. filiforme*, and *Z. marina* have been the species most often utilized. These four species probably constitute the majority of seagrass cover in the United States, but the others have not enjoyed similar scrutiny,

Figure 2. Bare root planting unit composed of a bundle of bare root Halodule wrightii *attached to metal anchor with paper twist-tie.*

and managers are not in a position to declare their relative importance. For example, it is now known that there are at least one million acres of seasonal *H. decipiens* off the west coast of Florida (Continental Shelf Associates, Inc. and Martel Laboratories 1985). This species has been transplanted on one occasion at depths over 10 meters (author, unpubl. data).

Site Selection. The selection of a planting site is the most critical aspect of the pre-project plan. Planting sites should be located on the impacted area, if the area has returned to conditions that will support seagrasses. Examples of such sites include backfilled access canals, pipelines, and power line crossings. If a site has been altered to the point where it will no longer support seagrass, then the restoration site should be located in the same water body, as near to the impact site as possible. If such a site cannot be found, then anthropogenically impacted areas in communicating water bodies should be selected to minimize local system losses.

A basic requirement in selecting a restoration site is addressing the point raised by Fredette et al. (1985): "If seagrass does not currently exist at the (chosen) site, what makes you believe it can be successfully established?" If no site history data exist (e.g., aerial photographs), the only alternative to this approach that may be employed is a commitment to a scientifically valid environmental monitoring project to evaluate temperature, salinity, currents, sediments, and especially light penetration to the bottom.

Sites should not be selected within existing, naturally patchy seagrass meadows (Figure 1; Fonseca et al. 1987b; Fonseca 1989). Some seagrass beds exist naturally in this configuration for any of several reasons: the existence of an annual population, their colonization of sediment-filled pockets in surface bedrock, a hydrodynamically active (high energy) setting, or some combination thereof. Although a seagrass may be transplanted in some of these areas and proliferate, the environmental factors acting to create the patchiness will shortly revert the area to pre-planting levels of patchiness. Thus, only a temporary pulse of productivity would be achieved and no persistent increase in seagrass acreage would be generated. To maintain a given level of seagrass coverage in high energy sites, a given level of unvegetated area must be anticipated. The seagrass cover and its associated inter-patch spaces must be regarded *in toto* as seagrass habitat because one does not exist without the other. Consistent definitions for seagrass habitat boundaries, especially in regard to patchy beds, have not yet been established (Fonseca 1989).

Apart from between-patch spaces, sites also should not be selected on naturally-occurring unvegetated areas. Such a selection displaces other habitats and assumes no intrinsic functional value of unvegetated bottom; nor does it address the question raised earlier by Fredette et al. (1985). It is likely that a natural balance of vegetation and non-vegetated areas is required for maintenance of specific marine resource values. Unvegetated areas are likely to be zones of significantly higher predation, which mobilizes energy up the food chain. Under this scenario, vegetated areas act as a reservoir of prey items that are less energetically costly for predators to obtain in unvegetated areas. Transplants into these areas in an attempt to reduce patchiness or alterations of seagrass species composition are hypothesized to negatively affect trophic transfer of energy for some fauna.

The following choices for restoration sites are given in order of preference: (1) areas previously impacted by poor water quality that once had seagrass, and the water quality has improved; (2) conversion of filled or dredged areas that were once seagrass meadows back to original elevation and transplant onto them (some subtropical areas, such as the Florida Keys, may colonize on their own with pioneer seagrass species, e.g., shoalgrass); (3) conversion of filled or dredged areas regardless of their previous plant community to a suitable elevation for seagrass; and (4) conversion of uplands zoned for development to seagrass habitat.

Timing of Construction. Timing of restoration is the key to determining timing of any site construction or modification. That timing is determined by the life history of the seagrasses in the area. For example, if a restoration operation uses many flowering shoots of eelgrass, then many shoots will die before giving rise to new, vegetative shoots. Seeds that may be set by transplanting the flowering shoots are not a reliable means of revegetating an area because of the variability of seed set and retention and low germination rates. Unless otherwise demonstrated through field surveys, natural seeding or vegetative fragment recruitment should not be counted on to provide significant coverage. Therefore, construction timing may also be affected by the availability of planting stock, especially in the case of salvage operations (a case where destruction of a seagrass bed is permitted, such as bridge construction, and the seagrass may be salvaged and used as planting stock at a new site).

Knowledge of the flowering factor may be a valuable asset in planning a seagrass restoration, though such data are not always available. The existence of annual forms (plants growing from seed and flowering the same year, as opposed to the typically perennial form) within an area may preclude seagrass restoration altogether unless seeding techniques are developed (Fredette et al. 1985). Research on *Z. marina* in Chesapeake Bay is continuing to resolve this problem for this species (R.J. Orth and K.A. Moore, Virginia Institute of Marine Science, pers. comm.).

Timing of construction, which is based on the actual timing of planting, should consider potential utilization of the planting site by fauna. For example, setting of herring eggs on eelgrass in the Pacific northwest or shorebird nesting on dredge material islands are natural phenomena that may preclude the disruptive activity of site engineering, donor site harvest, or even human presence.

Environmental Conditions. Relatively few sites have been created specifically for seagrass planting. Adequate depth of sediment over bedrock, exposure to waves and currents are primary site construction considerations. The obvious exception are the surf grasses which grow in crevices among rocks.

If a site is constructed for seagrasses, then in addition to considerations of wave energy, tidal currents, tidal range (for sufficient light penetration vs. exposure at low tide), flushing effects on temperature and salinity should be considered. Because light is of utmost importance in seagrass growth and survival, a site should be constructed to minimize sediment resuspension (i.e., reduce wave energy) while maintaining appropriate temperature regimes (author, pers. obs.). Environmental tolerance ranges for the various species are described in Phillips (1984), Thayer et al. (1984), Zieman (1982) and Zieman and Zieman (1989). These measures require extensive field monitoring to average out conditions at a given site. *In lieu* of these data, plans should be drawn up that provide depths equivalent to adjacent seagrass beds. If no beds are adjacent to the site, then a statistically valid monitoring of the light conditions is prudent.

In high current areas, patchy beds develop (Figure 1). While the patches and between-patch area may be correctly termed a seagrass habitat, performance of planting is measured by the actual area of seagrass cover (cover being the area where the long shoots overlap). Therefore, in high current or wave areas, a greater area will need to be planted to obtain the desired total area of bottom coverage. These areas are also prone to greater planting unit loss due to erosion. In quiescent areas, a relatively continuous meadow will often be attained, and planning of planting acreage is straightforward.

Prolonged freshwater inflow and lowering of salinities in seagrass beds is often fatal to the plants. Widgeongrass and shoalgrass are two euryhaline exceptions to this rule.

Sediments within the sand size range have not been shown to limit seagrass growth *per se.* Other covarying factors such as nutrient supply, currents and, in the case of fine-grained sediments, resuspension and light reduction, have been related to seagrass growth. Transplanting of bare-root sprigs (Derrenbacker and Lewis 1982; Fonseca et al. 1982, 1984) has been criticized as being of limited use in highly organic sediments (R. Alberte, Hopkins Marine Laboratory, pers. comm.) although successful restorations have been performed in sediments with 4% organic matter content (by combustion) in North Carolina (Kenworthy et al. 1980). Concomitantly low light conditions which do not allow the plant to generate and transport oxygen to the roots to fight sulfide intrusion may explain the poor performance of some restorations placed in sediments with high organic content (M. Josselyn, San Francisco State University, pers. comm.).

Fertilizer. Fertilization of plantings may be useful in areas with little or no organic matter in the sediment (<=2%). Depending on the sediment type, phosphorus (in carbonate sediments) or nitrogen (in siliceous sediments) may be limiting. This is an area of intense debate in seagrass research (Bulthuis and Woelkering 1981; Short 1983a, b; Pulich 1985; Williams 1987). Some studies have been completed specifically on fertilizer effects on seagrass transplants (Orth and Moore 1982a, b; Fonseca et al. 1987b) but the results are equivocal, based on inconsistent performance and assessment of the fertilizers used. At present, fertilization is considered to be acceptable in low (< 2%) organic sediments, but no reduction in planting intensity is recommended (Fonseca et al. 1987b). Phosphorus additions to plantings in sediments with 2% carbonate content has significantly improved population growth rates (author, unpubl. data).

Associated Impacts. A common problem that occurs in a seagrass mitigation project is associated damage to adjacent beds. For example, a permit is issued to allow destruction of a grass bed for laying of an underground pipeline with replanting to occur in the backfilled area over the pipe. But during the operation, maneuvering of barges, etcetera, causes erosion and burial of adjacent seagrass habitat not specifically identified with the permit. Also, excavation of the pipeline corridor could change sediment depths over bedrock in the planting area (S. Markley, Dade County Department Environmental Management, Miami, Florida, pers. comm.). Ideally, the permit should contain language identifying such potential impacts and develop contingency plans for the mitigation of the habitat loss.

Monitoring. Monitoring specifications have been laid out in at least two publications (Fonseca et al. 1987b; Fonseca 1989). Those documents point out that no one data type can stand alone in a monitoring program. Several factors must be considered and these may lead to an ecologically valid characterization of seagrass restoration success. Fonseca (1989) describes the following monitoring format:

"The number of planting units that survive should be recorded. This may be expressed as a percentage of the original number, but the actual whole number is critical as well. A random (as opposed to arbitrary) sample of the average number of shoots and area covered (m²) per planting unit should be recorded up until coalescence (the point where individual planting units grow together and the planting unit origin of individual shoots cannot be readily observed). The number of surviving planting units may then be multiplied by the average area per planting unit to determine the area covered on the planting site. The data from pre-coalescence surveys may be used to compare with existing data to assess performance relative to other, local plantings by plotting the average number of shoots (not area) per planting unit over time. The comparison may be statistical or visual (which often suffices to detect grossly different population growth rates). Shoot addition is recommended over area addition because areal coverage varies with the environmental setting of the planting. For example, in high current areas, shoots grow more densely, whereas studying shoot addition is a more accurate means of assessing the asexual reproductive vigor of the plantings. After coalescence, the area of bottom covered should be surveyed using randomized grid samples (Fonseca et al. 1985). These data may be used to assess persistence of the planting as well as total seagrass coverage."

Population growth and coverage data are scarce at this time. If collection of these data could be instituted, then managers would quickly develop the capability to objectively and efficiently deal with seagrass mitigation projects.

Designation of Responsibility. Because of the burden of permit loads relative to available staff, responsibility must be placed on the developer to execute a responsible (to the public) restoration project. However, because seagrass restoration is effectively an "end of the pipe management" option (Colby 1989), stringent safeguards must be incorporated into the project plan. Options (and resources) for replanting must be instituted. Because most developers lack the expertise to initiate the complicated procedure of seagrass restoration, experienced personnel should be involved at some level. Peer review of the plan prior to its adoption must be performed. These proactive steps not only protect the resource but act to clearly and fairly apprise the developer of the requirements which they face. Fonseca (1989) states:

"It is often necessary to secure funding from those responsible for the planting ahead of time. This is because experience has shown that after a permit is issued and a project completed, it is difficult short of legal action to get the planting done. Performance bonds or letters of credit from applicants to contractors have been used with success in this regard. In any event, points of finance, as well as the technical language of restoration technique, site selection and monitoring are critical elements of the 404 permit.

"Random samples should be collected on survival, number of shoots and area covered per planting unit. If a planting site is sufficiently small, all planting units should be surveyed for presence or absence (survival survey). The existence of a single short shoot on a planting unit indicates survival. If a site is large, then randomly selected rows or subsections (area in m²) should be sampled. Since each row or subsection is actually the level of replication, at least 10 replicates should be performed at the level which one wishes to generalize their findings (e.g., over the whole planting site). At the very least, stabilization of the running mean of survival or shoots and area per planting unit should be obtained as a measure of statistical adequacy.

"Presence or absence, and number of short shoots per planting unit are straightforward measures, although they usually require snorkeling or SCUBA diving to make them (a factor that surprisingly is not considered, or equipped for, by many attempting these data collections). The area covered by a planting unit may be measured by using a quadrat criss-crossed with string on 5 cm centers that is laid over the planting unit. In this case, the number of 5 × 5 cm grids (or half grids if there are only 1 or 2 shoots in the 5 × 5 cm grid) that have seagrass shoots are totaled and converted to meters square of cover for the planting unit. The quadrat method is more appropriate for seagrasses that propagate in long runners (e.g., shoalgrass), and do not form a clear radial growth pattern (e.g., eelgrass). An individual can be trained to perform these counts in a few hours, and can count individual planting units in 5-10 minutes or less at early stages of a restoration's development."

Duration of Monitoring. According to Fonseca (1989):

"Monitoring of shoot numbers and area covered per planting unit should proceed quarterly for the first year after planting and biannually thereafter for a minimum of two more years (a total of three years). After planting units begin to coalesce and the planting unit from which shoots originated can no longer be discerned, areal coverage data should be recorded and counts on a planting unit basis suspended."

Interpreting Monitoring Results. Fonseca (1989) writes:

"The population growth and coverage data may be compared periodically with published values (dependent on species and location) as a relative indicator of performance. More important, the computations described above allow a direct comparison on a unit area basis of planted versus lost acreage. Success may then be based on whether the appropriate ratio of coverage (e.g., 1:1, or 2:1) has been generated, a quantitative measure consummate with ecological function. If the restoration project is for mitigation, then compliance may thereby be interpreted as both acreage generated and the unassisted persistence of that acreage over time (the three year period). The persistence issue is also critical. If the planting does not persist, then resource values have experienced a net loss and the project has not been effective. Loss of plantings (e.g., bioturbation, storms, prop-scarring) is part of the risk in restoring seagrass systems. In all cases, it is often necessary to plan for at least a partial replanting of the site due to the above factors.

"If impacts occur early enough in the planting season, additional planting units may be added as a mid-course correction. If there are fewer than 90 days left before the first major seasonal decline of local, natural grasses, replanting should be postponed until the next year. If replanting is performed, then the monitoring clock should be reset to zero. Otherwise, a site could experience chronic planting failures without any impetus to change procedure.

"Another mid-course correction may be implemented upon examination of the population growth and coverage data. If the population growth is within expected limits for the species and location, then one may be assured that the observed coverage rate is the best that can be expected for that site. If the coverage rate is lower than expected while the shoot rate is as predicted, then the projected timetable for grow-out should be lengthened appropriately. Although this would not necessarily change the permit conditions or length of commitment of the applicant, it allows an objective evaluation that is fair to all parties. In other words, the restoration is performing as well as can be expected and the anticipated coverage should be reached, albeit at a later date. Both low shoot generation and coverage rates indicate that a restoration is in trouble. Timetables may again be altered, but replanting may be warranted or a new site sought if the rate of shoot addition to planting units was not significantly different from zero after the first year of growth. Chronic failures of plantings need to be carefully examined. It is important to distinguish between acts of nature and acts of incompetence or non-compliance. For example, if a permit applicant does everything asked of him by resource agencies and still cannot, after three years, come up with the acreage required, is the applicant required to finance planting in perpetuity? There needs to be some measure of agency responsibility in the site selection and approval process that prevents this. At what point has the applicant fulfilled the permit requirements? In reality, seagrass restoration has been and will continue to be a risky management option. This point becomes more profound if one considers that a terrestrial crop cannot be guaranteed, despite millennia of collective practise. The agencies involved should proceed with a clear realization that they are taking a calculated risk in their ability to prevent a net loss in habitat. If chronic failure is due to chronic non-compliance with established (in the permit) procedures, then existing, formal methods of ensuring compliance should be instituted."

Future Research Directions and Needs

Management of the seagrass resource suffers from a lack of purposeful objectives and long-term goals for meeting the challenge of no net loss of habitat. Those seeking information to guide projects involving seagrass suffer from a lack of centralized sources of information and opinion. Resource managers must realize that the loss of this habitat is often long-term and is manifested (as are most wetland losses) in many small, individual projects. The cost to resource management agencies to address projects piecemeal, without consistent intra- agency policy appears to be enormous, when measured in work-force hours. Because very few projects impact large acreages, there is little to attract attention to the cumulative problem. Only by taking a vigorous, proactive conservation stance on modifications to this habitat, and one that is actively supported by the highest offices of the respective agencies, will there be any hope of slowing, stopping and reversing the trend of seagrass habitat loss. In concert with such a revitalization of management philosophy, there are research gaps which need to be addressed.

1. A definition and evaluation of functional restoration of a seagrass bed must be made; is it just a floristic survey? Will faunal abundance follow? Will ecologically and economically valuable functions be realized, and if so, how long will this take? Coverage and persistence of seagrass plantings should be used along with numerical abundance and composition of macrofauna should take priority in functional assessment. By targeting animals higher up on the food chain, one would hope to evaluate the status of various, integrated resource dimensions provided by the seagrass bed in their combined ability to support these animals. These surveys should be

made in comparison to ambient natural beds. Evaluation of recruitment, growth and survival should be the next phase of functional evaluation.

2. Population growth and coverage rates should be compiled for seagrasses in all regions so that growth patterns may be better defined. Parent data sets on transplant performance should be centrally compiled on a regional basis as a measure of performance. Areas lacking in these data should have experimental plots initiated and monitored. Species lacking these data are *Enhalus acoroides*, all *Halophila* species, eelgrass on the west and northeast coasts, all seagrasses in the northern Gulf of Mexico (*S. filiforme*, *T. testudinum*, *H. wrightii*, *Ruppia maritima* and all *Halophila* species), *Z. japonica* and all *Phyllospadix* species.

3. The resource role of mixed species plantings should be evaluated.

4. The impact of substituting pioneer for climax species in "compressed successional" transplanting (Derrenbacker and Lewis 1982) on faunal composition and abundance should be investigated for resource maintenance.

5. Culture techniques for propagule development (seed and tissue culture) should be refined, bypassing the need to damage donor beds when salvage is not available.

6. Transplant optimization techniques should be explored, especially the use of fertilizers.

7. The importance of maintaining genetic diversity of restored seagrass beds should be investigated in terms of habitat resiliency and ensuring short-term transplant success.

8. The development of a consistent national policy on seagrass management must be developed wherein special conservation consideration should be given to artificially-propagated seagrass meadows. Site evaluation methodologies, especially for light availability, should be standardized and adopted by all resource agencies. The cumulative impact of small-scale, piecemeal loss of seagrass by de-

liberate (e.g., dredging) and accidental (prop-scarring) impacts should be considered in local and regional assessments for the protection and maintenance of this valuable habitat.

Acknowledgments

Thanks are extended to the NOAA for developing and supporting this Symposium. Comments from several anonymous reviewers improved the manuscript. Ann Manooch proofed the manuscript and references. A special thanks goes to Dr. Gordon Thayer for his hard work in marshaling the speakers and the subsequent papers.

Literature Cited

Addy, C.E. 1947. Eelgrass planting guide. Md. Conserv. 24:16-17.

Backman, T.W.H. 1984. Phenotypic expressions of *Zostera marina* L. ecotypes in Puget Sound, Washington. Ph.D. Dissertation, University of Washington, Seattle, 226 p.

Backman, T.W.H. 1985. Selection of *Zostera marina* ecotypes for transplanting, p. 1088-1093. *In* Oceans '85, Vol. 2., Marine Technology Society, Washington, D.C.

Bulthuis, D.A. and W.J. Woelkering. 1981. Effects of *in situ* nitrogen and phosphorus enrichment of the sediments on the seagrass *Heterozostera tasmanica* in Western Port, Victoria, Australia: a decade of observations. Aquat. Bot. 19:343- 367.

Colby, M.E. 1989. The evolution of paradigms of environmental management in development. World Bank Policy, Planning and Research Working Paper, WPS 313. 37 p.

Continental Shelf Associates, Inc. and Martel Laboratories, Inc. 1985. Florida Big Bend seagrass habitat study narrative report, a final report by Continental Shelf Associates, Inc., submitted to the Mineral Management Service, Metairie, Louisiana. Contract No. 14-12-0001-30188. 47 p.

Cowardin, L.M., V. Carter, F.C. Golet and E.T. LaRoe. 1979. Classification of wetlands and deep-water habitats of the United States. U.S. Fish and Wildlife Service, FWS/OBS-79/31.

Derrenbacker, J.A. and R.R. Lewis. 1982. Seagrass habitat restoration in Lake Surprise, Florida Keys, p. 132-154. *In* R.H. Stoval (ed.), Proceedings of the 9th Annual Conference on Wetlands Restoration and Creation, Hillsborough Community College, Tampa, Florida.

Fonseca, M.S. 1989. Regional analysis of the creation and restoration of seagrass systems, p. 175-198. *In* J.A. Kuslerand and M.E. Kentula (eds.), Wetland creation and restoration: the status of the science. Vol. 1: Regional reviews. U.S. Environmental Protection Agency Environmental Research Lab., Corvallis, Oregon. EPA/600/3-89/038a.

Fonseca, M.S., W.J. Kenworthy and G.W. Thayer. 1982. A low cost planting technique for eelgrass (*Zostera marina L.*). U.S. Army Engineer Coastal Engineering Research Center, Ft. Belvoir, Virginia, Coastal Engineering Technical Aid No. 82-6. 15 p.

Fonseca, M.S., W.J. Kenworthy, K.M. Cheap, C.A. Currin and G.W. Thayer. 1984. A low cost transplanting technique for shoalgrass (*Halodule wrightii*) and manatee grass (*Syringodium filiforme*). U.S. Army Engineer Waterways Experiment Station, Vicksburg, Mississippi, Instruction Report EL-84-1. 16 p.

Fonseca, M.S., W.J. Kenworthy, G.W. Thayer, D.Y. Heller and K.M. Cheap. 1985. Transplanting of the seagrasses *Zostera marina* and *Halodule wrightii* for sediment stabilization and habitat development on the east coast of the United States. U.S. Army Engineer Waterways Experiment Station, Vicksburg, Mississippi, Technical Report EL-85-9. 49 p.

Fonseca, M.S., W.J. Kenworthy and G.W. Thayer. 1987a. Transplanting of the seagrasses *Halodule wrightii, Syringodium filiforme,* and *Thalassia testudinum* for habitat development in the southeast region of the United States. U.S. Army Engineer Waterways Experiment Station, Vicksburg, Miss., Technical Report EL-87-8. 47 p.

Fonseca, M.S., W.J. Kenworthy, K.A. Rittmaster and G.W. Thayer. 1987b. The use of fertilizer to enhance transplants of the seagrasses *Zostera marina* and *Halodule wrightii.* U.S. Army Engineer Waterways Experiment Station, Vicksburg, Mississippi, Technical Report EL-87-12. 45p.

Fonseca, M.S., G.W. Thayer and W.J. Kenworthy. 1987c. The use of ecological data in the implementation and management of seagrass restorations. Fla. Mar. Res. Publ. 42:175-187.

Fonseca, M.S., W.J. Kenworthy and G.W. Thayer. 1988. Restoration and management of seagrass systems: A review, p.353-368. *In* D.D. Hook, et al. (eds.), The ecology and management of wetlands. Vol. 2, Management use and value of wetlands. Timber Press, Portland, Oregon.

Fonseca, M.S., W.J. Kenworthy D.R. Colby, K.A. Rittmaster and G.W. Thayer. 1990. Comparisons of fauna among natural and transplanted eelgrass *Zostera marina* meadows: criteria for mitigation. Mar. Ecol. Prog. Ser. 65:251-264.

Fredette, T.J., M.S. Fonseca, W.J. Kenworthy and G.W. Thayer. 1985. Seagrass transplanting: 10 years of Army Corps of Engineers research, p. 121-134. *In* F.J. Webb (ed.), Proceeding of the 12th annual conference on wetlands restoration and creation, Hillsborough Community College, Tampa, Florida.

Iverson, R.L. and H.F. Bittaker. 1986. Seagrass distribution and abundance in eastern Gulf of Mexico waters. Estuarine Coastal Shelf Sci. 22:577-602.

Kenworthy, W.J., M.S. Fonseca, J. Homziak and G.W. Thayer. 1980. Development of a transplanted seagrass (*Zostera marina*) meadow in Back Sound, Carteret County, North Carolina, p. 175-193. *In* D.P. Cole (ed.), Proceedings of the 7th annual conference on the restoration and creation of wetlands, Hillsborough Community College, Tampa, Florida.

Kenworthy, W.J. and D. Haunert. 1990. Results and recommendations of a workshop convened to examine the capability of water quality criteria, standards and monitoring programs to protect seagrasses from deteriorating water transparency. Workshop proceedings, Nov. 7-8, 1990, S.W. Fla. Water Manage. District, W. Palm Beach, Florida. 151 p.

Lewis, R.R. 1987. The restoration and creation of seagrass meadows in the southeast United States. Fla. Mar. Res. Publ. 42:153-174.

Lewis, R.R., M.D. Durako, M.J. Moffler and R.C. Phillips. 1985. Seagrass meadows of Tampa Bay: a review, p. 210-246. *In* S.F.Trent, J.L. Simon, R.R. Lewis and R.L. Whitman (eds.), Proceedings of the Tampa Bay area scientific information symposium (May 1982). Burgess, Minneapolis, Minnesota. 663 p.

Orth, R.J. and K.A. Moore. 1981. Submerged aquatic vegetation of the Chesapeake Bay: past, present and future. Trans. N. Am.Wild. Nat. Resour. Conf. 46:271-283.

Orth, R.J. and K.A. Moore. 1982a. The biology and propagation of *Zostera marina*, eelgrass, in the Chesapeake Bay, Virginia. Va. Inst. Mar. Sci. Spec. Rep. Appl. Mar. Sci.Ocean Engin. 265. 187 p.

Orth, R.J. and K.A. Moore. 1982b. The effect of fertilizers on transplant-ed eelgrass *Zostera marina* in the Chesapeake Bay, p. 104-131. *In* F.J. Webb (ed.), Proceedings of the 9th annual conference on wetlands res-toration and creation. Hillsborough Community College, Tampa, Flori-da.

Orth, R.J. and J. van Montfrans. 1990. Utilization of marsh and seagrass habitats by early stages of *Callinectes sapidus*: a latitudinal perspective. Bull. Mar. Sci. 46(1):126-144.

Phillips, R.C. 1960. Observations on the ecology and distribution of the Florida seagrasses. Florida State Board Conserv. Mar. Lab. Prof. Pap. Ser. 2. 72 p.

Phillips, R.C. 1982. Seagrass meadows, p. 173-202. *In* R.R.Lewis (ed.), Creation and restoration of coastal plant communities. CRC Press, Boca Raton, Florida.

Phillips, R. C. 1984. The ecology of eelgrass meadows in the Pacific northwest: a community profile. U.S. Fish and Wildlife Service, FWS/OBS-84/24. 85 p.

Pulich, W. 1985. Seasonal growth dynamics of *Ruppia maritima* Aschers in southern Texas and evaluation of sediment fertility types. Aquat. Bot. 23:53-66.

Riner, M.I. 1976. A study on methods, techniques and growth character-istics for transplanted portions of eelgrass (*Zostera marina*). M.S. The-sis, Adelphi University, Garden City, New York. 104 p.

Roberts, M.H., R.J. Orth and K.A. Moore 1984. Growth of *Zostera mari-na* seedlings under laboratory conditions of nutrient enrichment. Aquat. Bot. 20:321-328.

Short, F.T. 1983a. The response of interstitial ammonium in eelgrass (*Zostera marina L.*) beds to environmental perturbations. J. Exp. Mar. Biol. Ecol. 68:195-208.

Short, F.T. 1983b. The seagrass, *Zostera marina*: plant morphology and bed structure in relation to sediment ammonium in Izembek Lagoon, Alaska. Aquat. Bot. 16:149-161.

Thayer, G.W., W.J. Kenworthy and M.S. Fonseca. 1984. The ecology of seagrass meadows of the Atlantic Coast: a community profile. U.S. Fish and Wildlife Service, FWS/OBS-84/02. 147 p.

Thayer, G.W., M.S. Fonseca and W.J. Kenworthy. 1986. Wetland mitiga-tion and restoration in the southeast United States and two lessons from seagrass mitigation, p. 95-117. *In* The Estuarine management practices symposium, Baton Rouge, Louisiana. National Sea Grant College Program.

Thom, R.M. 1990. A review of eelgrass (*Zostera marina L.*) transplanting projects in the Pacific Northwest. Northwest Environ. J. 6:121-137.

Thorhaug, A. 1974. Transplantation of the seagrass *Thalassia testudinum* Konig. Aquaculture 4:177-183.

Williams. S.L. 1987. Competition between the seagrasses *Thalassia testudinum* and *Syringodium filiforme* in a Caribbean lagoon. Mar. Ecol. Prog. Ser. 35:91-98.

Wood, E.J.F., W.E. Odum and J.C. Zieman. 1969. Influence of sea grasses on the productivity of coastal lagoons, p. 495-502. *In* A. Ayala Castanares and F.B. Phleger (eds.), Coastal lagoons. Universidad Nacional Autonoma de Mexico, Ciudad Universitaria, Mexico, D.F.

Zieman, J.C. 1982. The ecology of the seagrasses of South Florida: a community profile. U.S. Fish and Wildlife Service, FWS/OBS-82/25. 126 p.

Zieman, J.C. and R.G. Wetzel. 1980. Productivity in seagrasses: methods and rates, p. 87-116. *In* R.C. Phillips and C.P. McRoy (eds.), Handbook of seagrass biology, an ecosystem perspective. Garland STPM Press, New York.

Zieman, J.C. and R.T. Zieman. 1989. The ecology of the seagrass meadows of the west coast of Florida: a community profile. U.S. Fish Wildl. Serv. Biol. Rep. 87(7.25). 155 p.

Zieman, J.C., R. Orth, R.C. Phillips, G.W. Thayer and A.T. Thorhaug. 1984. The effects of oil on seagrass ecosystems, p 36-64. *In* J. Cairns and A.L. Buikema (eds.), Restoration of habitats impacted by oil spills. Butterworth Publishers, Boston, Massachusetts.

CHAPTER 4

Large-Scale Restoration of Seagrass Meadows

Hugh Kirkman
CSIRO Division of Fisheries
Western Australia

Abstract

Seagrass meadows in Western Australia are probably the largest and most diverse in the world. Their importance as nursery areas, filters to overlying water and as sediment stabilizers are discussed with regard to Western Australia in particular and Australia generally. Examples of natural destruction are given with predictions of the rate of return under natural conditions. In some cases these natural responses to seagrass meadow damage are extremely slow and it is necessary for human intervention to quicken their recovery. Human-related destruction by eutrophication, raised sediment loads, development (dredging, land-fill, moorings) and the destruction by propellers and aquaculture are discussed.

Restoration of seagrass meadows has not been completely successful anywhere in the world. The history and current research on restoring seagrass meadows in Australia are reviewed and, where possible, likened to projects in the United States. Restoring seagrass meadows is hampered by such problems as: (1) finding a suitable propagule, (2) damaging donor beds, (3) preparing and modifying site, (4) keeping a propagule bank, (5) attaching the propagules, (6) the slow rate at which rhizomes spread and (7) difficulties of large-scale planting. Also discussed, the possible direction of future research and the means by which funding for the implementation of this research and the control of pollution in coastal marine habitats.

Habitat

Geographic Range

Temperate Western Australian seagrasses, of which there are seventeen species of eight genera (Table 1), grow from Shark Bay (26° S, 114° E) in the north, to the border with South Australia in the south (32° S, 129° E). Most species also grow further east in the gulfs of South Australia. *Posidonia*, which is represented by several species, is widely distributed in Western Australia, but only one species grows in Victoria, northern Tasmania and central New South Wales (Figure 1). Some genera that grow in Western Australia (*Zostera* and *Halophila*) also grow in estuaries on Australia's east coast. However, at Brisbane, Queensland (27.28° S, 153.03° E) tropical seagrass species first appear and their diversity increases to Torres Strait.

Characteristics

Seagrasses grow in sheltered sand or muddy-sand subtidal or intertidal estuaries and bays along the coast. The underground roots and rhizomes anchor plants to the substratum and take up nutrients from the surrounding sediments. The leaves may be very productive in terms of photosynthetic gain (Kirkman and Reid 1979). Two species, *Amphibolis griffithii* and *A. antarctica*, usually found growing on sand also grow on limestone reefs where they accumulate sand around their rhizomes. *Thalassodendron pachyrhizum* is only found on reefs and, as its name implies, has very large, strong rhizomes. Seagrasses have been found to 48 m depth in southern Western Australia (Manning CSIRO Division of Fisheries, pers. comm.) but the depth to which they grow varies according to the light-attenuating properties of local waters. While seagrass mead-

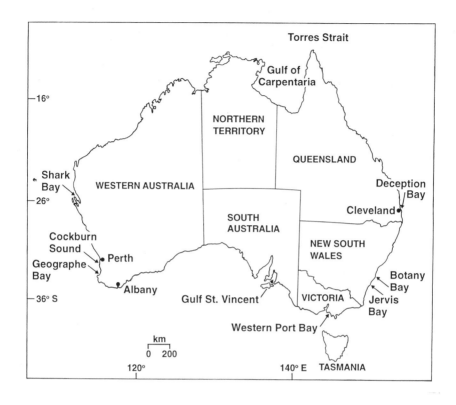

Figure 1. Map of Australia (Victoria, Tasmania and New South Wales).

ows on the east coasts of Australia are mostly found in estuaries, the seagrasses of Western Australia are usually found in open, but sheltered ocean embayments. Only *Zostera mucronata* and *Halophila ovalis* are found in estuaries in Western Australia. *Ruppia* spp. are also found in estuaries; however, they are not considered as seagrasses here. The species of seagrass found in Western Australia are listed in Table 1.

Table 1. Temperate Western Australian Seagrasses.[a]

Genus	Species
Zostera	*mucronata* den Hartog
Heterozostera	**tasmanica* (Martens ex Ascherson) den Hartog
Posidonia	* *australis* Hooker
	* *sinuosa* Cambridge and Kuo
	angustifolia Cambridge and Kuo
	den hartogii Kuo and Cambridge
	ostenfeldii den Hartog
	robertsoneae Kuo and Cambridge
	kirkmanii Kuo and Cambridge
	* *coriaceae* Kuo and Cambridge
Amphibolis	* *antarctica*
	* *griffithii*
Halophila	* *ovalis* (R.Br.) Hook.
	spinulosa (R.Br) Ascherson
Halodule	*uninervis* (Forsk.) Ascherson
Syringodium	* *isoetifolium* (Aschers.) Dandy
Thalassodendron	*pachyrhizum* den Hartog

[a] Major species considered in this paper are identified with an asterisk (*).

Functional Values

Seagrass meadows cover approximately 20,000 km^2 (4.9 million acres) in Western Australia and probably constitute the largest area of seagrass in the world (Kirkman and Walker 1989). This area is about equal to that occupied by rainforest in Australia and roughly equivalent to the area of seagrass meadows found along the coasts from North Carolina to Texas in the United States (Thayer and Fonseca, NMFS, Beaufort Laboratory, pers. comm.). South Australia has about 5,000 km² (Shepherd and Robertson 1989), while Victoria and New South Wales have only a small area (155 km² in New South Wales). Little is known of the area covered by seagrass on Tasmania's north coast and total area of seagrass in Queensland is not known, although it is very extensive.

The contribution of seagrasses to primary production of coastal waters has been estimated. The total carbon pool held by the Western Australian seagrass meadows is about 3 million tons, calculated by multiplying area by a conservative estimate of seagrass carbon biomass (150 gC m⁻²). The plants take up about 160,000 tons of carbon each year (assuming that the productivity is 0.4 mg g⁻¹d⁻¹). However, productivity varies in relation to depth and location. Phytoplankton productivity for the inshore areas of southern Western Australia has not been estimated but concentrations, measured by Chiffings (1979), were always less than 1 mg m⁻³.

The estimates of area presented above are approximate because no mapping or ground truth has been carried out. There are also only very loose approximations used in obtaining the above carbon pool and uptake figures. These figures have been derived using means of seagrass productivity and biomass gathered from work carried out in other Australian seagrass meadows. Little baseline data are available for the seagrass meadows of Western Australia although Kirkman and Walker (1989) and Kirkman (1985) have described the ecology of many of the meadows.

Links with Fisheries

Globally for the past sixty years (Dexter 1944; Cottam and Munro 1954; Adams 1976; Heck and Thoman 1984; Heck et al. 1989) and in Australia for twenty years (Young 1978; Coles et al. 1987), seagrass meadows have been recognized as nursery areas for commercially and recreationally important fish and crustacea. The fauna of seagrass meadows is more abundant and diverse than that of unvegetated areas (Young and Wadley 1979; Heck et al. 1989; Kirkman et al. 1991).

The decline of seagrasses in many bays near centers of human population in Australia has alerted fisheries managers and marine ecologists to the importance of seagrass meadows as fish habitats. These declines have resulted in serious alterations to the communities on which higher consumers depend. Abundances of many invertebrate species that form important components of the diets of commercial fish and crustacean species are strongly related to the amount of seagrass cover (Stoner 1980; Bell and Westoby 1986). Consequently, fundamental alterations to the structure of benthic communities seem inevitable once seagrass meadows are destroyed. It has long been suspected that serious declines in fisheries would result from significant losses of coastal and estuarine seagrass habitats. This is the logical conclusion to be drawn from evidence that seagrass meadows are important feeding and nursery habitats for fishes and fishery species. (Pollard 1976, 1981, 1984).

Prey depletion and food chain alterations or a loss of shelter for juvenile stages are factors likely to link fishes to declines in seagrass meadows. The fact that small epibenthic crustaceans, which are particularly important in the diets of seagrass-associated fishes (Pollard 1984; Watson et al.1984), appear to be highly sensitive to the loss of seagrass cover suggests that many fauna which utilize seagrass habitats are suffering severe interruption to their food supply. However, while it is clearly documented that many Australian estuarine and coastal fishery species utilize seagrass meadows, the degree to which their prey populations are altered by seagrass loss and the degree to which fishes are capable of compensating for any such changes are not understood. This is a problem that is common to many habitats worldwide and is not restricted to seagrasses. However, in some cases, the ecological significance of seagrass meadows can be directly assessed in economic terms, for example:

1. The cessation of all 21 commercial fishing operations in Western Port Bay, Victoria since 1974 has occurred at the same time that 71%, by area, of the seagrass meadow disappeared (Bulthius, Victorian Marine Laboratories, Queens-cliffe, pers. comm.; Bulthius et al. 1984).

2. Reduction in catch of tiger prawns (*Penaeus esculentus* and *P. semisulcatus*) in the Gulf of Carpentaria occurred after a cyclone had removed large areas of seagrass (Poiner, CSIRO Fisheries, Cleveland, pers. comm.).

Sediment Stabilization

Seagrass meadows have been shown to reduce the movement of sediment along coastal margins. In Geographe Bay in Western Australia, a dense *Posidonia* meadow covers about 70% of the sand sheet surface and provides a strong stabilizing influence on an otherwise mobile substrate. From 1941 and 1975 aerial photographs, it was determined that the net rate of accretion of sand over 34 years was more than 3.8 times the average for the preceding 4,600 years, as calculated from the average width and known age of the beach ridge system. This increase coincides with a significant decrease in seagrass cover (up to 45% in some locations) over the area within 3 km of the shore (Paul and Searle 1978). In this case the accretion had occurred inshore of the declining seagrass meadows and was a result of the added water movement caused by the disappearance of seagrasses.

Christiansen et al. (1981) found that the coastal morphology of Kyholm, Denmark was relatively stable from 1802 to 1933, but between 1933 and 1978 there have been two periods of drastic progradation correlated with die-back of *Zostera marina* in the 1930s.

Fonseca et al. (1982) found that shoot bending and subsequent in-meadow current velocity reduction are mechanisms that affect self-shading and photosynthesis as well as providing habitat stability in *Zostera marina*. Seagrasses may be used as a non-structural alternative to subtidal erosion control. Fonseca and Fisher (1986) showed the value of different species as sediment stabilizers and postulated that significant alteration of estuarine circulation patterns could result from removal of seagrass meadows or their creation through transplanting.

Water Filtration

Seagrass meadows suppress resuspension of suspended particulate material and enhance its deposition from the water column (Ward et al. 1984). In this way water is filtered of suspended material, allowing more light to get to the seagrass blades. Any organic matter in the suspended material adds to the food web of the seagrass meadow (Brasier 1975). Sedimentation also occurs more rapidly, sometimes leaving reefs of seagrass plants well above the unvegetated substrate below.

Food Chain Base

Detrital seagrass leaves supply food to the detritivores and filter feeders in the meadows (Zieman et al. 1984). Leaves provide substrate for epiphytic algae and associated communities. Leaves that are washed out of the seagrass ecosystem also provide organic material for off-shore ecosystems (Wolff 1976).

Impacts and Responses

Natural Destruction

For the smaller species of seagrass such as *Syringodium isoetifolium, Halophila ovalis* and *Heterozostera tasmanica,* storms may cause wave action to sweep out whole seagrass meadows including rhizomes (Kirkman and Kuo 1990). These areas may become quickly revegetated from sediment seed banks and by propagules from nearby undisturbed meadows. However, when the larger, climax community species such as *Posidonia* and *Amphibolis* are removed by severe, rare storms, they may take tens to hundreds of years to return (Clarke and Kirkman 1989). This slow recovery is a direct result of slow population growth rates and is an important consideration in both management and restoration of these habitats.

Greenhouse Effect

Changes brought about by greenhouse sea-level rise must be determined. If the predictions of greenhouse warming and sea-level rises of 1.5 m eventuate, seagrasses will be severely affected. The lower depth limit of seagrass meadows is governed by the amount of light received by the plants. Leaf density is usually lower at the deeper edges of seagrass meadows but these edges amount to considerable areas of seagrass and habitat for animals. If storms and subsequent river and land run-off increase, the sediment load over seagrass meadows will increase, thus further reducing the depth at which they grow. Severe storms are expected to become more frequent, thus destroying seagrass meadows more often.

Human Destruction

Human disturbances are now destroying seagrass meadows of the climax community plants faster than the natural spreading rate of these plants, resulting in a net loss in area covered by seagrasses. The most common type of impact on seagrass meadows in Australia in general and in Western Australia in particular is the reduction of photosynthetically active radiation to seagrasses. This reduction may be caused by phytoplankton blooms, an increase in epiphyte loads on seagrass leaves, smothering by free-living algae (all caused by excess concentrations of nutrients in the water) and/or by increased sediment loads in the water. This also appears to be a major factor impinging on seagrasses in the United States (W.J. Kenworthy, NMFS, Beaufort Laboratory, pers. comm.).

Here are some examples of large scale losses of seagrass meadows:

1. Between 1954 and 1978, 79% of the seagrass-covered area of Cockburn Sound (Western Australia) was lost (Cambridge and Mc-Comb 1984). The major cause was reduction in light available to leaves due to both heavy epiphyte loads and reduced water clarity from plankton blooms. These blooms of algae resulted from high concentrations of nutrients released from nearby factories (Silberstein et al. 1986).

2. About 50% of the area of seagrass meadow in Princess Royal Harbour and Oyster Harbour (near Albany, Western Australia) have been lost since 1972. The loss resulted from similar causes to those identified in Cockburn Sound, in addition to run-off of agricultural fertilizer from farmlands in the catchment of Oyster Harbour (Kirkman 1987).

3. Adelaide Beaches (Gulf St. Vincent, South Australia) have lost 365 ha of seagrass meadow due to sewage sludge. Much larger areas are now showing deleterious effects in areas adjacent to sewage discharge points in the Gulf (Neverauskas 1987).

4. Western Port Bay (Victoria) has lost about 71% of its seagrass meadows since 1974. While the causes are not clearly understood, land development in the upper reaches of the drainage-shed leading into the Bay has caused more sediment and nutrients to enter the Bay (Bulthius et al. 1984).

5. Botany Bay (New South Wales) seagrasses have been devastated by both excess sediment loads, which cause increased turbidity, and increased wave action resulting from human-made changes in the hydrology of the Bay (Larkum 1976).

6. Also causing reduction in area of seagrass meadows is direct or indirect destruction as the result of developments such as marinas or channel dredging. Developers, particularly in Queensland estuaries, are required to replace seagrass meadows they destroy as part of their operations. At the present time this is not possible, and governments, realizing the loss of valuable fish habitat, are controlling development.

Responses to Damage and Destruction

It has been shown that colonizing species of seagrass will recover more quickly than climax species (Kirkman and Kuo 1990). Because of their smaller size and less well-developed rhizome systems, however, they are much more readily removed by wave action so they are of less use in stabilizing sediment and do not develop the same faunal communities.

For climax communities of *Posidonia* and *Amphibolis*, response to damage caused by removal of plants including rhizomes is very slow, i.e., 60-100 years. There is evidence, in a number of *Posidonia* meadows around southern Australia, that they take tens of years to regenerate naturally. Four examples are given below.

1. In Jervis Bay on the southern coast of New South Wales in the early 1960s, seismic blasts in a *Posidonia australis* meadow caused about fifteen 20 m diameter circles to be cleared. There is no depression in the cleared areas and the sediment in the clearings appears to be similar in size distribution to that in the seagrass meadow. Nearly thirty years later the circles are well defined and have a few *Posidonia* rhizomes encroaching from the edges to a distance of about 30 cm (Figure 2).

2. Seismic testing in Gulf St. Vincent in South Australia was carried out thirty years ago, the clearings can still be easily seen from aerial photography in the *Posidonia angustifolia* meadows.

3. *Posidonia* fibre from the leaf sheaths of old shoots was used as a source of cellulose before and during the first World War. The fibre was found in the sediment and also came from the seagrass itself. Dredges removed the live seagrass from the sediment surface then dredged out the sand containing fibre. This industry had about fifteen dredges and two thousand people working in it at the peak of its activity (Winterbottom 1917). The scours from the dredges are still visible.

Figure 2. Seismic "holes" in seagrass meadow in Jervis Bay, New South Wales.

4. In a number of bays in southern Australia amphibious vehicles have driven into the seagrass beds and left tracks which have remained for years.

These areas make excellent experimental sites given that the history of the disturbance is known.

Restoration and Creation

Review of the History of Restoration in Australia

The global history of restoration is covered in other papers (see Fonseca et al. 1987; Fonseca, this volume). In Australia, CSIRO is the only organization actively developing methods to restore seagrasses on a large scale. Some previous work in Botany Bay (Larkum and West 1982) has failed and there have been several attempts to transplant seagrass for experimental purposes. Kirkman

(1978) planted *Zostera capricorni* in tanks and was able to sustain the plants by adding fish to remove amphipods which were eating the leaves. Cambridge (1979) transplanted *Posidonia australis* and *P. sinuosa* seedlings into four sites in Cockburn Sound which had varying degrees of pollution. She wanted to obtain information on their tolerance to disturbances such as chemical pollution, temperature and salinity changes. Controls, grown in an unpolluted bay nearby, showed the best survivorship while those in polluted areas fared badly.

The only efforts at restoration of seagrass meadows in Australia have been with transplanting vegetative material. Restoration attempts using *Posidonia* have been unsuccessful due to the slow growth of its rhizomes. Restoration attempts with other genera generally have been unsuccessful due to the unsuitability of the area chosen to restore.

Natural recovery of 15 square kilometers of seagrass meadow occurred between 1975 and 1980 in Deception Bay, Queensland with *Zostera capricorni.* This species was a dominant species before and after decline but has many of the characteristics of a colonizer. It is believed that excess sediment load from land-clearing in the catchment of a nearby river caused the original decline in this species. When this sediment load ceased the seagrasses returned, but even with the attributes of a colonizer it took five years for the recovery of 15 km² of *Zostera capricorni.* Other records from Queensland suggest that seagrass meadows do naturally establish, but the species that are successful are pioneers and never develop the complex above and below ground community of *Posidonia* and *Amphibolis.* No recovery of *Posidonia* or *Amphibolis* has ever been recorded.

It is not possible, at present, to restore seagrass meadows on a large scale in Australia nor can restoration attempts elsewhere be assured of success. Because of the importance of these systems for fisheries, the Australian Government is encouraging research on restoration of seagrass meadows.

Current Research on Restoration in Australia

CSIRO Division of Fisheries has been, for the past fifteen years, one of the leaders in seagrass research in Australia. As demonstrated by the examples above, there are a number of places (now, usually under government control) where extensive seagrass meadows have disappeared. There are also a number of coastal developments (government and private sector) which cannot proceed until provision can be made for replacement of seagrass meadows.

Research in Australia over the last decade has led to a greater understanding of the ecology and biology of seagrass meadows, particularly in the diverse meadows of Western Australia. Only now are we beginning to address restoration through application of this ecological information

The current studies by CSIRO at Cleveland, Queensland and in Perth, Western Australia are the only research on restoring seagrass meadows at present underway in Australia. At Cleveland, attempts are being made to move large turfs of *Zostera capricorni* from areas subject to development to other suitable areas.

In Western Australia the most likely species for successful restoration programs, marked in Table 1, are being individually screened for their ability to restore or create seagrass meadows. This screening consists of a series of steps. Sexually and vegetatively grown propagules are tested for their suitability for large scale collection, storage and planting; trial plots are then planted and closely monitored for leaf growth and spreading ability (Kirkman 1989). If plants perform well in trials, they will be planted on a large scale and monitored for coalescence.

The objective of restoration and creation of seagrass meadows on a large scale at this initial stage of the project in Western Australia is to plant propagules and assist them to the stage where planting units coalesce. Considerations of total system function and evaluation of its function, as has been incorporated into the work of Fonseca (Fonseca, this volume), are not part of the project. At this stage, finding an endemic species—the propagules of which can be planted over large areas and will coalesce with neighboring propagules—will be considered an initial success. Optimal spacing of planting units has to be determined and the size and developmental stage of propagules and time of planting are important considerations.

The most important research goal in this project has been identified as the need for a plant physiologist to develop methods to increase the growth rate of rhizomes of the dominant species. Once this has been achieved the engineering problem of mechanical planting of large areas with propagules should be more easily resolved.

Much of the biology of the seagrass species most likely to succeed in restoration is known. Below, a summary of the relevant biology and how it relates to possible restoration is given.

Seagrass Biology and Restoration

In order for us to develop a successful restoration program we have had to learn a great deal about the biology and ecology of the plants themselves. The fewer number of species and the wealth of research on biology and ecology of these species has made restoration experiments on seagrasses in the United States more common than in Australia. The ability of the majority of the seagrass species that have been experimented with in the United States, to spread rapidly by vegetative growth also makes the plants more amenable to experimental research (Fonseca, this volume).

Vegetative Propagules. Although, in terrestrial transplants, it is common practice for vegetative material to be used, little success has been achieved with these forms of seagrass transplants and rate of extension is slow (Fonseca et al. 1987). Seagrass revegetation trials have usually been attempted using vegetative parts such as plugs (plant cutting with sediment attached) or by sprigs (plant cutting without sediment).

In three Western Australian species there is another form of vegetative reproduction, i.e., the production of adventitious roots in the two species of *Amphibolis* and *Heterozostera tasmanica* (Cambridge et al. 1983; Kuo et al. 1987). These roots grow from the lignified stems of the plants and, if their production can be induced, they may be effective planting units (PUs) for transplanting.

The potential for use of terrestrial plant horticultural practices for enhancing growth is evident. The development of effective PUs may benefit from the application of methods such as hormone application, trimming, layering of nodes and application of root-promoting material and tissue culture to the PUs.

The PUs must be readily available as they will be required in large numbers to sustain commercial operations. Vegetative propagules, for example, meristematic rhizomes, may be collected at any time of the year but there may be optimum times for collection and for planting. Care has to be taken not to damage the donor sites. Some species (i.e., *Posidonia sinuosa, P. australis* and *P. coriacea*) have very few growing tips on their rhizomes and these generative shoots grow very slowly (less than 2.5 cm per year, West 1990). *Amphibolis* species may be useful in restoration as they produce many rhizome shoots; however, the rate of growth of shoot material is not known. There are limitations to where we may be able to transplant vegetative material of various species, however. *Heterozostera tasmanica* grows rapidly on cleared sediment and blowouts. It may be a fast growing and useful colonizer in sheltered bays, but will not tolerate vigorous water movement. *Halophila ovalis* is even more fragile but grows very rapidly once established (Kirkman, unpubl. data).

Reproductive Propagules. Sexual material may be more easily collected at times of the year and seed or seedling banks may be used

Figure 3. Posidonia australis *early seedling.*

to store large quantities of propagules. The collection of sexual material has the added advantage of not destroying donor meadows. Seedlings, however, are delicate and seeds may have poor germination rates.

Seagrass flowers are sometimes difficult to find and recognize. However, some species flower prolifically and have readily available fruit, for example, *Posidonia australis* and *P. coriacea,* while others have very few flowers (i.e., *P. sinuosa).* The mature fruit of *Posidonia* has a germinated seed inside; it can be planted in various planting media (Figure 3). The endosperm remains attached to the seedling and provides the seed with food for several months. Neither the uptake of nutrients, nor even photosynthesis, are important until this reserve is depleted.

Flowers produced by *Halophila ovalis* occur at or below sediment level and, although numerous, are difficult to find. The result-

Figure 4. Amphibolis antarctica *seedling soon after falling from parent plant.*

ing seeds remain in the sediment until disturbed and are probably stimulated to germinate by light, e.g., *H. engelmanii* (McMillan 1987, 1988).

Viviparous seedlings of the two *Amphibolis* species, are attached to the parent plant until they are well developed. Seeds are difficult to find and probably become non-viable once detached from the parent plants on which they rely for growth. The seedlings grow very slowly although their grappling apparatus (Figure 4) is useful in holding them down in the early stage of their development (Ducker et al. 1977). The plants grow new leaves rapidly but little is known of the rate at which their rhizomes spread. Adventitious roots form on *Amphibolis* plants (Kuo et al. 1987). Attempts will be made to develop these roots on *Amphibolis* stems to enhance attachment.

Heterozostera tasmanica produces many obvious inflorescences which can be collected when mature. These have potential to provide a plentiful supply of seed. Seedlings from *H. tasmanica* have not been found nor seeds germinated in the laboratory. In the

American Zosteraceae, *Zostera marina* seeds are plentiful and viable (Addy 1947); if the same treatments are applied to *H. tasmanica*, it may be possible to produce seedlings.

Seed Collection. Seed collection is the first part of any restoration project requiring sexually reproductive planting units. These seeds may be germinated in aquaria and then planted out as seedlings or they may be distributed on the restoration site. *Posidonia* seeds germinate inside fruit which can be collected when ripe. After dehiscence the seedling is ready for distribution or planting. At this stage, *Posidonia* seedlings like *Thalassia testudinum* cannot be considered a restoration stock because of the slow spreading rate of the rhizomes. There is a need for a plant physiologist to experiment with growth hormones to enhance rhizome growth.

Seed Bank. Sometimes it may not be feasible to plant seedlings or vegetative propagules as soon as they are large enough for transplanting or when they are collected, or to distribute seeds at a favorable time. In these cases a seed or propagule bank would allow their retention until a favorable time for planting. Once PUs have been collected they may need to be stored for considerable time, for various reasons, e.g., the planned planting area is not ready, weather is not suitable for planting, or man-power is not available. Large aquaria may be required for seedling retention in the storage facility.

Fertilizer. We intend to evaluate the importance of nutrient additions to enhance transplant growth. Excess nutrients will be given during the storage period to give the PUs an advantage once they are planted. These doses of fertilizer will be given either through the sediment, as slow release fertilizer, or by increasing the water nutrients. Increasing nutrient concentrations to assist PUs requires caution as algal epiphytes may also use the nutrients and overgrow the PUs. Algal spores or viable pieces of algae must be excluded during storage, it may be necessary to use artificial or sterilized seawater. Fonseca (this volume) has discussed the use of fertilizers.

Photosynthesis Vs. Irradiance Curves. Experiments are needed to determine the light climate required to keep PUs alive indefinitely in aquaria. As part of these experiments light compensation levels and the minimum amount of light required for maximum oxygen production will be determined at different temperatures. "P vs. I" curves for each species will be calculated. Routine use of the non-destructive "P vs. I" curve technique will be made to assess the well-being of plants, and respiration measurements will be used to determine whether rhizomes remain metabolically viable. Once optimum light levels are known for each species, experiments to determine optimum levels of secondary determining factors (i.e., nutrients) can be carried out. These experiments are required not only for laboratory rearing but also so we will know the conditions most conducive to growth in the field.

Selection and Preparation of Sites. Seagrass meadows will only grow and sustain themselves in suitable conditions. Before a seagrass restoration project can be effective, the marine environment must be suitable for seagrass establishment and growth. If the site previously had seagrass meadows growing on it, the vector or disturbance that caused the decline must be removed or ameliorated. Mitigation schemes that do not take this into account are doomed to failure. The environmentally-sound selection of the areas in which restoration is to take place is vital for the success of the project. It is useless to attempt to restore seagrass meadows if the conditions for seagrass growth are not suitable.

Monitoring of water quality, sediment stability and hydrology must also play a part in total ecosystem management. It would be unfortunate if the failure of a restoration project due to deteriorating physical or chemical properties at the site were blamed on the inability of the seagrass to spread.

If there were no seagrasses growing within living memory on the proposed site, it is probably unsuitable for seagrass meadow creation. Therefore, changes such as hydrodynamic ones would have to be made to this site before chances of successful transplants can be expected.

Where human-induced impacts have occurred restoration will be most difficult, particularly in the cases where the physico-chemical or hydrological environment has changed. The original environmental conditions must be re-established before restoration can be successfully carried out. This requirement is well illustrated in (1) the attempts to restore *Posidonia australis* in Botany Bay, where mechanical changes to the Bay had caused hydrological changes and restored plants were eroded away (Anink, State Pollution Control, Sydney, New South Wales, Australia) and (2) nutrient concentration must be reduced in Cockburn Sound and Princess Royal Harbour to levels similar to those over nearby healthy seagrass meadow areas.

The environmental conditions required for any of these restoration attempts to be successful need to be carefully evaluated. These conditions may be different from those required to sustain a dominant meadow. Water movement, in the early part of establishment, must be low so that sediment does not swamp the new plants or erode away from the roots and rhizomes. After establishment, fairly vigorous water movement may be necessary to supply sufficient nutrients to sustain plants (Fonseca and Kenworthy 1987). Other factors, e.g., light, may be limiting for the young seedlings or spreading PUs but at acceptable levels for the dominant meadow. The spacing of PUs is an important factor in determining the rate of regrowth and the cost.

Screening of Seagrass Species. The choice of the seagrass species to use in a restoration attempt must depend upon the characteristics of the site to be restored. It is reasonable to assume that *Heterozostera tasmanica* will spread faster than the *Posidonia* species. *H. tasmanica* could then be suitable to plant and quicker to grow in sheltered bays or estuaries. It would not, however, be suitable in areas where water motion is too vigorous and PUs would soon be washed out. Multiple-species planting may also be the most successful approach in some areas. Pioneer species may be planted with dominants and used to stabilize sediments while the dominants become established. Artificial seagrass, for instance, Monsanto's "Astroturf," could be useful in strips to stabilize sediments before a restoration program.

Transplanting. Probably the most difficult practical part of a restoration program is successfully planting out the PUs. Each PU must be stabilized in the sediment immediately because of the likelihood of its being washed away. Rhizomes must have growing tips capable of extending the plant over unvegetated sediment. PUs must be easily planted in large numbers over large areas and well able to be handled by mechanical planters or divers. We have grown *Posidonia australis, sinuosa* and *coriacea* in Growool, a product of spun rock wool, and used it to hold the plant down in plantouts (Figure 5).

Management of Restoration Sites. Management must play a large part in the successful restoration or creation of a seagrass meadow. First, management is responsible for ensuring that the site to be planted is in a suitable condition for planting. This is often difficult because the causal agent responsible for the decline may not have been eliminated. Once seagrass is gone the habitat changes in hydrological, chemical and physical ways. Second, the management must ensure that suitable conditions for seagrass

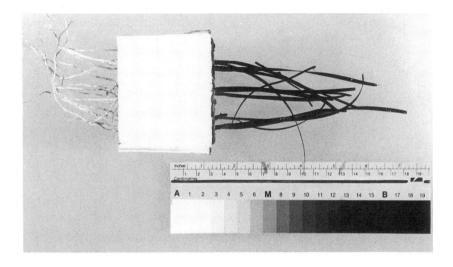

Figure 5. Posidonia australis *seedling after four months in Growool.*

regeneration continue after planting. Monitoring for physical and chemical parameters must continue either indefinitely or at least for a number of years while the seagrass grows and until it is certain that the deleterious effects that caused the original loss have gone.

Protocols

Coordination of Tactical and Strategic Research in Australia
CSIRO is a federally funded research organization whose charter is to provide the results of tactical and strategic research to the Australian public. CSIRO Fisheries is dedicated to the management of the coastal zone as it pertains to fisheries. This is considered a fairly wide environmental charter encompassing nursery areas and fish habitats. The organization has responsibility for research into restoration and creation of seagrass meadows and the practical applications may then be handed on to the Australian States.

In Australia, restoring seagrass meadows on a large scale, if it were possible, would be the responsibility of the states. They in turn should have levied the organizations responsible for the damage. Individual developers who destroy seagrass meadows will have to replace them. Evaluation of whether a seagrass restoration project has been successful should be carried out by the states or CSIRO with funding from polluters or developers. Currently, CSIRO is working on seagrass restoration with appropriation funding. Once a breakthrough has occurred the intellectual property will be passed on to the relevant state department. It is anticipated that long term management would be required to maintain seagrass meadows. CSIRO would be responsible to management for the upkeep and continued monitoring of new seagrass meadows.

Peer Review

Research into restoration of seagrass meadows has the temptation to short-cut sound scientific methods for quick results and remedies. Sound statistical methods and a defined, easily measured variable, such as rhizome length or number of shoots, are important criteria to include in experimental design. The technology of transplanting or of seeding areas for repair or mitigation is not so well known that every restoration project that applies it will be a success. Even though one would expect more success with restoring damaged sites with *Heterozostera tasmanica, Halophila ovalis* or one of the *Zostera* species, guarantees of success are not possible. Plantings of the dominant species *Posidonia, Amphibolis* and northern hemisphere tropical species like *Thalassia testudinum* will not be successful at this stage. Proposals for large expensive projects should be sent to relevant, informed scientists for review much as draft manuscripts are. Ultimately though, the evidence of coalesced propagules sustained over a period of some years (perhaps 10) is the real measure of success.

Future Needs and Directions

Much has been achieved in developing methods for large scale restoration of seagrass meadows. Propagule numbers are obtainable by either field collection of seeds or seedlings or tissue culture of vegetative material (Durako 1988). Planting on a large scale can be done by mechanical planters or broadcasting seed from planes (R. Orth, Virginia Institute of Marine Science, pers. comm). The biology of the seagrass plants and their communities is well enough known to ensure that conditions for satisfactory sustained growth are obtained before planting.

As stated earlier, the most important aspect of restoration on a large scale in Australia is to be able to obtain a propagule with a fast growing rhizome. The need for plant physiologists to determine the driving factors for rhizome growth is of paramount importance. Currently, if it were required to plant 2,000 ha of seagrass, and propagules were to be placed a meter apart, 20 million propagules would be required. If the rhizomes were to grow 2.6 cm per year (West 1990), it would take 20 years for propagules to coalesce. During this time, the chances of a storm occurring that could wash out the unprotected propagules is highly likely. Coalescence, at this stage, is the criterion for assessing success as it suggests some degree of stability to the system and can be easily assessed. If the time for coalescence could be shortened by an order of magnitude, the plants would have more chance to successfully coalesce and a restoration program would be feasible.

Most of the responsibility for judging success, evaluating, and managing restored areas should be the responsibility of the Environmental Protection Authorities in each state. Funds should come from the polluters, i.e., town sewage authorities, industry and perhaps even farmers or fertilizer firms. The value of seagrass ecosystems has been realized in a number of Australian states and demands are made on industry to clean up and lessen their effluent directly discharged onto seagrass meadows, for example, Environmental Protection Authority, Western Australia recommendations on Albany Harbours (EPA 1990).

In Australia the future needs of restoration programs are seen as basic physiology, hormone and tissue culture work. Engineering problems such as mechanical planting or seed harvesting can be overcome, but the basic problem of slow rhizome growth has yet to be solved.

Some consideration must be given to planting species that were not previously growing in the area; however, the problem of introductions and hybridization then arises. The communities associated with different seagrass species are different, and some of the beneficial uses of a particular species may be lost under the cultivation of another.

Acknowledgments

My thanks go to the National Oceanic and Atmospheric Administration for making it possible for me to visit Washington, D.C., participating in the Symposium this book is based on, and for assisting in making the Symposium so successful. I am grateful to Mr. Jeremy Fitzpatrick, Drs. Vivienne Mawson, Gordon Thayer, and an anonymous reviewer who made useful comments on the manuscript.

Literature Cited

Adams, S.M. 1976. The ecology of eelgrass, *Zostera marina* (L.), fish communities. I. Structural analysis. J. Exp. Mar. Biol. Ecol. 22: 269-291.

Addy, C.E. 1947. Germination of eelgrass seed. J. Wildl. Manage. 11, 279.

Bell, J.D. and M. Westoby. 1986. Abundance of macrofauna in dense seagrass is due to habitat preference, not predation. Oecologia 68: 205-209.

Brasier, M.D. 1975. An outline history of seagrass communities. Palaeontology 18: 681-702.

Bulthius, D.A., D.M. Axelrad, A.J. Bremner, N. Coleman, N.J.Holmes, C.T. Krebs, J.W. Marchant and M.J. Mickelson. 1984. Loss of seagrasses in Western Port. Progress Report 1, Dec. 1983 to March 1984. Victorian Marine Laboratories, Queenscliff, Victoria.

Cambridge, M.L. 1979. Cockburn Sound Environmental Study Technical report on seagrass. Dept. Conservation and Environment, Perth, Western Australia. Rept. No. 7, 100 p.

Cambridge, M.L., S.A. Carstairs and J. Kuo. 1983. An unusual method of vegetative propagation in Australian Zosteraceae. Aquat. Bot. 15: 201-203.

Cambridge, M.L. and A.J. McComb. 1984. The loss of seagrass in Cockburn Sound, Western Australia. 1. The time course and magnitude of seagrass decline in relation to industrial development. Aquat. Bot. 20: 229-243.

Christiansen, C., H. Christoffersen, J. Dalsgaard and P. Nornberg. 1981. Coastal and near-shore changes correlated with die-back in eel-grass (*Zostera marina, L.*) Sediment Geol. 28: 163-173.

Clarke, S.M. and H. Kirkman. 1989. Seagrass dynamics, p. 304-345. *In* A.W.D. Larkum, A.J. McComb (eds.), Biology of seagrasses, a treatise on the biology of seagrasses with special reference to the Australian region. Elsevier, Amsterdam.

Coles, R.G., W.J. Lee Long, B.A. Squire, L.C. Squire and J.M. Bibby. 1987. Distribution of seagrasses and associated juvenile commercial penaeid prawns in north eastern Queensland waters. Aust. J. Mar. Freshwater Res. 38: 103-119.

Chiffings, A.W. 1979. Cockburn Sound Environmental Study. Technical report on nutrient enrichment and phytoplankton. Dept. of Conservation and Environment, Perth, Western Australia, Rept. No. 3, 59 p.

Cottam, C. and D.A.Munro. 1954. Eelgrass status and environmental relations. J. Wildl. Manage. 18: 449-460.

Dexter, R.W. 1944. Ecological significance of the disappearance of eelgrass at Cape Ann, Massachusetts. J. Wildl. Manage. 8: 173-176.

Ducker, S.C., N.J. Foord and R.B. Knox. 1977. Biology of Australian seagrasses: the genus *Amphibolis* C. Agardh (Cymodoceaceae). Aust. J. Bot. 25: 67-95.

Durako, M.J. 1988. III 9. Turtle grass (*Thalassia testudinum* Banks ex Konig)—a seagrass, p. 504-520. *In* Y.P.S. Bajaj (ed.), Biotechnology in Agriculture and Forestry Vol. 6. Crops II. Springer-Verlag, Berlin.

EPA. 1990. Recommendations of the Environmental Protection Authority in relation to the environmental problems of the Albany Harbours. Environmental Protection Authority, Western Australia, Bulletin 442, Perth, Western Australia.

Fonseca, M.S. and W.J. Kenworthy. 1987. Effects of current on photosynthesis and distribution of seagrasses. Aquat. Bot. 27: 59-78.

Fonseca, M.S., W.J. Kenworthy and G.W. Thayer. 1987. Restoration and management of seagrass systems: a review, p. 353-368. *In* D.D. Hook, et al. (eds.), The ecology and management of wetlands, Vol. 2. Timber Press, Portland, Oregon.

Fonseca, M.S. and J.S. Fisher. 1986. A comparison of canopy friction and sediment movement between four species of seagrass with reference to their ecology and restoration. Mar. Ecol. Prog. Ser. 29: 15-22.

Fonseca, M.S., J.S. Fisher, J.C. Zieman and G.W. Thayer. 1982. Influence of the seagrass *Zostera marina* L. on current flow. Estuarine Coastal Shelf Sci. 15: 351-364.

Heck, K.L., Jr., K.W. Able, M.P. Fahay and C.T. Roman. 1989. Fishes and decapod crustacea of Cape Cod eelgrass meadows: species composition, seasonal abundance patterns and comparison with unvegetated areas. Estuaries 12: 59-65.

Heck, K.L., Jr. and T.A. Thoman. 1984. The nursery role of seagrass meadows in the upper and lower reaches of Chesapeake Bay. Estuaries 7: 70-92.

Kirkman, H. 1978. Growing *Zostera capricorni* Aschers. in tanks. Aquat. Bot. 4: 367-372.

Kirkman, H. 1985. Community structure in seagrass communities in southern Western Australia. Aquat. Bot. 21: 363-375.

Kirkman, H. 1987. Decline in seagrass beds in Princess Royal Harbour and Oyster Harbour, Albany. Environmental Protection Authority, Perth, Western Australia, Tech. Ser. 15. 11p.

Kirkman, H. 1989. Restoration and creation of seagrass meadows. Environmental Protection Authority, Perth, Western Australia, Tech. Ser. 30. 20 p.

Kirkman, H. and D.D. Reid. 1979. A study of the role of the seagrass *Posidonia australis* in the carbon budget of an estuary. Aquat. Bot. 7: 173-183.

Kirkman. H. and Kuo, J. 1990. Pattern and process in southern Western Australian seagrasses. Aquat.Bot. 37: 367-382.

Kirkman, H., P. Humphries and R. Manning. 1991. The epibenthic fauna of seagrass beds and bare sand in Princess Royal Harbour and King George Sound, Albany, south-western Australia. Proceedings of the 3rd international marine ecology workshop: the marine flora and fauna of Albany, Western Australia, Jan. 1989.

Kirkman, H. and D.I. Walker. 1989. Regional studies—Western Australian seagrass, p. 157-181. *In* A.W.D. Larkum, A.J. McComb and S.A. Shepherd (eds.), Biology of seagrasses, a treatise on the biology of seagrasses with special reference to the Australian region. Elsevier, Amsterdam.

Kuo, J., I.H. Cook and H. Kirkman. 1987. Observations of propagating shoots in the seagrass genus *Amphibolis* C. Agardh (Cymodoceaceae). Aquat. Bot. 27: 291-293.

Larkum, A.W.D. 1976. Botany Bay: national asset or national disaster. Operculum 5: 67-75.

Larkum, A.W.D. and R.J. West. 1982. Stability, depletion and restoration of seagrass beds. Proc. Linn. Soc. N.S.W. 106: 201-212.

McMillan, C. 1987. Seed germination and seedling morphology of the seagrass *Halophila engalmanii* (Hydrocharitaceae). Aquat. Bot. 2: 179-188.

McMillan, C. 1988. The seed reserve of *Halophila engalmanii* (Hydrocharitaceae) in Redfish Bay. Aquat. Bot. 30: 253-259.

Neverauskas, V.P. 1987. Accumulation of periphyton biomass on artificial substrates deployed near a sewage sludge outfall in South Australia. Estuarine Coastal Shelf Sci. 25: 509-517.

Paul, M.J. and J.D. Searle. 1978. Shoreline movements Geographe Bay, Western Australia, p.207-212. *In* 4th Australian conference on coastal and ocean engineering. Adelaide, South Australia. The Institution of Engineers, Australia.

Pollard. D.A. 1976. Estuaries must be protected. Aust. Fish. 35: 105.

Pollard, D.A. 1981. Estuaries are valuable contributors to fisheries production. Aust. Fish. 40: 7-9.

Pollard, D.A. 1984. A review of ecological studies on seagrass-fish communities, with particular reference to recent studies in Australia. Aquat. Bot. 18: 3-42.

Shepherd, S.A. and E.L. Robertson, 1989. Regional studies—seagrasses of south Australia, western Victoria and Bass Strait, p. 211-229. *In* A.W.D. Larkum, A.J. McComb and S.A. Shepherd (eds.), Biology of seagrasses, a treatise on the biology of seagrasses with special reference to the Australian region. Elsevier, Amsterdam.

Silberstein, K., A.W. Chiffings and A.J. McComb. 1986. The loss of seagrass in Cockburn Sound, Western Australia. III. The effect of epiphytes on productivity of *Posidonia australis* Hook. F. Aquat. Bot. 24: 355-371.

Stoner, A.W. 1980. The role of seagrass biomass in the organisation of benthic macrofaunal assemblages. Bull. Mar. Sci. 30: 537-551.

Ward, L.G., W.M. Kemp and W.R. Boynton. 1984. The influence of waves and seagrass communities on suspended particulates in an estuarine embayment. Mar. Geol. 59: 85-103.

Watson, G.F., A.I. Robertson and M.J. Littlejohn. 1984. Invertebrates of the seagrass communities in Western Port. Aquat. Bot. 18: 175-197.

West, R.J. 1990. Depth related structural and morphological variations in an Australian *Posidonia* seagrass bed. Aquat. Bot. 36: 153-166.

Winterbottom, D.C. 1917. Marine fibre. South Australian Dep. Chem. Bull. 4. 36p.

Wolff, T. 1976. Utilization of seagrass in the deep sea. Aquat. Bot. 2: 161-174.

Young, P.C. 1978. Moreton Bay, Queensland: a nursery area for penaeid prawns. Aust. J. Mar. Freshwater Res. 29: 55-76.

Young, P.C. and V. Wadley. 1979. Distribution of shallow-water epibenthic macrofauna in Moreton Bay, Queensland, Australia. Mar. Biol. 53: 83-97.

Zieman, J.C., S.A. Macko and A.L. Mills. 1984. Role of seagrasses and mangroves in estuarine food webs: temporal and spatial changes in stable isotope composition and amino acid content during decomposition. Bull. Mar. Sci. 35: 380-392.

CHAPTER 5

Restoring Coral Reefs with Emphasis on Pacific Reefs

James E. Maragos

Environment and Policy Institute
East-West Center, Honolulu, Hawaii

Abstract

Coral reefs near most of the populous centers in the United States, sovereign territories and affiliated Pacific nations are undergoing pronounced decline from a variety of anthropogenic activities. Under ideal circumstances coral reefs take decades or more to recover, and where pollution or other stresses are chronic, recovery is postponed indefinitely. Although a variety of mitigation activities are presently being used to protect coral reefs from additional decline, especially during coastal development, little effort has been directed at restoring reefs already degraded. In fact only one full-scale recovery has been accomplished (Kaneohe Bay, Hawaii) although several restoration techniques have been tested on a pilot basis elsewhere. There is also the good chance of a second full scale restoration being accomplished off south Florida (Molasses Reef, Key Largo). The cost of both full-scale restorations exceeds $6-25 million.

Given the high cost and limited progress to date on coral reef restoration, much more field and laboratory effort is needed on research and development of techniques. Management action needs to focus on preventing additional declines, until such time as coral reefs begin recovering on a national (and global) scale. Even then prevention should continue to be emphasized. This includes strengthening coastal resource management in the areas of coral reef inventorying, mitigation, monitoring, regulatory control, and multi-agency cooperation.

Overview of Habitat Type

Coral reefs are wave resistant structures harboring plants and animals in shallow tropical seas and consisting of the remains of calcium carbonate secreting organisms. Historically the most comprehensive definition is:

> Coral reefs are deterministic phenomena of sedentary organisms with high metabolism living in warm marine waters within the zone of strong illumination. They are constructional physiographic features of tropical seas consisting fundamentally of a rigid calcareous framework made up of the interlocked and encrusted skeletons of reef-building (hermatypic) corals and calcareous red algae. The framework controls the accumulation of sediments on, in, and around itself. These sediments are derived from organic and physical degradation of the frame and organism[s] associated with the reef constructors and have a bulk ten or more times as great as the frame itself (Wells 1957).

While most of the reef environment is depositional, the seaward growing portion of the system is essential for the survival and maintenance of the remainder of the system (Wiens 1962; Guilcher 1987). Coral reefs are successful and predominate in many tropical benthic environments because of their ability to grow or maintain structures in the face of heavy or prevailing wave action. Their ability to thrive in the clearest, most transparent of waters gives coral reefs an advantage over other benthic ecosystems which generally require greater concentrations of inorganic nutrients. Oceanic waters in the tropics are notoriously low in nutrients, and the greater water current, flushing, and circulation regimes associated with reefs may somehow stimulate more efficient uptake of the limiting nutrients (Atkinson 1988). Reef building corals and other reef organisms also contain symbiotic algae, termed zooxanthellae, which live within the tissues of corals to produce food and take up carbon dioxide and nutrients excreted by the coral animal. In turn the coral is able to secrete its carbonate skeleton much more quickly than non-reef building corals (*ahermatypes*) and use some of the food produced by the algae. Warm water and abundant sunlight, which characterizes much of the tropics, speeds up metabolism, calcification and coral growth.

Basic Types of Coral Reefs

There are several basic types of coral reefs, including fringing reefs, bank reefs, barrier reefs, table reefs, and atolls (Guilcher 1987). *Fringing reefs* grow immediately offshore, covering the uppermost marine slopes of high islands. As the reefs continue to grow offshore, the inner depositional zones tend to be deeper and dominated by sand deposits. If this process continues the outer growing edge becomes a *barrier reef* separated from the high island shoreline by a deeper lagoon. Darwin was the first to implicate subsidence as the cause for the development of barrier reefs from fringing reefs and atolls from barrier reefs (Darwin 1842). The reef maintains upward growth along the outer barrier while subsistence of the high island continues. Eventually the volcanic island sinks beneath the waves leaving only the *atoll* reef, generally a ring of peripheral reefs surrounding a lagoon. Deep drilling studies at Midway and the Marshall Islands confirmed that atolls are the subsided tops of drowned volcanoes which have sunk up to several thousand feet over a period of millions of years (Wiens 1962).

Not all reefs evolve in this simple sequence. For example, sea level rose more than 300 ft since the last ice age some 8,000 years ago. In response to the rising water levels, upward reef growth continued to maintain close proximity to the sea surface. Thus the living portions of all modern day reefs have been built during the past 8,000 years, in response to sea level fluctuations during the glaciations. *Bank reefs* are a special category of reefs that have grown upward on offshore banks or platforms in response to the rising sea levels over the past 10,000 years. These are not true barrier reefs, although they are separated from high islands or continental coasts by wide expanses of deeper water. *Table reefs* are like atolls in terms of evolutionary sequence but are simple in form and do not enclose a lagoon. Many table reefs occur among the atolls of Micronesia.

Global Distribution

Coral reefs are concentrated in tropical seas between the Tropic of Cancer and Tropic of Capricorn, ranging between 22.5° N and 22.5° S latitudes, respectively. However warm water masses and currents which predominate along the western sides of the major oceans allow reefs to extend into higher subtropical and temperate latitudes. For example reef corals occur as far north as 34° N latitude and are found off Japan, Bermuda and Virginia due to the warming effects of the Kuroshio and Gulf Stream currents. Conversely coral reefs are less developed along the eastern sides of the Atlantic and Pacific Oceans where prevailing currents are cooler and where freshwater discharges discourage coral growth. In contrast, the Indian Ocean, which has a seasonally reversing monsoonal ocean circulation system, supports good coral reef development along its eastern and western boundaries. The best reef development occurs within the Malay Archipelago, the complex equatorial island and continental area between the Indian and Western Pacific Ocean. Here currents tend to concentrate many types of reef life and habitat abundance and diversity achieve its highest development.

Distribution in the United States and Affiliated Islands

Although the contiguous 48 states occur in temperate latitudes, bank reefs occur offshore from the south coasts of Texas and Florida (Figure 1). The Gulf Stream and Florida Current allow the South Florida reefs to thrive, and the warm waters of the Gulf of Mexico support the Texas reefs. Among the states, Hawaii supports

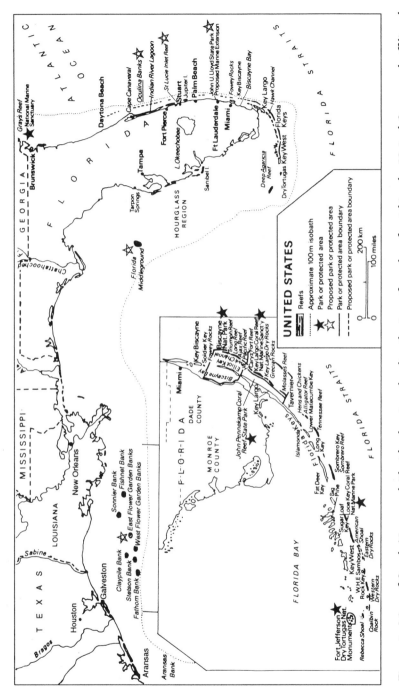

Figure 1. Map of South Texas and Florida showing the distribution of coral reefs within the United States and affiliated islands. After Coral Reefs of the World (IUCN and UNEP 1988).

145

the best developed reefs, stretching some 1,500 miles between the high island reefs to the southeast and the atolls to the northwest. The tropical high island portion of Hawaii supports better coral growth but generally "younger" and smaller reefs, compared to the older atolls at the opposite subtropical and temperate end of the archipelago near Midway and Kure Atolls (Table 1, Figure 2).

United States sovereign territories occur at lower latitudes where marine waters are warmer and support much larger and better developed coral reefs. Off Puerto Rico (Figure 3) and the U.S. Virgin Islands (Figure 4) in the Caribbean, these are mostly fringing reefs and some barrier reefs. Off the Northern Mariana Islands (Figure 5), Guam (Figure 6), and American Samoa (Figure 7) in the southwest Pacific, these are also generally fringing reefs and a few barrier reefs. One atoll (Rose) occurs off Samoa. Other isolated Pacific territories of the United States are atolls (Kingman, Palmyra, Wake, Johnston) or table reefs (Jarvis, Howland, Baker). All except one are located in the Line Islands with Wake being the northernmost of the Marshall Islands. Swains Island off Samoa is an elevated atoll. The best coral reefs occur within the United States affiliated island nations in the Pacific which consist of the Freely Associated States and Trust Territory of the Pacific Islands: Republic of the Marshall Islands (Figure 8), Federated States of Micronesia [FSM] (Figure 9), and the Republic of Palau (Figure 10). The Marshalls include the world's largest atolls, numbering 28 atolls and 5 table reefs. The FSM consists of 29 atolls, 7 table reefs and 5 volcanic island complexes with fringing and barrier reefs. Palau consists of one large barrier reef and volcanic island complex, numerous fringing reefs, 4 atolls and 4 table reefs (Table 1).

Major Species on Coral Reefs

Reef building corals are the dominant organisms in terms of bulk and framework growth, and are most concentrated along ocean slopes, lagoon slopes, lagoon pinnacles, and outer shallow

Table 1. Distribution of coral reefs in the United States and affiliated islands. Parentheses indicate number of major islands or atolls in the affiliated island governments.

Atlantic Ocean/Caribbean	Pacific Ocean
Florida	A. States and Territories
Texas	Hawaii (32)
U.S. Virgin Islands (3)	American Samoa (7)
Puerto Rico	Northern Mariana Islands (14)
	Guam
	Wake Atoll
	Johnston Atoll
	Palmyra Atoll
	Kingman Reef
	Jarvis Island
	Howland Island
	Baker Island
	B. Compact Nations and Trust Territory
	Republic of the Marshall Islands (34)
	Federated States of Micronesia (41)
	Republic of Palau (10)

reef flats (Wells 1957). Coralline algae may cover more reef surface than reef corals along the wave-exposed upper ocean reef slopes and are important in cementing corals and other bulky reef remains together to form the rigid, wave-resistant reef structure. Sand-producing green algae generally predominate in back reef and other depositional zones such as the lagoon slopes and floor, and atop the deeper seaward terraces below the wave zone. Among animals, molluscs are perhaps the next most important reef builders, especially the larger clams and snails. Other important organism groups contributing sediment materials to the reef include foraminiferans and echinoderms (sea urchins, sand dollars, and sea cucumbers in particular).

The diversity and abundance of reef dwelling organisms has been well documented and widely recognized. The most important

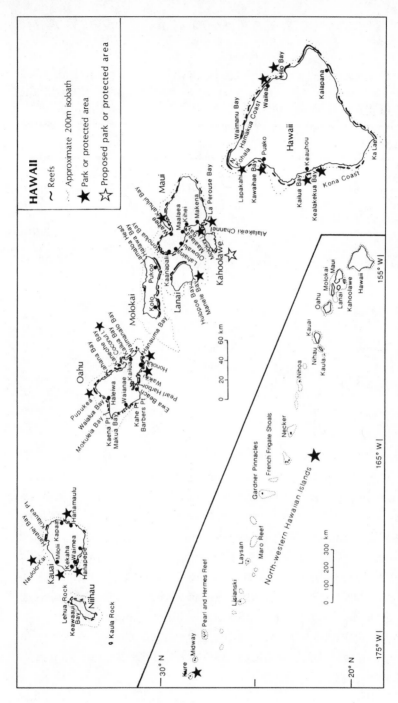

Figure 2. Map of Hawaii showing the distribution of coral reefs within the United States and affiliated islands. After Coral Reefs of the World (IUCN and UNEP 1988).

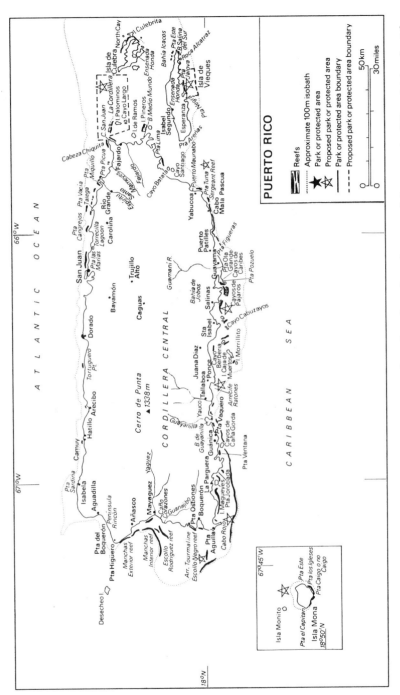

Figure 3. Map of Puerto Rico showing the distribution of coral reefs within the United States and affiliated islands. After Coral Reefs of the World (IUCN and UNEP 1988).

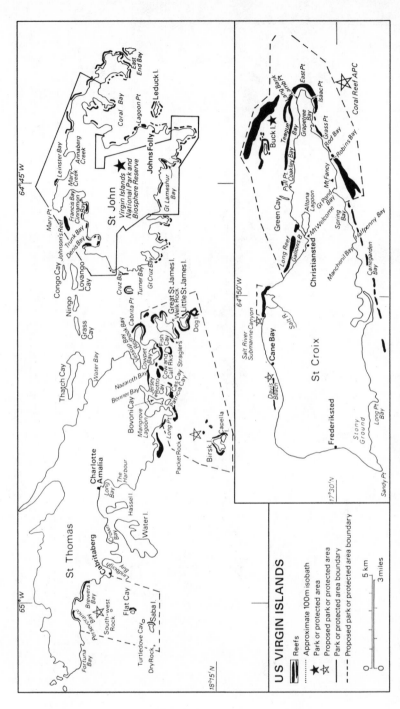

Figure 4. Map of the U.S. Virgin Islands showing the distribution of coral reefs within the United States and affiliated islands. After Coral Reefs of the World (IUCN and UNEP 1988).

150

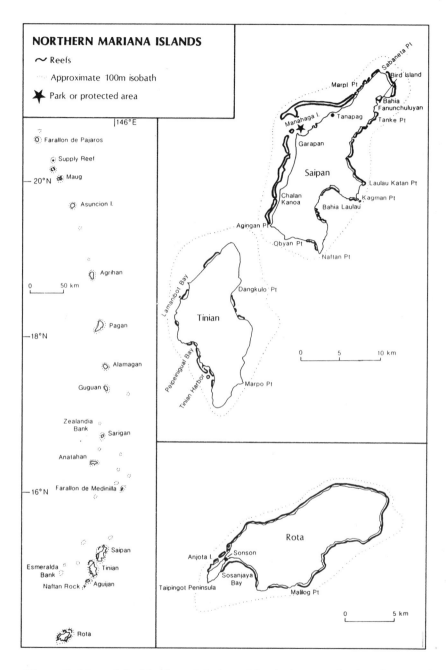

Figure 5. Map of the Northern Mariana Islands showing the distribution of coral reefs within the United States and affiliated islands. After Coral Reefs of the World *(IUCN and UNEP 1988).*

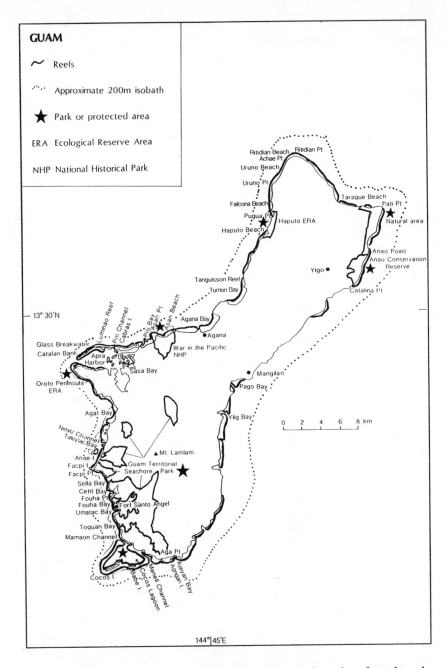

Figure 6. Map of Guam showing the distribution of coral reefs within the United States and affiliated islands. After Coral Reefs of the World *(IUCN and UNEP 1988).*

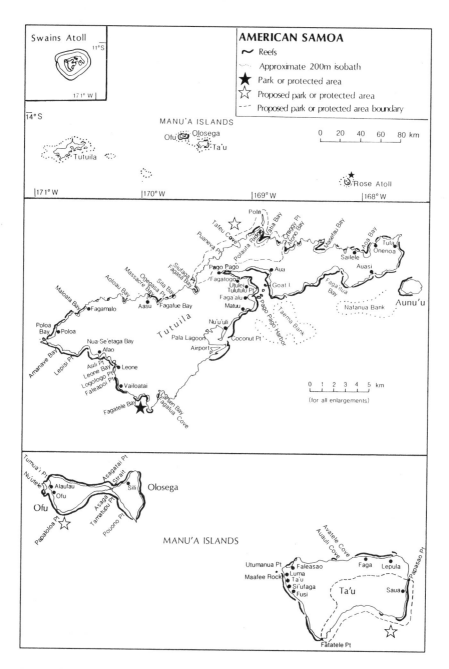

Figure 7. Map of American Samoa showing the distribution of coral reefs within the United States and affiliated islands. After Coral Reefs of the World *(IUCN and UNEP 1988).*

153

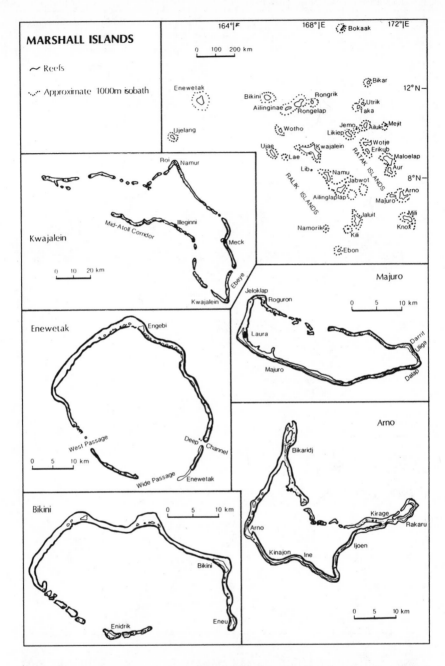

Figure 8. Map of the Marshall Islands showing the distribution of coral reefs within the United States and affiliated islands. After Coral Reefs of the World *(IUCN and UNEP 1988).*

154

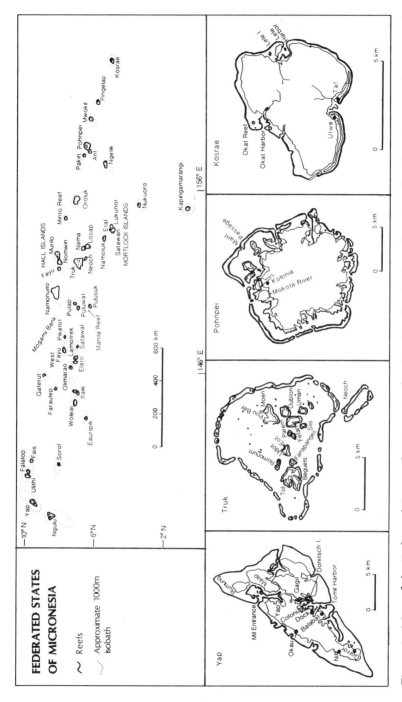

Figure 9. Map of the Federated States of Micronesia showing the distribution of coral reefs within the United States and affiliated islands. After Coral Reefs of the World (IUCN and UNEP 1988).

155

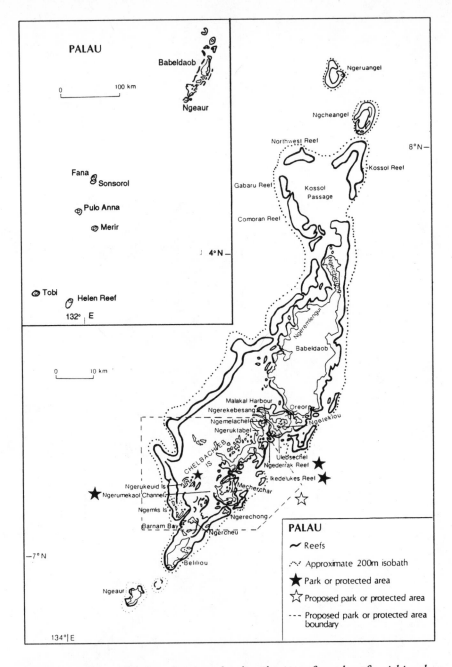

Figure 10. Map of Palau showing the distribution of coral reefs within the United States and affiliated islands. After Coral Reefs of the World *(IUCN and UNEP 1988).*

non-reef building biota of coral reefs include reef fishes and a variety of crustaceans (crabs, lobsters, shrimp). Marine vertebrates (sea turtles, dugongs, manatees, monk seals) are often found in the vicinity of coral reefs. Fleshy and other non-carbonate producing marine algae (seaweeds) are also important components, especially on reef flats. Many seabird species in the tropics are closely associated with reef environments and use the beach and land habitats of nearby islands, cays, or continental coasts for nesting and breeding.

Where reefs are wide, the back depositional portions of the reef flats often support seagrass meadows. Mangrove swamps are often found along the innermost reef flats adjacent to calm shoreline areas. Reef ecosystems are particularly important and diverse where corals, reef fish, seagrasses, and mangroves all occur within close proximity to one another. Many reef species migrate between these zones as part of diurnal and life stage cycles. Particularly good development of these three contiguous habitats occurs off the Caribbean and Caroline Islands. Seagrasses and mangroves are poorly developed in Hawaii, Guam, and the Northern Mariana Islands, and are less developed in Samoa and the Marshall Islands.

Sandy beaches are often associated with coral reefs, because reef derived sediments including the remains of coral, sand producing algae, echinoderms, foraminifera, and micro-molluscs are carried shoreward by currents and winds to accumulate along the coast. Reef derived "white" and sometimes "pink" sand beaches consist almost entirely of calcium carbonate and magnesium carbonate. Sand drifting down the coast is constantly replenished by sand carried in from the reef.

Functional Values of Coral Reefs

The principle values of coral reefs to man are:

- Source of many varieties of food
- Shoreline protection from waves and storms

- Replenishment source of sand for beaches
- Aesthetics and recreation
- Tourism attraction
- Mineral extraction (sand, coral rubble aggregate, quarry stone)
- Medicines
- Scientific research
- Educational resource
- Habitat for rare species
- Concentration of high biodiversity
- Physical support and foundation for coral islets on table, bank, barrier and atoll reefs

In addition, many "high" islands have major land areas consisting of reefs which have been raised above sea level, sometimes by tectonic forces. For example, the ground under most of Honolulu and all of Waikiki consists of elevated "fossil" reefs. The northern half of Guam, portions of Saipan and Tinian, most of the Rock Islands of Palau and large segments of the U.S. Virgin Islands consist of raised reefs. The upper Florida keys are fossil reefs.

In the United States and its sovereign territories, coral reefs are now generally protected from mineral extraction except for the dredging associated with port and airfield construction. However, in the Freely Associated States (Marshalls, FSM), and to a lesser extent Palau, coral reefs are important and convenient sources of construction materials, particularly for unpaved roads, concrete production, and fill land expansion (Maragos 1991).

Ecological Functions

Coral reefs are centers of high natural biodiversity and productivity, providing habitat for thousands of species of fish and shellfish and hundreds of species of corals, algae, sponges, echinoderms and many other organism groups. Coral reef systems provide food, shelter, breeding, and nursery areas for many reef and non-reef organisms. Coral reefs are also ecologically linked to mangroves and seagrasses where these systems co-occur. The upward growth of coral reefs along the tops and sides of subsiding volcanoes over the past 50 million years or more has made possible the existence of many thousand coral cays and islands and their associated terrestrial ecosystems on atolls and barrier reefs throughout the tropics.

Water Quality Functions

Healthy reefs both require and maintain good water quality. Better coral development occurs where water turbidity, suspended sediments, and nutrient concentrations are low and where dissolved oxygen concentrations, water temperature, light intensity, and water circulation is high. Corals also grow best where seawater salinities are near normal (35 ± 5 ppt).

Healthy reef metabolism does not alter ambient water quality substantially. As the "climax" communities for many coastal habitats in the tropics, living coral reefs exert few changes that would enable competitors to displace them. Reef communities are a mix of autotrophic and heterotrophic species. During bright daylight hours, reef metabolism may reduce slightly the seawater concentrations of carbon dioxide, dissolved nutrients, and carbonate ions and increase slightly dissolved oxygen, organic carbon and organic nutrient concentrations. During nighttime hours, carbon dioxide concentrations increase and dissolved oxygen levels decrease due to greater respiration and the cessation of photosynthesis. Besides corals, algae and coralline algae, some species of molluscs, soft corals and foraminiferans have developed symbiotic associations with algae to contribute to reef photosynthesis. Adequate water circulation and flushing is needed to remove waste products and promote high productivity of coral reefs.

Socio-Economic and Cultural Functions

Coral reefs serve as natural breakwaters, protecting coastal settlements from natural forces including tsunamis, storm surges, and large waves from storms, hurricanes, and typhoons. Many commercial activities depend upon coral reefs: sportfishing, aquarium fish collecting, sport diving, tourism, offshore sand mining, acquisition of construction materials, mariculture, collection of biochemical by-products for pharmaceuticals and cosmetics. Corals and other reef invertebrates are also collected for the curio and aquarium trades. Considerable scientific research is conducted on coral reefs, and many reef areas are used for educational purposes, including class field trips and field courses on marine biology. Subsistence fishing and gathering is the most important socio-economic function of reefs in the developing economies of the Pacific and Caribbean territories. Many varieties of reef fishes, crabs, lobsters, conchs, tritons, helmets, green snail, trochus, giant clam, seaweeds, sea turtles, clams, cockles, mussels, and oysters are collected from reefs for food or for ornaments. In the Pacific Islands certain reef flats and offshore reefs are sacred sites. The early Carolinians and Hawaiians also constructed stone fishtraps and fish corrals on reef flats. Many fishing shrines and rock walled fishponds were constructed on reefs by the early Hawaiians.

Rare Species and Biodiversity

Coral reefs are recognized for serving as habitat for many species of animals and plants including rare species. The Pacific region contains approximately 500 species of reef corals (Veron 1986), several thousand species of reef fishes and molluscs and hundreds of species of algae, echinoderms, and perhaps sponges. On the Caribbean side, only about 60 species of stony corals are found because of the younger geological age and smaller size of the region, but soft corals (including sea whips and fan corals) and seagrasses appear better developed. The highest centers for coral reef biodiversity among the United States sovereign territories, affiliated islands and states include Palau, followed by the FSM, Guam and the Northern Mariana Islands. The atolls of southern Palau may have the most diverse reef biota of any in the world. Biodiversity decreases towards the eastern side of the Pacific and further away from the equator. For example equatorial Palmyra Atoll in the Line Islands contains only about 100 species of corals (Maragos 1987). Further to the north and near the subtropical boundary, Hawaii supports only about 45 species (Maragos 1977; Grigg 1981; Maragos and Jokiel 1986). However, Hawaii is the only United States affiliated area to support large numbers of several endemic reef species, including corals (Maragos 1977; Maragos and Jokiel 1986; Jokiel 1987).

Many officially designated or proposed threatened or endangered species inhabit or use coral reefs in the United States areas. On the Atlantic/Caribbean side, several species of sea turtles either feed on shallow reefs or use nearby white sand beaches for nesting and breeding. On the Pacific side, the officially recognized species include sea turtles, the Hawaiian monk seal, the dugong and the salt water crocodile (Palau), and the largest giant clam (Western Pacific). Depleted Caribbean species include the conch, and depleted Pacific species include several smaller giant clams and the coconut crab. In addition the humpback whale, an endangered species, uses deeper reef areas as calving grounds in Hawaii, other Pacific Islands and the Virgin Islands in the Caribbean. Several endangered sea bird species also nest on sand cays and beaches next to coral reefs in the Pacific and Atlantic.

Natural Limitations of Reefs

Coral reefs are generally confined to tropical areas due to the need for warm waters and abundant sunlight. In subtropical areas such as Hawaii, cooler winter temperatures retard growth and allow the destructive forces of storms and waves to exact a greater toll (Grigg 1983; Dollar 1982). For example the seasonal approach of large surf from the north (Bering Sea and Aleutians) during the winter months can cause extensive damage to corals and has been the cause preventing coral reef development along northern facing coasts. At depth, light limits reef growth to the upper 100 meters or less. The deepest reef corals were observed at Johnston Atoll at a depth of 180 m (Maragos and Jokiel 1986). The fastest reef growth occurs at depths less than 30 m. A combination of lower temperature and greater competition from benthic algae may limit corals at high latitudes (Coles 1988).

Due to slow coral recolonization and recovery rates on reefs, coral reefs may require decades for full recovery from damaging storms and typhoons (Endean 1976). Based upon recolonization of corals on historic lava flows (Grigg and Maragos 1974), boat harbors (Maragos 1983; Schuhmacher 1988), and coral reefs recovering from pollution in Hawaii (Maragos 1985; Evans 1986), full recovery takes 15 to 50 years or more.

Geological instability can also control coral reef growth. For example, the youngest island in Hawaii, less than 800,000 years old, is rapidly subsiding as it grows, and corals cannot grow upward and offshore fast enough to create a reef structure except along a two mile stretch of the oldest most sheltered coast (Puako, Island of Hawaii). Similarly, islands in the volcanically active northern Mariana Islands fail to support reef growth. At the other extreme, tectonic activity has pushed much of the Palau Islands above sea level during recent geological time, converting patch reefs and barrier reefs into "rock islands". The large earthquake (7.2 on the Richter scale) off Hawaii in 1975 also caused massive slumping of shallow coastal habitats, exposing reefs to heavy wave action. Any upward or downward movement of sea level relative to the islands, whether from tectonic forces or global warming, is disruptive to coral reef growth, requiring the slow processes of recolonization to establish reefs on new surfaces.

Geographic isolation has also indirectly limited reef growth in Hawaii (Maragos 1977). Many of the wave resistant, fast growing corals (especially *Acropora*) found elsewhere in the Pacific have failed to reach Hawaii, which is 500 to 2,000 miles upstream of its nearest coral reef neighbors to the south and west (Grigg 1981). Spur-and groove formation, a key to healthy reef growth in heavy wave action, is poor in Hawaii, and many reef flats are submerged well below low tide levels. Mangroves and most seagrass species have also failed to colonize Hawaii naturally, resulting in greater mobility of detrimental sands and muds. To a similar degree, geographic isolation at nearby Johnston Atoll has also probably limited reef growth along the windward sides of the atoll.

Benthic algae compete for space with corals. The massive die-off in the Caribbean of sea urchins which feed on algae, demonstrates that coral reefs may depend heavily upon the activities of herbivores to keep benthic algae in check (Sammarco 1982a, b 1985; Bak 1984; Lessios 1984; Carpenter 1985; Hughes 1987).

Natural predators on corals include the crown-of-thorns starfish (*Acanthaster*) on Pacific reefs (see Birkeland 1989a, b), the cushion star *Culcita*, and several species of butterfly fishes (*Chaetodon*, *Forcipiger*). By far the most dramatic predator is *Acanthaster* which was reported as major infestations during the past two decades in the following United States areas: Guam, Hawaii, American Samoa, Ponape, Majuro Atoll, Arno Atoll and Saipan. Although the starfish is a natural predator on corals, some earlier researchers believe that major infestations were linked to human-caused disturbances and pollution (Endean 1976, 1987). The present consensus now is that the starfish undergoes boom-and-bust cycles which are common among other predator-prey relationships. Heavy starfish predation on corals is a major concern, however, regardless of the cause, because coral recovery takes so long. Pollution or a series of large storms during the intervening recovery phase can damage reef structures, reduce habitat and expand the adverse ecological consequences to other groups of reef organisms, especially fish.

Coral reef diseases have also been reported in recent years, particularly on Caribbean reefs (Antonius 1981, 1988; Gladfelter 1982; Peters 1983, 1984). These include coral bleaching, white band disease, black band disease, shut down reaction, and other possible bacterial attacks (Mitchell and Chet 1975). Pacific reefs have been generally spared these problems except for bleaching. However, tumorous growths on corals have just been reported off Hanauma Bay, Hawaii's most popular coral reef site (Cynthia Hunter of Honolulu, pers. comm.). The causes for these maladies are not well understood. Bleaching is a stress response by corals to expel their (pigmented) symbiotic algae. Higher water temperatures and seawater dilution from tropical storms have been implicated as the cause for bleaching on Caribbean and Pacific reefs (Glynn 1983, 1988). Bleached corals sometimes recover (reestablishing their symbionts) or sometimes die. Slower growth of corals during the bleaching stage renders corals more vulnerable to attack from other corals and algae which compete against corals for bottom space.

Human Impacts on and Responses by Coral Reefs

On a global scale the principal human impacts to coral reefs include sedimentation (Rogers 1990), dredge-and-fill (Bak 1978; Salvat 1987; Maragos 1991), and eutrophication (Pastorak and Bilyard 1985; Kinsey 1988) from a total of 14 major impact categories. In the United States and its sovereign territories, dredging and filling on reefs is now largely controlled (Maragos 1986), but impacts from sedimentation, eutrophication, fresh water kills (Banner 1968), overfishing, overuse by visitors (Gladfetter 1977; Tilmant 1987; Rogers 1988), ship groundings and anchor damage (Smith 1985; Gittings 1988) exert a heavy toll on United States reefs. Impacts to reefs from thermal effluents are also largely controlled by

United States environmental regulations (Coles 1984; Neudecker 1987; Jokiel and Coles 1990). Except in Hawaii where fishing with bleach is still widely used, destructive fishing practices are mainly confined to reefs in developing countries (Gomez 1981; Polunin 1983). Nuclear weapons testing by the United States during the post World War II years caused major damage to Enewetak and Bikini atolls in the northern Marshalls (Maragos 1986) and perhaps also at Kiritimati (Christmas) Atoll in Kiribati. Impacts to reefs have been documented for the Middle East reefs from oil spills and oil exploration (see Loya 1976), but have been essentially confined to Panamanian waters, the Gulf of Mexico, and the bank reefs off southern Texas near United States waters. As noted earlier bacterial and viral attacks on corals are mostly reported from Caribbean reefs and may be exacerbated by other human derived impacts. A few researchers, most notably Grigg and Dollar (1990) believe there has been little documentation of the global decline of coral reefs in the published scientific literature. Nevertheless, there is considerable documentation in non-journal publications (see Carpenter and Maragos 1991; Maragos 1991, for example).

Sedimentation

Sedimentation from the discharge of upland eroded soils causes the greatest damage to coral reefs (Johannes 1975). Coastal construction can also resuspend marine sediments and damage reefs as has been documented in the Philippines (Hodgson and Dixon 1988). Upland and coastal mineral extraction and logging activities also generate discharges of slurry and tailings which may reach coastal reefs, although this last source of input is not prevalent in the United States affiliated areas except in the FSM. Sediments accumulating on the bottom can smother corals and other reef organisms in low lying areas (Rogers 1990). Suspended sediments can also damage many corals by reducing light penetration needed for photosynthesis, requiring excessive mucus secretion by corals to rid its tissues of sediment, clogging of respiratory and reproductive functions, and rendering corals more susceptible to bacterial or algal attacks (Rogers 1990). Important case studies for sedimentation effects on reefs include Kaneohe Bay and Pago Pago Harbor, American Samoa (Maragos 1972) and Puerto Rico (Acevedo and Morelock 1988).

Dredging and Filling

The creation of land from fill activities permanently converts coral reefs to terrestrial habitats. Dredging mechanically destroys the upper-most living layers of coral reefs (Salvat 1987; White 1987), and reef organisms are often slow to recover if the newly dredged habitats are not well flushed by currents or tides or if sediments accumulate on bottom surfaces (Maragos 1991). Under optimal circumstances corals tend to recolonize only on wave absorbers and elevated hard surfaces such as channel and basin walls where water quality and water flushing is suitable for coral settlement and survival (Maragos 1983; Schuhmacher 1988). Dredging or filling on coral reefs can greatly expand the zone of impacts to adjacent areas if currents and tidal regimes are modified (see Kaly and Jones 1990), such as occurred at Palmyra (Figure 11; Dawson 1959; Carpenter and Maragos 1989; Maragos 1991). Military construction at Palmyra and other forward bases in the Pacific during the World War II era (Johnston, Wake, Midway) was responsible for much of the impacts to reefs.

In the United States sovereign areas, dredging and filling have been mostly associated with the construction of coastal transportation projects such as aircraft runways, docks, ports, navigation channels, roadways, and causeways. Micronesia dredging and filling are also conducted to obtain construction materials for non coastal (upland) road construction and coastal land fill expansion. Here political and development interests consider reefs to be "free" and unencumbered resources available for exploitation. Offshore

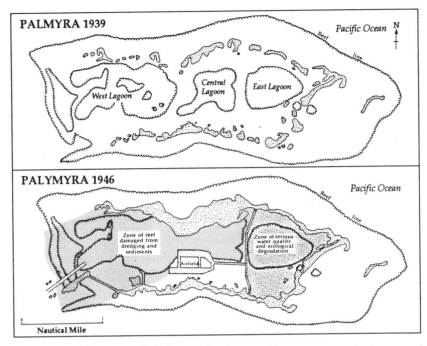

Figure 11. Palmyra Atoll before and after World War II era dredging and filling. After Dawson (1959) and Carpenter and Maragos (1989).

sources of materials are definitely more attractive in these areas where environmental regulations are lax. Otherwise, developers are required to pay landowners for removal of upland sources of fill. Dredging and filling have expanded during the past half century with the advent of explosives and heavy machinery use (Maragos 1991). A considerable amount of coastal construction was also conducted before and during World War II in the Pacific. More recently, new resorts in many Pacific areas are advocating dredging and filling to construct marinas, sandy beaches and to expand land holdings (see Salvat 1987; White 1987).

Eutrophication

On a global basis, agricultural sources of fertilizers and soil run-off from clear-cut forested areas can dump large quantities of nutrients into coastal waters, which in turn can stimulate benthic algal or phytoplankton growth and cause eutrophication (Pastorak and Bilyard 1985, Fujita 1991) on coral reefs. In United States affiliated areas, the gradual shift from agriculture to resort and coastal urban development also generates nutrients from sewage and golf courses. Nutrient discharges into confined water bodies with long turnover or flushing times aggravates eutrophication and stress to coral reefs (Maragos 1985). The introduction of "excess" nutrients to coral reef environments allows potential benthic competitors, (e.g., benthic algae) to establish and displace corals and other reef benthos (Maragos 1972; Banner 1974) as occurred in Kaneohe Bay. In worse case situations phytoplankton growths in overlying waters can reduce sunlight needed for coral photosynthesis, increase organic loading to sediments (Smith 1981) and possibly cause sediments to go anaerobic and become toxic to reef organisms (Sorokin 1970; Maragos 1972). In Caribbean areas, the maintenance of herbivorous sea urchins which feed on the excessive algal growths may be inhibited by poor water quality, including eutrophication. Increased nutrients can also increase bacteria and attacks on corals (Mitchell and Chet 1975).

Freshwater Kills and Flooding

Urban development (Maragos 1972) and some forms of irrigated agriculture can increase freshwater runoff to coastal areas which can be detrimental to coral reefs, sometimes resulting in freshwater "kills" (Banner 1968). Rainwater which would otherwise percolate through the humus soil layers of vegetated lands is now trapped by pavement and impervious open agricultural lands, collected by flumes and lined channels, and carried to the coast. Although corals can tolerate temporary lowering of salinities during large rainstorms, prolonged dilution can kill some reef organisms with subsequent decomposition leading to oxygen reductions and secondary kills of reef organisms (Banner 1968), especially within embayments which are slow to mix or flush out freshwater discharges.

Overfishing

Fish and shellfish are major components of coral reef ecosystems, and their depletion can reduce some important ecological functions and disrupt the balance between important predators and prey. For example, over-exploitation of herbivorous reef species may reduce algal cropping and lead to greater than normal growths of benthic algae that could out compete corals and coralline algae for bottom space (Sammarco 1985; Glynn and Colgan 1988; Hay 1988; Woodley and Clark 1989). Over-exploitation of reef predators may increase the proportion of prey species that are coral predators or whose feeding habits can damage corals (Bohnsack 1982 cited in Woodley and Clark 1989). There are few documented case studies of the effects of overfishing; however, see Munro and Williams (1985) and Salvat (1987).

Overfishing of reefs on Oahu, Hawaii by recreational fishermen is such a problem that one can rarely see any fish 12 inches or longer anywhere except in the few marine sanctuaries. These latter (such as Hanauma Bay) are so crowded with visitors and tourists that the reefs in the sanctuaries themselves may be stressed. With roughly 100,000 recreational fishermen and sport fishermen on the island, it is politically difficult to support stronger regulatory controls over recreational fishing. In developing countries, depletion of fishing stocks often leads less skilled fishermen to resort to illegal or destructive fishing practices to put food on the table (Gomez 1981; Polunin 1983; Salvat 1987).

Disruptive Fishing Practices

The most damaging of illegal or destructive fishing practices includes the use of explosives (including dynamite), poisons (such as Clorox and sodium cyanide), pounding the reef bottom with bamboo sticks (muro-ami) or with stones attached to lines (kayakas) as occurs in the Philippines, and long or large nets, which catch excessive quantities of desirable and undesirable species. The use of explosives and poisons, in particular, can damage adjacent reef habitat including corals. Explosives generate shock and concussion that stuns or kills fish, ruptures their air bladders, shatters or dislodges corals and also kills shellfish. Many of the fish affected are unwanted and many that are desired are not collected because they sink to the bottom (see Polunin 1983). Corals or coralline algae dislodged by blast fishing can be later moved about by waves and strong currents, effectively destroying them, as probably occurred off Barbers Point, Oahu (Maragos 1991). Other documented sites for reef damage from fish blasting include Truk Lagoon, FSM (Devaney 1975), Palawan, and other areas in the Philippines (Gomez 1981).

A recent ominous trend has been the collection of "live rock" for the aquarium trade. These rocks consist of stony corals, soft corals and other attractive reef substrates. Although United States sovereign areas have regulations to control the collection of live corals, commercial operations are well established in the Philippines with some "live rock" companies now trying to obtain approval in Pohnpei (FSM). Removal of rocky substrate from reef habitats is particularly damaging because many commensal and infaunal reef organisms are now lost to home aquariums. Live corals are also collected as curios or souvenirs by resort divers. In one bizarre case, an entrepreneur on Yap (FSM) has resorted to manual collection of live corals to build up the foundation for a proposed house because the government had not given him permission to dredge reefs to obtain construction and fill material (Maragos unpubl.).

On some islands, chicken wire enclosures have been placed on reef flats to trap coral reef fishes, replacing the traditional stone weirs and reed netting used for centuries by the islanders (Zann in Helfrich 1979). However the rusting or corroded wire is thought to be injurious to juvenile fish. Enclosures left unattended or wire fish traps left or lost on the reef bottom can trap and kill fish for years.

Overuse by Recreationists and Visitors

Recent publicity and monitoring of reefs off the Florida keys and southern coasts have documented widespread damage to fragile reefs within protected parks and "sanctuaries" (Gladfelter 1977; Tilmant 1987; Rogers 1988). Damage results from standing on corals, accidental coral breakage by swimming divers and illegal collection of corals. In the U.S. Virgin Islands, visitor use of snorkeling trails at Trunk Bay and Buck Island in a National Parks has damaged many fragile branching corals. In contrast, corals at Hanauma Bay, Hawaii are more robust and noticeable coral damage is still minor. Reef walking tours can also damage corals and resuspend sediments that are then carried to adjacent reefs.

Thermal Stress

Reef organisms live near their upper sublethal temperatures in the tropics (Johannes and Betzer 1975). Thus, even the slight elevation of ambient water temperatures from thermal effluents, especially during the warmer summer months, can lead to coral kills (Jokiel and Coles 1990; Coles 1984). Fish kills have also be associated with the use of toxic chemicals in power plant effluents (McCain and Peck 1977). Power plants often use seawater for once-through cooling systems, discharging heated effluents onto reef environments (Neudecker 1987). In one case off Kahe, Oahu (Coles 1984) and in another off Piti Channel in Guam (Marsh and Gordon 1974), corals were killed by buoyant thermal effluents impinging on the bottom. These kinds of impacts can be avoided through proper dilution of heated water to control temperature, and through proper placement of outfalls (to avoid plume impingement on the bottom). United States Federal regulations have been strengthened in recent years to avoid thermal impacts to coral reefs from heated effluent discharges.

The periodic rise in seawater temperatures in recent years has been thought to cause massive coral bleaching off the Gulf of Panama (Glynn 1983) and globally (Williams and Bunkley-Williams 1988). Many researchers believe that the rising sea water temperatures may be related to global climate change, the warming of the atmosphere by the "greenhouse effect" caused by increased carbon emissions. Frequent or prolonged temperature conditions triggering bleaching responses in corals may cause massive damage to reefs. Some researchers (Nunn 1991), however, believe that recent warming trends are the most current stages of "natural" warming attributed to the ongoing warming of the earth from the last ice age.

Ship Groundings and Anchor Damage

The recent grounding of the *Wellwood* off the reef at Key Largo National Marine Sanctuary (Hudson and Diaz 1988) and ship groundings on Bermuda (Smith 1985) are examples of an increased frequency of ship groundings on coral reefs worldwide. Reefs near navigation channels and ships trapped in typhoons are the most serious risks. Aside from the physical reef damage from the collisions, groundings often lead to oil spills from ruptured hulls and the dragging of cables and anchors and the ditching of cargoes to lighten the ships load to facilitate removal and salvage operations. The propeller motion of vessels attempting to free grounded vessels may resuspend sediments and cause "sand blasting" of corals and other reef organisms. Also, reef materials loosened by ship collisions can be later shifted by winter storms and typhoons, causing damage to adjacent reefs (Hudson and Diaz 1988).

Oil Pollution

Although there have been no major oil spills near coral reefs in United States Pacific waters, there have been major oil spills in Puerto Rico, the Red Sea and Persian Gulf. Oil spills can occur anywhere at any time, and there have been numerous minor oil and fuel spills in United States waters especially at offshore moorings and harbors. Floating oil that contacts the reef at low tide constitutes the greatest risk (Johannes 1972) but oil spills can also worsen the adverse effects of other sources of water pollution on reefs (Loya 1976).

Response of Reef Habitat and Communities to Impact

Under general circumstances, coral reefs can take more than one to several decades to achieve full ecological recovery, depending on the extent of damage, the type and duration of impact, and the complexity of the affected reef communities. Fish, fleshy algae, and coralline algae and some invertebrates appear to begin recolonization within a year, but corals often take longer to achieve a visible start (Table 2; Grigg and Maragos 1974; Endean 1976; U.S. Army Corps of Engineers 1983). Corals represent the climax stage in coral reef ecological succession on outer reef flat and reef slope habitats, and substrates need to be stable and properly conditioned prior to coral larval settlement and attachment.

One factor that tends to accelerate coral and reef recovery is the presence of surviving adult colonies nearby, which provide live coral fragments and larval recruits for recolonization (Maragos 1972; Shinn 1976). Abundant elevated rocky substrates and good water flow also favor coral colonization, as noted in reef flat quarry holes throughout Micronesia (Maragos 1991).

On the other hand, eutrophication or excessive sedimentation can slow down coral recovery rates or postpone colonization altogether until water quality conditions improve, as noted in Kaneohe

Bay (Maragos 1985; Evans 1986), Sand Island, Hawaii (Dollar 1979, 1986), and also in the Red Sea (Loya 1976). Sediment accumulation on hard surfaces and resuspension such as in boat harbor basins may prevent coral colonization and result in the transformation of bottom habitat to soft bottom communities. Sand in suspension, in the presence of heavy wave action can scour boat channels, the toe of breakwaters, and the walls of navigation channels, retarding recovery (Helfrich 1975; Maragos 1983).

Most documented examples of coral reef recovery have resulted entirely from natural processes. No large-scale restoration of a reef has occurred with human assistance except for the reefs in Kaneohe Bay. Many techniques for active rehabilitation have been tested (Woodley and Clark 1989), but to date natural processes have accounted for most *in situ* coral reef recovery. The grounding of the *Wellwood* at Key Largo led to an out-of-court, 6 million-dollar settlement against the shipping company (and owner of the grounded vessel) that will contribute to the actual restoration of the damaged Molasses Reef site. Studies have been done to analyze rehabilitation options and scope out future recovery work (Hudson and Diaz 1988).

Aside from these two examples, little else is on the horizon to advance the state of the art and success for human-aided coral reef restoration. Research institutions continue to evaluate future possible options and monitor recovery rates of previously denuded reef areas. Some innovative artificial reef development has been accomplished by Hudson et al., (1990), and the unique mooring systems developed in the Florida keys to reduce anchor damage by Halas (1985) are now being expanded to other reefs.

Restoration and Creation of Coral Reefs

In the literature, one recent paper is devoted to the review of techniques to rehabilitate coral reefs (Woodley and Clark 1989).

Table 2. Main impacts to coral reefs worldwide.

Impact	Effects
Sedimentation	upland soil erosion coastal construction mining
Dredging and Filling	transportation military installations urban development marine mining resorts
Eutrophication	sewage discharges agricultural discharges soil erosion golf courses
Fresh Water Kills and Flooding	increased freshwater runoff changing agricultural practices urbanization
Overfishing	
Destructive Fishing Practices	explosives poisons pounding (muro-ami, kayakas) nets and wire enclosures coral harvesting
Overuse by Recreationists and Visitors	
Bacterial and Viral Infections	
Thermal Stress	power and desalination plant effluents global climate change/bleaching
Ship Groundings	
Anchor Damage	
Weapons Testing	
Oil Development and Spills	
Pesticide and Heavy Metal Toxicity	

They divided rehabilitation approaches into three categories: impact mitigation, passive rehabilitation, and active rehabilitation. The first involves reducing the level of loss or compensating for the loss of coral reefs attributed to anthropogenic factors. The adverse impacts have already occurred or are expected to occur, and mitigation is exercised to reduce these in part or completely. Woodley and Clark (1989) defined passive rehabilitation as impact mitigation which promotes or accelerates the process of natural recovery. They defined active rehabilitation as a more aggressive approach of direct intervention or manipulation to influence the course of recovery. For purposes of the present discussion, I define restoration of coral reefs to include all three of these approaches; all tend to reduce adverse effects or increase beneficial responses through a combination of natural and human-assisted actions.

Selection Criteria for Coral Reef Restoration

It is essential to focus on the particular components of reef systems warranting preferential attention; and three criteria may be important in focusing restoration efforts. First, man views some reef resources to be more important than others. Secondly, some reef components are "essential" to coral reefs. Finally, some reef organisms are less able to reestablish on reefs or do it more slowly. Restoration efforts, thus, could be directed to those groups that rate relatively high in all three of these criteria or attributes. Table 3 is an analysis involving the rating of different reef organism groups against these three criteria. Although admittedly subjective in approach, the table offers a means of focusing attention on those fewer coral reef groups that warrant restoration assistance.

Based on this analysis, reef corals are the most important target group, closely followed by edible invertebrates and fish. Coralline algae have the lowest priority. Other reef components are also generally not targeted for restoration. Although all coral reef organisms probably play a role in the maintenance and survival of reefs, cer-

Table 3. Criteria for targeting reef organisms for coral reef restoration. Priorities: 1=highest, 3=lowest.

Reef Component	Importance to Man	Slow or Difficult Recovery	Essential to the Reef	Overall Priority
reef corals	2	1	1	1
coralline algae	3	2	1	3
fleshy algae	2	3	3	3
edible shellfish	1	1	3	2
other invertebrates	3	2	3	3
edible reef fish	1	2	2	2
other reef fish	2	2	2	3

tain groups play a more important role. These groups, if threatened by competitors, man, predators, or if they are slow to recover from impacts, should be the targets of restoration efforts.Reef corals and coralline algae are the critical components of the reef structure, particularly in wave exposed environments, but reef corals take much longer to recolonize and recover on reefs if denuded or removed by human or natural factors. Reef fish and edible shellfish are generally important to man as a source of food and income.

In addition, marine reptiles (sea turtles, sea snakes, marine lizards, marine crocodiles) and marine mammals (monk seals, humpback whales, dugongs, manatees) are visitors to reefs and are either very rare, threatened or endangered species. Whether or not eaten by man, their rare status elevates their importance for protection. However, it is beyond the scope of this paper to address restoration of marine mammals and reptiles.

Strategies to restore coral reefs vary among these target groups but can generally be categorized as in Table 4.

Most of the above strategies are theoretical since very few have actually been exercised as part of coral reef restoration or research efforts. Actually restoration efforts that have been practiced on coral reefs, experimentally tested in the laboratory, or used in the field are listed (Table 5). Additional techniques have promise but are untested (Table 6).

The rest of the discussion focuses on the restoration of higher priority target groups: the corals, edible fish, and shellfish. Important spawning and nursery grounds for reef fish and shellfish (especially shrimps, crabs, molluscs) need to be identified and protected from water pollution or physical modification. An inventory of such areas on specific reefs is the most practical first step in protecting these sites. Designation of these areas as sanctuaries is an important later step. However, inventory and sanctuary designation do not in themselves constitute restoration without adequate on-site monitoring and enforcement.

Transplantation of Reef Corals

Several workers have transplanted corals on an experimental basis to assist coral reef recovery (Maragos 1974; Shinn 1976; Bouchon 1981; Auberson 1982; Harriott and Fisk 1988; Hudson and Diaz 1988). Maragos (1972, 1974) used non-toxic coated wire to fasten corals to metal frames. He concluded the technique was feasible only in lagoon environments where surge and wave action would not overturn the metal platforms and where water pollution did not degrade coral growth and survival and promote algal competitors. He also concluded that coral transplantation may be an important means to allow colonization in sand dominated reef environments.

Other techniques have been used to establish coral transplants. Over a period of many years, Peter Frey of Fiji (pers. comm.) has

Table 4. Strategies to consider in restoration of coral reefs for selected target organisms.

Strategy	Target Group
Increase Recruitment	
improve dispersal	corals
improve fecundity	fish, shellfish
increase larval settlement	corals
increase larval survival	corals, fish, shellfish
increase spawning stock biomass	fish, shellfish
protect spawning sites	corals, fish, shellfish
Accelerate Recovery Rates	
increase abundance or cover	all
increase growth rate	all
increase food supply	all animals
Reduce Mortality to Adults	
maintain environmental quality	corals, shellfish
reduce pollution	all
reduce predation by man	all
reduce natural predation or disease	corals
improve shelter	fish, shellfish
Reduce Competition	
reduce benthic (fleshy) algae	corals
increase herbivores that feed on benthic algae (urchins, fish)	corals
reduce undesirable invertebrates (sponges, tunicates)	corals

Table 5. Coral reef restoration techniques already attempted and which should be expanded.

Action	Effect
transplantation of reef corals	accelerate coral recolonization and provide shelter for other reef life
removal of sewage discharges	reduce anaerobiosis and competition from benthic algae and invertebrates
construction of artificial reefs	provide shelter for fishes and other reef life
construction of reef flat quarry holes	provide suitable habitat for corals, fishes, and edible invertebrates
rubblemound revetments and break-waters	
modification of current regimes	promote better flushing and water quality
replanting and restoration of mangrove fringe	control and buffer sediment and nutrient discharges from land, and provide biological recruits
replanting of seagrass beds	stabilize offshore sand deposits, and provide biological recruits
mooring buoy systems on coral reefs	eliminate anchor damage to fragile corals by tourism boating and fishermen
extend sewage and thermal outfalls and expand diffuser ports further offshore	reduce thermal and sewage stress to fish and corals by dilution and offshore and deep water discharge
cementation of reefs damaged by ship groundings	stabilize hard substrates to promote coral and other reef recolonization
remove disease bearing organisms and coral predators	reduce mortality to corals and other desirable reef species
establish buffer zones or barriers around important reefs	keep potential damaging activities away from coral reefs

Table 6. Additional coral reef restoration techniques that should be pursued.

Action	Effect
mariculture recruitment of larvae (fishes, giant clams, green snail, pearl oysters, sea urchins, corals, etc.)	accelerate recolonization of reef life and eliminate competitors
ban on fishing for herbivorous species	maintain adequate predation on benthic algae competitors
hydraulic or mechanical removal of sediments	reestablish suitable settling habitat for coral larvae and shelter for other reef species
mechanical modification of current wave and tidal regimes	improve water quality, flushing and removal of deleterious sediments
habitat creation (dredged channels quarry holes, artificial reefs, etc.)	establish suitable additional environments for coral reef life
reducing eutrophic stress by controlling soil erosion, runoff, agriculture, and golf courses	eliminate nutrients which stimulate the growth of algae competitors
transplantation of herbivorous fish and sea urchins	increase foraging on benthic algal competitors
remove benthic algae which compete against corals	reduce smothering, shading, and mortality to corals and other desirable reef species
long range monitoring program	basis for response planning and action to restore/protect reefs
fisheries reform over use of destructive poisons, muro-ami, kayakas, explosives, fine mesh nets and chicken wire enclosures	protect habitat and reduce unneeded mortality to reef life
reduced fishing pressure on herbivorous fish	increase the herbivores feeding on benthic algae which compete against corals

successfully transplanted black corals from deeper to shallower water, wedging them into crevices to promote their growth and "sustainable harvesting" for his jewelry business. Wedging of stony corals has also been used (Woodley and Clark 1989). The most successful technique appears to be the actual cementation of live coral colonies to hard reef substrate (Hudson and Diaz 1988). A variation of this technique involves the cementation of precast concrete domes and other shapes resembling corals that eventually serve as colonization sites for living stony and soft corals (Hudson 1990). The advantage of cementation is that corals can be attached in wave exposed environments as was successfully accomplished on a small scale at the *Wellwood* grounding site at Key Largo (Hudson and Diaz 1988). However, on soft bottom environments the use of frames and pegs may be the only feasible techniques, unless there are elevated outcrops upon which to cement corals.

The major disadvantage of coral transplantation is the required time and expense of underwater work for full-scale reef restoration. In addition, the corals must be collected from some other reef environment, and there must be some assessment conducted there to establish that removal of transplants will not cause more harm than they solve. Thus, coral transplants may do little to contribute additional abundance to the total coverage of corals. However, transplantation can greatly accelerate the recovery of corals in environments where they have been denuded or where sedimentary regimes may postpone or inhibit natural recovery.

Removal of Sewage Discharges

In 1977-78 two major sewage discharges were removed from the confined southern lagoon in Kaneohe Bay, which immediately led to a decline in nutrients and related biomass levels in overlying lagoon waters. By 1983, benthic algal population levels in the lagoon also declined substantially resulting in the dramatic recovery of reef coral populations (Maragos 1985; Evans 1986). The lowered nutrient levels were thought to slow down the growth of benthic algae, allowing corals a chance to compete successfully for space against the algae. The long residence time of lagoon waters exacerbated the effects of elevated nutrients by allowing more time for phytoplankton and benthic algae to grow and degrade water quality and benthic habitats. In contrast, the new sewage outfall sites are located outside the bay where water resident time is less than one day, insufficient to cause eutrophication beyond the immediate boundaries of the outfall site. The greater flushing of waters by wave action has also maintained existing coral reef environments.

In 1990 there has been a slight reversal in the gains reported in 1983 in Kaneohe Bay for corals and a slight increase in abundance of the competing benthic alga (Chris Evans of Honolulu, pers. comm.). Some suspect that unauthorized sewage discharges in the lagoon during power outages or increased discharges from cesspools, septic tanks, golf courses, etc., may be causing the reversals. The matter is presently being investigated for a masters thesis research (Chris Evans of Honolulu, pers. comm.).

The cost of relocating the sewage outfalls outside the bay was high because construction involved the placement and anchoring of an outfall pipe in open seas to depths of 100 ft up to one-half mile offshore. The project was more expensive because the original outfalls were improperly sited inside the lagoon. Considerable expense can be avoided by careful analysis and site selection prior to construction. Sewage outfalls placed in deeper open waters offshore are better than sewage outfalls, even those including sewage treated to the secondary level, placed inside lagoons or protected embayments (Dollar 1979, 1986). In fact, the discharge of raw sewage off the ocean reef slope at Majuro Atoll was shown to cause only minor or localized effects on coral reefs (Maragos unpubl.). Although proper initial siting and planning can reduce costs and impacts, there may be a number of coral reefs presently stressed by eutrophication from sewage discharges that would benefit from relocation of sewage outfalls to less damaging coastal sites.

Construction of Artificial Reefs

Artificial reefs can help restore coral reefs in two principal ways: providing shelter or secluded hiding places for reef fish and shellfish and providing suitable substrates for settlement and colonization by corals and other sessile reef biota (Mueller 1988). Sheehy and Vik (this volume) give considerable background material on artificial reefs and, thus, this technique will not be addressed in detail here. The most benefit to coral reefs results from artificial reefs constructed of long lasting stable materials placed in environments where fish and corals are not locally common, such as on barren sandy lagoon terraces or flats (see Hudson 1990). In some circumstances, artificial reefs can provide needed shelter for reef fishes or suitable settling sites for coral larvae.

Artificial reefs that rely on natural recruitment of fish, shellfish, and corals may slow down the net benefit to coral reefs for years because the recruits are derived from other reef environments. Artificial reefs may aggravate overfishing by concentrating an already depleted resource. The maximum benefit of artificial reef structures will be proportional to the amount of new habitat established and eventually fully occupied. The cost of artificial reefs is often expensive but there may be low cost options appropriate for coral reefs (Sheehy and Vik, this volume; Mueller 1988). Reefs constructed of concrete last the longest and are most suitable for coral colonization, but artificial reefs of metal may also be suitable (Fitzhardinage and Bailey-Brock 1989). The dramatic colonization of World War II shipwrecks in Chuuk (Truk) lagoon is testament to the potential of relatively large metallic "reefs" to provide abundant and diverse habitat to many reef organisms. In contrast, artificial reefs made of tires, although suitable for reef fish, appear unattractive to coral settlement (Fitzhardinage and Bailey-Brock 1989).

Artificial reef development could rely on mariculture stocks to supply some inhabitants, provided that laboratory and tank rearing of fish and shellfish larvae is safe and feasible.

Construction of Reef Flat Quarry Holes

The outer shallow reef flats of many Pacific reefs consist of extremely hard pavement or "flagstone" suitable as armor rock for protective structures when quarried. Whether these same types of habitats are as common on Atlantic and Caribbean reefs is not clear. Although the actual quarrying operations can cause some damage to coral reef communities (Maragos 1991), shallow outer reef flats in the "pavement" zone are generally devoid of much live coral. Excavating the quarry holes, however, deepens this habitat to well below a mean-low-water depth allowing recolonization by reef corals, fish and shellfish. Quarry holes that are located near good wave action but which are not too deep to trap sand appear to be the most favorable environments for coral (Titgen 1988; Lamberts and Maragos 1989; Maragos 1991). Quarry holes that are too shallow (less than 1-2 m depth) or too close to sand or mud beaches are less suitable for corals due to sediment accumulation or scour.

Although no quarry hole has actually been constructed for the purpose of coral reef enhancement, many such holes excavated for their quarry rock have in fact promoted dramatic and rapid coral colonization, such as at Kwajalein Atoll (Titgen 1988), Majuro Atoll (Lamberts and Maragos 1989) and Enewetak Atoll (Losey 1973) in the Marshall Islands. It is also clear that quarry holes can be designed and sited to maximize natural colonization by reef organisms, and some quarry holes would also be suitable sites for culturing shellfish such as giant clams, conches, lobsters, oysters, green snails and trochus (Berg 1976; Gillett 1988; Prescott 1988; Price 1988; Sims 1988; Yamaguchi 1988). Quarry holes at 2-4 m in depth on the outer half of wave-exposed reef flats away from sandy beaches, or coral islets, appear to be the most optimal environments for coral and fish colonization, based upon observations in the Marshalls and Caroline Islands. Recovery can occur within 20 years in some cases (Maragos 1991).

Rubblemound Revetments and Breakwaters

Quarry stone from reef flats, from upland (volcanic) sources or prefabricated on land can be designed and placed for revetments and wave absorbers in a manner that can promote coral and reef fish colonization (Schuhmacher 1988; Maragos 1991). In Hawaii and the Red Sea, boat harbor protective structures have been monitored for 5 to 20 years and indicate that several families of reef corals will colonize and show normal growth rates on certain types including rubblemound structures (Maragos 1983; Schuhmacher 1988). In addition, excavated channel walls and basin walls may also attract coral recruits. Reef fishes of many types can quickly colonize the interstices of rubblemound revetments, moles and breakwaters (Brock 1983). Channel slopes are also preferred habitat for many reef fishes. Although harbor construction often results in greater impact to coral reefs than is mitigated from the newly created habitat in protective structures, inland harbors can also promote good coral colonization in essentially new marine habitats, as demonstrated in Honokohau Harbor, Hawaii (Maragos 1983, 1991). Water quality conditions on the outer harbor basin and entrance channel walls were sufficient to support colonization by most of the coral species found outside the harbor, while damage from construction outside the harbor was minimal.

Outfall Extensions for Thermal Effluents

Ocean outfalls have been extended for thermal effluents to protect coral reefs in Hawaii (Coles 1984; Jokiel and Coles 1990). In response to U.S. Environmental Protection Agency regulations, the ocean outfall off the Kahe Power Plant, Oahu was extended offshore to prevent the impingement of the buoyant surface water plume on the bottom. The thermal effluent was generated by the intake of seawater used in the once-through cooling system for the power plant. The heated effluent was also diluted with normal (unheated) seawater to reduce water temperatures. As a result of both the extension and dilution, corals naturally recolonized bottom environments previously exposed to lethal seawater temperatures (Coles 1984). Furthermore, the temperature increases after dilution appeared to favor greater colonization by some species and more vigorous growth rates.

Cementation of Reefs Damaged by Ship Groundings

Experimental rehabilitation techniques of the damaged reef at Key Largo has been tested on a pilot scale as part of a proposed full scale restoration program (Hudson and Diaz 1988). Approximately 644 m² of the underlying reef framework was flattened and fractured by the grounding of a 122m long freighter, the M/V *Wellwood* in August 1984. A total of 1,282 m² of the reef sustained 70-100% loss of live coral. A one square meter area of the reef framework was reestablished by grouting with quick setting underwater cement (8 parts Portland II cement mixed with 2 parts molding plaster). Hardening was achieved within 3-6 minutes after mixing. This same mixture was used to fix transplanted corals and to reattach still-living large coral heads dislodged during the grounding. These techniques were successful and may be used during the planned restoration of the remainder of the damaged reef. When accomplished this restoration will constitute only the second full-scale restoration of a coral reef.

The cost of reef framework restoration is probably high, comparable to that of coral transplantation, due to need for extensive underwater work by divers and the use of airlift bags to move large intact head corals. Recementation of the framework is essential to protect adjacent undamaged parts of the reef from the movement of the large dislodged corals and reef blocks and to promote coral recolonization on the flattened reef framework. Recovery of reef corals would also be inhibited by the movements. Removal or reattachment of loose materials from the damaged reef site is required prior to cementation. Based upon the practical and effective experience gained at the *Wellwood* site, Hudson and Diaz (1988) conclude that each grounding will present a unique set of problems and solutions to rehabilitate damaged coral reefs.

Removal of Disease Bearing Organisms, Coral Predators and Competing Benthic Algae

Removal of benthic organisms detrimental to corals has been accomplished on several reefs. Harold Hudson (pers. comm.) of the U.S. Geological Survey in Miami Beach has tested two techniques to treat corals infected with "black-band" disease (Antonius 1988), an infestation caused by a type of *cyanophyte* or blue-green algae which results in progressive destruction of coral tissue and the death of the coral. One technique involves the use of an aspirator hooked to a SCUBA tank to suck off the algae from the coral. Sloughed off suspensions of the algae were collected and stored in barrels for later disposal on land to prevent the collected blue-green algae from infecting other corals. The other technique involves the application of molding clay or cement over the infected area. In both techniques, treatment retarded the spread of the algae. However, the main drawback of either technique is the extensive diving effort needed to treat each coral colony, especially branching corals. Competing species of benthic algae can also be removed from reefs either using rakes (Woodley and Clark 1989), or through manual removal.

The physical removal or killing of the coral eating starfish (*Acanthaster*) has been attempted on several Pacific reefs (Dick Wass of Honolulu, pers. comm.). After the major starfish infestation in American Samoa in 1978, the government offered a bounty to citizens for each starfish collected from the reef and taken to a land disposal site. Poisons were injected into starfish on reefs in Guam, other parts of Micronesia, Australia, and Hawaii (Cheney 1973; Endean 1987; Henry Sakuda of Honolulu, pers. comm.). Two effective poisons were commercial hypochlorite (Clorox) and ammonium hydroxide. The injected starfish die in situ without regeneration. For a twenty-year period following eradication of the starfish infestation off South Molokai, there has been little sign of starfish recovery (Maragos unpubl.). Another technique for starfish control involves picking up the starfish and carrying them offshore where they are dropped into deep water. This last technique is effective only off reefs with very steep dropoffs such as the ocean side of barrier reefs and atolls. Biological control of the starfish has also been investigated, including the culturing and release of natural predators of the starfish (triton's trumpet, *Charonia tritonis*, and a shrimp, *Hymenocera picta*). The crown-of-thorns starfish is absent from the Atlantic-Caribbean Ocean. However Caribbean predators on corals including the snail *Coralliophila abbreviata*, and the fireworm *Hermodice carunzulata* may be worthy of control such as removal by divers (Woodley and Clark 1989).

Fisheries Controls

Controls over spearfishing on heavily fished reefs can result in dramatic recovery of targeted species of reef fish. This is no better illustrated than in Hawaii at Hanauma Bay, one of only a few sites where spearfishing is banned. Here large schools of reef fish can be seen swimming among the thousands of daily visitors to the park. Outside the sanctuary a diver rarely sees a large reef fish, and only from a "safe" distance beyond the reach of a speargun. Reef fish have also responded favorably to controls over spearfishing in southern Florida (Bohnsack 1982, cited in Woodley and Clark 1989).

The principal drawback of regulating spearfishing is finding the political will among government officials and legislators. In Hawaii, for example, fishermen have great political power. Even today adequate data are not yet collected on the catch from recreational fishermen, despite years of having depleted reef fish stocks along Oahu, the most populous island of Hawaii.

Equally important is regulating destructive fishing techniques, including use of poisons, explosives, large gill nets, and the pounding of reefs with sticks and stones. Nevertheless, there has been little success in achieving controls where they are most needed such as in the developing island nations, including the Federated States of Micronesia among the United States affiliated areas. Education, vocational retraining of fishermen, and greater restrictions over the transportation and sale of explosives are the most likely long term remedies for removing destructive fishermen or their techniques from reefs.

Coastal Resources Management: Zoning of Coral Reefs

The Great Barrier Reef Marine Parks Authority has established a zoning scheme to control activities potentially destructive to reefs and to promote research and educational activities potentially beneficial to reefs. The establishment of buffer zones to separate potentially incompatible use levels on reefs is also included (Great Barrier Reef Marine Park Authority annual reviews). The zoning concept for the park has widespread application to many other coral reefs, but very few other island nations have adopted the comprehensive approach adopted by the authority. The Great Barrier Reef itself benefits by its location offshore where people do not live on or near the reef. Elsewhere in the Pacific Islands, zoning is an alien concept to many governments where it is believed to conflict with "traditional" controls. Zoning also undermines political power among elected officials. Zoning schemes have also been developed for a few small reef sanctuaries in Hawaii, but the level of enforcement and size of the sanctuaries are very limited by comparison to what is actually needed and the scale of sanctuaries on the Great Barrier

Reef. Zoning is now apparently used in Bermuda and the Cayman Islands to control potentially destructive activities on coral reefs.

Replanting of Seagrasses and Mangroves

As noted, coral reef ecosystems are often ecologically linked to adjacent seagrass and mangrove areas. For example, species may spend different parts of their life-cycles in separate systems. Thus, restoration of seagrass and mangrove areas via transplantation can indirectly benefit adjacent coral reef areas. Companion papers by Fonseca, Kirkman and Cintron-Molero in this volume treat the restoration of seagrasses and mangroves respectively in greater detail, including the efficacy of transplanting schemes. Large-scale transplantation schemes have not been implemented near coral reefs. The author notes mangrove transplanting schemes have been initiated in Bali (Indonesia) and Palawan (Philippines) and seagrass replanting schemes in Rota (Northern Mariana Islands) adjacent to coral reef areas. Long-term monitoring of these small scale projects is needed to determine the degree of success and identify improvements to existing schemes.

Coral Reef Mitigation for Federally Approved or Funded Projects

To comply with several United States environmental laws and implementing regulations, federal agencies (either sponsors or regulators) often impose measures to reduce or avoid adverse environmental impacts to coral reefs and other environmentally significant resources. On coral reefs these techniques have included: water quality standards, turbidity control basins and curtains, relocation of corals from dredge and fill sites to safer areas, monitoring for the outbreak of ciguatera fish poisoning, tagging and monitoring of individual corals or reef transects, design changes to maintain water flushing/residence time, and design precautions to prevent pollution discharges, beach erosion, etc. (Maragos 1991). In virtually all of these cases, the goal of mitigation is to reduce the effects or compensate for anticipated losses in environmental values and functions. However, there is still little established federal policy to enhance ecological habitat and areas. Such a policy would lead to greater research and development of techniques to restore or rehabilitate coral reefs.

Additional Techniques That Could Be Pursued to Restore Coral Reefs

There are at least ten additional approaches that could be pursued to restore coral reefs (Table 6), and most of these serve as logical extensions of research and development from restoration techniques used to date.

Culturing and Stocking of Larvae and Juveniles of Target Species

There have been great advances in mariculture research on the culturing and growth of desirable species such as giant clams (*Hippopus, Tridacna*) in Australia and Palau (Price 1988; D. Hopley of Townsville and G. Heslinga of Palau, pers. comm.). Mariculture systems are also producing abundant recruits of the Caribbean queen conch, *Strombus gigas* (Berg 1976) and success is being reported for other Pacific species such as spiny lobsters (Prescott 1988), top shell (Gillett 1988), green snail, Turbo (Yamaguchi 1988) and the black-lipped pearl oyster, *Pinctada margaritifera* (Sims 1988). To benefit reef restoration initiatives, mariculture research could be extended to develop the protocols for optimal health, size, habitat, and other requirements for the release of cultured larvae and juveniles to promote restocking of specific reef sites. Mariculture research should also extend to "essential" reef species including other shellfish, reef corals and reef fishes. The culturing and release of coral larvae in restoration sites may be important for the survival of degraded reef environments (R. Richmond of Guam 1990, pers. comm.). Mariculture of seaweeds is also being advocated as an alternative vocation for fishermen to reduce over-fishing and eutrophication stress to coral reefs (Rodney Fujita of New York, pers. comm.)

The life cycles of the larval stages of reef fishes appear to be more complex, and larvae of target fish may be more difficult to rear in laboratory conditions. In addition, overfishing of predatory reef fishes (groupers, snappers, barracudas, jacks, etc.) may lead to an overabundance of "prey" species such as planktivorous reef spe-cies which in turn may feed more heavily on the larvae of more de-sirable reef organisms (Gaines and Roughgarden 1987, cited in Woodley and Clark 1989). Clearly, more field and laboratory re-search is needed on the survival, behavior, and interactions of lar-vae of desirable reef fishes.

The laboratory culturing and release of organisms is not without controversy. Some scientists maintain that spreading hatchery-reared larvae may alter genetics of natural populations, spread dis-eases like viruses, or be a waste of time by releasing organisms with inappropriate behavior for survival (i.e., they produce fat barracu-da). To minimize these problems, brood stock would need to be collected from near the target reefs and include a variety of individ-uals from different populations and races.

Increased Abundance of Herbivorous Species

An important strategy for controlling the damage to reef corals and other important reef benthos caused by the overabundance and growth of competing algae is to promote greater populations of and grazing pressure by herbivores that feed on the algae (see Sammarco 1985). The mass mortality of the herbivorous Caribbe-an sea urchin *Diadema antillarum* (Sammarco 1982a; Lessios 1984; Carpenter 1985) was followed by increased abundance of benthic algae on the reefs, and Sammarco (1982b) demonstrated the potential importance of the grazing sea urchin in controlling algae during experiments in which he increased the numbers of urchins on study sites. On the other hand, other evidence and observations suggest that high densities of sea urchins can also erode the framework of coral reefs and retard coral larval settlement, growth and survival (Maragos unpubl.). Hence, it seems prudent at this time to pursue additional site specific research on sea urchin and related coral reef population dynamics before advocating the propagation and release of sea urchins to control benthic algal growth.

Herbivorous reef fishes are also important grazers on reefs and if overfished, benthic algae populations can increase and outcompete corals. Some researchers for example speculate that the excessive growths of the algae *Dictyosphaeria cavernosa* in Kaneohe Bay may have been attributed, in part, to the lack of herbivores due to overfishing. Thus, restocking of herbivorous reef fishes and restrictions on the capture of these species (especially surgeon fish and parrot fish) serve as important potential strategies for controlling benthic algae. However, care should be taken to avoid damage to "host" reefs resulting from the transfer of herbivorous fish to the restoration sites.

Hydraulic or Mechanical Removal of Sediments

Accumulated sediments on reef environments can be removed to promote coral larval settlement and recolonization on degraded reefs. Suction dredges or similar pipeline devices could be floated over reef sites to pump off sediments. A major drawback in these types of operations is the high cost. If reef restoration actions can be combined with more "conventional" uses of the dredging equipment nearby (such as to restore safe navigation depths in a nearby harbor), it may be possible to reduce costs, especially for mobilization and demobilization of equipment used for both the harbor and restoration work.

Two other potential drawbacks of mechanical removal of sediments must be noted: (1) the sources of the reef sedimentation must be eliminated beforehand and (2) selection of appropriate disposal sites and procedures may be challenging. Often upland soil erosion from logging, ranching, and large-scale agriculture and irrigation contribute the sediments to reefs. Thus, before removal of sediments can be considered economically or technically feasible, land use controls and management would be needed to stop further soil erosion. Hydraulic removal of sediments using a pipeline or suction dredge involve the mixing of the sediment with water to form a slurry which is easier to transport off the dredging site. The sand to water mixtures in the slurry approximate 20-80%, respectively, indicating that the disposal of the slurry can pose a major problem if not done properly. The construction of diked sedimentation basins near land may be needed to dewater the slurry (see Maragos 1991).

Removal of sediments by mechanical means, such as using a clamshell dredge, may cause damage to adjacent reef areas unless the dredging site is separated from them using rigid berms or turbidity curtains. Barges may be needed alongside the floating dredge plant to stockpile excavated material. A less desirable option, which avoids the need for barges and floating equipment, is to build causeways out from the shoreline so that crane operated clamshells or draglines can remove sediment from them and stockpile them in dry environments. Then loaders and dump trucks can transport the sediment to the disposal sites. All of these procedures are costly and can cause additional impacts to the reef sites being restored.

Mechanical Modification of Current, Wave and Tidal Regimes

Another strategy is to modify substrate and shoreline characteristics of reefs to allow stronger water currents to carry away undesirable sediments from reefs. Such modifications can also reduce water residence time in lagoons, enhance water quality, and reduce eutrophication stress. Although such manipulations have not been accomplished to date for the purpose of restoring reefs, several situations indicate this approach could be feasible. For example, reefs degraded by causeways constructed across them which block circulation and water movement can be improved by installing culverts and bridge openings to restore circulation. New harbors could be designed with "circulation culverts" such as accomplished at Agana, Guam to promote better circulation and reduce adjacent impacts (Maragos 1991). Entrance channels and navigation channels could be designed and sited to increase water circulation on degraded reefs. In one instance, a failed incomplete small boat channel at Okat, Kosrae has turned out to be an inadvertent success for funneling strong wave-driven water currents through the channel to the harbor where increased circulation is benefiting water quality and conditions for coral reefs (Maragos 1991). Although this circumstance constitutes a fortuitous accident, other such channels could be intentionally designed to enhance water quality conditions on reef sites targeted for restoration.

Modifications to current and circulation regimes can cause major damage to coral reefs, and so such proposals to fulfill reef restoration goals must be approached with the utmost caution. One rule of thumb would be to avoid current or circulation modifications at existing reef sites in good or excellent condition and to confine such modifications to sites which truly need to be restored. Cutting channels through the perimeter reefs of lagoons, for example, should not be pursued for pristine or healthy atolls. However, reestablishing bridges or culverts on atolls degraded by causeways may be appropriate (see Kaly and Jones 1990; Maragos 1991).

Habitat Creation on Shallow Reef Flats
Other physical modifications of coastal environments are possible to promote reef enhancement and restoration, especially on

shallow reefs. Underwater observations in reef flat quarry holes in Micronesia, for example, demonstrate that such environments can be favorable for coral, reef fish and economically important invertebrates (Titgen 1988; Lamberts and Maragos 1989; Maragos 1991). Such holes can be specifically designed to maximize colonization and development of desired reef organisms. The holes could also be used for mariculture, ranching, grow out, holding "pens" or refugia for sought after species including conchs, trochus, green snail, giant clams, and possibly pearl oysters and helmet shells. Since the original intention for most existing quarry holes was to obtain quarry rock, specifically tailoring the design and siting of future holes should lead to greater opportunities for reef enhancement. Reef fishes may use the sites for spawning aggregations and to take shelter from larger predators or during low tide conditions.

Reducing Eutrophic Stress to Coral Reefs

A key future strategy on coral reefs in populated areas is to reduce the influx of nutrients and organics, which in turn may degrade a reef through eutrophication (see Dollar 1986). From a cost standpoint, it is very likely that outfall extension may be a more desirable option in lieu of or in addition to enhanced treatment. In particular, the discharge of anything less than tertiary-treated sewage into lagoons or protected embayments should be greatly discouraged. Conversely, desired sites for outfalls should be located in open ocean environments where water mixing and flushing is maximized; if feasible, the diffuser ports for the outfall should be placed in deeper water, below the level of most photosynthesis (e.g., below a depth of 30-60 m), or below the thermocline. This placement would further dilute sewage or discourage its upward movement to surface waters.

In addition to sewage, soil runoff from agriculture, fertilized lawns and golf courses are also major sources of nutrients. Land use controls must be exercised to control runoff or to locate these activities away from embayments or other confined water bodies where discharged nutrients can have greater adverse effects to coral reefs.

Transplantation of Reef Fishes, Sea Urchins, Other Herbivores and Predators

In some emergency situations, it may be justifiable to remove organisms from healthy reefs to sites where reef restoration is needed. If large scale transplantation is to be exercised, it should be the last resort and aim toward restoration of important coral reef areas or sites destined to serve as refugia or sanctuaries. Regular monitoring and surveillance of recipient sites would also be needed to protect desired transplants from poaching.

Long-Range Monitoring

Although there is considerable contemporary interest in reef monitoring, there is still no organized large-scale reef monitoring network and program established anywhere within the United States and affiliated islands. The recently initiated Environmental Monitoring and Assessment Program (EMAP), being managed by the U.S. Environmental Protection Agency (Hunsaker and Carpenter 1990), although nationwide in eventual scope, is not presently focusing on monitoring marine waters; nor does there appear to be any indication that the program will extend to tropical marine waters and coral reefs.

Monitoring is especially important on coral reefs because so many appear to be declining without the full knowledge or understanding by scientists and resource managers (see Grigg and Dollar 1990). The priorities for reef restoration will be dictated by the condition of coral reefs as a whole and the rate to which they are being degraded on a regional and national level. A systematic coral reef monitoring program might be the best means to accomplish this assessment and lead to priorities for which specific reef sites should be ear-marked for protection and restoration. At present, the state legislature in Hawaii, possibly assisted by resort and diving operations, will be sponsoring a small initial reef monitoring effort (Cynthia Thielen and Cynthia Hunter of Honolulu, pers. comm.). Resort and diving interests in Micronesia are also interested in monitoring nearby coral reefs. The National Park Service is sponsoring a 3-year long monitoring program in the U.S. Virgin Islands and South Florida (Dry Tortugas and Biscayne National Park) (Caroline Rogers of the U.S. Virgin Islands, pers. comm.).

Coastal Resource and Other Regional Management

Several of the previously discussed approaches, such as monitoring, land use controls, regulation of overfishing, and reef zoning can constitute separate components of an integrated regional coastal resource management program. At present, NOAA sponsors federally funded coastal zone management programs in several states and territories where coral reefs are found. NOAA could encourage greater emphasis on the need to protect or restore coral reefs through its policies and funding. In addition, coastal resource management (CRM) programs could be extended to United States affiliated islands which presently lack but which have requested United States sponsorship of CRM or Coastal Zone Management programs, including the Republic of the Marshall Islands, Federated States of Micronesia and the Republic of Palau.

Discussion of Current or Recommended Experimental Approaches That Should Be Pursued

Cementation of Reefs Damaged by Ship Groundings

If not yet implemented, NOAA should continue to press for full restoration of Molasses Reef, damaged by the 1984 grounding of the M/V *Wellwood* at Key Largo. Not only are the funds available, but restoration of this site will be the first full-scale mechanical restoration of a reef and will yield important practical experience in dealing with future restoration of reef frameworks. For example, if the restoration works, NOAA and other agencies will be provided with actual expenses and practical experience on which to base fines and costs of restoration. Regardless of whether restoration works or not, NOAA or other regulators need to establish appropriate policies, monitoring and penalties that keep shipping lanes far from valuable reefs. Prevention of groundings is clearly preferred from both the economic and ecological perspective.

Reducing Coral Mortality
The most feasible approaches to reducing coral mortality include:

1. Eliminating the causes of the mortality.
2. Increasing the populations of herbivores that feed on benthic algae that compete for space against corals.
3. Pursuing efficient procedures to remove disease bearing organisms from corals.
4. Manual removal of coral predators by divers.
5. Physical removal of competing benthic algae by divers.

A feasible national level approach would be the redesign and retrograding of sewage treatment plant and disposal systems near coral reefs. The emphasis on treatment should be matched with one that also considers ocean outfall extensions to remove sewage, treated or otherwise, from the vicinity of coral reefs. Although secondary treated sewage may be important in protecting public health, this level of treatment does little to discourage the growth of benthic algae stimulated by the nutrients in the treated sewage. Improvements in land management practices that will reduce soil erosion and the discharge of fertilizers and other non-point sources of pollution should also be implemented for coastal areas near valuable reefs. Tertiary sewage treatment plants, although costly and technically difficult to operate and maintain, may be the preferred approach where other sewage disposal options are not feasible. Tertiary treatment essentially removes all nutrients and pathogens from effluents. Regulators may also wish to require tourism proponents to finance tertiary treatment plants or ocean outfalls where other sewage disposal options may be damaging to coral reefs.

Hydraulic Removal of Sediments and Reestablishment of Hard Elevated Surfaces for Coral Colonization

Hydraulic removal of sediments may be feasible along coastal areas where walled sedimentation basins can be constructed on land to accept the collected sediments or sediment slurry mixture. Floating suction equipment seems more feasible and less potentially damaging to coral reefs compared to the use of draglines or clamshell dredges, the operation of which could fragment or fracture corals.

The placement of elevated rocky structures on sandy bottoms could also stimulate larvae coral settlement and provide shelter for fish. This may be an acceptable option in environments where sediment removal is not desired or feasible.

Optimizing Sites and Designs for Coastal Structures

A recent review of Pacific coastal construction (Maragos 1991) reveals that in some cases structures or features actually enhance rather that discourage coral reef development. A systematic retrospective survey and analysis of completed structures on coral reefs could identify the success stories as well as the failures and lead to better guidance on designing and siting channels, quarry holes, protective structures, and basins to enhance coral reef restoration.

Improving Water Circulation on Reefs

A retrospective analysis could also help to design structures to enhance circulation on reefs that may be damaged by construction, sedimentation, or eutrophication. Model tests may also be needed prior to final design. Constructing channels, culverts, bridge spans, and trestles over reefs should be pursued where circulation has been reduced to the level of causing water quality and ecological degradation. Engineering manipulations can take advantage of waves, tides, and winds to drive desired water movement.

Seagrass and Mangrove Planting

Both seagrass and mangrove replanting schemes could help to stabilize sediments and improve biological diversity and productivity on adjacent coral reefs. Design and implementation procedures should be detailed for planting operations.

Buffer Zones and Other Protective Designations

Heavily used coral reefs may benefit from zoning that could remove damaging activities and better separate incompatible uses of coral reefs. The Great Barrier Reef system of reef zoning can be used as a model for many other reef areas, especially offshore reefs that are not permanently occupied.

Carrying Capacity and Density Controls over Reef Use

Existing marine parks need to be evaluated on a case-by-case basis to assess the extent of reef damage from visitor use. For each coral reef park, a field assessment of damage from a variety of activities needs to be differentiated, such as from anchoring, spear fishing, boat contact, sewage discharges, litter, coral breakage by snorkelers or divers, unauthorized collection, other fishing, etc. This information then needs to be correlated to statistics on the number of visitor user days and type of reefs. On the basis of these analyses, educational programs and controls over density and duration of use can be implemented, and enforcement actions can be taken to limit unauthorized fishing, collecting, and other undesirable behavior in the more vulnerable reef environments. Carrying capacity studies have been attempted at Hanauma Bay, Hawaii's most popular coral reef park. Carrying capacity studies have also been conducted on the Great Barrier Reef Marine Park (Wendy Craik of Townsville, pers. comm.) and are in the planning stages for the National Marine Sanctuaries in the Florida keys.

Long-Range Monitoring and Response

Heavily used and important coral reefs need to be regularly monitored to assess temporal trends and the regional extent of potential problems and issues. Regional level monitoring programs are in the planning stages in Hawaii, (Hunter and Maragos 1991) on the Great Barrier Reef (Lassig 1988 Gillies and Craik In press) and in French Polynesia (Salvat 1991). Monitoring is an urgent and important strategy to get early warning of problems since reefs are generally fragile and slow to recover.

In order for monitoring to be implemented at a large enough scale to be effective, efficient communications networks between various regional monitoring projects need to be established, and many non-scientists need to be involved. There are simply not enough marine scientists working on coral reefs to actually do the monitoring, and it may be best to have scientists train others to do the monitoring, provide oversight and interpretation of monitoring results, do follow-up investigations on impacts to reefs uncovered during monitoring programs, and to establish a data library for efficient storage and retrieval of monitoring information. There is considerable interest among commercial dive operators, diving clubs, and educational institutions in volunteering the time to conduct the bulk of field monitoring. Many of these groups visit coral reefs regularly and are in the ideal position of accomplishing monitoring efficiently and providing early feedback to scientists.

Federal assistance in support of monitoring is warranted. For one, the U.S. Environmental Protection Agency's Environmental Monitoring and Assessment Program (see Hunsaker and Carpenter

1990) should be extended to coral reef environments in both the Atlantic and Pacific Ocean, and perhaps NOAA could initiate the dialogue with EPA to achieve this end. Also, NOAA could sponsor additional field training, library development, and monitoring activities in federally significant coral reef environments such as in National Parks, Sanctuaries, estuaries and critical habitat for federally listed endangered and threatened species. State and territorial governments could provide matching funds especially to extend monitoring to State and County coral reef parks.

Monitoring also needs to include a combination of high tech and low tech procedures, particularly if non-scientists are to carry much of the burden. Underwater photography and video procedures (see Maney 1990) are now available which may be very useful in coral reef environments. In addition, tagging and inspection of individual corals, quadrat, transects and relevant water quality sampling may also be warranted (see Gillies and Craik In press). Depending upon the type of stresses likely to occur in specific reef areas, protocols to detect the effects of sedimentation, eutrophication, anchor damage, visitor damage, destructive fishing, coral bleaching, coral diseases, predation, etc., may need to be developed and exercised.

Fisheries Reform and Control

Controls over fishing intensity, fishing gear, and techniques need to be strengthened, especially for recreational fishing. Spearfishing and aquarium fish collecting activities need to be strictly controlled and confined to areas where these activities will not jeopardize other reef functions or values. Meaningful statistics on gear use and catch also need to be regularly collected and differentiated for each individual reef area. Enforcement for controlling destructive fisheries techniques needs strengthening, including mandatory confiscation of gear and boats used in intentional illegal fishing. If more coral reefs *outside* marine parks were in good condition and supported a variety of attractive reef fish and other attractions, then many more reef enthusiasts would visit them, relieving pressure *within* the few established coral reef parks.

Medical Cures for Bacterial and Viral Infections

Great strides have been made in detecting and documenting many reef diseases (Peters 1983, 1984). As advances continue to be made, more research emphasis should be placed on developing "cures." *In situ* cures and preventive measures that can be accomplished by diving communities and other reef enthusiasts should also be pursued. Research on diseases will also be essential for the possible culturing and release of coral reef species.

Establishing Priorities for Reef Restoration

Establishing priorities for restoration of reefs involves a sequence beginning with general inventories and leading to actions taken at specific sites. Some coral reef areas are inherently more valuable than others, and if reef restoration funding is limited, then the more valuable sites need to be given earlier attention or higher priority.

Inventories

Coral reef inventories of the type sponsored by the U.S. Army Corps of Engineers, NOAA Sea Grant and other sponsors (Maragos and Elliott 1985) are an efficient means to evaluate coral reefs on a large scale or regional basis and identify sites that serve as candidates for protection, parks, restoration, or other management actions.

Monitoring and Assessment

Following identification of candidate sites, a monitoring and assessment program would be needed to determine the cause for the original degradation of the reef, and whether that cause(s) has been or can be eliminated as part of a restoration effort. Also a judgment must be made on cost and feasibility of restoration and whether natural recovery would be sufficient. As noted earlier, established techniques for restoration can be very expensive, and in many cases there may be no choice except to rely on natural recolonization and recovery.

Restoration Plans

Once reef restoration sites are selected, restoration planning procedures must be developed on a case-by-case basis. These must identify the target species and other desired outcomes of the restoration.

Interaction Between Researchers and Managers

Given the limited number of coral reef scientists presently engaged or experienced in coral reef restoration, a great deal of cooperation is needed between "researchers" and "managers." For example, the development of coastal zone or coastal resource management plans and programs should involve input by coral reef scientists and include incentives or provisions for reef restoration. Also, coastal resource inventories would benefit significantly from the availability of aerial photography and the availability of reef scientists to interpret them for purposes of identifying candidate sites for restoration and other management options.

Scientists will also need to work with park agencies and other conservation organizations to identify, evaluate and set aside coral reef areas as marine parks (see Salm and Clark 1984). And finally, coral reef scientists will need to work with regulatory agencies and proponents during environmental impact assessments to identify sites and measures to reduce or avoid potential impacts to coral reefs (see Carpenter and Maragos 1989). All of the above "preventive" strategies are seen as precautions to reduce the need for coral reef restoration, which is now a mostly unproven experimental approach to reef management.

Preconstruction and Construction Phase Mitigation

A number of procedures in conjunction with proposed coastal construction has been demonstrated to be successful in reducing impacts to coral reef environments (Kaly and Jones 1990, Maragos 1991):

1. Water quality performance standards.
2. Silt curtains and other barriers to block sediment movement.
3. Specifying less damaging dredging techniques.
4. Siting and design of dredge and fill sites away from reefs.
5. Requiring water dependency for any construction project proposed in the water or near a coral reef.
6. Preplacement of protective structures and seawalls and then back-filling behind them to prevent open water sedimentation.
7. Use of tagged and monitored reef corals in conjunction with environmental or water quality standards.

Steps to Ensure the Success of Restoration and Mitigation

In the case of mitigation, post-construction environmental audits should be accomplished after completion of construction and during operation of projects near coral reefs to determine whether the mitigation worked as planned. These after-the-fact surveys will also identify unanticipated "success" stories, impacts, and additional remedial measures that are needed.

In the case of restoration, the same sort of procedures should also be applied. More importantly, however, more full-scale restoration projects need to be implemented in the first place, to see what works, what does not work, and what the costs are. Only through trial and error can the confidence and success of reef restoration technologies be established.

Follow-up Needs and Considerations

Until meaningful and systematically developed and applied coral reef restoration projects are actually achieved, it is difficult to speculate what is needed beyond the initial phase. The following are best guesses on follow-up needs:

1. Long term monitoring and periodic evaluation of reefs actually restored (starting with Kaneohe Bay and Key Largo).

2. National policies (especially NOAA and EPA) and state and territorial policies that encourage, authorize and fund restoration projects.

3. Refine the criteria for documenting the success of restoration, which at this stage focuses on recovery of target groups of reef organisms described earlier.

4. Extend quantitative or field criteria to include coverage or biomass, species diversity, lack of disease, rapid growth, rapid resettlement and otherwise rapid restoration of reef health indicators.

5. Enlist independent scientists familiar with coral reef functioning, health and recovery to document the success of restoration.

Aside from federal, state and territorial governments, counties, villages and conservation organizations should participate in and fund restoration projects.

Future Needs and Directions

Scientific Research

Coral reef restoration is still so much in its infancy that the most important initial step is the sponsoring of pilot restoration projects and field trials for many techniques. Potential candidate sites should be identified and research and development on restoration techniques vigorously pursued. Pilot projects on advanced techniques for coral transplantation, reef framework repair, artificial reef development, sediment removal, eutrophication removal and benthic algae control should be pursued. Physical oceanographic and ocean engineering studies including modeling can also be applied to improve flushing and circulation on degraded reefs. In the laboratory several related projects should be pursued, including coral reef pathology and larval rearing of desirable species (corals, reef fishes, herbivores, desirable predators and anti-viral or anti-bacterial agents). Perhaps research centers for the culturing of coral reef target species can be established or expanded at existing tropical marine laboratories.

Management

Coral reef inventories need to be completed for all coral reef regions to identify those sites most worthy of special protection as well as sites worthy of restoration efforts. Monitoring and assessment programs can then be established at a number of candidate and control or reference sites. Aside from developing policies and legislation to support restoration, similar guidance is needed to expand or institutionalize successful mitigation procedures and post-construction and operational environmental audits. These can all be essentially national level initiatives.

Prevention of Further Loss of Coral Reefs

The decline of coral reefs on a global basis is so alarming, that until tried and tested restoration programs are successfully completed on actual full-scale coral reefs, continued loss of coral reefs under federal jurisdiction from anthropogenic causes should be *prohibited.* This means that further destruction of coral reefs from ports, roads, airfields, effluent discharges, other dredging, fill land expansion, and commercial fishing in federally controlled waters should be strictly minimized.

At this stage restoration should not be put forth as a successful technology that justifies a "license to kill" more reefs. Coral reef restoration should instead be viewed as a potential untested technology that may eventually succeed in helping to reverse the worldwide trend in coral reef degradation.

Role of Mitigation

At the present time only some mitigation techniques on coral reefs have been demonstrated to work in reducing the potential level of coral reef damage associated with coastal construction and other forms of development on coral reefs. However, the net result of mitigation is either no net loss or decline of reefs from impacts that cannot be avoided or mitigated. Thus, mitigation has not reversed the decline in coral reef environments. Nevertheless, retrospective analysis of a variety of completed projects and structures on coral reefs may lead to identification of potentially valuable techniques or designs for either mitigation and restoration.

Mooring buoy anchoring designs developed by Halas (1985), applied off Florida at Looe Key (William Causey, pers. comm.), and now advocated globally by Greenpeace (Jeanne Kirby of Washington, D.C., pers. comm.) should be widely sponsored and supported at other reefs. This technique can reduce or eliminate anchor damage from boats visiting coral reefs, and it serves as an important mitigation and restoration method for any marine park or destination site accessible by boat (Halas 1985).

Priority Regions

On the basis of visitor use and development pressure, coral reef restoration is especially needed among the following United States or United States affiliated areas: Hawaii, South Florida, Guam, Saipan, South Texas, U.S. Virgin Islands, Puerto Rico, and the

Pago-Pago Harbor region of American Samoa. In addition, coral reef restoration should be investigated at other United States possessions, including the following Pacific atolls: Johnston, Wake, Midway and Palmyra.

Priority Species

Restoration efforts should focus on the "target species" of coral reefs discussed earlier: reef corals, reef fishes, herbivores that feed on benthic algae, and edible shellfish. Other species or groups may warrant restoration assistance as assessed on a site-by-site basis.

Costs

Coral reef restoration is very expensive, and for any reef area, restoration costs will probably exceed one million dollars. The cost of retrofitting the sewage treatment plant system and outfall at Kaneohe Bay exceeded $25 million dollars, for example. During court proceedings on the *Wellwood* grounding, the cost of restoring the damaged reef at Key Largo was estimated at $30 million. A total of $6 million agreed to in an out-of-court settlement is to be used for the actual restoration, and it will be important to determine exactly how much that restoration will cost in the end. The high cost of restoration only serves to emphasize the importance of preventing further damage to coral reefs under United States jurisdiction.

Responsibility for Costs

The United States government should take the primary responsibility for sponsoring coral reef restoration. Through incentive arrangements, the Federal government may also be able to attract state, county, and private funding. Government sponsorship of restoration should especially be directed at areas where the actual individuals, groups, corporations, etc., responsible for the initial damage cannot be identified. The fine system used in the *Wellwood* case is a good model to use in requiring restoration where responsibility has been clearly established and assigned. Volunteer assistance is also likely among marine research laboratories, dive clubs, commercial diving operators and some educational institutions.

Coastal Zone Management Programs

Coastal zone or coastal resource management programs are partly sponsored by NOAA and established in all of the states and territories of the United States where priority attention is needed to restore coral reefs. Thus, NOAA may be able to exert more pressure leading to positive restoration activities and policies preventing further degradation of coral reefs. Revised language on coral reef restoration in the next reauthorization of the Coastal Zone Management Act may be one of the more fruitful and early steps in focusing attention on the restoration of coral reefs and other valuable marine habitats in the United States coastal zone.

Acknowledgments

I thank Sue Wells, Gordon Thayer, Ann Manooch and the anonymous reviewers for helpful criticisms on the draft manuscript, and Angelina Lau for word processing support.

Literature Cited

Acevedo, R. and J. Morelock. 1988. Effects of terrigenous sediment influx on coral reef zonation in southwestern Puerto Rico. Proc. 6th Int. Coral Reef Symp. 2:189-194.

Antonius, A. 1981. The "band" diseases in coral reefs. Int. Coral Reef Symp. 2:7-14.

Antonius, A. 1988. Black band disease behavior in Indo-Pacific reef corals. Proc. 6th Int. Coral Reef Symp. 3:145-150.

Atkinson, M.J. 1988. Are coral reefs nutrient-limited? Proc. 6th Int. Coral Reef Symp. 1:157-166.

Auberson, B. 1982. Coral transplantation: An approach to the re-establishment of damaged reefs. Kalikasan 11:158-172.

Bak, R.P.M. 1978. Lethal and sublethal effects of dredging on reef corals. Mar. Pollut. Bull. 9:14-16.

Bak, R.P.M., M.J.E. Carpay and E.D. De Ruyter Van Steveninck. 1984. Densities of the sea urchin *Diadema antillarum* before and after mass mortalities on the coral reefs of Curacao. Mar. Ecol. Prog. Ser. 17:105-108.

Banner, A.H. 1968. A freshwater kill on the coral reefs in Hawaii. Hawaii Inst. Mar. Biol. Tech. Rep. 15. 29 p.

Banner, A.H. 1974. Kaneohe Bay, Hawaii: Urban pollution and a coral reef ecosystem. Proc. 2nd Int. Coral Reef Symp. 2:685-702.

Berg, C.J., Jr. 1976. Growth of the queen conch, *Strombis gigas* with a discussion of the practicality of its mariculture. Mar. Biol. 34:191-199.

Birkeland, C. 1989a. The Faustian traits of the crown-of thorns starfish. Amer. Scientist. 77:154-163.

Birkeland, C. 1989b. The influence of enchinoderms on coral reef communities, p 1-79. *In* M. Jangoux and J.M. Lawrence (eds.), Echinoderm studies, Vol. 3. A. A. Balkema, Rotterdam.

Bohnsack, J.A. 1982. Effects of piscivorous predator removal on coral reef fish community structure, p. 258-267. *In* G.M. Caillet and C.A. Simenstad (eds.), Gutshop '81: fish food habit studies. Washington Sea Grant Publication, University of Washington, Seattle.

Bouchon, C., J. Jaubert and Y. Bouchon-Navarro. 1981. Evolution of a semi-artificial reef built by transplanting coral heads. Tethys 10:173-176.

Brock, R.E. 1983. A eleven year study of the structure and stability of the coral reef fish community in Honokohau Harbor, Kona, Hawaii, p. 74-85. *In* A decade of ecological studies following construction of Honokohau small boat harbor, Kona, Hawaii. U.S. Army Corps of Engineers, Honolulu District, Honolulu.

Carpenter, R.A. and J.E. Maragos. 1989. How to assess environmental impacts on tropical islands and coastal areas. Environment and Policy Institute, East-West Center, Honolulu. 345 p.

Carpenter, R.C. 1985. Sea urchin mass mortality: effects on reef algal abundance, species composition, and metabolism and other coral reef herbivores. Proc. 5th Int. Coral Reef Congr. 4:53-60.

Cheney, D.P. 1973. An analysis of *Acanthaster* control programs in Guam and the Trust Territory of the Pacific Islands. Micronesica 9:171-180.

Coles, S.L. 1984. Colonization of Hawaiian reef corals on new and denuded substrata in the vicinity of a Hawaiian power station. Coral Reefs 3:123-130.

Coles, S.L. 1988. Limitations on reef coral development in the Arabian Gulf: temperature or algal competition? Proc. 6th Int. Coral Reef Symp. 3:211-216.

Darwin, C. 1842. The structure and distribution of coral reefs. 3rd ed. D. Appleton and Co. (1898), New York, 332 p.

Dawson, E.Y. 1959. Changes in Palmyra Atoll and its vegetation through the activities of man, 1913-1958. Pac. Natur. 1:1-51.

Devaney, D., G.S. Losey and J.E. Maragos. 1975. A marine environmental survey of proposed construction sites for the Truk runway. Prepared for R.M. Parsons Co., Honolulu. 72 p.

Dollar, S.J. 1979. Ecological response to relaxation of sewage stress off Sand Island, Hawaii. Wat. Resour. Res. Center, Honolulu, Tech. Rep. 124.

Dollar, S.J. 1982. Wave stress and coral community structure in Hawaii. Coral Reefs 1:71-81.

Dollar, S.J. 1986. Response of the benthic ecosystem to deep ocean sewage outfalls in Hawaii: Benthic fluxes at the sediment-water interface. Ph.D. thesis, University of Hawaii, Honolulu.

Endean, R. 1976. Destruction and recovery of coral reef communities, p. 215-255. *In* O.A. Jones and R. Endean (eds.), Biology and geology of coral reefs. Academic Press, New York.

Endean, R. 1987. *Acanthaster planci* infestations. *In* B. Salvat (ed.), Human impacts on coral reefs: facts and recommendations. Antenne Museum Ecole Pratique des Houtes Etudes, French Polynesia. 253 p.

Endean, R., A.M. Cameron and L.M. De Vantier. 1988. *Acanthaster planci* predation on massive corals: the myth of rapid recovery of devastated reefs. Proc. 6th Int. Coral Reef Symp. 2:143-148.

Evans, C.W., J.E. Maragos and P.F. Holthus. 1986. Reef corals in Kaneohe Bay six years before and after termination of sewage discharges (Oahu, Hawaiian Archipelago), p. 76-90. *In* P.L. Jokiel, R.H. Richmond and R.A. Rogers (eds.), Coral reef population biology. Sea Grant Coop. Rep. Honolulu.

Fitzhardinge, R.C. and J.H. Bailey-brock. 1989. Colonization of artificial reef materials by corals and other sessile organisms. Bull. Mar. Sci. 44:567-579.

Fujita, R.M. 1991. Protecting ecosystem integrity in the Florida Keys National Marine Sanctuaries. Environmental Defense Fund, New York, 15 p.

Gaines, S.D. and J. Roughgarden. 1987. Fish in offshore kelp forests affect recruitment to intertidal barnacle populations. Science 235:479-481.

Gillett, R. 1988. Pacific islands trochus introductions. Workshop on Pacific Inshore Fisheries Resources, Noumea, New Caledonia, 1988, BP 61. 7 p.

Gillies, J. and W. Craik. In press. Environmental monitoring programs associated with developments in the Great Barrier Reef Marine Park. Ocean and Shoreline Management.

Gittings, S.R., T.J. Bright, A. Choi and R.B. Barnett. 1988. The recovery process in a mechanically damaged coral reef community: Recruitment and growth. Proc. 6th Int. Coral Reef Symp. 2:225-230.

Gladfelter, W.B. 1982. White band disease in *Acropora palmata*: Implications for the structure and growth of shallow reefs. Bull. Mar. Sci. 32:639-643.

Gladfelter, W.B., E.H. Gladfelter, R.K. Monahan, J.C. Ogden and R.F. Dill. 1977. Coral destruction. Environmental Studies of Buck Island Reef National Monument. U.S. National Park Serv. Rep. 144 p.

Glynn, P.W. 1983. Extensive "bleaching" and death of reef corals on the Pacific coast of Panama. Environ. Conserv. 11:133-146.

Glynn, P.W. and M.W. Colgan. 1988. Defense of corals and enhancement of coral diversity by territorial damselfish. Proc. 6th Int. Coral Reef Symp. 2:157-164.

Glynn, P.W., J. Cortez, H.M. Guzman and R.H. Richmond. 1988. El Niño (1982-83) associated coral mortality and relationships to sea surface temperature deviations in the tropical eastern Pacific. Proc. 6th Int. Coral Reef Symp. 3:237-244.

Gomez, E.D., A.C. Acala and A.C. San Diego. 1981. Status of Philippine coral reefs. 1981. Proc. 4th Int. Coral Reef Symp. 1:275-282.

Grigg, R.W. 1983. Community structure, succession and development of coral reefs in Hawaii. Mar. Ecol. Prog. Ser. 11:1-14.

Grigg, R.W. and S.J. Dollar. 1990. Natural and anthropogenic disturbance on coral reefs, p. 439-452. *In* D. Dubinsky (ed.), Ecosystems of the world 25, Coral Reefs. Elsevier, New York.

Grigg, R.W., J.W. Wells and C. Wallace. 1981. *Acropora* in Hawaii, Part 1. History of the scientific record, systematics, and ecology. Pac. Sci. 35:1-13.

Grigg, R.W. and J.E. Maragos. 1974. Recolonization of hermatypic corals on lava flows in Hawaii. Ecology 55:387-394.

Guilcher, A. 1987. Coral reef geomorphology. Wiley, New York. 228 p.

Halas, J.C. 1985. A unique mooring system for reef management in the Key Largo National Marine Sanctuary. Proc. 5th Int. Coral Reef Symp., 4:237-242.

Harriott, V.J. and D.A. Fisk. 1988. Coral transplantation as a reef management option. Proc. 6th Int. Coral Reef Symp. 2:375-380.

Hay, M.E., P.E. Renaud and W. Fenical. 1988. Large mobile versus small sedentary herbivores and their resistance to seaweed chemical defenses. Oecologia 75:246-252.

Helfrich, P. 1975. An assessment of the expected impact of a dredging project for Pala Lagoon, American Samoa. University of Hawaii Sea Grant Publication, UNIHI-SEAGRANT-TR-76-02. 76 p.

Helfrich, P. 1979. Utilization and management of inshore marine ecosystems of the tropical Pacific Islands, Proceedings, Institute of Marine Resources. University of the South Pacific (Fiji) and International Sea Grant Program, University of Hawaii (Honolulu). 116 p.

Hodgson, G. and J.A. Dixon. 1988. Logging versus fisheries and tourism in Palawan. East-West Environment and Policy Institute, Honolulu, Occ. Pap. No. 7. 95 p.

Hudson, J.H. and R. Diaz. 1988. Damage survey and restoration of M/V Wellwood grounding site, Molasses Reef, Key Largo National Marine Sanctuary, Florida. Proc. 6th Int. Coral Reef Symp. 2:231-236.

Hudson, J.H., D.M. Robbin, J.T. Tilmant and J.L. Wheaton. 1989. Building a coral reef in southeast Florida: Combining technology and aesthetics (abst.). Bull. Mar. Sci. 44:1067.

Hughes, T.P., D.C. Reed and M.J. Boyle. 1987. Herbivory on coral reefs: community structure following mass mortalities of sea urchins. J. Exp. Mar. Biol. Ecol. 113:39-59.

Hunsaker, C.T. and D.E. Carpenter (eds.). 1990. Ecological indicators for the Environmental Monitoring and Assessment Program. U.S. Environmental Protection Agency, Off. Res. Devel., EPA 60013-901060.

Hunter, C. and J.E. Maragos. 1991. Methodology for involving the recreational and educational diver community in long term monitoring of coral reefs (abst.). Pacific Science Congress, 27 May - 3 June 1991, Honolulu.

International Union for the Conservation of Nature and Natural Resources and United Nations Environment Programme (IUCN and UNEP). 1988. Coral reefs of the world. Conservation Monitoring Centre, Cambridge, U.K. 3 vols.

Johannes, R.E. 1975. Pollution and degradation of coral reef communities, p. 13-51. *In* E.J. Ferguson Wood and R.E. Johannes (eds.), Tropical marine pollution. Elsevier, New York.

Johannes, R.E. and S.S. Betzer. 1975. Marine communities respond differently to pollution in the tropics than at higher latitudes, p. 1-13. *In* E.J. Ferguson Wood and R.E. Johannes (eds.), Tropical marine pollution. Elsevier, New York.

Johannes, R.E., J.E. Maragos and S.L. Coles. 1972. Oil damages coral exposed to air. Mar. Pollut. Bull. 3:29-30.

Jokiel, P.L. 1987. Ecology, biogeography and evolution of corals in Hawaii. Trends Ecol. Evolut. 2(2):175-178.

Jokiel, P.L. and S.L. Coles. 1990. Response of Hawaiian and other Indo-Pacific reef corals to elevated temperatures. Coral Reefs 8:155-162.

Kaly, U.C. and G.P. Jones. 1990. The construction of boat channels across reefs: An assessment of ecological impact. Final Report (No. 4) of an environmental assessment of the impact of reef channels in the South Pacific. Prepared for New Zealand Ministry of External Relations and Trade, Auckland. 192 p.

Kinsey, D.W. 1988. Coral reef response to some natural and anthropogenic stresses. Galaxea 7:113-128.

Lamberts, A.E. and J.E. Maragos. 1989. Observations on the reef geomorphology and corals of Majuro Atoll. *In* Majuro Coastal Resource Inventory. University of Hawaii Sea Grant Program.

Lassig, B.R., C.L. Baldwin, W. Craik, S. Hillman, L.P. Zann and P. Ottesen. 1988. Monitoring the Great Barrier Reef. Proc. 6th Int. Coral Reef Symp. 2:313-318.

Lessios, H.A., D.R. Robertson and J.D. Cubit. 1984. Spread of *Diadema* mass mortality through the Caribbean. Science 226:335-337.

Losey, G.S. 1973. Study of environmental impact for Kwajalein Missile Range. Prep. for the U.S. Army Corps of Engineers, Pacific Ocean Div., Hawaii.

Loya, Y. 1976. Recolonization of Red Sea corals affected by natural catastrophes and man-made perturbations. Ecology 57:278-289.

Maney, E.J., J. Ayers, K.P. Sebens and J.D. Witman. 1990. Quantitative techniques for underwater video photography, p. 255-265. *In* Proceedings of the American Academy of Underwater Science, 10th annual science diving symposium. St. Petersburg, Florida.

Maragos, J.E. 1972. A study of the ecology of Hawaiian reef corals. Ph.D. thesis, University of Hawaii, Honolulu. 292 p.

Maragos, J.E. 1974. Coral transplantation: A method to create preserve and manage coral reefs. University Hawaii Sea Grant Publication UNI-HI-SEAGRANT AR-74-03. 30 p.

Maragos, J.E. 1977. Order Scleractinia stony corals, p. 158-241 *In* D.M. Devaney and L.G. Eldredge (eds.), Reef and shore fauna of Hawaii. Bernice P. Bishop Museum Special Pub. 64(1).

Maragos, J.E. 1983. Status of reef coral populations in Honokohau Small Boat Harbor 1971-1981. *In* A decade of ecological studies following construction of Honokohau Small Boat Harbor. U.S. Army Corps of Engineers, Honolulu District. 91 p.

Maragos, J.E. 1986. Coastal resource development and management in the U.S. Pacific Islands: 1. Island-by-island analysis. Prepared for Office of Technology Assessment, U.S. Congress. 81 p.

Maragos, J.E. 1987. Observations of corals and reefs at Palmyra Atoll, Line Islands. Unpubl. MS. U.S. Army Corps of Engineers, Pacific Ocean Division, Honolulu.

Maragos, J.E. 1991. Impact of coastal construction on coral reefs in Oceania: a review. Submitted MS.

Maragos, J.E., C. Evans and P. Holthus. 1985. Reef corals in Kaneohe Bay six years before and after termination of sewage discharges (Oahu, Hawaiian archipelago). Proc. 5th Int. Coral Reef Congr. 4:189-194.

Maragos, J.E. and M.E. Elliott. 1985. Coastal resource inventories in Hawaii, Samoa and Micronesia. Proc. 5th Int. Coral Reef Congr. 5:577-582.

Maragos, J.E. and P.L. Jokiel. 1986. Reef corals of Johnston Atoll: one of the world's most isolated reefs. Coral Reefs 4:141-150.

Marsh, J.A. and G.D. Gordon. 1974. Marine environmental effects of dredging and power-plant construction in Piti Bay and Piti Channel, Guam. Univ. Guam Mar. Lab. Tech. Rep. 8: 55 p.

McCain, J.C. and J.M. Peck, Jr. 1977. The toxicity of selected chemicals used in power generating stations to Hawaiian fishes. University of Hawaii Sea Grant Program Technical Report, UNIHI-SEAGRANT TR-77-01. 17 p.

Mitchell, R. and I. Chet. 1975. Bacterial attack of corals in polluted seawater. Microb. Ecol. 2:227-233.

Mueller, E.L. 1988. Managing inter-reefal environments and resources by artificial constructions. Proc. 6th Int. Coral Reef Symp. 2:387-392.

Munro, J.L. and D. M.C.B. Williams. 1985. Assessment and management of coral reef fisheries: Biological environmental and socio-economic aspects. Proc. 5th Int. Coral Reef Congr. 4:543-581.

Neudecker, S. 1987. Environmental effects of power plants on coral reefs and ways to minimize them, p. 103-118. *In* B. Salvat (ed.), Human impacts on coral reefs: facts and recommendations. Antenne Museum, Ecole Pratique des Hautes Etudes, French Polynesia.

Nunn, P.D. 1991. Keimami sa vakila na li ni Kalou (Feeling the hand of God): Human and nonhuman impacts on Pacific Island environments. East-West Environment and Policy Institute, Occ. Pap., Vol 13. 68 p.

Pastorak, R.A. and G.R. Bilyard. 1985. Effects of sewage pollution on coral reef communities. Mar. Ecol. Prog. Ser. 21:175-189.

Peters, E.C. 1984. A survey of cellular reactions to environmental stress and disease in Caribbean scleractinian corals. Helgol. Meeresunter. 37:113-137.

Peters, E.C., J.J. Oprandy and P.P. Yevich. 1983. Possible causal agent of "white band disease" in Caribbean acroporid corals. J. Invert. Pathol. 41:394-396.

Polunin, N.V.C. 1983. The marine resources of Indonesia. Oceanogr. Mar. Biol. Annu. Rev. 21:455-531.

Prescott, J. 1988. Tropical spiny lobster: an overview of their biology, the fisheries and the economics with a particular reference to the double spined rock lobster *P. penicillatus*. Workshop on Pacific Inshore Fishery Resources, Noumea, New Caledonia, 14-25 March 1988, WP 18. 36 p.

Price, C.M. 1988. Giant clam ocean nursery and reseeding projects. Workshop on Pacific Inshore Fishery Resources, Noumea, New Caledonia, 14-25 March 1988, BP 7. 3 p.

Rogers, C.S. 1990. Response of coral reefs and reef organisms to sedimentation. Mar. Ecol. Prog. Ser. 62:185-202.

Rogers, C.S., L. McLain and E. Zullo. 1988. Damage to coral reefs in Virgin Islands National Park and Biosphere Reserve from recreational activities. Proc. 6th Int. Coral Reef Symp. 2:405-410.

Salm, R.W. and J.R. Clark. 1984. Marine and coastal protected areas: a guide for planners and managers. International Union for the Conservation of Nature and Natural Resources, Gland, Switzerland. 302 p.

Salvat, B. 1987. Dredging in coral reefs, p. 165-184. *In* B. Salvat (ed.), Human impacts on coral reefs: facts and recommendations. Antenne Museum, Ecole Pratique des Hautes Etudes, French Polynesia.

Salvat, B. 1991. World coral reef site network project. A long term global plan to monitor coral reefs on indices of natural and anthropogenic changes (abst.). Pacific Science Congress, 27 May - 3 June 1991, Honolulu.

Sammarco, P.W. 1982a. Effects of grazing by *Diadema antillarum* Philippi (Echinodermata, Echinoidea) on algal diversity and community structure. J. Exp. Mar. Biol. Ecol. 65:83-105.

Sammarco, P.W. 1982b. Echinoid grazing as a structuring force in coral communities: whole reef manipulations. J. Exp. Mar. Biol. Ecol. 65:31-55.

Sammarco, P.W. 1985. The Great Barrier Reef vs. the Caribbean: Comparison of grazers, coral recruitment patterns and reef recovery. Proc. 5th Int. Coral Reef Congr. 4:391-398.

Schuhmacher, H. 1988. Development of coral communities on artificial reef types over 20 years (Eilat, Red Sea). Proc. 6th Int. Coral Reef Symp. 3:379-384.

Shinn, E.A. 1976. Coral reef recovery in Florida and the Persian Gulf. Environ. Geol. 1:241-254.

Sims, N.A. 1988. Pearl oyster resources in the South Pacific, Research for management and development. Workshop on Pacific Inshore Fishery Resources, Noumea, New Caledonia, 14-15 March 1988, WP 4. 25 p.

Smith, S.H. 1988. Cruise ships: A serious threat to coral reefs and associated organisms. Ocean Shoreline Manage. 11:231-248.

Smith, S.R. 1985. Reef damage and recovery after ship groundings on Bermuda (abst.). Proc. 5th Int. Coral Reef Congr. 2:354.

Sorokin, Y.I. 1970. Microbiological aspects of productivity in coral reefs. Res. Rep. Hawaii Inst. Mar. Biol. 40 p.

Tilmant, J.T. 1987. Impacts of recreational activities on coral reefs, p. 195-214. *In* B. Salvat (ed.), Human impacts on coral reefs: facts and recommendations. Antenne Museum, Ecole Pratique des Hautes Etudes, French Polynesia.

Titgen, R.H., A.M. Orcutt and P.J. Rappa. 1988. Marine environmental assessment report on United States Army leased lands at Kwajalein Atoll. Prepared for U.S. Army Corps of Engineers, Pacific Ocean Division by University of Hawaii Sea Grant Extension Service, Honolulu.

U.S. Army Corps of Engineers. 1983. A decade of ecological studies following construction of Honokohau Small Boat Harbor, Honolulu District. 91 p.

Veron, J.E.N. 1986. Corals of Australia and the Indo-Pacific. Angus and Robertson Publishers, London and North Ryde, Australia. 644 p.

Wells, J.W. 1957. Coral reefs. Geol. Soc. Am. Mem. 67(1):609-631.

White, A.T. 1987. Effects of construction activity on coral reef and lagoon systems, p. 185-193. *In* B. Salvat (ed.), Human impacts on coral reefs: facts and recommendations. Antenne Museum, Ecole Pratique des Hautes Etudes, French Polynesia.

Wiens, H.J. 1962. Atoll environment and ecology. Yale University Press, New Haven. 532 p.

Williams, E.H., Jr. and L. Bunkley-Williams. 1988. Circumtropical coral reef bleaching in 1987-1988. Proc. 6th Int. Coral Reef Symp. 3:313-318.

Woodley, J.D. and J.R. Clark. 1989. Rehabilitation of degraded coral reefs. p. 3059-3075. *In* O.T. Magoon *et al.* (eds.), Proceedings, coastal zone 1989, Charleston, South Carolina.

Yamaguchi, M. 1988. Biology of the green snail (*Turbo marmoratus*) and its resource management. Workshop on Pacific Inshore Fishery Resources, South Pacific Commission, Noumea, New Caledonia, 14-25 March 1988, WP 11. 9 p.

CHAPTER 6

Restoring Mangrove Systems

Gilberto Cintron-Molero
Department of Natural Resources
Commonwealth of Puerto Rico

Abstract

This review covers structural and functional ecology of mangrove systems, patterns of recovery and restoration after perturbation and the potential for artificial restoration and creation. Stand development depends on the availability of freshwater, nutrients and subsidiary energies such as tides. Stressors drain energy, reducing the amount of forest structure and ability to recover from damage but mangroves exhibit great tolerance and adaptability to environmental conditions. They possess rapid growth rates and generally have a high capacity for recovery. Management options include restoration or creation which may be required to compensate for mangrove areas lost as a result of development activities, or because it is desirable to create new mangrove habitats. Prior to a restoration/creation project, the site's potential for natural regeneration must be assessed. In high biomass, high nutrient areas it may be more cost-effective to leave the planting to nature. In low biomass, low nutrient areas plantings may be used to enhance natural recolonization. Site selection and preparation are critical. Plantings should be made using the species dominant at nearby locations with similar tidal elevations and flooding regimes. Planting sites must be protected. Uprooting by waves and erosion is often the most important factor impairing the success of planting efforts. By utilizing the characteristics of mangrove species and the natural environment, restoration efforts can be made more successful and less costly and can contribute to a greater degree to the productivity and aesthetics of the landscape.

223

Introduction

Mangroves are woody plant communities that occupy sheltered tropical and subtropical coastal estuarine environments. Although not closely related, mangrove plants have common morphological, physiological and reproductive adaptations that allow survival and development in very saline, waterlogged, reduced soils which are often poorly consolidated and subject to rapid change. They receive inputs of matter and energy from both land and sea. These inputs include freshwater, sediments and nutrients from land, and tidal flushing and saline intrusions from the sea. They act as energy subsidies that increase the performance of the system and maintenance of high rates of organic matter fixation and active ecosystem processes. The efficient allocation of organic production by mangroves allows the accumulation of a persistent structure in areas subject to active geomorphic processes and change (Thom 1967, 1984), climatic stresses like storms (Craighead and Gilbert 1962; Tabb and Jones 1962), frost (Lugo and Patterson-Zucca 1977), and hypersalinity (Cintron et al. 1978).

Definition

The word mangrove is used to describe the forest type or association, individual plants, or the complex assemblage of plants and animals associated with this ecosystem. Mangroves are limited on a global scale by low temperature and their lack of tolerance to frost. They extend latitudinally beyond the tropics to 29-30°C in both hemispheres. At those latitudes low temperatures (mean temperature for the coldest month between 12-16°C [Tomlinson 1980]) inhibit growth, and periodic frosts cause widespread mortality (Lugo and Patterson-Zucca 1977). At the latitudinal limits, forests are poorly developed and grade into salt marshes, the dominant landscape feature of temperate coasts.

Factors that Determine Degree of Development

Within the warmer portions of their range, mangrove development and cover is controlled by hydrology, physiography and climate. Mangrove coverage is most extensive on:

1. Low relief coasts. Here inland tidal penetration helps prevent the competition from non-salt tolerant species.

2. Macrotidal regimes. These force salt water inland, create suitable landforms and provide subsidies in terms of flushing, and the transport of materials and propagules.

3. Areas influenced by fluvial inputs. Rivers are geomorphic agents that create some of the substrates where mangroves develop, are sources of nutrients and freshwater inputs which dilute sea water to levels (<5-10 ppt) that promote best development.

4. Moist or wet environments. Rainfall in excess of evapotranspiration also promotes development, since runoff leaches accumulated salt. Mangroves develop best where rainfall is >2,000 mm y^{-1} and there is no pronounced dry season (Macnae 1968), but where other subsidies are available they may develop in arid environments.

5. Sheltered environments. Most mangrove species are shallow-rooted; both trees and seedlings are subject to uprooting, and the landforms over which they develop can be eroded and scoured by waves and currents. Mangroves develop best in sheltered environments or in the lee of protective structures.

6. Terrestrial sediment inputs. The availability of sediments promotes land building and establishment, and terrestrially-derived sediments are important sources of inorganic nutrients. Mangroves may grow in areas with little or no allochthonous sediment inputs, but their development is not as extensive.

The Vegetation

Tomlinson (1986) distinguishes three components of the mangrove plant community: (1) major elements, (2) minor elements and (3) mangrove associates. The major elements are species that possess specialized morphological and physiological adaptations that allow them to become established and compete successfully with other species, and, eventually, to dominate the structure of the ecosystem. Major elements have complete fidelity to the mangrove environment and do not intrude into terrestrial communities. These species are taxonomically isolated from their terrestrial relatives at the generic, subfamily or family level.

Only three species comprise the major elements of mangrove communities in Florida, Puerto Rico and the U.S. Virgin Islands. These are *Rhizophora mangle* (red mangrove), *Avicennia germinans* (black mangrove) and *Laguncularia racemosa* (white mangrove). Red mangroves generally occupy fringe or riverine environments characterized by active water flow and high flushing. The other two species tend to dominate in stagnant environments where water flows are reduced and often seasonal. The black mangrove is the most tolerant species and is dominant in areas subject, for example, to high substrate salinity or frost events.

The minor elements according to Tomlinson (1986) are those species that occupy peripheral habitats and only rarely form pure communities such as the fern, *Acrostichum* spp. Species belonging to this genus can be a normal component of the secondary succession pattern in some transitional environments or at disturbed sites, especially at the highest elevations where there are fresh water inputs.

Mangrove associates are plants usually found in transitional environments such as sandy areas (strand environments), areas influenced by large exchanges of fresh water or those subject to extremely saline conditions. The best known associates are *Conocarpus erectus* (buttonwood), *Hibiscus tiliaceus* (mahoe) and the succulents *Batis maritima* (saltwort) and *Sporobolus virginicus* (seaside rush-grass).

Only the species making up the major elements are generally used for restoration purposes, but it should be mandatory under some circumstances to use the minor elements and associates because of their affinity for unusual or restrictive sites within the mangrove environment, their contribution to landscape diversity or for aesthetic purposes.

Functional Values

As a result of the relatively high primary productivity and the active biological processes, mangrove wetlands can provide many goods and services which may be of direct or indirect public benefit both economically and to the urban and industrial environment. In Asia and South America, mangroves have been managed for their yields of lumber, fuelwood and charcoal. Mangroves are also known to be important to estuarine fisheries because they provide shelter for larval fish and crustaceans and contribute detritus and dissolved organic carbon to estuarine food webs (Heald 1969; Odum 1971; Twilley 1982). Because of their coupling to other systems, they contribute to the maintenance of landscape level structures and processes. This coupling is in the form of high quality exports such as those of migratory species of fish and shrimp to coastal shelves, support to migratory bird populations and services, and linkages between mangroves, seagrass beds and coral reefs, where these systems coexist.

Mangroves grow over landforms created and shaped by local geomorphic processes, creating a complex woody structure (trunks, branches, aerial roots, pneumatophores) that varies in degree of development and architecture in response to physiographic and climatic conditions. The heterogeneity of landforms and forest architecture gives rise to a variety of habitats that provide shelter, foraging grounds and nursery areas to many associated animals. The production of large amounts of litterfall leads to large exports of particulate and dissolved organic carbon. High levels of secondary productivity are supported by the production of phytoplankton and benthic autotrophs (seagrasses and algae), and detritus inputs from upland sources. The relative importance of each of these functional elements of the mangrove ecosystem is a function of the geomorphic, hydrologic and climatic characteristics of the area. Collectively these elements are the source of food that allows the establishment of a complex and diverse food web that may support large resident and migratory populations of mammals, reptiles, birds, fish, crustaceans, molluscs and other associated animals.

Mangrove wetlands perform other services as well. They dominate in depositional environments, and in certain settings play a role in controlling water quality and turbidity. The intricate network of roots binds the substrate, while aerial roots, trunks and pneumatophores dissipate water energy and promote accretion and reduce erosion. They contribute to bank and sediment stabilization. Mangroves are valuable for their contribution to education, research and the varied recreational opportunities they provide as well as for their scenic and landscape values. The nature and magnitude of the goods and services provided by a given mangrove wetland is a function of the region and the forcing functions which drive it.

Physiographic Types

Mangrove forests vary greatly in structural and functional characteristics (Lugo and Snedaker 1974; Pool et al. 1977; Cintron and Schaeffer-Novelli 1985). This variability is due to vegetation responses to forcing functions that vary in quality, time and intensity. At the landscape level, geomorphic processes provide an array of landform types (Hayes 1975; Galloway 1975; Thom 1984), a mold over which mangroves establish. The form and frequency of these landforms is a function of the geomorphic, hydrologic, oceanographic and climatic processes acting in the area. Thom (1984) recognized eight such major coastal landscape types colonized by mangroves. Each region can be characterized by typical landforms and processes that determine or influence mangrove growth and development patterns. Mangroves are also strongly influenced on a local scale by underlying micro-topography, as described by Thom (1967, 1984).

Lugo and Snedaker (1974) developed a classification scheme that relates forest physiognomy and functional characteristics to local hydrology and geomorphology. In each category, stands are subject to similar tidal characteristics and hydroperiods and reach similar levels of development, while sharing similar structural and functional attributes. This classification system is useful as a first approximation for rating mangrove structure and functional characteristics. The system, modified by Cintron et al. (1985), recognizes three forest types: riverine, fringe/overwash and basins. Dwarf (scrub) and hammock communities are recognized as special subtypes determined by local edaphic conditions. Riverine stands, developing under the influence of flowing fresh water, runoff, and moderate to high nutrient availability, accumulate the greatest biomass and exhibit the highest rates of litterfall production. The productivity of riverine stands is a result of more optimal growing conditions. Basin forests and fringe types, which develop under more rigorous conditions, follow in terms of structural development and productivity. Basin forests develop in relatively still water and low oxygen areas and may be affected by salt accumulations and anoxic sediments. Fringe/overwash mangroves, usually found farther away from terrigenous nutrient inputs and where wave and current energy levels are higher, accumulate less biomass and have lower rates of litterfall production.

The degree of development of a system is not a measure of its ecological value. Each forest type is coupled to the local environment, part of the landscape mosaic. It contributes to the maintenance of landscape diversity and functions.

Ecosystem Responses to Disturbance

Mangroves develop in areas subject to active geomorphic change. These disturbances include global, long term sea level changes, such as those that occurred during the Holocene Transgression. Late Pleistocene coastlines were 100-150 m lower; sea level rose at rates >100 mm y^{-1} for more than 10,000 years as the glaciers melted. This forced mangroves to migrate to their present position as sea level advanced. This rapid rate of sea level rise only began to decline during the last 3,000-5,000 years (Enos and Perkins 1979). Thus, modern mangrove stands are underlain by peat less than 5,000 years old.

Other disturbances include rapid subsidence in deltaic environments (Thom 1967), cyclic erosional episodes and change in hydrology due to long term oscillations in sea level (Wells and Coleman 1981), cyclonic disturbances (Wadsworth and Englerth 1959; Craighead and Gilbert 1962; Craighead 1964; Stoddart 1969) and tsunamis (Sachtler 1973). At their latitudinal limits, mangroves suffer massive mortalities due to frost events (Lugo and Patterson-Zucca 1977). Mangroves persist in spite of these disturbance regimes. Under some conditions they behave as perturbation-dependent systems, where the destructive effects of the disturbance are quickly countered by beneficial changes. Between disturbance events there may be declines in structure and vigor.

The resiliency shown by mangroves is an adaptive response to their occupation of rapidly changing landscapes and the selection of species with traits that allow the exploitation of resource rich but ephemeral habitats. Species with exploitive growth and reproductive strategies are selected where disturbances increase resource fluxes, since evolution produces effective resource users (Bazzaz and Sipe 1987).

Mangroves have pronounced characteristics of "pioneer species" in their reproductive biology and "mature phase" attributes in their community structure and vegetative growth (Tomlinson 1986). Some of the "r strategist" characteristics of mangroves include: (1) broad tolerances to environmental factors, (2) light-demanding species, (3) rapid growth, (4) rapid maturity, (5) continuous or almost continuous flowering and propagule production, (6) high propagule outputs in a wide range of environmental conditions, and (7) adaptations for short and long distance dispersal by an abiotic dispersal agent (tides) (Vogl 1980; Tomlinson 1986). These attributes have led many to consider mangroves as successional systems when they are in fact self-maintaining ecosystems in environments where these characteristics are required for survival (Lugo 1980).

Inertia and Resilience Properties of Mangroves

Ecosystem response to a disturbance is related to (1) energy flow through the system, (2) nutrient availability and (3) its stability and resilience properties (Holling 1973; Orians 1975; Westman 1978). In this paper we will examine the mangrove ecosystem's response to disturbances based on some of the properties as described by Orians (1975), Cairns and Dickson (1977) and discussed by Westman (1978). These properties are inertia or resistance and resiliency.

Inertia is the resistance to a disturbance. It is related to the presence of feedbacks which prevent deformation and provide a high degree of protection against environmental change. Among the factors that contribute to ecological resistance are (1) redundancy in function of component species, (2) tolerance to environmental fluctuations, (3) physical and chemical buffering capacity or flushing characteristics of the environment and (4) proximity of the system to its ecological limits (Cairns and Dickson 1977).

Mangroves may be considered to have a great resistance because of their redundancy of function. For example, all three species may be found occupying fringes under different environmental conditions. In the black mangrove there is a great difference between the physiological optimum for development and its ecological optimum. As a result this species can successfully occupy a wide range of environments, developing optimally in very low salinity environments (trees as high as 40 m) or as a shrub in highly saline environments.

Mangrove stands near the latitudinal limits of species tolerance will be more fragile than those in more benign environments. It is not as obvious that as the stand develops, the stand becomes more highly tuned to the local energy signature and, therefore, may become very sensitive (less resistant) to alterations in rates of subsidiary energy delivery or intensity. Changes beyond those that normally occur can induce massive mortalities in these areas (Jimenez et al. 1985).

Resilience, on the other hand, is related to the recovery of the system and a return to its earlier configuration after the stress is removed. Resiliency may be considered to have four components: elasticity, amplitude, hysteresis and malleability (Westman 1978).

Elasticity is the time required to restore a particular characteristic of a system to an acceptably close limit of its pre-impact level. Elasticity can be related to the slope of the biomass exponential build-up curve. It may be described by the reciprocal of the time constant of the biomass restoration curve (1/TC). The time constant, or relaxation time, is a measure of how long it takes the system to "settle down" or build up to a maximum level. Sixty-three percent of the equilibrium state biomass accumulates within one time constant following exponential restoration.[1]

The larger the value of the time constant, the slower the re-

---

[1] In this model, biomass builds up to a maximum level (B_{max}) in a fashion analogous to the accumulation of electrical charge in a capacitor. The fraction of the maximum biomass accumulated at time t, (B_t/B_{max}), is given by the equation:

$$\frac{B_t}{B_{max}} = 1 - e^{\frac{-t}{TC}}$$

sponse of the system and the smaller its elasticity. Mangrove growth may be very rapid in nutrient rich environments, such as those of south-east Asia where above ground biomass accumulations in excess of 200 tons ha^{-1} may be attained within 30 years. These data indicate that 60% of this biomass is attained within 14-15 years, closely following an exponential return to its pre-harvest state.

Amplitude is the area within which ecosystem restoration to the pre-stressed state can occur. In mangroves it may be related to the inhibition of recovery with increased frequency and intensity of stress. In this respect, the amplitude is a function of the elasticity, since a system subjected to a disturbance regime greater than the time required for total restoration (more than three time constants) will not be able to attain its maximum potential biomass accumulation. However, mangroves can persist in such an environment because they produce propagules and reproduce at an age much earlier than the time required for maximum biomass accumulation to take place. Red, black and white mangrove trees 5 years or younger are known to produce flowers (Davis 1940; Holdridge 1940). Mangroves may not return to their original state if the site's factors are altered by the disturbance.

Hysteresis is the degree to which the path of the restoration process is an exact reversal of the retrogression path. The work that is required for a system to become restored is a function of this path. In mangroves various mechanisms exist, such as the maintenance of an advance regeneration in "suspended growth" and coppicing, that speed up the restoration process by shunting normal establishment patterns.

Malleability is the capacity of the system to persist in a de-

---

where *TC* is the time constant. At a time interval equal to *TC*,

$$\frac{B_t}{B_{max}} = 1 - e^{-1} = 1 - 0.37 = 0.63$$

Thus, after a time interval equal to a time constant a value of 63% of the state is reached.

formed or altered state. It is the degree of plasticity of the system. Mangroves exhibit a high degree of phenological plasticity in growth form. They can develop as trees >30 m in benign environments or as dwarfed stemless shrubs in nutrient-limited environments. Microtopography plays an important role in influencing mangrove structure. Small changes in terrain elevation (and, therefore, in hydrology, and physical and chemical characteristics) determine to a great degree the standing biomass that will develop. This is true where flooding frequency and physico-chemical gradients develop, such as in arid areas where strong structural and floristic gradients are typical features of the landscape.

Patterns of Biomass Accumulation and Disturbance Regime

The occupation by mangroves of a newly accreted substrate leads to an accumulation of biomass at maturity which reflects the fertility of the site, predominant geomorphic and hydrologic processes, and the disturbance regime of the region. This biomass accumulation has been termed the maximum persistent biomass and may be less than the biomass that could be accumulated in a fertile and benign environment, where, ideally, the maximum potential biomass could be accumulated (Reichle et al. 1975). Because of their great malleability, biomass accumulations vary greatly in mangroves, as evidenced in the different degrees of structural development in each of the physiographic types. Biomass accumulations range from 14 tons ha^{-1} in dwarf stands to >200 tons ha^{-1} in the moist, hurricane free mangrove forests of Darien, Panama.

Figure 1, based on Ewel (1983), shows the paths of biomass accumulation in habitats subject to various degrees of natural stress. The maximum amount of persistent biomass decreases greatly along the gradient of increasing environmental stress as a result of the amplitude and malleability of the mangrove system. The amount of biomass is shown increasing hyperbolically as environmental quality increases. It takes a greater increment in environmental quality to produce a given increase in structure at "maturity" in a benevolent environment than it does in a harsh one. The recovery of structure is shown as a logistic growth curve. Considerable delays are shown in initial establishment in the very harsh environments; this is a result of low propagule recruitment and high mortality as well as slow growth. System recovery can follow three kinds of paths: (1) recovery by retracing the retrogression path and return to a condition much like the pre-disturbance state; (2) survival in a deformed state where the disturbance reduces the suitability of the site; or (3) increase in biomass (compared to the pre-disturbance state) if the environment has been made more suitable. Within a mangrove-dominated landscape all three processes may be occurring simultaneously, as landforms are created, modified or destroyed by natural processes.

The impact of a disturbance is scale dependent. A disturbance creating a gap is normal, since gaps are part of the landscape mosaic. The mosaic remains at steady state whereas the site is undergoing a recovery process. As a result, it is convenient to recognize two broad disturbance types based on the amount of overstory removed (Oliver and Larson 1990): (1) major or "stand-replacing" disturbances (those that remove or kill all the trees) and (2) minor disturbances (those that leave some of the predisturbance trees standing). The structure of forest stands is a function of the scale and periodicity of disturbances.

Sensitivity and Vulnerability to Disturbance

The intensity of natural disturbances is inversely related to their recurrence rate; the magnitude of storms or intensity of frost events as well as floods, increases as the recurrence rate of these events decrease. At the same time the sensitivity (decreased resistance) of mangrove stands increases with structural complexity. The larger, taller trees and higher proportion of standing dead biomass become increasingly vulnerable to windthrow. The more developed forests

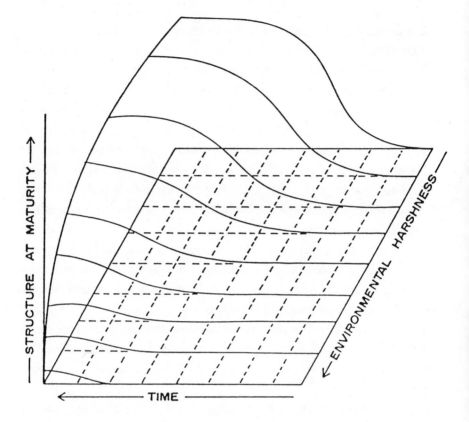

Figure 1. Paths of biomass accumulation in habitats subject to various degrees of environmental harshness. The maximum amount of potential biomass decreases along the gradient of environmental stress. It takes a greater increment in environmental quality to produce a given increase in structure at maturity in a benevolent environment than it does in a harsh one. Based on Ewel (1983).

are extremely sensitive to minor modifications of their hydrologic regime (Blasco 1983). Large-scale changes may be triggered as the susceptibility of the system converges with a disturbance event greater than the threshold magnitude necessary to disturb it (Jimenez et al. 1985; Oliver and Larson 1990). At these times, widespread mass mortalities may occur, since members of single cohort stands are equally susceptible (Wadsworth 1959; Jimenez et al. 1985).

Synchronous death and rapid reoccupation of available substrates leads to the establishment of even-aged stands. The age range of the cohort is limited by the brief interval that the growing space is available for colonization, due to rapid recruitment and competition. Later in the development of the forest, minor disturbances can lead to the formation of multicohort stands as new trees regenerate gaps. If there is a large time interval between disturbances, gaps may be created by natural death, windthrows, or lightning. These stands are characterized by inverted-J diameter distributions, where smaller diameter trees occur as random clumps whose size is closely related to the size of existing or former light gaps (Jimenez 1988b). Truly successional processes are not common in high disturbance environments; rather, differential establishment dominates and species zonation is not a reflection of succession but rather of the differential establishment on substrates of different microtopography, substrate type and hydrologic regime.

Restoration Mechanisms

To persist within a given disturbance regime, component species must complete their life cycles and have the capacity to repair and restore function within the disturbance recurrence interval. Mangroves have different mechanisms to deal with disturbance. These mechanisms reduce the time required for the restoration of vegetative cover and reproductive capacity and allow maximal recovery in the shortest time. This elasticity provides great resiliency. Vegetative recovery mechanisms are best developed in black and white mangroves, species that are dominant in the more rigorous basin environments. The red mangrove, dominant at the fringes, is sensitive to many of the same disturbances but because of its normal habitat it is less vulnerable to many of those stressors. It is more vulnerable to major types of disturbances, and its regeneration mechanisms are adapted to such events.

Recovery from disturbances follows two different reproductive strategies: (1) the maintenance of a large pool of tidally dispersed propagules of two types, small easily transported propagules which are dispersed throughout the area, reaching inland where tidal energies are weak, and large propagules which may become established where wave or currents are stronger; and (2) maintenance of an "advance regeneration," composed of saplings and young individuals that grow very slowly but may be released after a gap develops. In most cases this advance regeneration is of the same species as the canopy dominants. This is in contrast to many "pioneer" species of upland forests, which are unable to develop under their own shade. Although this "advance regeneration" may suffer from selective mortality due to competition (Jimenez et al. 1985), continuous recruitment substitutes for those losses. Regeneration by red mangroves is helped by the prolonged viability of its propagules during the dispersal phase.

Basin forests, dominated by black and white mangroves, are vulnerable to a greater variety of small and medium scale disturbances (drought, hypersalinity, windthrow). They cope with disturbances through their greater tolerance and tighter coupling with environmental fluctuations, such as propagule production peaks associated with optimal establishment conditions (Jimenez 1988a). They also have a great capacity for recovery from total defoliation and loss of branches, or even main trunks, through adventitious sprouts, coppicing and layering (production of roots from leaning or fallen stems; Snedaker et al. 1981).

These modes of vegetative regeneration may be dominant in stressed basins, bypassing reseeding, which relies on the establishment of propagules and may be less reliable in a rigorous environment due to mortality, predation, and damage. Ewel (1977) reported that in dry upland sites most of the regeneration after a disturbance originated as coppice from stumps and underground roots and stems. He suggested that, since seed germination and seedling establishment are critical stages in the life cycle of most woody plants, these stages are bypassed by plants found in harsh, unpredictable environments. The ability to coppice is a characteristic of black and white mangroves, which are the dominant species in basin forests. Because of these adaptive mechanisms, the restoration path may be different from the original establishment path and may occur rapidly if the disturbance is not prolonged or leaves no severe residual effects. Wadsworth (1959) reported the rapid regeneration by coppicing in a white mangrove stand in Puerto Rico. Within three years of clear-cutting, 60% of the white mangrove stems were of sprout origin. Similarly, Detweiler et al. (1976) reported that much of the regeneration of a disturbed site in Tampa Bay was due to the regeneration of existing trees.

In areas where seedling establishment is difficult (such as where there is too much turbulence), red mangroves have the ability to reorient their branches readily (Tomlinson 1986). Red mangrove fringes may contain recumbent branches which extend into waters too deep for seedling establishment (Kolehmainen and Hildner 1975).

I have summarized adaptive strategies that allow mangroves to survive and regenerate rapidly after disturbances. The restoration of mangrove environments should take advantage of these characteristics. The most efficient and rapid restoration effort will be that which takes advantage of these natural recovery processes.

Major Causes of Habitat Loss and Alterations in the Region

Historically, the major causes of mangrove wetland losses in the region may be attributed to: (1) development and changes in land use; (2) changes in hydrology and water quality; and (3) marine transportation activities (ports, navigation channels and marinas). Between 1981 and 1985, 193 ha of wetlands (including non-mangrove wetlands) were involved in NOAA/NMFS conservation efforts in Puerto Rico and the U.S. Virgin Islands. Activities were allowed in 114 ha and 295 ha were requested for restoration/creation activities (Mager and Thayer 1986). In Puerto Rico, 80% of the permit request area was for impoundment and fill activities, whereas in the U.S. Virgin Islands dredging was the most common proposed activity.

Development and Changes in Land Use

Urban and industrial development in the coastal zone is one of the primary factors leading to mangrove losses. Mangrove lands have been reclaimed for residential developments, industrial parks, and transportation infrastructure (roads, causeways, bridges). Reclamation for agriculture is no longer a cause of wetland losses, but the potential demand for conversion of mangrove lands to aquaculture exists. In the past, significant wetland losses came from the conversion of mangrove lands to solid waste dumps. Perhaps this reflected the common perception of mangrove lands as so worthless or noxious that reclamation, even by garbage, was considered preferable to leaving them in their original state.

The conversion of mangrove wetlands for urban and industrial development causes significant losses of fish and wildlife habitat and exposes the converted substrate to natural hazards such as storm surge flooding, erosion, subsidence, and earthquake damage. Structures built near mangrove areas often need flood or erosion control structures (seawalls, revetments, bulkheading, channelization) which often result in further habitat alterations.

Reclamation for agriculture was a leading cause of mangrove losses in the past. In Puerto Rico, more than 4,166 ha of fresh water and mangrove wetlands were drained for conversion to agriculture by the early 1950s (Koenig 1953); almost all of these former swamplands were later abandoned because of low productivity. Reclamation programs require structures for flood control, drainage, water table and salinity management. These are costly to maintain and agricultural yields are often poor. The acidification that follows drainage and exposure of some types of mangrove substrates makes the economic feasibility of agriculture in these areas questionable (Moorman and Pons 1975). Productive agricultural lands reclaimed from mangrove areas are rare (Hamilton and Snedaker 1984).

The use of mangrove lands for solid waste disposal was common in the past, and some are still used for that purpose. Waste deposits leach potentially hazardous chemicals, as well as viral and bacterial agents. Although mangrove sediments are reduced and may serve as sinks for metals, some may be incorporated into plant tissues and the food web. Filter feeders which are common in mangrove areas, and which may have considerable commercial value, are particularly vulnerable to becoming vectors of serious diseases. "Open" dumps may eventually become recolonized by mangroves as the solids disintegrate and sink into the soft substrate, but sanitary landfills cause irreversible damage and loss of mangrove areas. Leaching of hazardous and toxic substances from these dumps may continue long after the site is closed and abandoned.

Changes in Hydrology and Water Quality

Changes in hydrology cause irreversible damage to or degradation of mangrove areas. Channelization and water diversion schemes are particularly damaging. Diking and impoundment also quickly lead to degradation because they deprive mangroves of nutrient sources and external subsidies. Impoundment can raise water levels and cause mass mortality in the impounded area due to overtopping of gas exchange organs. In dry areas, impounding can reduce water flows that are necessary to dilute accumulated salts. Impoundment can be caused by poorly planned roads (Patterson-Zucca 1978) or as a result of well intentioned but poorly executed habitat enhancement actions. On the other hand, there are examples that show how awareness, good planning, and design can minimize impact. One such example is the dredging and bulkheading of a mangrove channel in metropolitan San Juan for navigation purposes. A porous bulkhead allows free water movement into the large mangrove area lining the channel, while retaining the sediments. No adverse effects on the mangroves have been observed as a result of this specific construction activity.

Pollution of poorly flushed water bodies (such as coastal lagoons with low turnover rates) is also a serious problem. Mangrove trees are not always harmed by the indirect discharge of sewage effluents (Clough et al. 1983), but their associated faunal communities are devastated, and receiving water bodies become eutrophic, with frequent massive fish kills due to de-oxygenation. Health hazards destroy the value of these areas for recreation or fisheries.

Marine Transportation Activities

In sheltered environments mangroves occupy areas which may be in demand for port development. Dredging of navigation channels can kill trees or degrade water quality. In the past mangrove areas were used as dredge spoil disposal sites, to reduce dredging costs and reclaim land. The expansion of port areas to include marshalling yards for containerized cargo, warehouses, docks and piers has caused significant mangrove losses in industrial areas. Losses of this type may be reduced by careful planning (avoidance). Unavoidable losses would require mitigation through replacement or other measures.

In recent times the rapid development of water sports has increased the demand for marina sites. Small boat marinas are often built in mangrove-lined, sheltered bays or within mangrove forests. Both Puerto Rico and the U.S. Virgin Islands have lost mangrove areas to marinas. Again, judicious planning and proper siting can eliminate mangrove losses and could incorporate the mangroves in the marina design.

Restoration and Creation

Restoration is defined by Webster as "a putting or bringing back into a former, normal, or unimpaired state or condition." Restoration as a policy presumes to take the system back to its initial state, presumably even accepting undesirable features of that state (Magnuson et al. 1980). A practical definition of restoration is given by Morrison (1990): Restoration is the reintroduction and re-establishment of community-like groupings of native species to sites which can reasonably be expected to sustain them, with the resultant vegetation demonstrating aesthetic and dynamic characteristics of the natural communities on which they are based.

In terms of the National Environmental Policy Act (NEPA) restoration is an element of mitigation. In the context of the NEPA regulations, mitigation includes: (1) avoidance of direct and indirect impact or losses, (2) minimization of the extent or magnitude of the action, (3) rectification or repair, rehabilitation or restoration of integrity and function, (4) reduction or elimination of impact by preservation and maintenance and (5) compensation by replacement (construction of a new wetland in another location) or by providing a substitute resource or environment.

Restoration Goals and Objectives

The goals and objectives of restoration may be varied. The primary goal of federal wetlands restoration is the reestablishment of habitats and functions unavoidably lost through the issuance of permits by the U.S. Army Corps of Engineers (USACOE). In this respect, the goal of the restoration/creation effort is the conservation of landscape diversity and functions and compensation for ecosystem services that have been or would otherwise be lost. The goals of such an effort must be:

- Replace the system that is being altered or destroyed by one of an equal type in terms of structural and functional characteristics.
- The surface area should generally be equal to or greater than the area to be developed.
- Landforms and configuration should be similar to those of the area to be developed.
- The restored/created unit must be in the same landscape unit as the area to be developed.
- The restored/created unit must managed for multiple uses and protected for perpetuity.

Recently there has been increasing concern about the use of "compensatory mitigation" to reduce wetland losses. This concern arises out of doubts about the ability of present day technology and know-how to create wetlands successfully, and about whether these created wetlands are identical to naturally occurring ones. Wetland managers must be cautious about accepting creation as a mitigation strategy in exchange for a permit to degrade or destroy an existing natural system (U.S. OTA 1984; Kusler and Kentula 1990). This is especially true of the complex, diverse and site-specific mangrove and freshwater forested wetland systems, where there is still insufficient knowledge about most basic ecosystem properties and processes to assure that they can be replicated successfully.

It has been said that mangrove wetlands are easy to create. There is little doubt that the most easily managed systems are those that are the most resilient (Holling and Clark 1975). Mistakes or misinterpretations of available information, leading to inappropriate restoration actions, may be absorbed by the resiliency of the system; but the great malleability of mangrove wetlands does not guarantee that the "restored" system will develop the structure and function of the original system, or will perform its natural functions. This is especially true where there is no natural model on which to base the "recreated" mangrove stand, as when new mangrove wetlands are created in areas which did not contain mangroves originally.

Other restoration/creation objectives may be related to specific uses and may be less demanding in terms of the quality of the created system. Some examples of these goals are: (1) landscape enhancement, (2) enhancement of recreational values and habitats for wildlife, including endangered species, (3) enhancement of water quality, and/or (4) provision of protection against erosion.

Restoration provides an opportunity to improve or enhance the landscape and increase environmental quality. It is particularly valuable in densely populated areas or in areas of industrial blight where plantings can enhance the environment even if no other tangible benefit may be obtained from such action. In these cases, the goal of the restoration may be primarily aesthetic, but many collateral benefits accrue from these actions. The basic premise is that there are many opportunities in human-dominated landscapes where the skillful use of wetland restoration or creation techniques can help preserve, enhance or maintain the integrity of the indigenous landscape. One such possibility is the provision of created mangrove corridors between natural stands that could provide for conduits between isolated stands and provide pleasant views or opportunities for recreation. Restored or created plant cover may be simultaneously functional, economical, ecologically suitable and aesthetically appealing (Morrison 1990).

Historic Restoration/Creation

In Puerto Rico, concern for the protection and restoration of mangrove areas dates from last century, when the Spanish Crown granted "concessions" or long-term leases of some mangrove stands for wood, charcoal and salt extraction, while retaining title. In the early part of this century, the government became concerned about protection of mangroves because of their overexploitation for firewood and charcoal production. In 1918, the Governor of Puerto

Rico incorporated all former Crown mangroves into the Insular Forest System. By the 1930s the U.S. Forest Service's Institute of Tropical Forestry had already undertaken silvicultural research on natural and artificial regeneration patterns of mangrove stands (Swabey 1939). Holdridge (1940) described the results of planting red mangrove propagules and black and white mangrove seedlings in the San Juan area. Mangroves were planted along canals where there were clear patches and natural regeneration was absent. Establishment and survival was >90% at those sites. Holdridge (1940) also described sapling transplants, and reported excellent growth and survival of white mangroves after two years.

Bowman (1917) described the Florida East Coast Railway's plantings of red mangrove along its railway line to the Keys, to protect the embankments and road bed from erosion. Davis (1940) reported planting more than 4,000 red mangrove seedlings in the Keys to determine if mangroves could be established in an exposed location. Although more than 80% survived one year after planting, a check of the area by Teas (1977) 32 years later found no survivors.

In the New World, interest in mangroves and their restoration dwindled until the late 1960s. Savage (1972) reported the results of plantings of red mangroves in Florida. Since that time a fairly extensive literature pertaining to mangrove and coastal community restoration has appeared (Teas 1977, 1981; Lewis 1982, 1990a, 1990b; Kusler and Kentula 1990).

A review of the available literature shows disturbingly mixed success of restoration efforts, even though many proponents of restoration still claim that these wetlands can be restored or created easily.

Factors Affecting Success. Failures of artificial plantings are related to three basic factors: (1) failure to recognize factors limiting establishment (need for shelter from wave and wind action, tides and currents), (2) lack of provision for proper hydrologic regime, and (3) failure to provide follow-up, including replacement for mortality and lack of consideration of stand maintenance.

Artificial planting efforts associated with permit actions are often hampered by their "one-shot" approach. Under these circumstances, conditions for establishment must be made extraordinarily optimal. When nature does the planting, the propagules disseminate over broad areas and only those that reach the proper locations develop into mature trees. Healthy mangrove stands produce massive quantities of propagules over space and through time, so that opportunistic plantings can take place during unusual or brief periods when conditions favor establishment (Jimenez 1988b). Propagule production and release may have adaptive significance related to establishment requirements. Nature's propagule pool is available to exploit favorable periods, and provides for a continual replacement of lost propagules. Artificial plantings may lack these properties if not properly designed.

The factors that hinder restoration efforts are those that affect natural colonization; therefore, naturally blank areas should be examined carefully before planting to ascertain the reasons for lack of natural establishment. Planting in these areas is often wasted effort, since the same forces that prevented or eliminated natural establishment can remove artificial plantings.

It is possible to identify ten broad types of problems or factors that control establishment and development. These are:

- Wrong planting elevation for desired species
- Excessive wave or current exposure
- Unsuitable substrate
- High salinity
- Excessive substrate temperature
- Damage by flotsam and wrack accumulations
- Disease, isopod infestations
- Grazing, trampling, vandalism

- Undesirable provenance
- Isolation from natural propagule sources

Wrong Planting Elevation. Creating the correct elevation and grade is one of the most critical aspects of any restoration project. Incorrect elevation is one of the most common reasons for poor performance or total failure of plantings, since a small change in elevation may represent a major difference in flooding frequency and duration. This can be caused by improper grading to a suitable elevation (poor project execution) or erroneous selection of elevation (wrong technical specifications), or both. In cases where the planting is done on spoil material, consideration must be given to consolidation of the material or foundation settlement so that the ultimate elevation does not fall below or remain above the elevation requirements. Kinch (1976) reported the failure of plantings due to settling of dredge spoil islands.

Planting failures can occur as a result of the transfer of elevation data from different geographic regions with different tidal ranges and characteristics (tide form) and degrees of non-tidal residuals (water levels unrelated to tide generating forces). Thus, although many planting elevations are prescribed in the literature for particular species (Teas et al. 1975; Teas 1981; Stephen 1984; Beever 1986; references in Lewis 1990a, b), the relationship of a vegetation unit to a tidal datum differs from one locality to another. This is why Lewis (1990a) states that good experimental evidence, which establishes the best zone of planting, is rare.

Mangroves are tidally dispersed. As a result propagules may be carried as far inland as the highest tides. Ideally, if a plane is projected from this point seaward, red mangroves become established on substrate elevations as far down as 30-60 cm below this line, but usually above the mean sea level datum. Seedlings do not normally tolerate complete flooding, and only very seldom occur where more than 30 cm of water covers their tops at high tide (Banus and

Kolehmainen 1976). Black mangroves, on the other hand, become established where the depth of tidal flooding is less than 5 cm (McMillan 1971). In reality, the situation is not so simple. Seaward mangrove rooting is constrained by turbulence, and inland colonization influenced by competition from mangrove associates or hypersaline conditions. Mangroves respond to very slight elevational differences (Thom 1967) but plantings may survive in less than ideal topographic elevations. However, they may grow too slowly and eventually lose in competition with other species (Ball 1980). In some instances, the plantings may eventually die and herbaceous vegetation predominate.

Lewis (1990a) suggests trial plantings before specifying the desired restoration/creation project elevation. Whereas this may be desirable, it can be impractical in terms of project development time and costs and may have limited value. A pilot study may contribute information about recruitment potential at the site, but because of the low mortality of red mangrove seedlings during the first year, even in marginal environments, the success of the plants in a pilot study of such short duration would have little meaning. For this reason it is best to specify project elevations similar to those at well-developed stands of the desired species near the selected site.

Lewis (1990a) also suggests that a slope be provided but does not specify a figure. Slopes are critical since very small gradients determine establishment and stand performance. Transverse gradients of mangrove areas are very small: in the order of 1:100 (steep gradients) to 1:1000 (low gradients). Gradients reflect the geomorphology of the site and tidal characteristics.

Excessive Wave or Current Exposure. Excessive exposure to high wave energy is another major cause of planting losses (Teas 1977). It arises from efforts to plant mangroves where there are none, without providing suitable compensating subsidies such as shelter. Empirical methods are available to assess fetch (USACOE

1980) and more formal methods appear in the Shore Protection Manual (USACOE 1977) for wave regime hindcasting, but the literature does not report on the application of these techniques to mangrove plantings. Where wind-waves or current may be a problem, the planting site should be relocated to reduce fetch and exposure to the prevailing wind and waves or behind the shelter of shoals or islands. Exposed locations will require the provision of protective structures. These should not impair circulation or cause stagnation. If improper consideration is given to wave and wind action, the planting effort is likely to be futile.

Erosion by boat wakes has been reported to inhibit seedling establishment and can cause the loss of plantings on channel margins.

Unsuitable Substrate. Mangroves develop best on silts and clays, usually overlain by veneers of organically-rich material. The quality of the soil depends on the source of the alluvium (Macnae 1968). Macnae (1968), citing Schuster (1952), states that some biological processes are needed before fresh alluvium can be utilized by mangroves. These processes may include the activities of bacteria, blue-green algae, diatoms and green algae and the activities of nitrifying bacteria and sulphate reducing bacteria. During this "aging" process, a continuous accumulation of organic matter enriches the substrate and increases its suitability for establishment.

Seedlings planted in sandy substrates or extremely compact muds derived from fill materials show poor establishment and growth. The substrate should be loose enough to facilitate root penetration. This is especially true at higher wetland elevations, where the substrate cannot be rapidly modified by the tidal deposition of organic layers. Non-mangrove substrates may need a time period to become suitable for establishment. Mangroves planted on poor substrates are usually stunted and grow very slowly.

High Salinity. Salt accumulations stunt growth. Red mangrove seedlings less than 50 cm tall can tolerate interstitial salinities in excess of 80 ppt; however, at this size they may still be surviving on stored reserves, because salinity tolerance decreases dramatically with further development. McMillan (1971) reported up to 80% rooting in black mangrove propagules at 65 ppt interstitial salinity; at 75 ppt, only 10% of the propagules rooted. Growth at high salinities is poor and trees are stunted; high mortality leads to scattered distribution. Survival is limited to those areas where the microtopography allows an escape from the high interstitial salinities. High salinities are the result of poor drainage and failure to provide for proper flushing, especially in environments where strong water deficits develop. The upper limit to mangrove establishment and survival appears to be near 90 ppt (Cintron et al. 1978).

Excessive Temperatures. Excessive temperatures may be important constraints to successful establishment. Craighead (1964) reported that areas where the temperature may exceed 43°C remain barren. Temperatures over 37°C inhibited rooting of black mangrove and exposures to temperatures of 39-43°C for 48 hours caused the death of rooted but stemless seedlings (McMillan 1971). Red mangroves are vulnerable to high temperatures. Barth and Lieth (1982) reported that 50% of seedlings exposed to 24-42°C outflow from a desalinization plant (salinity 40-60 ppt) survived after two years, but all plants died at temperatures greater than 45-48°C.

High temperatures become a problem in poorly-flushed areas or where the substrate elevation is so high that it dries and loses its capacity for evaporative cooling. The dark color of mangrove substrates is conducive to high heat absorption. In highly saline areas, very high temperatures may occur, because of the decreased specific heat of brines.

Excessive Damage by Flotsam, Wrack Accumulations. Floating debris will damage or destroy plantings by its movement when flooding occurs. Seagrass wrack accumulations can impair circulation and cause hypersaline conditions and severe damage to stands in dry areas. Debris accumulations occur in exposed locations and should be avoided when selecting planting sites. Debris deposits caused by unusual storm events must be removed to avoid damage to stands or plantings.

Disease, Isopod Infestations. The diseases affecting mangroves have received little attention. Olexa and Freeman (1975) reported three fungal diseases of mangroves in Florida. Two of those were pathogenic on black mangroves and one on red mangroves, where it is thought to cause galls. This disease appears to preferentially affect oldest trees (Jimenez et al. 1985) and has not been reported on mangrove seedlings or saplings. Infestations of the aerial roots of the red mangrove by the isopod *Sphaeroma* spp. have been reported (Rehm and Humm 1973). *Sphaeroma* is known to attack black and white mangroves (Rehm 1976). There is some controversy about the significance of faunal predators on aerial roots of the red mangrove since these roots branch opportunistically and *Sphaeroma* attacks may induce branching which is considered beneficial (Simberloff et al. 1978). Hannan (1976) reported the loss of all plantings low in the tidal zone due to *Sphaeroma*. None of the plantings at high tide level were damaged.

Grazing, Trampling, Vandalism. Severe herbivory is often visible on young seedlings developing under the canopy where they may be weakened by starvation (food storages depleted) or subject to physiological stresses (low light). Red mangrove propagules may suffer from severe grazing by crabs and other animals. Trampling (by animals such as cattle) may be a problem in some areas. Teas

(1977) reports damages due to vandalism. This problem is more severe close to urban areas.

Undesirable Provenance. Propagules may be obtained from nearby donor sites or from commercial nurseries. If commercial sources are used, the origin of the stock should be considered. Propagules from different latitudes have varying tolerance to chilling temperatures (McMillan 1975; Markley et al. 1982). Other differences may exist between individuals of the same species growing in different geographic areas in terms of growth form and resistance to drought and disease. Ideally, plant material should come from nearby sources (Lewis 1990a).

There appears to be a relationship between the size of the seedling and the trees which produce them. Larger trees produce larger seedlings (Banus and Kolehmainen 1976), so that the largest and best shaped propagules should be collected from trees of good form and development. These propagules should yield the best performance.

Isolation from Natural Propagule Sources. Upon establishment, a planting will loose individuals due to mortality from various causes: uprooting, predation and loss of viability. In nature, these losses are compensated for by constant recruitment. Planting sites should be selected such that they receive the greatest natural propagule subsidies and require the least intervention by man. This reduces maintenance costs and damage by trampling.

Considerations for Restoration Design

Natural systems are self perpetuating (persistent), autonomous, self repairing and highly reliable. These are properties that should be designed into a restoration or creation project. The restoration effort: (1) must be technically feasible; (2) must be self organizing, self repairing, self perpetuating; (3) must have no negative impact on nearby natural systems; (4) must be compatible with local natural processes; and (5) efforts should not be redundant with natural processes. The most efficient, economical and rapid restoration efforts are those that supplement rather than duplicate natural recovery processes (Detweiler et al. 1976).

Ecosystem Resilience and Natural Restoration. Given the high resiliency of mangroves, designers of creation/restoration projects should assess a site's potential for natural regeneration (proximity to propagule sources). In high biomass, high nutrient areas and for large scale projects it may be desirable for economic reasons to leave the planting to nature. Actual plantings may be used where the natural availability of propagules is reduced or where it is desirable to speed the rate of natural recolonization. In all circumstances, it is desirable to provide for natural recruitment to compensate for losses and mortality. Restoration efforts at these sites can be limited to preparing the terrain. Undoubtedly the most critical design factor is providing a proper elevation and unrestricted coupling to the local hydrology. Once this is done, nature's self-repairing properties will reduce construction and maintenance costs.

Current State of the Art

Wetland restoration is an art. There is still insufficient information about the most fundamental aspects of tropical wetland hydrology, soil chemistry and plant physiology. In addition, little is known about the dynamics of reproduction, dispersal and establishment in mangroves under all conditions. An understanding of these processes requires knowledge of the couplings and interactions between physiological processes and the environment. Poor knowledge of mangrove dynamics is partially compensated for by the resistance and resilience of the system, which simplify its management as long as desirable natural processes are allowed to operate unhampered.

Viability of Restoring Natural Functions. The fact that creation and restoration projects are often intended to substitute for natural systems requires that the created or artificial systems reach a functional equivalency to the replaced system. The poor performance of most restoration projects examined indicates that great caution should be exercised. It is clear that the state-of-the-art restoration/creation has not reached the maturity required to become a primary tool for compensatory mitigation.

Other Biologically Acceptable Goals. Although careful planning and execution can increase the success rate of restoration efforts, even within its limitations, restoration remains a powerful technique for less critical applications where the goal is to replace systems which had been destroyed or degraded and to provide for limited specific uses such as aesthetic or amenity purposes. When properly designed, creation and restoration efforts can contribute significantly to the replacement of lost or degraded landscape functions and can have a high degree of social acceptability.

Natural Restoration as an Alternative Strategy. The primary consideration in the design of a restoration effort should be the efficient use of natural subsidies. Natural regeneration is feasible where propagule sources are near. It is economical and simple. It is especially desirable in large projects which could otherwise be limited by available funding. During the 1950s all of the northern shore of Martin Pena channel, a tidal creek in the heart of metropolitan San Juan, Puerto Rico, was occupied by a large shantytown. In the 1960s an urban renewal program began the removal of the slum. By 1980 the last remaining houses had been removed. Mangrove colonization of the shoreline was rapid and now this area is covered by a vigorous growth of white and black mangroves. Government action was limited to the removal of the houses; nature took care of planting more than 30 ha of mangroves, which today are one of the most successful examples of mangrove restoration in Puerto Rico.

Developing a Restoration Strategy

Preconstruction Considerations. The primary criterion for a restoration project is the possibility of connecting the area to the local hydrology and providing adequate shelter. Other pertinent considerations are proximity to potential sources of plant propagules, animal colonizers and potential linkages to natural systems. The impact of the project on nearby systems must be assessed. Another factor relates to engineering: accessibility for equipment is a critical consideration since excavating and surface shaping equipment, if needed, has trafficability and reach limitations.

Ownership must be transferred to the state or a non-government bona fide conservation group, or a conservation easement restricting the use of the property be recorded on the Property Title. Unfortunately, this mechanism is inconsistently applied in Puerto Rico and it is not uncommon for "mitigation" sites to be encroached upon by the same project in a later development phase, especially when the mitigation planting fails and the land remains bare.

A field investigation is required to determine the characteristics and desirability of candidate sites. From this assessment alternate sites can be chosen and costs and benefits may be compared.

Although the possibility of "transplanting" animal populations has been mentioned in the literature, this appears impractical, inefficient and probably would be a trivial contribution since a site with unrestricted hydraulic/tidal access should become quickly invaded by larval dispersal and immigration from nearby source areas. A restoration site that is well coupled to adjacent waters will be colonized according to natural rates of immigration, developing by internal sorting processes its own co-adapted set of species. Since the site is initially undersaturated with species, the imbalance between immigrations and losses would lead to rapid restoration. Further, the number of species and population density will increase at a rate proportional to the habitat's increase in complexity and productivity. It may be inappropriate to introduce animal species until a habitat suitable for them has time to develop.

Steps to Ensure Success

Current Technology. The restoration or creation of mangrove systems requires that detailed attention be given to site conditions. Wave and wind action, tides and currents, salinity, and substrate types are factors that may limit establishment even if other conditions are suitable. The art of restoration is in the ability of the restorer to recognize, assess and provide for natural processes such as those that take place in adjacent mangrove stands. The success of a restoration/creation project depends on the accuracy with which natural forcing functions are provided.

Project Timing. Site preparation should be finished by the time that propagule availability is peaking, usually late summer. If these schedules do not match, then it may be desirable to delay planting until good quality propagules become available. If the project is to rely on natural restoration it is best to allow normal recruitment to take place before planting (one to two years). In such cases, planting may not be necessary or can be limited to the blank areas where regeneration has been unsatisfactory. Often when sites are planted, natural recruitment eventually dominates, and may out compete the planted species.

Site Preparation. The success of a restoration project depends on the accuracy of the substrate elevation. It is important that an "as built" survey be made before the construction equipment is removed from the site (Lewis 1990a). It is desirable that the survey be tied to the level of adjacent, well developed stands of the desired species, since the mean tide level may differ from the land datum by different amounts in different locations due to tidal gradients. Very small differences in elevation can cause great changes in the outcome of the project in terms of recruitment or success of the desired species. Verification of the accuracy of the grading may be made by monitoring tidal flooding of the area.

Planting Materials, Sources and Considerations for their Use. Red mangrove is the most common species used for restoration projects because it is the easiest and most economical species to obtain. It characteristically exhibits a high viability, high establishment success and low mortality. It is readily available from natural stands or from commercial sources. Propagules may be collected and stored or shipped for use for more than 20 days in moist containers without damage or significant loss of viability. The freshly collected hypocotyl is the unit of choice for planting.

Seedlings may be allowed to develop primary roots before planting, but there appears to be no significant advantage to allowing initial growth of the propagule at the nursery. The only advantage cited for such an approach is that the seedling is allowed to develop a healthy root system before implantation. However, this adds costs and increases the difficulty of transportation and planting, offsetting any advantage.

According to Goforth and Thomas (1980), there is no advantage in planting seedlings 12-18 months old over propagules. After 23 months, survival of the transplanted seedlings was no greater than that of the planted propagules.

During the planning phase of the project, efforts should be made to find propagule sources from nearby, readily accessible high quality stands. Depending on the size of the project, various locations should be identified. Propagule collection must be closely supervised. It is preferable to collect propagules from trees, avoiding low stunted trees that may have undesirable genetic qualities (these are unfortunately the most easily reached). Only mature, easily detached (abscission layer well developed) propagules should be used.

If fallen propagules are collected, then they should satisfy the following criteria (based on Goforth and Thomas 1980):

• Presence of undamaged green terminal growth bud
• Absence of evidence of dehydration
• Absence of developing roots and leaves
• Absence of isopod or insect damage
• Have the general characteristics of a recently released propagule
• Avoid malformed or damaged propagules

Seeding black and white mangroves has not been considered practical in the United States. Although these propagules may be broadcast by hand, they have an absolute requirement for a stranding period previous to rooting of about seven days for black mangrove and five days for white mangrove (Rabinowitz 1978). Thus, to allow establishment, they must be broadcast when the area is not expected to be flooded by the tides for at least a week. It is expected that a much lower percentage of the propagules will become established. Lewis and Haines (1981) report low success using this method. Because of the low rate of establishment sowing should be generous and it may be desirable to do it repeatedly.

Greater success has been obtained by planting black mangrove seedlings grown in nurseries to a 10-15 cm size. This material is easily grown in small plastic bags; it can be easily transported and planted. Potted seedlings have good survival in areas where the propagules may be disturbed by flooding during the establishment period. Large scale plantings using this technique are done successfully in the United Arab Emirates.

Goforth and Thomas (1980) reported that transplanted small red mangrove trees (2-3 years old, 0.4-0.8 m tall) had the best survival (98%) after a 23 month monitoring period. Leaf and prop root development was greatest at exposed sites using this material. Root balls were 25-50% of the tree height. There are reports (Pulver 1976) of the successful replanting of small trees (5 or more years old, 0.5-1.5 m tall). Planting of small trees of this size may be desirable where seedlings may be uprooted and washed away. According to Pulver (1976) their large size and more extensive root system offers a greater chance and faster shore protection than can be achieved using seedlings.

This technique offers another advantage. Since mangroves produce propagules from an early age, these small trees may be used to assure propagule availability at the site. Pulver (1976) provides guidelines for replanting small trees.

Planting Techniques. As described earlier, the red mangrove is the easiest, most economical species for propagation. Propagules are inserted into the substrate deep enough so that they will not fall over (Watson 1928). For a 20-30 cm seedling, planting depth is about 4-7 cm. Spacing is selected to provide a rapid coverage. Planting in parallel rows at 1.0 m spacing appears to provide the advantage of economy and rapid full cover. However, if the cost of planting is a limiting factor there is no reason why wider spacing may not be used. Closer spacings are desirable to control erosion or compensate for scouring losses (0.5 m vs. 1.0 m). The number of propagules required is the reciprocal of the square of the spacing times the planting area; therefore, small increases in spacing reduce the requirements for propagules substantially.

Because of poor establishment of broadcast propagules (Lewis and Haines 1981), it may be preferable to plant black and white mangroves as potted seedlings or saplings, where natural recruitment is deficient. Propagules may be established on potting media in a nursery under controlled conditions and transferred to the field when they reach 10-15 cm. The size of the container must be large enough to allow good root growth.

The planting of larger trees as propagule sources in low recruitment environments may foster rapid coverage of the area. In a restoration site in Puerto Rico, red and black mangrove trees only three years old were producing large numbers of propagules. Black mangroves were aggressively spreading over a red mangrove planting area this way.

There are inconclusive reports in the literature about the use of fertilizers on mangroves (Kinch 1976; Lewis 1990a), but the use of slow release fertilizers has been recommended (Teas 1977, 1981; Darovec et al. 1975). It may be desirable under some circumstances (when planting on low fertility substrates) to use slow release fertilizers to promote growth during the time that the substrate is being slowly enriched by microbial activity and organic deposition. They may be useful when rapid establishment is needed to reduce losses to wave action (Goforth and Thomas 1980).

Pilot Propagation Studies. Lewis (1990a) suggests a pilot propagation study prior to planting. It may be more practical to design the project in such a way that it has elevational characteristics identical to those of nearby stands of the desired species, and then to allow natural restoration to occur. The area should be checked at the end of at least one fruiting and dispersal season after site preparation; at this time adequacy and type of natural regeneration would be assessed. Blank areas would then be planted. This would avoid redundancy with natural processes and would allow the establishment of the most appropriate species, rather than forcing the establishment of a species that may perform unsatisfactorily at the site or be out competed by faster growing, better adapted volunteers.

Air Layering. Air layering has been suggested as a technique to provide stock plants for transplantation without removing mangroves from a donor area. It is a horticultural technique in which short segments of bark and phloem are stripped away to the cambium, promoting the development of roots at the girdled area. Later the branch (layer) is removed from the parent plant and planted. All three mangrove species produce roots using this method (Carlton and Moffler 1978). The technique, although successful, has limited application due to the ease of obtaining propagules and seedlings in most locations and the current preference to plant seedlings rather than saplings.

The use of cuttings may be a practical means to propagate some species such as *Conocarpus*, which have a high regenerative ability and horticultural value (silver leaf buttonwood). *Conocarpus*, which roots readily from stems bent into the soil, roots readily from both cuttings and air layering.

Planting Area Diversity. Although not usually considered in mangrove restoration planning, large scale projects should provide species and habitat diversity similar to that of the altered wetland. A greater diversity represents an increase in potential users. Restoration projects should favor the establishment of more than one species by providing elevation gradients as well as open water and shallow wading areas. Open water areas should be connected to the main water bodies through channels to provide for circulation and to act as conduits for colonization by estuarine organisms.

Post-Restoration Maintenance and Monitoring. A great deal has been written about post-restoration maintenance and monitoring. The point of view expressed in this paper is that the project should be tightly coupled to local hydroperiod through unrestricted tidal circulation, and have elevations identical to those of stands of the desired species in the vicinity of the restoration/creation site.

In other words, the project has to be designed to receive all available natural subsidies. If this is done successfully, then the stand will be self-maintaining and will require little if any intervention. In administering systems such as mangroves which have high regenerative power and are highly resilient, management is optimized when intervention is minimized (Lugo and Cintron 1975). Management can be limited to the protection of the processes that drive the system and the intervention to extract or make use of the desired products or services. This is the basis of mangrove silviculture, which when properly executed makes use of the regenerative powers of the system for its self maintenance.

System Attributes and Components that Require Evaluation. The stand is a dynamic entity; it changes as it develops due to plant competition for growing space. During development the density of the stand decreases, but the biomass accumulation and structural complexity increase. A good measure of the functional equivalency of the restored wetland is how close the structural attributes of the created site will resemble well developed natural stands in the area.

It is expected that monitoring of the stand may be limited to determining that there has been adequate regeneration and to planting in those areas that have remained blank or where there has been loss or mortality of planted material.

If a detailed documentation of the restoration process is desired, then standard techniques to assess the structural and functional characteristics of stands should be used (UNESCO 1981).

Vegetation Management Practices. Artificial thinning is expensive and requires considerable care and experience but natural thinning and increases in spacing will occur without human intervention. Early stagnation, and the need for thinning, may be delayed or avoided by using wider initial spacings than those now used. When crown closure occurs at wider spacing, the trees are larger and may be more vigorous. In areas which receive large rates of natural recruitment, it may be desirable to avoid stagnation and accelerate development by removing late arrivals or young trees showing poor growth. Heavy thinning of mangrove stands should be avoided since it makes the trees even more vulnerable to windthrow.

Site Management and Monitoring—Minimum Monitoring Duration. Ideally, the burden for management costs of the restoration site should be borne by the project proponent until maturity and functional equivalency are reached. In mangroves, this represents a period of 20 to 30 years (time to reach maximum persistent biomass), although over 60% of that amount will have accumulated by the 15th year, assuming exponential growth. Protection of the site to ensure the persistence of the system is just as important. This can only be guaranteed through title transfer to the state resources management agency or through the use of other legal instruments that provide protection in perpetuity (perpetual conservation easements). Management of the area can then be guided by site assessments, performed or contracted out by the conservation agency. Maintenance costs would be borne by the developer as part of the project, and fiscal responsibility could be assured at the time of permitting by requiring a performance bond. If these measures seem draconian, it must be remembered that the purpose of restoration projects is to stop the net overall loss of wetlands acreage; it seems justified to pass on to developers the full cost of restoration, if they are permitted to alter or destroy a functional wetland.

Presently the greatest difficulty with restoration/creation as a mitigation alternative is that, in many state jurisdictions, there is no established procedure to track initial compliance, much less eventual persistence. Uniform monitoring procedures must be established to ensure that projects comply with the objectives originally set. The restoration/creation plan must be successful as a condition of the permit, and the applicant must be able to set aside funds to ensure that the project meets the required success criteria. A 75% minimum initial persistence of at least two years duration should be required; the developer should submit quarterly inspection reports during this period, and the monitoring agency should inspect and endorse at least once yearly during this period. A "successful" restoration project should continue to receive inspection visits, at least once every five years, after the initial period is completed successfully. Seedling or propagule mortality in excess of the allowable 25% during the first two years should be replaced at the developer's cost.

Success Criteria. The criteria used to determine long term success are now highly subjective. There are no detailed quantitative assessments of structural or functional attributes of the restored/created wetlands. There are no quantitative data on the longer term survival rates of the planted material nor the degree to which success is due to natural regeneration. In part, this is because structural attributes take decades to become fully restored in forested wetlands.

A project might be considered successful if the coverage of the area is as intended and remains so during the development of the stand. It is possible to quantify the development of the stand through time in terms of: (1) species composition, (2) stem density, (3) basal area, (4) vegetation height, (5) percentage canopy cover, (6) leaf area index, (7) mean diameter of the stand and (8) above ground biomass. Structural attributes change with time as the stand develops to fully utilize the available site, but the performance of the stand may be readily assessed on the basis of height and above ground biomass accumulation at different ages. Above ground biomass is easily estimated by non-destructive allometric techniques. Although presently there are few data of this type, data

accumulated from studies at different restoration sites can eventually be built into a data base. The general pattern of mangrove stand
development is described by Cintron and Schaeffer-Novelli (1984,
1985) and Jimenez et al. (1985).

Of course, the ultimate test is that species composition and biomass accumulation at maturity should be similar to those of natural stands nearby.

The Project Success Evaluation Process

Evaluations of project success should be done by the state natural resources agency based on its own studies or contracted assessments. Although the primary permits for restoration projects
almost always are granted by the U.S. Army Corps of Engineers,
this agency is not set up to do long-term monitoring, nor is it the
lead natural resource stewardship agency, even at the federal level.
Lacking structural data that can be used to judge performance and
success, the evaluation process must initially be restricted to the assessment of the success of the planting during the second year, or if
the site was to be allowed to self-regenerate, the degree and nature
of recruitment.

Future Needs and Direction

If restoration/creation is to be accepted as a true tool for mitigation, it must become much more reliable, and reliability will come
only as a result of greater scientific knowledge. The restoration and
creation of mangrove wetlands depends on basic information about
the processes that determine establishment patterns that have been
described earlier.

Information Needs to Ensure Success or Enhancement

There is a need for fundamental information concerning the dynamics of propagule production and dispersal, and establishment

requirements. These processes are intimately linked to hydrologic, edaphic and light factors. Although recovery is dependent on those process, little is known about them and their relationships and interactions with the environment. Information is required regarding (1) propagule production: crop size per tree in relation to tree size and crown class, flooding frequency, soil salinity; (2) causes of losses of reproductive effort: abortion, herbivory, predation; (3) dispersal: losses on dispersal, viability; and (4) establishment: rates of establishment, mortality, growth.

High priority should be assigned to studies of physiological ecology of establishment, and to mangrove soil chemistry. At present we know that both soil type and microtopography are extremely important to early survival, but cannot point to specific mechanisms causing mortality in many cases. Too little effort has been devoted to understanding the structural and functional changes that take place during stand development. These data are required to monitor performance of the stand and determine success, and to prepare stand management or enhancement schemes (such as thinning schedules). Finally, there are significant differences in the responses of mangrove stands to large and small scale disturbances. Information is needed on gap dynamics and how the stand changes after it reaches maturity.

Viability of Replacement of Functional Values

This issue is complex, since systems must have a critical size (minimum size for viable ecosystem function) to carry out certain functions. Piecemeal fragmentation of a large system and compensation by an array of small dispersed units would certainly degrade many landscape functions. This is avoided by the requirement that the restoration or creation action be done within the same landscape unit which was altered or degraded. However, this policy is not often part of the mitigation strategy, although it is now recognized as desirable in a federal interagency memoranda of agreement. There is no reason why a well designed project (designed to be highly coupled to natural processes) will not eventually function as well as nearby natural systems.

Funding Sources for Long-Term Research

NOAA has the resource management responsibility for the nation's living marine resources. One way of exercising this responsibility is through the support of scientific research aimed at providing the tools for managing these resources for sustainable use. One of NOAA's central research issues is related to the functional equivalency of restored habitats. It is clear that functional equivalency can only be attained by restoration projects based on a thorough understanding of natural wetland processes. Long-term research programs in these areas need to be established and supported by multi-year funding if the Administration's policy of no-net-loss of wetland habitats is to be successful.

Incorporation of Research Results into Restoration Plans

In order to improve the ability to restore or create wetland habitats and to manage wetland resources in a more effective manner, results from monitoring and research must be made available to technical and scientific audiences in a timely manner. This requires the provision for disseminating findings as they become available in symposia, workshops and technical memoranda as well as in the peer-reviewed literature. Research as well as actual restoration or creation efforts would benefit from the availability of the following products: (1) new protocols for quantifying mangrove hydrology (upon which successful restoration could be based); (2) technical reports containing improved methodologies for restoration or creation; and (3) standardized techniques to measure the structural and functional characteristics of restored areas. Research programs should take advantage of the opportunities to monitor natural restoration patterns after events such as hurricanes or human activities. Granting agencies should support research on the development of some restoration sites, and should more directly support restoration conferences or symposia in which the papers presented would be peer-reviewed. The Sea Grant network may be used to develop and distribute information related to restoration or creation opportunities and methodologies to local environmental managers and project developers.

Establishing a Restoration/Creation Ethic

We have witnessed society's growing awareness of environmental matters. People are beginning to realize that the quality of our lives, and even our survival, depends on the health and viability of the natural environment. The conference from which this book originates was a result of an emerging trend, a new ethic, of human-environmental interaction. We have realized, albeit slowly, that we cannot take perennially without giving back. Restoration of wetlands is a belated attempt to "give back", at least in part, what we have been destroying during the past century, because now we understand that we need wetlands.

It is possible to use restored areas as educational tools, to foster the development of this ethic at an early age. Science teachers and students may be involved in field oriented workshops on restored areas, sponsored by the federal government (Coastal Zone Program), state conservation agencies, or the private sector (business or other non-governmental organizations). Through these activities, a more balanced view of the human role in the biosphere may be presented.

Literature Cited

Ball, M.C. 1980. Patterns of secondary succession in a mangrove forest in south Florida. Oecologia (Berl.) 44:226-35.

Banus, M. and S.E. Kolehmainen. 1976. Rooting and growth of red mangrove seedlings from thermally stressed trees, p. 46-53. *In* G.W. Esch and R.W. McFarland (eds.), Thermal ecology II. CONF 750424, Technical Information Center, ERDA, Springfield, Virginia.

Barth, H. and H. Lieth. 1982. Applicability of mangroves for the development of ecologically based mariculture systems, p. 235-39. *In* J.J. Symoens, S.S. Hooper and P. Compere (eds.), Studies on aquatic vascular plants. Royal Botanical Society of Belgium, Brussels.

Bazzaz F.A. and T.W. Sipe. 1987. Physiological ecology, disturbance and ecosystem recovery, p. 203-227. *In* E.-D.Schulze and H. Zwolfer (eds.), Potentials and limitations of ecosystem analysis. Springer-Verlag, Berlin.

Beever, J.W. 1986. Mitigative creation and restoration of wetland ecosystems—a technical manual for Florida. Draft Report. Florida Dept. of Environmental Regulation, Tallahassee.

Blasco, F. 1983. Mangroves du Senegal et de Gambie, statut ecologique-evolution. Institut de la Carte Internationale du Tapis Vegetal. Centre National de la Reserche Scientifique E.R.73, Universite Toulouse III. Universite Paul Sabatier, Toulouse, France. 86 p.

Bowman, H.H.M. 1917. Ecology and physiology of the red mangrove.- Proc. Am. Philos. Soc. 56:589-672.

Cairns, J., Jr. and K.L. Dickson. 1977. Recovery of streams and spills of hazardous materials, p. 24-42. *In* J. Cairns, Jr., K.L. Dickson and E.E. Herricks (eds.), Recovery and restoration of damaged ecosystems. University of Virginia Press, Charlottesville.

Carlton, J. M. and M.D. Moffler. 1978. Propagation of mangroves by air-layering. Environ. Conserv. 5(2):147-150.

Cintron, G., A.E. Lugo and R. Martinez. 1985. Structural and functional properties of mangrove forests, p. 53-66. *In* W.G. D'Arcy and M.D. Correa A. (eds.), The botany and natural history of Panama. Monographs in Systematic Botany, 10. Missouri Botanical Gardens, St. Louis.

Cintron, G., A.E. Lugo, D.J. Pool and G. Morris. 1978. Mangroves of arid environments in Puerto Rico and adjacent islands. Biotropica 10(2):110-121.

Cintron, G. and Y. Schaeffer-Novelli. 1984. Methods for studying mangrove structure, p. 91-113. *In* S.C. Snedaker and J.G.Snedaker (eds.), The mangrove ecosystem: Research methods.UNESCO, Paris.

Cintron, G. and Y. Schaeffer-Novelli. 1985. Caracteristicas y desarrollo estructural de los manglares de Norte y Sur America. Ciencia Interamericana. Organizacion de los Estados Americanos, Washington, D.C. 25(1-4):4-15.

Clough, B.F., K.G. Boto and P.M. Attiwill. 1983. Mangroves and sewage: a re-evaluation, p. 151-161. *In* H.J. Teas (ed.), Tasks for vegetation science, Vol. 8. Dr. W. Junk Publishers, The Hague.

Craighead, F.C. 1964. Land, mangroves and hurricanes. Fairchild Trop. Garden Bull. 19:5-32.

Craighead, F.C. and V.C. Gilbert. 1962. The effects of hurricane Donna on the vegetation of southern Florida. Q. J. Fla. Acad. Sci. 25:1-28.

Darovec, J.E., Jr., J.M. Carlton, T.R. Pulver, M.D. Moffler, G.B.Smith, W.K. Whitfield, Jr., C.A. Willis, K.A. Steidinger and E.A. Joyce, Jr. 1975. Techniques for coastal restoration and fishery enhancement in Florida. Fla. Dep. Natur. Resour. Mar. Res. Publ. 15.

Davis, J.H. 1940. The ecology and geologic role of mangroves in Florida. Carnegie Inst. Wash. Publ. 32: 305-412.

Detweiler, T.E., F.M. Dunstan, R.R. Lewis and W.K. Fhering. 1976. Patterns of secondary succession in a mangrove community, p. 52-81. *In* R.R. Lewis (ed.), Proc. 2nd Annu. Conf. Restoration of Coastal Vegetation in Florida. Hillsborough Community College, Tampa, Florida.

Enos, P. and R.D. Perkins. 1979. Evolution of Florida Bay from island stratigraphy. Geol. Soc. Am. Bull. 90 (1):59-83.

Ewel, J. 1977. Differences between wet and dry successional tropical ecosystems. Geo-Eco-Trop. 1:103-117.

Ewel, J. 1983. Succession, p. 217-223. *In* F.B. Golley (ed.), Tropical rain forest ecosystems, A. Structure and function. Elsevier, Amsterdam.

Galloway, W.E. 1975. Process framework for describing the morphologic and stratigraphic evolution of depositional environments, p. 87-98. *In* M.L. Broussard (ed.), Deltas: Models for exploration. Houston Geological Soc., Houston, Texas.

Goforth, H.W. and J.R. Thomas. 1980. Planting of red mangroves (*Rhizophora mangle L.*) for stabilization of marl shorelines in the Florida Keys, p. 207-230. *In* D.P. Cole (ed.), Proceedings of the 6th annual conference on wetlands restoration and creation. Hillsborough Community College, Tampa, Florida.

Hamilton, L.S. and S.C. Snedaker. 1984. Handbook for mangrove area management. East-West Center/IUCN/UNESCO/UNEP. East-West Center, Honolulu, Hawaii. 123p.

Hannan, J. 1976. Aspects of red mangrove reforestation in Florida, p. 112-121. *In* R.R. Lewis (ed.), Proceedings of the 2nd annual conference on restoration of coastal vegetation in Florida. Hillsborough Community College, Tampa, Florida.

Hayes, M.O. 1975. Morphology of sand accumulations in estuaries, p. 3-22. *In* L.E. Cronin (ed.), Estuarine research, Vol. 2. Geology and Engineering. Academic Press, New York.

Heald, E.J. 1969. The production of organic detritus in a south Florida estuary. Ph.D. Dissertation, University of Miami, Florida.

Holdridge, L.R. 1940. Some notes on the mangrove swamps of Puerto Rico. Caribb. Forest. 1(4):19-29.

Holling, C.S. 1973. Resilience and stability of ecological systems. Annu. Rev. Ecol. Syst. 4:1-24.

Holling, C.S. and W. Clark. 1975. Notes toward a science of ecological management, p. 247-251. *In* W.H. van Dobben and R.H. Lowe-McConnell (eds.), Unifying concepts in ecology. Dr. W. Junk. Publishers, The Hague.

Jimenez, J.A. 1988a. The dynamics of *Rhizophora racemosa* Meyer, forests on the Pacific coast of Costa Rica. Brenesia 30:1-12.

Jimenez, J.A. 1988b. Floral and fruiting phenology of trees in a mangrove forest on the dry Pacific coast of Costa Rica. Brenesia 29:33-50.

Jimenez, J.A., A.E. Lugo and G. Cintron. 1985. Tree mortality in mangrove forests. Biotropica 17(3):177-185

Kinch, J.C. 1976. Efforts in marine revegetation in artificial habitats, p 102-111. *In* R.R. Lewis (ed.), Proceedings of the 2nd annual conference on restoration of coastal vegetation in Florida. Hillsborough Community College, Tampa, Florida.

Koenig, N. 1953. A comprehensive agricultural program for Puerto Rico. U.S. Government Printing Office, Washington, D.C. 299 p.

Kolehmainen, S.E. and W.K. Hildner. 1975. Zonation of organisms in Puerto Rican red mangrove (*Rhizophora mangle L.*) swamps, p. 357-369. *In* G.E. Walsh, S.C. Snedaker and H.J. Teas (eds.), Proceedings of an international symposium on the biology and management of mangroves. Inst. Food Agric. Sci., University of Florida, Gainesville.

Kusler, J.A. and M.E. Kentula. 1990. Executive summary, p. xvii-xxv. *In* J.A. Kusler and M.E. Kentula (eds.), Wetland creation and restoration. Island Press, Washington, D.C.

Lewis. R.R. 1982. Mangrove forests, p. 153-171. *In* R.R. Lewis (ed.), Creation and restoration of coastal plant communities. CRC Press, Boca Raton, Florida.

Lewis, R.R. 1990a. Creation and restoration of coastal plain wetlands in Florida, p. 73-101. *In* J.A. Kusler and M.E.Kentula (eds.), Wetland creation and restoration. Island Press, Washington. D.C.

Lewis, R.R. 1990b. Creation and restoration of coastal wetlands in Puerto Rico and the U.S. Virgin Islands, p. 103-123. *In* J.A. Kusler and M.E. Kentula (eds.), Wetland creation and restoration. Island Press, Washington. D.C.

Lewis, R.R. and K.C. Haines. 1981. Large scale mangrove restoration on St. Croix, U.S. Virgin Islands-II. Second year, p.137-148. *In* D.P. Cole (ed.), Proceedings of the 7th annual conference on restoration of coastal vegetation in Florida. Hillsborough Community College, Tampa, Florida.

Lugo, A.E. 1980. Mangrove ecosystems: Successional or steady state?, p. 65-72. *In* J. Ewel (ed.), Tropical succession. Biotropica 12(suppl.):65-72.

Lugo, A.E. and G. Cintron. 1975. The mangrove forests of Puerto Rico and their management, p. 825-846. *In* G.E. Walsh, S.C. Snedaker and H.J. Teas (eds.), Proceedings of an international symposium on the biology and management of mangroves. Inst. Food Agric. Sci., University of Florida, Gainesville.

Lugo, A.E. and C. Patterson-Zucca. 1977. The impact of frost stress on mangrove ecosystems. Trop. Ecol. 18:149-61.

Lugo, A.E. and S.C. Snedaker. 1974. The ecology of mangroves. Annu. Rev. Ecol. Syst. 5:39-64.

Macnae, W. 1968. A general account of the fauna and flora of mangrove swamps and forests in the Indo-West-Pacific region. Advan. Mar. Biol. 6:73-270.

Mager A., Jr. and G.W. Thayer. 1986. National Marine Fisheries Service habitat conservation efforts in the Southeast Region of the United States from 1981 through 1985. Mar. Fish. Rev. 48(3):1-8.

Magnuson, J.J., H.A. Regier, W.J. Cristie and W.C. Sonzogni. 1980. To rehabilitate and restore the Great Lakes, p. 95-112. *In* J. Cairns, Jr. (ed.), The recovery process in damaged ecosystems. Ann Arbor Science Publishers, Ann Arbor, Michigan.

Markley, J.L., C. McMillan and G.A. Thompson Jr. 1982. Latitudinal differentiation in response to chilling temperatures among populations of three mangroves, *Avicennia germinans, Laguncularia racemosa* and *Rhizophora mangle*, from the western tropical Atlantic and Pacific Panama. Can. J. Bot. 60:2704-2715.

McMillan, C. 1971. Environmental factors affecting seedling establishment of the black mangrove on the central Texas coast. Ecology 52:927-930.

McMillan, C. 1975. Adaptive differentiation to chilling in mangrove populations, p. 62-68. *In* G.E. Walsh, S.C. Snedaker and H.J. Teas (eds.), Proceedings of an international symposium on the biology and management of mangroves. Inst. Food Agric. Sci., University of Florida, Gainesville.

Moormann, F.R. and L.J. Pons. 1975. Characteristics of mangrove soils in relation to their agricultural potential, p. 529-547. *In* G.E. Walsh, S.C. Snedaker and H.J. Teas (eds.), Proceedings of an international symposium on the biology and management of mangroves. Inst. Food Agric. Sci., University of Florida, Gainesville.

Morrison, D. 1990. Landscape restoration in response to previous disturbance, p.159-172. *In* M.G. Turner (ed.), Landscape heterogeneity and disturbance. Springer-Verlag, New York.

Odum, W.E. 1971. Pathways of energy flow in a south Florida estuary. University of Miami Sea Grant Bull. 7. 162 p.

Olexa, M.T. and T.E. Freeman. 1975. Occurrence of three unrecorded diseases on mangrove in Florida, p. 688-692. *In* G.E. Walsh, S.C. Snedaker and H.J. Teas (eds.), Proceedings of an international symposium on the biology and management of mangroves. Inst. Food Agric. Sci., University of Florida, Gainesville.

Oliver, C.D. and B.C. Larson. 1990. Forest stand dynamics. McGraw-Hill, New York. 467 p.

Orians, G.H. 1975. Diversity, stability and maturity in natural ecosystems, p. 139-150. *In* W.H. van Dobben and R.H. Lowe-McConnel (eds.), Unifying concepts in ecology. Dr. W. Junk, The Hague.

Patterson-Zucca, C. 1978. The effects of road construction on a mangrove ecosystem. M.S. Thesis, University of Puerto Rico, Rio Piedras, Puerto Rico. 77 p.

Pool, D.J., S.C. Snedaker and A.E. Lugo. 1977. Structure of mangrove forests in Florida, Puerto Rico, Mexico and Costa Rica. Biotropica 9(3): 195-212.

Pulver, T.R. 1976. Transplant techniques for sapling mangrove trees, *Rhizophora mangle, Laguncularia racemosa,* and *Avicennia germinans* in Florida. Fla. Mar. Res. Publ. 22.

Rabinowitz, D. 1978. Dispersal properties of mangrove propagules. Biotropica 10(1):47-57.

Rehm, A.E. 1976. The effects of the wood-boring isopod *Sphaeroma terebrans* on the mangrove communities of Florida. Environ.Conserv. 3:47-57.

Rehm, A.E. and H.J. Humm. 1973. *Sphaeroma terebrans:* a threat to the mangroves of southeastern Florida. Science 182:173-174.

Reichle, D.E., R.V. O'Neill and W.F. Harris. 1975. Principles of energy and material exchange in ecosystems, p. 27-43. *In* W.H. van Dobben and R.H. Lowe-McConnell (eds.), Unifying concepts in ecology. Dr. W. Junk, The Hague.

Sachtler, M. 1973. Inventario y fomento de los recursos forestales: Republica Dominicana, inventario forestal. PNUD/FAO. FO:SF/DOM 8, Informe Tecnico 3.

Savage, T. 1972. Florida mangroves as shoreline stabilizers. Fla. Dep. Natur. Resour. Prof. Pap. Ser. 19. 46 p.

Schuster, W.H. 1952. Fish culture in brackish water ponds of Java. Indo-Pacific Fish. Counc. Spec. Publ. 1. 143 p.

Simberloff, D., B.J. Brown and S. Lowrie. 1978. Isopod and insect borers may benefit Florida mangroves. Science 210:630-632.

Snedaker, S.C., J.A. Jimenez and M.S. Brown. 1981. Anomalous aerial roots in *Avicennia germinans* (L) in Florida and Costa Rica. Bull. Mar. Sci. 31:467-470.

Stephen, M.F. 1984. Mangrove restoration in Naples, Florida, p. 201-216. *In* F.J. Webb, Jr. (ed.), Proceedings of the 10th annual conference on the restoration of coastal vegetation in Florida. Hillsborough Community College, Tampa, Florida.

Stoddart, D.R. 1969. Post hurricane changes on the British Honduras reefs and cays: Re-survey of 1965. Atoll Res. Bull.131: 1-25.

Swabey, C. 1939. Forestry and erosion in Haiti and Puerto Rico. Jamaica Dep. Sci. Agric. Bull. 21.

Tabb, D.C. and A.C. Jones. 1962. Effects of hurricane Donna on the aquatic fauna of North Florida Bay. Trans. Am. Fish. Soc. 91:375-378.

Teas, H.J. 1977. Ecology and restoration of mangrove shorelines in Florida. Environ. Conserv. 4:51-58.

Teas, H.J. 1981. Restoration of mangrove ecosystems, p. 95-103. *In* R.C. Carey, P.S. Markovits and J.B. Kirkwood (eds.), Proceedings of a workshop on coastal ecosystems of the southeastern United States. U.S. Fish and Wildlife Service, Office of Biological Services, FWS/OBS-80/59. 257 p.

Teas, H.J., W. Jurgens and M. C. Kimball. 1975. Plantings of red mangrove (*Rhizophora mangle L.*) in Charlotte and St. Lucie counties, Florida, p. 132-161. *In* R.R. Lewis (ed.), Proceedings of the 2nd annual conference on the restoration of coastal vegetation in Florida. Hillsborough Community College, Tampa, Florida.

Thom, B.G. 1967. Mangrove ecology and deltaic geomorphology: Tabasco, Mexico. J. Ecol. 55:301-343.

Thom, B.G. 1984. Coastal landforms and geomorphic processes, p.3-17. *In* S.C. Snedaker and J.G. Snedaker (eds.), The mangrove ecosystem: research methods. UNESCO, Paris.

Tomlinson, P.B. 1980. The biology of trees native to tropical Florida. Harvard University Printing Office, Allston, Mass. 480p.

Tomlinson, P.B. 1986. The botany of mangroves. Cambridge University Press, Cambridge. 413 p.

Twilley, R.R. 1982. Litter dynamics and organic carbon exchange in black mangrove (*Avicennia germinans*) basin forests in a southwest Florida estuary. Ph.D. Dissertation, University of Florida, Gainesville.

United Nations Educational, Scientific and Cultural Organization(UNESCO). 1981. The mangrove ecosystem: Research methods. Monographs on oceanographic methodology. UNESCO, Paris. 251 p.

U.S. Army Corps of Engineers. 1977. Shore protection manual. U.S. Government Printing Office, Washington, D.C.

U.S. Army Corps of Engineers. 1980. Erosion control with smooth cordgrass, gulf cordgrass, and saltmeadow cordgrass on the Atlantic Coast. Coastal Engineering Research Center, TN-V-2, Washington, D.C.

U.S. Office of Technology Assessment (OTA). 1984. Wetlands: Their use and regulation. Office of Technology Assessment. OTA-O-206. U.S. Government Printing Office, Washington, D.C.

Vogl, R.J. 1980. The ecological factors that produce perturbation-dependent ecosystems, p. 63-94. *In* J. Cairns, Jr. (ed.), The recovery process in damaged ecosystems. Ann Arbor Science Publishers, Ann Arbor, Michigan.

Wadsworth, F.H. 1959. Growth and regeneration of white mangroves in Puerto Rico. Caribb. Forest. 20:59-71.

Wadsworth, F.H. and G.H. Englerth. 1959. The effects of the 1956 hurricane on forests in Puerto Rico. Caribb. Forest. 20:38-57.

Watson, J.C. 1928. Mangrove forests of the Malay Peninsula. Malayan Forest. Rec. 6:1-275.

Wells, J.T. and J.M. Coleman. 1981. Periodic mud flat progradation, northeastern coast of South America: A hypothesis. J. Sedimen. Petrol. 51(4):1069-1075.

Westman, W.E. 1978. Measuring the inertia and resilience of ecosystems. BioScience 28:705-710.

CHAPTER 7

Restoring Kelp Forests

David R. Schiel
Department of Zoology, University of Canterbury
Christchurch, New Zealand

Michael S. Foster
Moss Landing Marine Laboratories
Moss Landing, California

Abstract

Kelp forests, shallow subtidal communities dominated by large brown algae such as Macrocystis pyrifera *that form surface canopies, are probably the most productive marine communities in temperate waters. The high productivity and habitat complexity provided by these plants contribute to the formation of diverse communities with considerable ecological, aesthetic and economic value. Moreover, food and habitat are exported from kelp forests to associated communities such as sandy beaches and the deep sea. Losses of kelp, on spatial scales from localized sites to hundreds of kilometers, have occurred because of increased turbidity, sedimentation, and perhaps toxic substances associated with sewer discharges, and also because of low nutrients, high temperatures and severe water motion during El Niño oceanographic events. Fishing and hunting of animals associated with kelp forests have caused the extinction of one mammal and the severe reduction of another, as well as numerous reductions in populations of crabs, lobsters, abalone and fishes. Decreases in predators may indirectly lead to losses of kelp by allowing increases in grazing by sea urchins.*

More kelp is not necessarily more "natural." Kelp forests exhibit a dynamic range of structure, and restoration must be designed and evaluated in the context of current natural variation as well as historical changes induced by human activity. The environmental conditions necessary for natural recovery or restoration of kelp and associated algae are the presence of stable substrata for attachment (usually rock), suitable water quality for the growth and maintenance of kelp populations and, in some cases, reduction of sea urchin populations. If these conditions can be met and the area is near a mature stand of kelp, natural recovery is often rapid. Without nearby natural populations, however, recovery may be slow and highly variable, depending in part on dispersal to the area, the size of the habitat to be restored, and the presence and activity of herbivores. Recovery rates of kelp, abalone and other species can be increased by transplanting or "seeding" various life stages into natural habitats, and new communities can be established (at the expense of soft bottom communities) by placing hard substrata into suitable areas. Although some species can be restored, restoration may be difficult or impossible for other species that have undergone large scale deterioration due to human exploitation. If these species have life history characteristics that hinder rapid recovery (e.g., episodic recruitment, slow growth, etc.), the only effective restoration technique may be significant reductions in commercial and recreational exploitation over large regions.

Introduction

Research during the past decade has significantly advanced the knowledge of how kelp communities work. In one sense, this research can be divided into the traditional categories of examining physical and biological factors that affect the growth, survival and reproduction of kelp and associated organisms. Of crucial importance, however, is the methodology that has been a common theme to many recent studies. Meticulous, field-based, manipulative experiments have provided the strongest tool for understanding community processes. Not only main factors have been investigated but, probably more importantly, the interactions among factors.

The single most important feature of good experimental designs is the ability to partition variation into "explained" and "unexplained" categories.

We introduce this paper with a preamble about variation and the methods used to partition it because they are the essence of managing kelp communities. We believe it was not until research focused on variation, rather than on average properties of kelp forest organisms, that significant progress was made in understanding the important processes that structure these communities. As elaborated in our review below, past attempts at kelp forest restoration have primarily been "trial and error" experiments; claims of positive results have frequently been qualitative and equivocal, the causes of failure are often unknown, and most variation has remained "unexplained." We argue that the majority of restoration efforts in kelp forests are necessarily experimental. Their success, however, will be greatly enhanced by a thorough understanding of the causes of degradation, clearly stated restoration objectives, rigorous sampling and experimental designs, peer review at all stages of the restoration effort, and publication of the results in the scientific literature. Anything less will be a poor use of public funds and will inhibit the development of the science of restoration.

This paper focuses primarily on communities dominated by the giant kelp, *Macrocystis pyrifera*, in California. Although there are other kelps that form surface canopies, the literature on them is not so extensive. We present an overview of how kelp communities are structured and go on to discuss impacts on these communities, attempts to restore them and efforts to create new kelp forests. Finally, we suggest guidelines for future work on restoration and creation of kelp forests.

Overview of Kelp Forests

Distribution

The term kelp "forest" is usually reserved for subtidal marine communities dominated by large brown algae (kelps) that form floating canopies on the surface of the sea. These kelps are highly productive and also provide a three-dimensional aspect to the near-shore environment, with a consequent increase in habitat and food for other species. The major species that form floating surface canopies along the west coast of North America are *Macrocystis pyrifera*, generally found south of Santa Cruz, California, *Nereocystis luetkeana*, found mostly north of Point Conception, California, and *Alaria fistulosa*, in Alaska (Figure 1; Druehl 1970). Kelps with floating canopies do not occur along the east coast although plants can obtain heights of over 6 m above the bottom (R. Vadas, University of Maine, pers. comm.).

Virtually any transect through a *Macrocystis pyrifera* forest in California would encounter hundreds of species of plants and animals, and several spatial patterns of species abundances would be evident (Figure 2). Along much of California, seagrass occurs in the immediate subtidal zone, where sand patches are common. As hard reef becomes predominant, several species of smaller kelps, including *Egregia laevigata*, are often found. There is usually a lush understory of various brown and red algae as well as mobile and encrusting invertebrates. In slightly deeper water (ca. 2-5 m), stipitate kelps such as *Eisenia arborea* and *Laminaria* spp. can be abundant. Beyond that depth, *Macrocystis pyrifera* forms a surface canopy over understory kelps such as *Pterygophora californica* and *Laminaria farlowii*. *Macrocystis* can be found in depths beyond 30 m, but this depends on water clarity and the availability of hard substrata. Throughout the kelp forest are hundreds of species of fish, while benthic areas are usually completely covered by low-lying algae and invertebrates. This is a very simplified summary that overlooks the

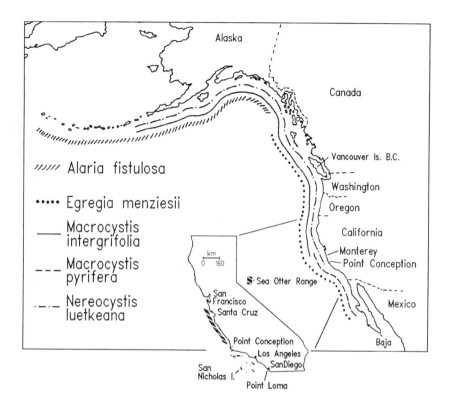

Figure 1. Map of west coast of North America showing the general distribution of the major canopy-forming species of kelp. Inset shows more detail of California, including the distributional range of sea otters. Adapted from Druehl (1970) and Foster and Schiel (1985).

considerable variation along a coastline in the occurrence and distribution of particular species (Foster and Schiel 1985).

Some species are generally considered to be "important," either in the context of community processes or commercial and recreational interests. In *Macrocystis* forests these species are the giant kelp itself, the understory kelps, sea urchins, sea otters, abalone, lobsters (south of Point Conception; Figure 1), and several species of fish.

In addition to the depth-related patterns of distribution there is clearly a vertical structure to kelp forests (Figure 2). A line from the hard substratum at the floor of a kelp forest running vertically to the sea surface would pass through possibly an encrusting invertebrate, one or more species of low-lying algae, understory kelp and giant kelp (e.g., Foster 1975a). Each of these layers has associated with it particular physical and biological conditions that may affect organisms in the other layers.

Kelp Life Histories

An understanding of kelp forest dynamics is necessarily based on the life history characteristics of the major species. *Macrocystis* and other kelps in the order *Laminariales* have an alternation of generations, with a microscopic gametophyte stage followed by the much larger sporophyte (Figure 3; Neushul 1963; Abbott and Hollenberg 1976; Lobban 1978). *Macrocystis* spores are released from specialized blades (sporophylls) at the base of the plant. These settle and grow into male and female gametophytes (1N) that produce eggs and sperm. Mobile sperm are released from males and fertilization takes place on the female gametophyte attached to the substratum. These microscopic 2N zygotes then develop into the much larger sporophytes. While kelp gametophytes can be maintained in lab cultures for years (e.g., Luning 1980), they probably do not survive more than a few months in the field (Deysher and Dean 1986a; Reed 1990). The timing of the life history patterns is species-specific but, generally, reproductive maturity of the sporophytes occurs during the first or second year of growth, depending on the species, and maximum longevity is from one to several years. *Macrocystis* will live for up to 10 years, while *Nereocystis* is an annual (Abbott and Hollenberg 1976). *Macrocystis pyrifera* is capable of elongating at a rate of 50 cm per day (North 1971a). When conditions are most favorable for growth, over 50% of the biomass

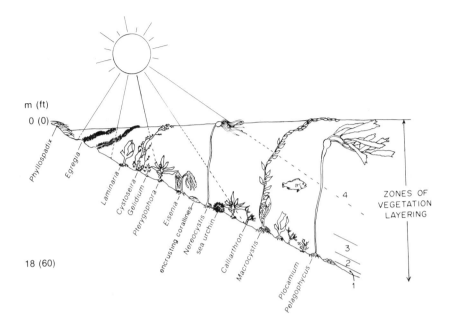

Figure 2. The distribution of common seaweeds along a depth gradient of a kelp forest. The four vertical layers of vegetation include (1) filamentous species and encrusting coralline algae, (2) low-lying algae, such as Gelidium, Calliarthron *and* Plocamium, *that may form canopies a few cm above the bottom, (3) understory, canopy-forming kelps such as* Eisenia, Laminaria *and* Pterygophora, *and (4) kelps that form canopies in mid-water or on the sea surface such as* Egregia, Macrocystis *and* Nereocystis. *This is stylized and some species do not co-occur at the same sites. Diagram adapted from Dawson and Foster (1982) and Foster and Schiel (1985).*

of older plants is within 1 m of the sea surface (D.C. Barilotti, Kelco Company, pers. comm.).

Kelp Forest Environments and Economic Values

Food webs of kelp forest communities are very complex (North 1971b; Rosenthal et al. 1974; Foster and Schiel 1985), but they are based primarily on the productivity of kelp. Kelp fronds are essentially belts of tissue that continuously erode (Mann 1973; Chapman and Craigie 1977) or, as in *Macrocystis*, decay or are broken off the parent plant, contributing organic material and recycling nutrients to coastal communities. There is usually a considerable amount of drift material that stays within the kelp forest, providing food for grazing invertebrates such as sea urchins, abalone, and gastropods (Mattison et al. 1977; Harrold and Reed 1985). Drift material accumulating on the sea surface (kelp paddies), on the sea floor and on beaches provides additional food and habitat to associated communities and contributes significantly to the overall productivity of coastal waters (Gerard 1976; Cailliet and Lea 1977; Harrold and Lisin 1989).

The physical presence of kelp also has a significant impact on communities. A kelp canopy can reduce surface light by over 90%

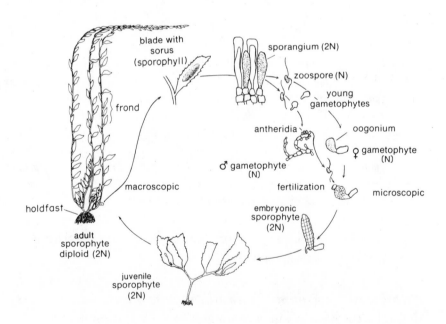

Figure 3. The life history of the giant kelp Macrocystis pyrifera. *Modified from Dawson and Foster 1982.*

to areas below, thereby affecting the composition and growth rates of organisms in the understory (Reed and Foster 1984). Giant kelp harbors many species of small invertebrates, which in turn are fed on by a wide variety of fishes (Quast 1971). The combined effects of the kelp canopy and predation of larvae by fishes may act as filters and severely deplete the number of larvae drifting inshore from pelagic areas, with possible consequences on the composition of inshore areas (Bernstein and Jung 1979; Gaines and Roughgarden 1987).

Severe water motion can alter kelp communities by removing the kelp plants (Cowen et al. 1982; Dayton and Tegner 1984a) but in less extreme conditions the floating canopy acts as an offshore damper that reduces the force of seas.

Economically, kelp forests are the most important marine communities in the western United States. *Macrocystis* is harvested commercially, and there are large commercial and recreational fisheries for many species of fishes, lobsters, abalone and sea urchins (North and Hubbs 1968; Frey 1971; Tegner 1989). These communities also have value for recreational sports such as scuba diving, kayaking, and viewing nearshore marine life such as birds, sea otters and seals.

Finally, it should be mentioned that kelp forests serve as a natural laboratory and classroom to train marine scientists, many of whom go on to do the type of research discussed here.

Variability in Kelp Forests

While kelp forests in different localities have many features in common, there is no such thing as a "typical" kelp forest. As knowledge of distribution and abundance is extended to an increasing number of sites, it is clear that there is significant variation in kelp community composition at every spatial scale from geographic to extremely localized (Foster et al. 1983; Dayton et al. 1984; Dayton and Tegner 1984a; Foster and Schiel 1985, 1988; Holbrook et al. 1990a). The examination of variation within sites, often called "patch dynamics", has been useful in determining important processes (Dayton et al. 1984).

Perhaps uniquely among marine communities, the temporal variation in giant kelp abundance has been assessed many times since the early 1900s because of the economic importance of kelp harvesting (e.g., North and Hubbs 1968; North 1971b; Foster 1982). There has been considerable fluctuation in the extent of kelp canopies along the coast of California during the 20th Century. This was accentuated in the 1980s with the El Niño events that reduced the abundance of kelp in many areas (Wilson and Togstad 1983; Tegner and Dayton 1987). Recognition of this natural variation is essential in determining how kelp forests should be managed, if restoration is appropriate, and whether or not various management efforts are successful.

Impacts and Responses

Natural Impacts

There is now a reasonable knowledge about the factors affecting the distribution and abundance of kelp. These must be the focus in any discussion about broader community effects because there is considerable evidence that changes in kelp abundance can have a major impact on most other organisms in the community (Dayton 1985; Schiel and Foster 1986; Holbrook et al. 1990b). Because this point is central to any attempts at management of kelp-dominated communities, it must be elaborated.

Many species are affected by low light levels beneath canopies of kelp. For example, Foster (1982) and Carr (1989) found an inverse correlation between the abundance of *Macrocystis* and the cover of understory algae, a relationship that presumably results from canopy shading. Overstory canopies also greatly affect algal recruitment (Reed and Foster 1984) as well as the distribution of understory species (Kastendiek 1982). Of particular importance to those concerned with commercial and recreational fishing is the relationship between the abundance of fishes and the character of nearshore habitats. Carr (1989) showed experimentally that providing kelp habitat in areas otherwise devoid of kelp significantly changes the species composition and recruitment of nearshore fishes. Other studies have shown that different species of fish can finely partition habitats (Choat 1982; Ebeling and Laur 1985; Holbrook et al. 1990b). Studies of a number of sites in southern California, however, have shown a weak relationship between overall fish assemblage structure and vegetation structure (Holbrook et al. 1990a). Holbrook et al. (1990a) suggest that this general lack of specialization on habitats produced by the vegetation may reflect the high temporal variability of these habitats.

Given that associated organisms are affected by the abundance of the kelp plants themselves, factors that directly cause kelp mortality or a decrease in the cover of canopies are particularly important. The natural factors can be categorized into physical and biological. Storms have the most visible effects on kelp forests. In most years, winter storms in central California remove much of the surface canopy of *Macrocystis* so that an annual cycle of canopy presence and absence can be seen (Figure 4; Cowen et al. 1982; Kimura and Foster 1984). Exceptionally severe storms, however, can have effects more devastating than just canopy removal. Storms producing nearshore wave heights >6 m have been rare during most of the 20th Century, but the frequency of such events increased along the southern California coast during the 1980s (Seymour et al. 1989). Kelp forests growing on sand were virtually destroyed, while others showed depth-related effects. In a large kelp forest near San Diego, for example, 66% of *Macrocystis* plants were killed at a depth of 12 m, while only 13% of those at 18 m were killed (Dayton and Tegner 1984; Seymour et al. 1989). Further north in Carmel Bay, central California, a dense kelp forest was removed to such an extent that few *Macrocystis* plants could be found

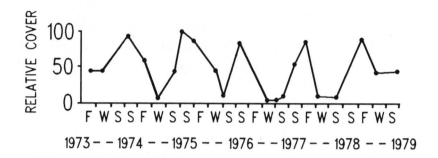

Figure 4. Change by season (F, W, S, S = autumn, winter, spring, summer) in the percentage cover of the Macrocystis *surface canopy in a kelp forest at Carmel, central California. From Kimura and Foster (1984).*

after the 1982-83 winter storms (Schiel and Foster unpublished data). A common means of kelp removal is that some large plants become detached and then drift through the forest, entangling and removing other plants (Rosenthal et al. 1974). The effects of severe storms are therefore exacerbated in dense kelp forests.

Kelp forests are generally restricted to areas with predominantly hard substrata. The type of substratum can affect the character of algal stands. For example, the stipitate kelp *Pterygophora californica* occurs in dense aggregations of large, old plants on hard conglomerate and sandstone substrata, but only small, young plants are found on soft mudstone in exposed sites (Foster 1982; Reed and Foster 1984).

Sediment load also affects kelp forests. Scour and abrasion are directly related to water motion and, within kelp forests, there is an inverse relationship between sediment load and the abundance of kelp (Figure 5; Cowen et al. 1982; Weaver 1977). Sediment effects on adult plants generally occur during periods of severe water motion when an influx of sediment may bury the lower portions of the thallus. Otherwise the effects are seen on the early life stages. Along with the associated effects of reduced light and nutrients, sedimentation can prevent spores from settling on the substratum

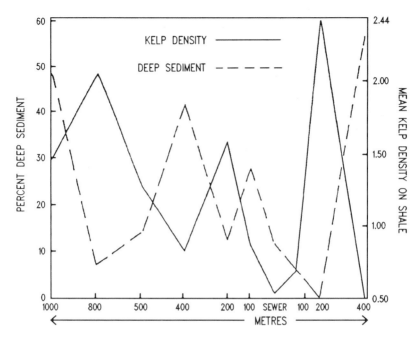

Figure 5. Diagram showing the inverse relationship between the density of adult kelp plants and the percentage cover of sediment of at least 2 cm deep in a kelp forest in central California. X-axis is expressed as distance from a sewer outfall. From Weaver (1977).

and will drastically reduce the growth and survival of gametophytes (Devinny and Volse 1978; Norton 1978; Deysher and Dean 1986a).

Sea water temperature and nutrients are inversely correlated along the coast of California (Jackson 1977; Zimmerman and Kremer 1984; Deysher and Dean 1986a, b) so their effects cannot be treated separately. It is generally believed that prolonged temperatures greater than about 20°C, along with low nutrients, can set the limit of *Macrocystis* distribution (North 1971b). These factors particularly affect early life stages. Deysher and Dean (1986b) concluded that a factor correlated with temperature, probably the level of nutrients, could be the most important determinant of recruitment success of *Macrocystis*.

There are also relationships between temperature and irradiance that affect sporophyte production of *Macrocystis* and other kelps (e.g., *Nereocystis*; Vadas 1972). Dense recruitment of *Macrocystis* (>50 sporophytes m^{-2}) occurs only when temperatures are below 16.3°C and irradiance is above 0.4 E m^{-2}d^{-1} (Deysher and Dean 1986a). As temperatures approach 16°C, higher irradiance levels are required to achieve dense recruitment. It appears there may be regional variability in the temperatures at which sporophyte development will occur. In laboratory studies, Luning and Neushul (1978) found that the upper limit for gametogenesis was between 17° and 20°C in *Macrocystis* from the Santa Barbara region, while Deysher and Dean (1986b) found 100% production of sporophytes at 20°C, when light levels were high, from plants obtained 300 km further south. As pointed out by Deysher and Dean (1986b), however, there is variation in gametogenesis between experiments controlled in a laboratory and those in nature, where several factors interact simultaneously.

In addition to effects on early life stages, giant kelp canopies usually deteriorate during summer when inorganic nitrogen is low and temperature is high, particularly in southern California (North 1971b; Zimmerman and Kremer 1986). This deterioration can be delayed because of the ability of kelps, including *Macrocystis*, to store nitrogen when concentrations are high in the surrounding water and then use these reserves to maintain growth when environmental nitrogen is low (Wheeler and North 1981; Gerard 1982). Water motion up to a current speed of 2-4 cm s^{-1}, such as that caused by wave surge, enhances the uptake of inorganic nitrogen by increasing nutrient transport through the diffusion boundary layer of fronds (Wheeler 1980; Gerard 1982).

When the fronds of adult plants deteriorate during late summer, they become particularly vulnerable to encrustation by sessile invertebrates such as bryozoans. These encrustations, combined with the effects of fish grazing on them, can accelerate the deterioration of kelp canopies (Bernstein and Jung 1979).

Of the biological factors impacting kelp forests, intensive grazing by sea urchins has received the most attention in California and worldwide (Lawrence 1975; Harrold and Pearse 1987). Mobile aggregations of sea urchins can rapidly remove large tracts of kelp, leaving dramatically altered communities in their wake. The effects have been such a prominent focus of research that grazing by echinoids is considered by many to be a "controlling" or "regulating" factor in kelp communities generally (reviews in Foster and Schiel 1985, 1988). We (Foster and Schiel 1988) reviewed all available surveys of California kelp forests to provide spatial and temporal contexts to intensive grazing by echinoids in giant kelp forests. We found that only about 9% of sites along the California coast (outside the range of the sea otter) were affected by large scale (ca. 103 m²) deforestation. This indicates that the sea otter/sea urchin paradigm is not generally applicable in this area.

It is also apparent that storms may influence the effects of grazing by sea urchins. When drift algae are abundant, echinoids tend to feed passively or move only small distances to capture accumulating drift material (Mattison et al. 1977). When intense storms remove much of a kelp forest, however, drift algae often become scarce. Sea urchins then become active grazers, may form large aggregations, and remove large numbers of attached plants (Ebeling et al. 1985; Harrold and Reed 1985).

Human-Induced Impacts

The impact of humans on kelp forests has been difficult to quantify in most cases. This is not necessarily for lack of effort but because the effects are often subtle, occur over a long time within a context of considerable natural variation, and there may be no historical reference points. An issue that has been hotly debated in several parts of the world is the impact of various forms of fishing on kelp communities. For example, commercial fisheries for lobsters have drastically lowered their numbers in most coastal areas. Because lobsters are known predators of sea urchins, it has been argued in California (Tegner and Dayton 1981; Tegner and Levin 1983) and Nova Scotia (Mann 1973; Breen and Mann 1976) that the reduction in lobster numbers may have allowed increases in sea urchin populations, with consequent deleterious effects on kelp abundance. Others have argued that the evidence is at best equivocal for this (Miller 1985; Pringle 1986), and the recent review by Elner and Vadas (1990) highlights the effects that lack of proper experimentation has had on resolving the issue.

It is likely that more heat than light will continue to be generated by this debate in California, which is essentially unresolvable because historical data on the finer structure of kelp communities is lacking. Nevertheless, it seems clear that some species, particularly prominent residential ones, are far less numerous in kelp forests than they once were. California sheephead (*Semicossyphus pulcher*), which are able to consume large numbers of sea urchins (Nelson and Vance 1979), are relatively scarce in areas accessible to humans along the coast of California (Cowen 1983). Giant sea bass (*Stereolepis gigas*), a large (150 kg) and once numerous fish in kelp forests, was severely fished, is now rarely seen, and specimens are invariably small (Frey 1971). The extinction in the 1700s of the Stellar's sea cow (*Hydrodamilis gigas*), a giant (up to 6,000 kg, 7 m in length) passive herbivore resident in nearshore kelp communities, is the only known extinction in kelp communities (Domning 1978). It was hunted by aboriginal man, and the last sea cow is believed to have died in 1741. We will never know the effects many of these alterations to species composition and size structure have had on the character of kelp forests.

Having pointed out how many factors and organisms can affect the distribution and abundance of kelp and the importance of the kelp plants to the greater community, it is perhaps surprising that kelp harvesting appears to have few deleterious effects on kelp communities. The surface canopies of *Macrocystis* forests have been harvested commercially since the early 1900s in California (North 1971b). Around 54,000 metric tons are harvested annually with some forests being harvested up to three times per year (McPeak and Glantz 1984). The concern about the possible impacts of kelp harvesting on the wider biological community stimulated a considerable amount of research during the 1950s and 1960s. Investigations were made into the potential loss of kelp stands, harmful effects on fish populations due to loss of food and habitat, and a potential increase in beach erosion due to the loss of kelp (North 1971b). These studies found no adverse effects of harvesting (North and Hubbs 1968). Cut fronds usually deteriorate slowly and sink to the bottom while uncut fronds replace them in the canopy (Rosenthal et al. 1974). Light is increased to the understory after harvesting which enhances recruitment of *Macrocystis* (Kimura and Foster 1984). More recent studies have also concluded that current harvesting practices do not negatively impact giant kelp or associated species (Barilotti and Zertuche-Gonzalez 1990). No overall reduction in fishes or invertebrates has been found (North and Hubbs 1968; Miller and Giebel 1973) except perhaps for turban snails (*Tegula* spp.) that tend to be removed with kelp canopies (Hunt 1977).

It has been argued that the historical hunting of sea otters (*Enhydra lutris*) and the depletion of populations along the coast of California has had a large impact on the character of kelp forests. Sea otters consume up to 25% of their body weight (up to 36 kg) per day, feeding on a range of invertebrates (Kenyon 1969; Estes et al. 1981; Kvitek and Oliver 1988). They are able to remove virtually all large grazers such as sea urchins and abalone from open reef areas, leaving only smaller individuals in cryptic habitats (Estes et al.

1981; Hines and Pearse 1982). Sea otters in California are currently listed as endangered and have full legal protection.

More direct effects of human habitation involve the deposition of materials into coastal habitats. This includes run-off from coastal farms and agricultural areas and nearshore construction; the alteration of currents and sand movement by construction along and offshore; increased sedimentation from dredging; increased turbidity from power plant discharges; and the sedimentation, organic and nutrient enrichment, and addition of toxic substances resulting from sewage discharge. These impact kelp communities in several ways. Changes in current flows due to the construction of structures such as piers and breakwaters can result in sand movement and sedimentation or burial of nearshore habitats. The most important effects, however, are probably the reduction in light transmission due to turbidity, and the sedimentation that covers rocky substrata. These can reduce the light necessary for photosynthesis and developmental processes, and severely reduce or impede altogether the attachment and gametogenesis of kelps (Norton 1978; Devinny and Volse 1978; Deysher and Dean 1984, 1986a). The best-known example of this is the sewage outfall on the Palos Verdes Peninsula near Los Angeles.

The decline of one of the largest kelp forests in California, at Palos Verdes, began in the early 1950s as sewage discharge rates increased. After the period of warm oceanic conditions during the late 1950s, the kelp forest did not recover (Grigg and Kiwala 1970; Meistrell and Montagne 1983). Contributing to the decline and persistent lack of recovery were increased water turbidity (Eppley et al. 1972), benthic sludge which may have inhibited gametophytes (Grigg and Kiwala 1970), toxic chemicals such as DDT in the sewage (Burnett 1971), and metals such as copper which even at low levels can inhibit gametophytes (Smith 1979).

Probably the most immediate and dramatic man-induced impacts on kelp forests were the Tampico tanker diesel fuel spill in

Baja California, Mexico in 1957 and the Santa Barbara, California offshore oil well blow-out and spill in 1969. In the Mexico spill there was considerable mortality of invertebrates, such as sea urchins, abalone, lobsters and sea stars, but damage to the kelp plants was less obvious (North et al. 1964). In the case of the crude oil spill at Santa Barbara, numerous birds associated with the kelp forest were killed and mysid shrimp abundance declined (Ebeling et al. 1971), but overall there was little obvious damage to kelp forest algae, invertebrates or fishes, despite the large amount of oil that fouled the surface canopy (Foster et al. 1971). The crude oil was partially weathered, stayed mostly on the surface of the sea, and did not stick to the kelp fronds. Determining the full effects of this spill was complicated by the record storms that occurred at about the same time (Foster et al. 1971).

These are examples of events that have been studied and for which there is reasonable information on impacts. Other influences such as sedimentation from coastal run-off, change in current flow, loss of canopy from boat traffic and numerous others can produce localized effects. There will always be problems in quantifying the effects of these impacts, especially given the often large seasonal and year-to-year natural variability of kelp communities. This inability to quantify effects will always leave ample room for speculation about impacts. This "window of uncertainty," often combined with a misunderstanding of scientific evidence, provides the opportunity for populist statements about upsetting the "balance of nature." As one of the great early ecologists, Charles Elton (cited in Connell and Sousa 1983) pointed out, the concept of the balance of nature is appealing, but "It has the disadvantage of being untrue." Variation and interactions are the hallmarks of natural communities. The answers to questions about the effects of various activities lie in better data and not in simplified paradigms about how natural ecosystems may work.

Natural Responses to Impacts

The responses of kelp communities to episodes of destructive grazing by sea urchins have been well-studied. During the 1960s, there was considerable concern that populations of sea urchins were increasing and that this was resulting in serious deforestation of kelp communities (North 1963a, b; North and Pearse 1970). Consequently, there were many observational and experimental studies on grazing by sea urchins and the responses of other members of the forest community to removal of *Macrocystis*.

There are often large numbers of sea urchins in a kelp forest but they normally have little impact on attached plants (Lowry and Pearse 1973; Foster 1975a; Cowen et al. 1982). This is because the sea urchins are usually widely dispersed and feed on drift algae (Mattison et al. 1977; Vadas 1977; Duggins 1980). What have been termed "feeding fronts" (Leighton 1971), large aggregations of mobile sea urchins, are occasionally formed, possibly as a result of the unavailability of drift algae (Dean et al. 1984; Harrold and Reed 1985). Aggregations of *Strongylocentrotus franciscanus* (red sea urchins) and *S. purpuratus* (purple sea urchins) of up to 90 animals per square meter have been documented as removing thousands of square meters of *Macrocystis* during a single year (Leighton et al. 1966). These sorts of effects are seen worldwide. For example, grazing episodes by *Strongylocentrotus drobachiensis* during the late 1970s to early 1980s in Nova Scotia removed kelp from hundreds of kilometers of coast (Wharton and Mann 1981; Miller 1985).

Grazing by sea urchins can have many community effects, especially through changes in algal species composition and abundance. Kelps are usually removed first, followed by lower-lying species. In some kelp forests, the largest plants are removed through detachment of holdfasts by grazers (Vadas 1977). Because all organisms are not equally vulnerable to grazing, however, more subtle results may occur by their differential removal from the community (Schiel 1982).

The removal of dense aggregations of sea urchins both naturally and artificially usually results in a large recruitment of algae. This has been demonstrated on a small experimental scale of a few square meters (e.g., Duggins 1980; Andrew and Choat 1985), and over hundreds of kilometers of coastline where there was a mass mortality of sea urchins (Miller 1985). In experiments where *Macrocystis* was removed to simulate the effects of grazing, several species of understory kelps recruited first, followed by recruitment of *Macrocystis* (Pearse and Hines 1979). This sequence may vary with seasonal differences in recruitment and other factors (Foster 1975a, b). The community reverted to one dominated by a *Macrocystis* canopy within 1-2 years.

These studies demonstrate that on all spatial scales, kelp communities become re-established within a few years of sea urchin removal, provided there is a source of algal propagules nearby. The reduction of kelp canopies allows increased light to areas below, and previously grazed areas provide increased primary substratum for attachment of other species, resulting in good conditions for the successful recruitment of algae. In giant kelp communities, *Macrocystis* becomes readily re-established as a dominant species because it grows quicker and larger than other algae and, despite fewer numbers per given area, can form a dense canopy over other species and suppress further recruitment (Pearse and Hines 1979; Dayton et al. 1984).

Given these interactions between sea urchins and kelp, three major features have emerged to the point where they can be validly termed "generalizations." The first is that areas dominated by sea urchins comprise distinct habitats that may persist for many years (e.g., Ayling 1981; Chapman 1981). These habitats have their own characteristics of encrusting organisms, mobile invertebrates and fishes that are quite distinct from nearby kelp-dominated habitats (Ayling 1981; Choat 1982; Choat and Kingett 1982; Ebeling et al. 1985; Harrold and Reed 1985; Ebeling and Laur 1988). The sea urchin-dominated and kelp-dominated habitats are so distinct that they have been called "alternate stable states" by some authors (Simenstad et al. 1978) and occur in kelp forests worldwide (Schiel and Foster 1986). The second general feature is that the removal of sea urchins will usually result in kelp recruitment, the rapidity and composition of which depends on the availability of propagules. Kelps are often seasonal in their periods of reproduction and recruitment (Foster 1975a; Dayton et al. 1984; Schiel 1988) and recruits often appear within several meters of fertile adult plants (Anderson and North 1966; Deysher and Norton 1982) although dispersal can be much further during storms (Reed et al. 1988). The third general feature is the patchiness of kelp forest communities. Grazing, along with many other processes, ensures that species assemblages and size structure vary throughout any kelp forest (Dayton et al. 1984; Foster and Schiel 1985).

Responses of species other than kelp have not been studied so extensively. Abalone, and large reef fish such as giant sea bass and California sheephead are examples of species severely impacted by fishing. These species are now drastically reduced from their historical abundance (although good quantitative studies are often lacking) and their size-frequencies are skewed towards small individuals (Cowen 1985). These species are long-lived, have relatively slow growth rates, have low recruitment success in any one year, and are continuously vulnerable to removal by commercial and recreational fishing. Because of this disturbance and their life history features, these species tend to remain relatively scarce. Other species that are fished heavily, such as kelp and blue rockfish, are very abundant, relatively fast growing, and produce large numbers of recruits annually (Miller and Geibel 1973). Because of these characteristics, present fishing activity appears to have minimal impact on their populations, and their main requirement seems to be the presence of *Macrocystis* or other habitats with high relief.

Finally this section would be incomplete without a mention of the sea otter, *Enhydra lutris*. This animal was hunted to near extinction by the turn of the century (Kenyon 1969). Despite complete legal protection and a considerable range expansion (Figure 1) since 1914, it appears that population numbers in California have not greatly changed from the approximately 1,300 individuals recorded in the 1960s (Riedman and Estes 1988). It is clear that sea otters can have drastic effects on kelp forest communities by removing large invertebrates, particularly sea urchins and abalone (Estes and Harrold 1988). If populations expand southward beyond Point Conception into southern California, they could have a significant impact on invertebrate fisheries. There is evidence that Pismo clam populations in central California were severely reduced by sea otters (Estes and VanBlaricom 1985). Whether the expansion and re-establishment of sea otters within their historical range in California will return these areas to their former "natural" state is a moot point (VanBlaricom and Estes 1988). It is fairly certain, however, that they will change the modern structure of kelp communities, have a negative impact on commercial and recreational fisheries of abalone, sea urchins, and probably crabs and lobster, and a positive impact on kelp abundance. However, these impacts will vary from site to site depending on the abundance and behavior of sea otter prey prior to otter foraging (see sea urchin deforestation discussion above).

Restoration and Creation

Since the 1950s, there have been numerous attempts to restore kelp forests. These have largely centered on giant kelp and how to increase its abundance. The major projects have been the Kelp Habitat Improvement Project (documented in annual reports, e.g., North 1963a, 1974), continuing studies by Kelco Company (the largest kelp harvesting company in the State of California), and by the California Department of Fish & Game. While not directed towards restoration, the Kelp Ecology Project to study the San Onofre kelp forest near the San Onofre Nuclear Generating Station did considerable work on *Macrocystis* life history regulation and population dynamics (Table 1). Many techniques have been used to en-

Table 1. Major projects aimed at restoring or expanding *Macrocystis* forests in California.

Project	Location	Years	Selected Publications
Kelp Habitat Improvement Project	Mostly Point Loma, San Diego	1956-70	North (1971b, 1976) Numerous unpublished reports
Kelco Co.	Mostly Point Loma	ongoing	Wilson and McPeak (1983)
Kelco Co.	Santa Barbara	1987-ongoing	Barilotti (Kelco Co., pers. comm.) and unpublished report
California Department of Fish and Game	Southern California	1971-ongoing	Wilson et al.(1978) Wilson and McPeak (1983) Numerous unpublished reports
Kelp Ecology Project	San Onofre	1978-87	Dean (1985) Dean and Deysher (1983) Dean and Jocobsen (1986) Deysher and Dean (1984, 1986a, b) Numerous unpublished reports

hance the abundance of *Macrocystis*, including the killing of sea urchins, removing understory kelps, transplanting adult *Macrocystis* from other areas, and seeding areas with spores and small, laboratory-reared sporophytes. Success has been claimed in the restoration of several kelp forests but, as will become apparent below, the effects of human intervention are often obscured by inadequate study designs and natural events. The reasons that some restoration attempts have been successful are, therefore, frequently open to debate.

The Kelp Habitat Improvement Project was the first long-term study into *Macrocystis* communities. North (1971b) interacted with many scientists from different organizations, including Kelco Co. and the California Department of Fish and Game, to determine if the abundance of *Macrocystis* in southern California could be enhanced in places where forests had become depleted. This depletion in the 1950s was related to the interacting events of increased sewage discharge, increased sea urchin grazing and El Niño oceanographic changes (discussed above). There also appears to have been significant increases in the abundance of sea urchins throughout the ensuing decades that contributed to the near elimination of kelp from previous forests off the Palos Verdes Peninsula near Los Angeles and Point Loma near San Diego (Leighton et al. 1966; Tegner and Dayton 1981; North 1983; Tegner 1989). Both of these forests were near major sewer outfalls.

Removal of Grazers

Efforts at restoration have fallen into four categories: grazer control, kelp transplantation, competitor control and monitoring (Wilson and McPeak 1983). Sea urchins were at densities of >50 m^{-2} in many areas of Point Loma and Palos Verdes (Leighton 1971) and it was considered necessary to remove them from several thousand square meters in order to establish a permanent stand of kelp. Several removal techniques were used. Smashing sea urchins with hammers wielded by scuba divers was effective, but labor-intensive. Kelco biologists used this technique in 50 hectares of Point Loma in 1981 and there was a successful recruitment of *Macrocystis* during the following year (Wilson and McPeak 1983). A mechanized technique was used by Kelco to control sea urchins at high densities (>30 m^{-2}). Divers dislodged sea urchins which were then sucked through a diver-held hose to a suction dredge on the surface. This macerated the sea urchins and then discharged them back into the sea. This technique is considered not to be cost-effective when sea urchins are at low densities. However, urchins can be concentrated by taking advantage of their feeding behavior. Kelp attached by divers to the sea floor attracts large numbers of sea urchins, which can then be removed.

A relatively inexpensive method that has been used to control sea urchins is the use of quicklime (CaO), which produces lesions in the sea urchin epidermis that eventually kills the animal (North 1963a; Wilson and McPeak 1983). Early methods used pebblized quicklime dispersed from a ship. Later, the quicklime pebbles were pumped from the sea surface through a hose held by a diver, who directed it over concentrations of sea urchins.

Each of these methods of grazer control has serious drawbacks. The first two are labor-intensive and costly to implement over a large area, while the use of quicklime also affects other echinoderms such as sea stars. K. Wilson (California Department of Fish and Game, pers. comm.) states, however, that the diver dispersal method can avoid most damage to other species. None of these control methods are currently being used because, with the advent of a major commercial market for red sea urchin roe in Japan, sea urchins are no longer a pest, but a valuable resource. Various techniques are now being considered to enhance sea urchin populations (Tegner 1989). The fishery has probably aided the expansion of kelp forests in some sites off central and southern California (Wilson and McPeak 1983; A. Chess, National Marine Fisheries Service, pers. comm.).

The results of sea urchin removals in deforested areas generally show natural recruitment of kelp if there are fertile adults nearby. Large areas of Point Loma reverted to a kelp forest in the early 1960s. Prior to the period of restoration, however, there were changes in waste discharge and dredging in the San Diego area which decreased sedimentation and turbidity, giving better conditions for kelp recruitment and growth (Wilson and McPeak 1983). It is therefore impossible to determine how effective the restoration program was. Similar results occurred at Palos Verdes, where the expansion of the kelp forest occurred in the mid-1970s at the same time that waste water quality was improved (Meistrell and Montagne 1983).

An Ethical Consideration

We believe that the driving force for restoring kelp forests, indeed even the concept of "restoring," is the widely-held notion that areas with abundant kelp represent the "natural" state of these nearshore communities. The reasoning is exemplified by Jacques Cousteau in an advertisement for the Cousteau Society: "An ecosystem of classic simplicity is the sea otter, the kelp, and the sea urchin. Years ago the charming sea otter was abundant along the California coast, but now it has been almost wiped out. So the urchins it used to feed on gnaw at the roots of the kelp, and what were once fecund marine jungles are now scrubby deserts." While not wishing to embark on a "Save the Sea Urchin" campaign, it should at least be pointed out that sea urchins comprise a normal part of kelp communities (Foster and Schiel 1988). Transitions from kelp to sea urchin to kelp communities can occur naturally in as little as five years (Ebeling et al. 1985). Viewing the "natural" state of these communities as being dominated by kelp may be a form of selective myopia—and the most likely primary causes of drastic changes to kelp cover were a combination of pollutants, sedimentation, turbidity and natural variation in environmental conditions. Rather than hammers, mechanical maceration and infusions of chemicals to kill sea urchins, an obvious approach to consider is to reduce human input. This, of course, is often considered too costly and difficult, but it does strike at the causes rather than the symptoms.

Transplantation of Kelp

Considerable effort was expended during the 1970s in transplanting kelp to areas where it formerly occurred. The aim was to establish nuclei from which kelp stands would expand by "seeding" surrounding areas with spores (Wilson et al. 1978; Wilson and McPeak 1983). Three main methods were used: (1) plants were brought from kelp forests and tethered to chains spread along the substratum in transplant areas; (2) small plants were tied by rubber inner tubes to rocks; and (3) young *Macrocystis* were tied to the cut stipes of understory kelps. About 35,000 plants were transplanted in the San Diego area between 1973 and 1976 (Wilson and McPeak 1983). While new recruits were noted around some transplants (K. Wilson, California Department of Fish and Game, pers. comm.), these techniques have proved to be costly and labor-intensive, and their success at expanding kelp forests over more than a few meters has not been documented.

A recent method of kelp forest restoration is the seeding of ropes or other physical structures and the placement of these into appropriate subtidal areas. The techniques for large-scale rearing of *Macrocystis* sporophytes in the laboratory are now well-developed (e.g., Deysher and Dean 1986a), but methods of successful planting in the field are still experimental, and will no doubt vary depending on local conditions such as substratum characteristics and water motion. For example, Barilotti (Kelco Company, pers. comm.) has developed several methods to establish kelp forests on sand bottoms near Santa Barbara. In this habitat the plants originate from growth centers, large accumulations of holdfast material from many generations of plants. One method to produce growth centers uses so-called "mushroom" anchors made from concrete (Figure 6). *Macrocystis* juveniles are attached to the anchors before placing them on the bottom. The anchors provide a surface for hapteral (= holdfast branches) development, while the shape of the anchor helps to prevent its movement as well as sand abrasion and

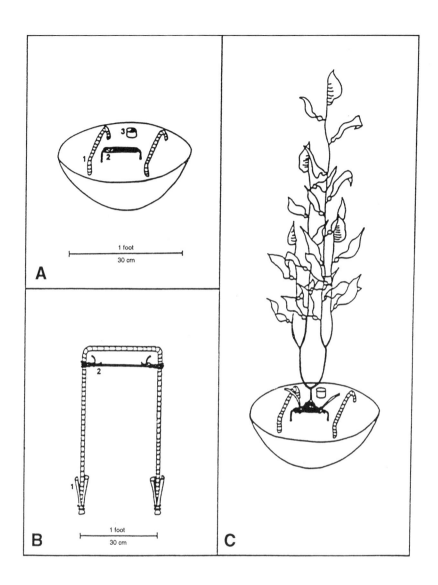

Figure 6. "*Mushroom*" *anchors used by Barilotti (Kelco Co. pers. comm.) for establishing* Macrocystis *plants on the sandy floor of the devastated kelp forest at Goleta, California. A = the concrete anchor with handles of reinforcing steel (1), an attachment site for plants (2) and a PVC sleeve that a rod is slid through when deploying anchors from a boat (3). B = a staple made of reinforcing bar, used to secure anchor to sea floor, with barbs that improve stability in the sand (1) and nylon line that helps secure plant (2). C = anchor with young sporophyte attached.*

scour. The success of this enhancement program is still being evaluated.

These kelp forests are unique in that they occur on sand bottoms. Efforts are continuing to refine the mushroom anchor technique as well as others that anchor adult plants in sand with "staples" of reinforcing bar, and use them over a wider area. Harger and Neushul (1983) successfully planted a biomass test farm of *Macrocystis* by placing holdfasts in mesh bags filled with gravel. These anchored the plants to a sand bottom until hapteral growth made the holdfast itself an anchor. A variety of transplant methods are possible, but their effectiveness and cost will no doubt vary from site to site.

Management of Sea Otters

Sea otters are one of the most high profile and emotive features of kelp forests. Otters are intimately associated with kelp surface canopies and commonly feed on invertebrates in the kelp forests. The preferred habitat of otters in California is the surface canopy in proximity to rocky substrata (Woodhouse et al. 1977). Otters sleep in the canopies, which provide protection from predators and severe water motion (Kenyon 1969). Canopies also act as nurseries for pups. While population size has increased since the early 1900s (discussed above), animals have proved to be vulnerable to entanglement in gill nets, the use of which was banned inshore in the mid 1980s. Nevertheless, it has been argued that sea otter populations along the coast of California may also be vulnerable to a large disaster such as an oil spill (VanBlaricom and Jameson 1982) and that active means should be used to extend their range (reviewed in Levin 1988).

These concerns prompted a translocation program to establish a separate population on San Nicolas Island, 90 km west of Los Angeles. This island has extensive *Macrocystis* forests, would provide ideal habitat for the sea otters, and is within the historical range of otters (Woodhouse et al. 1977; Harrold and Reed 1985). Since 1987, 138 otters, captured along the mainland coast of central California, have been moved to San Nicolas Island. Fourteen of these presently (October 1990) remain around the island, plus three young that were born there (G. VanBlaricom, U.S. Fish and Wildlife Service, pers. comm.). The fate of 80 of the 124 animals that left the island is unknown. Of the remaining 44, many eventually returned to the vicinity of their capture. While success has been low, the remaining animals appear to have become residents (VanBlaricom, pers. comm.). There is concern for the loss of animals as a result of this program, as well as continuing concern that translocation of otters may assist their southward expansion into southern California and have severe effects on shellfish and sea urchin fisheries. Whether or not translocations should continue is currently being debated.

Enhancement of Abalone

Several species of abalone (*Haliotis* spp.) are fished commercially in California. These have been overfished and, in some cases, affected by sea otters so that species have tended to be depleted sequentially (Tegner and Butler 1989). The most valuable of these is the red abalone, *Haliotis rufescens*, for which several attempts have been made to enhance populations in southern California. Hatchery-raised juveniles have been released into natural habitats in several kelp forests. There has generally been poor survival due to predation by octopus, crabs and fish (Tegner and Butler 1985, 1989). There is considerable evidence that hatchery-raised abalone are more vulnerable to predators than natural juveniles, possibly due to behavioral differences (Schiel and Welden 1987; Schiel In press). Other attempts at enhancement have taken advantage of the short larval life and the consequent short dispersal of planktonic abalone. Mature red abalone were transplanted to the Palos Verdes kelp forest, an area that has a natural gyre where larvae may be entrained

(Tegner and Butler 1985). A relatively large recruitment occurred during the following year. Approaches similar to those tried for abalone are now being considered for red sea urchins (Tegner 1989).

Reserves

Reserves are a type of preservation that entail various forms of non-extraction and could aid restoration. Despite the fact that reserves are often heavily used recreationally, a non-extraction policy can have discernible effects on kelp forest organisms over a period of several years. Changes in species diversity usually are not seen but some fish and invertebrate species may increase in abundance. The most noticeable, documented effect of subtidal reserves, however, is that resident, long-lived species may reach larger sizes than in nearby areas where extraction is allowed (Cole et al. 1990). Unfortunately, there is not much direct information on the effects of marine reserves in *Macrocystis* forests because not many reserves exist, and possible changes in the few that do have not been well investigated.

Creation of New Kelp Forests with Artificial Reefs

Artificial reefs have been established primarily to satisfy fisherman, but some have been constructed as a means of compensation where human activities have threatened to destroy existing kelp communities (Grove 1982; Sheehy and Vik this volume). Pendleton Artificial Reef was constructed in 1980 near San Diego by Southern California Edison Company and the California Department of Fish and Game to determine the potential to mitigate possible losses of kelp forests due to the operation of coastal power plants (Grant et al. 1982; Carter et al. 1985a, b). This project ran for several years, but was unsuccessful in establishing *Macrocystis* until 1988. K. Wilson (California Department of Fish and Game, pers. comm.) reports that *Macrocystis* now fully covers the reef. The probable reasons for this long successional history are instructive.

There was an early and extensive covering of encrusting inverte-brates and small algae (Carter et al. 1985a). *Cryptoarachnidium argilla* (a sediment-fixing, encrusting ectoproct) and barnacles cov-ered 66% of the understory after three years while small filamen-tous and foliose red algae formed an overstory. These apparently inhibited colonization by other species.

The reef itself was composed of eight slightly separated rock modules that were fairly steep on the sides. Carter et al. (1985a) suggest that the resulting high water motion on the module crests, and low light on the sides, may have reduced kelp growth and de-creased survivorship. The reef attracted large numbers of the herbiv-orous fishes, halfmoon (*Medialuna californiensis*) and opaleye (*Girella nigricans*). Over 600 *Macrocystis* and 200 *Pterygophora* (an understory kelp) plants were transplanted to the reef, but the blades and growing tips were intensively grazed by the fish and no plants survived (Carter et al. 1985b). Grazing by herbivorous fish has been a problem in several other kelp transplanting attempts (discussed in Carter et al. 1985b). Gill nets have been used to re-move grazing fishes and mesh has been used to protect transplant-ed kelp, but neither has been successful in the long term.

Attempts at establishing red abalone on Pendleton Artificial Reef were also unsuccessful. Over 18,000 laboratory-raised juve-niles (2-4.5 cm shell length) were placed onto the modules during 1981 (Grant et al. 1982) and by May 1982 none could be found. As for other attempts at transplanting red abalone juveniles, mortal-ity was most likely caused by predators, especially crabs (*Cancer* spp.).

Finally, Pendleton Artificial Reef was situated on a sand floor sev-eral kilometers from the nearest stand of kelp. As dispersal of spores and larvae of many species, including giant kelp, usually de-creases rapidly with distance (e.g., Reed et al. 1988), there was a di-minished chance of natural colonization of the reef by kelp forest species. The isolation of the reef, its topography, and the flow of water around the modules were probably the major factors influencing the presence of low-lying species and large numbers of grazing fishes (Carter et al. 1985a, b; Ebeling et al. 1985).

K. Wilson (California Department of Fish and Game, pers. comm.) suggests that *Macrocystis* eventually established on the reef because of improvements in oceanographic conditions suitable for kelp growth throughout the region where the reef is located. When kelp began to establish in abundance at Pendleton Artificial Reef there was also considerable recruitment and growth on several local reefs, perhaps diluting populations of herbivorous fish formerly concentrated around transplants at Pendleton Artificial Reef. In addition, the appearance of juvenile kelp in areas where mature plants had been placed indicates that these may have acted as spore sources

Another artificial reef was installed in 1984 by Pacific Gas and Electric Company along the coast of San Luis Obispo County near the Diablo Canyon Nuclear Power Plant in central California. The modules of this reef were placed on sand floors but were within 50 meters of natural rocky substrata supporting kelp communities. These reef modules were each approximately 100 × 50 m of fairly low relief large boulders and rubble, and were therefore much larger than Pendleton Artificial Reef as well as of a different configuration. The San Luis reef developed a kelp forest (*Nereocystis*) within six months of placement, and this kelp is now being replaced by the understory kelp, *Pterygophora*. Juvenile rockfish also began to recruit soon after the reef was placed, and continue to do so (S. Krenn, Pacific Gas and Electric Company, pers. comm.).

Restoration Techniques

As illustrated by the examples above, there has been mixed success in restoring kelp forests. This is due to the physical rigors and variability of nearshore environments, their relative inaccessibility, and complex interactions that occur among associated organisms in rocky reef communities. The context for this is the physical environment of light, temperature, nutrients and water motion. Sometimes these factors can be controlled, at least within certain bounds, but often they are uncontrollable, broad-scale natural phenomena. From the restoration work done to date, it is unclear whether pro-active efforts have resulted in positive and lasting results or if natural phenomena would have produced similar results without human intervention.

Despite the uncertainty about the results of human intervention there is still a good rationale for attempts at restoration. Because of the complexities of interactions among nearshore communities, human activities, and the natural environment, the causes of changes to kelp forest community structure will often be obscure. Past research into restoration has clearly defined some of the important processes and interactions affecting the presence of *Macrocystis*, and techniques have evolved to make use of this knowledge. If there is a consensus among monitoring agencies that the presence of kelp is desirable, it is valid to attempt to circumvent some of the vagaries of natural processes and create favorable conditions to facilitate the formation of kelp stands.

Clean Up the Environment

The most obvious means of enhancing kelp forests that have been affected by human activities is to reduce or cease these activities. The Palos Verdes kelp forest is a good example of positive results. Sedimentation associated with the sewage outfall clearly had an effect on the ability of kelp spores to settle, and on gametophytes to grow and germinate. Studies on kelp gametophytes have shown that light is one of the most important resources for gametophytes (Luning and Neushul 1978; Deysher and Dean 1986a, b) and that up to 99% of incident light can be removed by even a fine layer of silt, thereby preventing gametogenesis altogether (Norton 1978). Although there was an extensive program of grazer removal

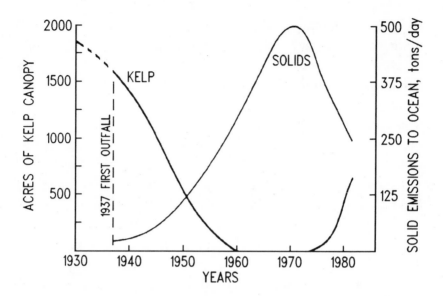

Figure 7. Diagram showing the inverse relationship between the extent of the Macrocystis pyrifera *surface canopy and the emission of solids from the sewer outfall at the Palos Verdes Peninsula. From Meistrell and Montagne (1983).*

and kelp transplantations at Palos Verdes, the re-establishment of kelp forest habitat coincided with (and was most likely due to) the change in waste water quality and consequent reduction in turbidity, sludge discharge and discharge of substances potentially toxic to kelp and associated species (Figure 7). It seems most likely that the removal of high densities of sea urchins and transplantation of adult plants facilitated the process of kelp recovery once the sewage was more thoroughly treated before discharge, especially by reducing sludge. The most reasonable interpretation of the return of kelp to Palos Verdes, therefore, is that natural recovery occurred, but only after environmental conditions were improved and grazers were altered. Further expansion of the kelp forest at Palos Verdes may be inhibited by accumulated sediments and the metals and other toxins bound in them.

Transplantation of Kelp and Seeding with Spores

The transplanting of adult or juvenile kelp plants has not been very successful in restoring kelp to isolated, rocky sites. The idea is appealing in that providing nuclei of fertile plants to serve as a proximal source of propagules makes sense. The reality, however, has been somewhat different. Transplanted kelp are prone to detachment and tend to attract large numbers of benthic and mobile grazers. Furthermore, transplanting is expensive in time and labor.

Seeding habitats with spores, gametophytes or small sporophytes holds some promise for restoration of kelp-depleted areas (North 1976). There is now a considerable knowledge of the conditions and techniques necessary to raise *Macrocystis* from spores to sporophytes (Neushul 1963; Deysher and Dean 1984, 1986 a, b; Dean and Jacobsen 1986; Reed 1990). Various seeding techniques have been used. Placing fertile sporophytes of *Macrocystis* into mesh bags and then attaching these to the substratum has produced successful recruitment of kelps (Dayton et al. 1984). However, this has only worked on the scale of a few square meters. The potential problems with seeding by spores fall into several categories. First, kelp gametophytes have fairly specific requirements of light, temperature and nutrients for reproductive development. Deysher and Dean (1986b) have shown that these "windows" of favorable conditions may occur only once every few years in particular coastal areas. Reed (1990) demonstrated that the availability of "safe sites" for successful recruitment can vary on the order of weeks. Since gametophytes appear to be short lived in nature (discussed above), it may be necessary to repeat spore releases several times in order to achieve successful reproduction at high enough densities to produce a successful recruitment of sporophytes. If this is to be done over large areas, as restoration attempts would require, the procedure may be prohibitively expensive in time and money. Furthermore, hard substrata may have to be added to sandy habitats such as those recently denuded by storms near Santa Barbara.

One promising technique is to raise small sporophytes in a laboratory, where physical conditions can be controlled and there is a high chance of success (Deysher and Dean 1986a). The small sporophytes, attached to ropes or other surfaces, are then transplanted to appropriate localities. This technique has been successful on a small scale but not over larger areas. The main problems have been the vulnerability of small sporophytes to grazers (e.g., Harris et al. 1984) and the vertical stratification of physical conditions of light, temperature and nutrients that may render the area immediately above the substratum unsuitable for sporophyte growth (Deysher and Dean 1986b). These negative scenarios are really cautionary tales about the importance of the physical environment, even when the biological environment is suitable. Laboratory-rearing of sporophytes significantly reduces environmental uncertainties, but successful restoration will occur only when "windows of opportunity" exist in the natural environment.

There are special areas in which some form of seeding seems to be the only chance for the replenishment of a kelp forest in the short term. A recent example is the natural loss of kelp forests growing on sand bottoms near Santa Barbara. The forests persisted because old and heavy *Macrocystis* holdfasts served as substrata for recruitment, and as anchors for new plant growth. Low nutrients and high temperatures during 1982-1983 reduced the kelp canopy and the severe storms during 1987-1988 finally removed these growth centers. Since that time, there has been little new recruitment of kelps because stable substratum is limiting (D.C. Barilotti, Kelco Company, pers. comm.).

In assessing how to restore kelp to this area, a reasonable starting point is to ask what conditions allowed the formation of the original forest. Although the specifics cannot be known for certain due to the age of the original forest, it is clear that some form of hard substrata suitable for spore attachment was available, at least temporarily, that water motion was reduced so that sand scour and the direct physical removal of young plants was minimal, and that light, temperature and nutrient conditions fell within the range suitable for kelp growth. Once large plants became established, the formation of growth centers for perpetuation of the forest was possible.

Successful anchors that will also form growth centers will most likely provide the key to restoring these kelp forests. D.C. Barilotti (Kelco Company, pers. comm.) developed several models before deciding on a concrete, mushroom-shaped anchor, weighing about 30 kg. Three methods were used in seeding the anchors with kelp: (1) securing natural juveniles to the anchors before transplanting; (2) raising small sporophytes in a laboratory, transplanting these to field sites until they grew into larger juveniles (>1 m), and attaching these plants to the anchors for transplanting; and (3) putting out anchors near adult plants to act as attachment sites for natural spore fall. The success of these methods is still being evaluated. The problems encountered, however, are similar to other restoration exercises: the loss of plants from grazing, loss of anchors, and poor growth and survival in some sites due to changing physical conditions.

Because of the commercial interest of Kelco Company in restoring the kelp forests in Santa Barbara, D.C. Barilotti (Kelco Company, pers. comm.) provides costings for restoration there (the only recorded attempt at doing so). The cost of constructing and deploying anchors at the rate of four hectares a day is $1,517 a hectare. Fastening juveniles to anchors and then deploying them costs $3,074 a hectare.

It is interesting that even in this case where most of the important factors appear to be understood, restoration has been a slow process over several years. It seems likely that this may be the take-home message: even with the best techniques and a good understanding of critical factors, persistence of effort or at least patience (see Pendleton Artificial Reef discussion above) may be required for large-scale restoration. It also seems likely that some form of synergy will occur whereby beyond some critical size the kelp forest will expand naturally if suitable substrata can be provided.

Artificial Reefs

Techniques for the construction and placement of artificial reefs are necessarily evolving, and there are only a few examples on which to judge effects. Nevertheless, the main message seems to be to avoid placing small reefs in isolated sandy areas if the desired result is the rapid appearance and persistence of a stand of kelp. These sorts of reefs are unlikely to provide the surface area necessary for a kelp forest since physical conditions are often unsuitable, dispersal to the reef may be limited, and transplanted kelp are overrun by grazers (contrast development on Pendleton Artificial Reef and the artificial reef in San Luis Obispo discussed above).

Natural kelp forests are commonly bounded by sand that limits their expansion (Foster and Schiel 1985). It seems the creation of kelp forests could be best achieved by placing hard substrata on these soft bottom habitats near natural stands. Natural processes of spore production, dispersal and kelp growth would then probably save the need for more intervention by humans. A single fertile *Macrocystis* plant can put out millions of spores (Anderson and North 1966). These natural processes far exceed human capabilities of restoration.

Whatever the future use of artificial reefs is in the expansion or restoration of kelp forests, it should be pointed out that sand substrata provide distinct habitats that are a normal part of general coastal ecosystems. Therefore, the use of artificial reefs is a form of habitat substitution. Interstitial fauna, echinoderms such as sand dollars and sea stars, tubeworms, sea anemones, sea pens and fishes such as sole and halibut are typical residents of sand habitats. The dumping of hard substrata onto these sorts of areas represents a loss of habitat and a significant alteration to a natural community (Davis et al. 1982). Decisions to "restore" kelp using artificial reefs must therefore be made with due care for the other sorts of habitats present along a coastline.

Guidelines for Future Restoration and Creation

This section will be structured as a series of questions that should be considered before any attempts are made at restoration or creation of kelp forests.

Has the Kelp Forest Been Degraded?

There have been so many research and monitoring programs in the *Macrocystis* forests of the West Coast that changes in kelp abundance should be able to be placed into a spatial and temporal context. If kelp abundance seems to be reduced, monitoring should be done, especially using underwater surveying techniques (Foster et al. 1985) in order to determine: (1) the numbers of kelp; (2) the size structure; and (3) the spatial scale over which abundances are reduced (localized or more general?).

The spatial context is important to determine because of the inherent patchiness of kelp communities. There are many processes that affect communities on a small scale such as grazing, competition for space, and stratification of physical conditions. It is therefore important to determine if changes are greater than this inherent patchiness, i.e., are changes real?

It is much more difficult to put changes into a temporal context unless effects are devastating. This is because there can be considerable seasonal variation in kelp forest structure, particularly due to winter storms and variable recruitment. There is also inter-annual variation in the severity of storms and in oceanographic conditions of temperature and nutrients. Years when canopies are particularly lush may simply reflect a period of mild winters and may actually be the exception.

"Degradation," therefore, requires reference points both spatially and temporally. Variables of numbers and sizes of plants and animals in a structured sampling design can provide the spatial context, e.g., is this area of coast different from that area of coast? The temporal context can only be provided by comparisons among years, preferably using the same sampling design. One of the problems with the temporal context is that most of the long-term data on *Macrocystis* abundance are maps of the areal extent of surface canopies. This does not give an accurate representation of the underlying community structure. When canopies are reduced, all the components of a kelp community may still be present (e.g., Foster 1982; Wilson and Togstad 1983).

What Is Causing the Degradation?

The answer to this is often obvious. Coastal construction and waste outfalls are examples of point-source items that can affect kelp forests. The major uncertainties in ascribing causes are that changes to kelp community structure may occur slowly over several years (e.g., declines due to fishing) and may be difficult to discern against the background of natural variation. It is always tempting to implicate environmental degradation through human activities as the cause of changes in natural systems, especially when the changes occur over long periods of time. However, the history of kelp forest monitoring in California suggests that where human activities are responsible for alterations to the environment, such as increases in turbidity and sedimentation, the source is readily identifiable, e.g., the sewage outfall at Palos Verdes. It must also be recognized that the major documented changes in kelp forests throughout the 20th Century have been due to oceanographic events. As monitoring of coastal areas becomes more extensive, experimental work in kelp forests increases, and a better understanding of El Niño effects is gained, it is becoming increasingly clear that increases in temperature, decreases in nutrients and severe storms are the most important factors affecting kelp forests over a wide geographic scale. While these factors may be exacerbated by localized human activities, they are still part of the long-term natural variation of coastal ecosystems.

Biotic interactions, especially grazing by sea urchins, have frequently been cited as being major agents in the demise and lack of recovery of kelp forests (reviews in Foster and Schiel 1985; Schiel and Foster 1986). Grazing by sea urchins becomes more intense as the amount of drift algae decreases (Mattison et al. 1977; Harrold and Reed 1985). Grazed substrata provide good habitat for recruitment of sea urchins (Tegner and Dayton 1977; Andrew and Choat 1985), which may then come to dominate. However, domination by sea urchins is a natural feature of kelp forests (discussed above). The range of community structures on shallow reefs and the processes that control them are now understood reasonably well, and it is only through well-designed sampling and monitoring that any suggested sea urchin problems caused by man can be evaluated (Elner and Vadas 1990). This becomes particularly complex when changes in sea urchin populations are linked to changes in their predators, and both predators and prey vary in abundance naturally.

Can the Causes of Degradation Be Eliminated?

There are ecological, economic and political considerations in this decision. Ecological factors are usually the proximal causes for the lack of *Macrocystis* recruitment into local areas where kelp forests once existed. For example, sea urchins and understory kelps may effectively prevent successful settlement of *Macrocystis*. The considerations are whether or not it is appropriate to tamper with these species, which comprise natural habitats.

Economic and political factors are important in that moving sewage outfalls, tertiary treatment of sewage, cessation of coastal construction, and reduction in commercial and recreational activities in kelp forests all have attendant costs and other impacts on various interest groups. Changes in sewage treatment and discharge can be enormously expensive in direct costs. Changes in other activities have many inherent costs such as loss of present and potential employment and even loss of enjoyment. The decisions, therefore, necessarily fall into the political realm. If a cause of degradation can be identified, a decision must ultimately be made about the relative merits and costs of removing the source of contamination, who is responsible, and who is able or willing to pay the price. The decision is further complicated because there is often conflict between commercial and recreational interests. This normally requires some form of intervention by a regulatory agency which must consider the evidence for degradation of the environment, whether altering human activities is likely to lead to restoration of the kelp forest, and whether potential benefits are commensurate with the costs.

Regulation of activities and monitoring of the coastal environment are now the norm in the United States. These requirements can be an effective means of identifying causes of degradation of the environment and reducing or ceasing them altogether.

Are Mitigation and Compensation Desirable?

Decisions have frequently been made to allow activities that are likely to lead to degrading of nearshore kelp forests. These include the construction of piers and breakwaters and the discharges from power plants. In some cases, mitigation and compensation for the potential loss of kelp forest habitat have been considered. The major decisions are whether the provision of hard substrata is likely to increase the size of a kelp forest to compensate for the portion lost, whether it is advisable to sacrifice some other habitat such as a sand floor by the construction of a reef, and whether the financial costs are justified.

These questions are not always easily answered. Besides the ecological, economic and political factors involved in the decision-making process, the results of creation of new habitats have not been predictable in the past. While guidelines now exist (see previous section) for the installation of reefs, the end point of establishing kelp stands on them is far from certain. Moreover, even if kelp is established, it may not persist.

Will Degraded Areas Recover Naturally?

In cases where the causes of degradation have been identified and reduced or eliminated, the major decision is whether or not to rely on natural processes to effect recovery of kelp forests. If the degradation occurred over a wide area of coastline, say many kilometers, it will probably be expensive and futile to attempt to restore the kelp forest directly. Over smaller spatial scales, such as specific kelp stands, transplanting kelp and other species, seeding with small sporophytes or larvae, or creating new hard substratum may be viable options.

One of the criteria in deciding how pro-active to be is the length of time it is likely to take to achieve the desired result, i.e., a kelp forest. Given the right physical conditions, successful recruitment of kelp, and no overgrazing by fish and sea urchins, it will take at least a few years for a mature stand of kelp to develop. Of course, the process will take longer if some of the physical and biotic conditions are unfavorable. The decision to be pro-active, therefore, must be weighed against the costs, the length of time that significant restoring efforts may be required, and the likelihood that "restoring" will actually achieve more than natural recovery.

Most evidence to date suggests that natural recovery swamps efforts at restoring. The exceptions are localized areas where high densities of grazers were removed (allowing the natural recovery of kelp) and perhaps the kelp forests near Santa Barbara where natural substrata for kelp attachment were lost.

Is Biological Manipulation Futile?

Based on past experience the answer to this has mostly been yes. Several factors, however, suggest that this option may be warranted in some cases in the future. There is now a considerable storehouse of knowledge about the important processes in kelp forests. This has increased considerably during the past decade and will probably continue to do so. In a statistical, sense this understanding means that the source of the previously unexplained variance in numbers of kelp now can be ascribed to particular factors and their interactions (e.g., temperature, nutrients, light, grazers). Being able to reduce the unexplained variance is another way of saying that results of manipulations are becoming more predictable than they previously were.

Spatial scale affects this interpretation. As the size of manipulations increases, more patchiness (i.e., variation) in results will occur. Although this is a natural feature of kelp forests, the types of patches that result from manipulations may not be the ones hoped for.

We believe that manipulations should be attempted only when it appears unlikely that natural recovery will occur in a short time, say 2-3 years, and that pro-active means are likely to achieve a positive result (e.g., sand bottom kelp forests). This, therefore, necessarily involves the detailed consideration of each coastal area proposed for restoration: How far is it from a source of kelp propagules? Have benthic grazers become so abundant that natural recovery of kelp is unlikely in the short term? Has the physical environment recovered to such an extent that kelp recovery, either natural or manipulated, is likely? What are the likely costs of intervening, especially compared to the real and intrinsic costs of having no kelp forest in that area?

Do Reserves Work?

The answer is a qualified yes. A comparison of areas that are protected from exploitation either by regulation or inaccessibility shows that prominent resident species like lobsters, large benthic-feeding fishes and abalone are more abundant and reach a larger size in protected areas (Cowen 1983; Cole et al. 1990). An interesting feature, however, is the lack of evidence that protection from exploitation has any effect on the habitats themselves. The structure of kelp forests such as at Hopkins Marine Life Refuge (a scientific reserve) and Stillwater Cove (relatively inaccessible) in central California falls well within the range of *Macrocystis* forests throughout that region (Foster and Schiel 1985). A ten-year study in a marine reserve in New Zealand showed no significant changes in kelp habitats after protection, even though some reef species were more abundant and larger than in nearby areas (Cole et al. 1990). In contrast, there are reports of dramatic changes in heavily exploited rocky intertidal habitats after protection (Duran and Castilla 1989).

One often-cited value of reserves is that they serve as a reservoir of propagules that will assist in seeding exploited areas nearby. While this makes intuitive sense, the evidence for it is equivocal. For example, the use of reserves to protect lobsters has shown no discernible effect, against a background of natural variation, on recruitment outside the reserves (Conan 1986).

Reserves clearly have merit in protecting reef species that are prone to commercial and recreational exploitation. It seems unlikely, however, that their value extends to having a significant impact on the kelp forest themselves, and their effects on areas outside the reserves are, at present, questionable.

The Science of Kelp Forest Restoration

Past Approaches

We have reserved this section to discuss the scientific aspects of past restoration attempts and what might be a useful approach in the future.

The approaches to restoration work in the past have varied from entirely ad hoc ("Let's have a go and see what happens") to the carefully planned ("Let's pose hypotheses, then test them"). A general indication of which studies fall into each category is given by a scan through the reference list to see which studies are published in refereed journals.

Although it is hardly novel to stress the use of scientific processes (Fisher 1971; Green 1979; Underwood and Denley 1984), it is nevertheless worth repeating for kelp restoration work. It is also worth emphasizing the necessity of peer review at each stage of a program. While organizations and agencies are often short of people and money, it is false economy to embark on any field research without testing ideas on colleagues and having proposals reviewed. It is much cheaper to bear the initial costs of review than to discover later that an entire field program does not stand up to scrutiny because of faulty planning.

Established Protocols

Posing Testable Hypotheses. Every aspect of the work, from initial sampling to manipulations, should require an explicitly-stated and testable hypothesis. This point is considerably more than pedantic. If sampling, monitoring, or experiments are to be done, there is always a spatial and temporal context that must be part of the design. Simple examples are:

- *Macrocystis* abundance at Site A = *Macrocystis* abundance at Site B. This involves a variable (number of *Macrocystis* plants) that can be measured.

- *Macrocystis* size structure at Site A = *Macrocystis* size structure at Site B. The variable here is some measure of plant size, such as the number of fronds.

Hypotheses may get considerably more complicated where multiple factors are involved. Lack of attention to formulating hypotheses usually results in the collection of data that is inadequate or inappropriate for the questions being asked.

Adequate Sampling Design. What is the proposed structure of the data set? How will the data be collected? How will they be analyzed? Has a cost-benefit analysis been done to determine the most efficient way of sampling (see Andrew and Mapstone 1987)?

Proper Experimental Design. Any attempt at restoration, whether it involves manipulation or natural recovery, is an experiment. As such, it is necessary to have replication and controls. By definition, these must be pre-planned. Literature reviews during the 1980s have highlighted the large number of major flaws in published experimental papers in all fields of ecological research (Underwood 1981; Hurlbert 1984; Underwood and Denley 1984; Andrew and Mapstone 1987). Research in subtidal areas has also been particularly prone to lack of replication and suitable controls (Schiel and Foster 1986). The excuse is often that subtidal work is logistically difficult (it is), that shortcuts are sometimes necessary (they usually are not), or that since restoration is not research (in most cases it is) normal scientific procedures need not be followed (they should be).

There have been many attempts at kelp forest restoration where something was merely tried to see what would happen. For example, kelp plants may be transplanted or sea urchins smashed. A proper experimental approach would be to mark out similar areas in one (or preferably more than one) experimental site, randomly assign treatments and controls (e.g., kelp addition, grazer removal), then compare treatments and controls for particular variables such as kelp recruitment.

Of course, explicitly stating this seems trivial. Our only excuse for it is that experimental designs have tended to be the poor relation of restoration work, often being either confused or ignored altogether. The penalty to pay for this is that if a form of restoration works, the reasons for success are not clear (e.g., there may be no unmanipulated controls for comparison), and if failures occur few lessons are learned and, thus, considerable amounts of money are wasted. The logic often seems to be that if a few tries do not work, then more tries should be made, instead of using suitable replication and spatial scales in the first place.

Proposals. In most cases, it is now mandatory to produce a written proposal outlining all sampling, experimental and analytical procedures. After peer review, this then forms the blueprint of the work to be done. Peer review is particularly critical at this stage because it is the final step before the field program begins.

Any sort of restoration work in kelp forests requires a broad range of expertise in natural history, experimental ecology, implementing field work, laboratory work and data analysis. These should be reflected in the reviewers of proposals.

Progress Reports. As illustrated in previous sections, restoration work (this includes monitoring as well as manipulations) usually occurs over several years. Regardless of the duration of funding, annual reports detailing sampling, experiments, data analysis and interpretation should be peer-reviewed. It would be the rare field program that did not require some modification as it proceeded.

Publication. Either through incentives or contractual arrangements, publication of results in widely-disseminated, peer-reviewed journals should be required. The system of journal publication may have its flaws but it is still generally recognized as an arbiter and disseminator of quality scientific work, and there are several journals that publish applied research. A great proportion of restoration-related work, much of it funded by public agencies, is relatively inaccessible in papers and reports that are difficult to obtain. This work may be of high quality but nevertheless contributes little to the corpus of literature on kelp forests because it is little known and generally not available for critical analysis. There is usually nothing proprietary about the work, especially if it is funded by public agencies. Perhaps those providing funds should require final reports (or at least relevant parts of them) to be written in the form of scientific papers which can then be readily submitted to appropriate journals. Whatever the method of persuasion, it is certain that providers of funds for restoration have considerable say in the requirements for scientists to distribute information and ensure it is of high quality.

Steps to Ensuring Success

The history of kelp forest restoration work shows that "success" is measured not only in endpoint results (i.e., is a kelp forest reestablished) but also in what is learned about the processes affecting kelp forests. This is not an academic sidestep. The endpoint of having a lush kelp forest in a particular place in a given time cannot be ensured, nor is it likely to be in the future. With the information gained from past scientific work, however, the balance of probability has shifted towards better prediction of likely results of natural processes and manipulative techniques.

We believe that "ensuring success" comes only in the adherence to the principles and procedures outlined above. Outcomes are then measured in whether a process has an effect, and the magnitude of that effect, against the background noise of natural variability.

Within the context of kelp forest restoration, we also include monitoring programs. These can be thought of as mensurative experiments: hypotheses are posed, factors are tested, variables are measured (Underwood 1981; Hurlburt 1984; Andrew and Mapstone 1987). Without monitoring programs, it is difficult or impossible to put changes in kelp forests into a spatial or temporal context.

Because of the vastness of kelp forests along the West Coast of North America, public agencies, businesses and academics will have to continue to cooperate, and cooperate more fully (Anon. 1990) in monitoring and experimental programs. These groups will continue to have their own agendas and objectives, but an underlying theme of cooperation is achieved through the process of peer review. This already exists to a great extent, but cannot be overemphasized. If peer review is required and if funding agencies require scientists to take heed of recommendations from reviews, there is no better way to ensure, or at least increase the odds, that our understanding of kelp forests and how to restore them will improve.

Acknowledgments

We thank D.C. Barilotti (Kelco Company, San Diego, California), S. Kren (Pacific Gas and Electric Company, Avila, California), G. VanBlaricom (U.S. Fish and Wildlife Service, Santa Cruz, California) and K. Wilson (California Fish and Game, Long Beach, California) for providing unpublished information, and D.C. Barilotti, G. Thayer, K. Wilson and R. Vadas for their helpful comments on the manuscript. Special thanks to G. Thayer for inviting us to present this paper and for organizing the symposium.

Literature Cited

Abbott, I.A. and G.J. Hollenberg. 1976. Marine algae of California. Standford University Press, Stanford, California 827 p.

Anderson, E.K. and W.J. North. 1966. *In situ* studies of spore production and dispersal in the giant kelp, *Macrocystis pyrifera.* Proc. Int. Seaweed Symp. 5: 73-86.

Andrew, N.L. and J.H. Choat. 1985. Habitat related differences in the survivorship and growth of juvenile sea urchins. Mar. Ecol. Prog. Ser. 27: 155-161.

Andrew, N.L. and B.D. Mapstone. 1987. Sampling and the description of spatial pattern in marine ecology. Oceanogr. Mar. Biol. Ann. Rev. 25: 39-90.

Anon. 1990. Managing troubled waters. Nat. Acad. Press, Washington, D.C. 125 p.

Ayling, A. 1981. The role of biological disturbance in temperate subtidal encrusting communities. Ecology 62: 830-847.

Barilotti, D.C. and J.A. Zertuche-Gonzalez. 1990. Ecological effects of seaweed harvesting in the Gulf of California and Pacific Ocean off Baja California and California. Hydrobiologia 204/205: 35-40.

Bernstein, B.B. and N. Jung. 1979. Selective pressures and coevolution in a kelp canopy community in southern California. Ecol. Monogr. 49: 335-355.

Breen, P.A. and K.H. Mann. 1976. Destructive grazing of kelp by sea urchins in eastern Canada. J. Fish. Res. Board. Can. 33: 1278-1283.

Burnett, R. 1971. DDT residues: distributions of concentrations in *Emerita analoga* (Stimpson) along coastal California. Science 174: 606-608.

Cailliet, G.M. and R.N. Lea. 1977. Abundance of the "rare" zoarcid, *Maynea californica* Gilbert, 1915, in the Monterey Canyon, Monterey Bay, California. Calif. Fish Game 63: 253-261.

Carr, M.H. 1989. Effects of macroalgal assemblages on the recruitment of temperate zone reef fishes. J. Exp. Mar. Biol. Ecol. 126: 59-76.

Carter, J.W., A.L. Carpenter, M.S. Foster and W.N. Jessee. 1985a. Benthic succession on an artificial reef designed to support a kelp-reef community. Bull. Mar. Sci. 7: 86-113.

Carter, J.W., W.N. Jessee, M.S. Foster and A.L. Carpenter. 1985b. Management of artificial reefs designed to support natural communities. Bull. Mar. Sci. 37: 114-128.

Chapman, A.R.O. 1981. Stability of sea urchin-dominated barren grounds following destructive grazing of kelp in St. Margaret's Bay, eastern Canada. Mar. Biol. 62: 307-311.

Chapman, A.R.O. and J.S. Craigie. 1977. Seasonal growth in *Laminaria longicruris*: relations with dissolved inorganic nutrients and internal reserves of nitrogen. Mar. Biol. 40: 197-205.

Choat, J.H. 1982. Fish feeding and the structure of benthic communities in temperate waters. Annu. Rev. Ecol. Syst. 13: 423-449.

Choat, J.H. and P.D. Kingett. 1982. The influence of fish predation on the abundance cycles of an algal turf invertebrate fauna. Oecologia 54: 88-95.

Conan, G.Y. 1986. Summary of Session 5: Recruitment enhancement. Can. J. Fish. Aquat. Sci. 43: 2384-2388.

Cole, R.G., A.M. Ayling and R.G. Creese. 1990. Effects of marine reserve protection at Goat Island, northern New Zealand. N.Z. J. Mar. Freshwater Res. 24: 197-210.

Connell, J.H. and W.P. Sousa. 1983. On the evidence needed to judge ecological stability or persistence. Am. Natur.121: 789-824.

Cowen, R.K. 1983. The effect of sheephead (*Semicossyphus pulcher*) predation on red sea urchin (*Strongylocentrotus franciscanus*) populations: an experimental analysis. Oecologia 58: 249-255.

Cowen, R.K. 1985. Large scale pattern of recruitment by the labrid, *Semicossyphus pulcher*: causes and implications. J. Mar. Res. 43: 719-742.

Cowen, R.K., C.R. Agegian and M.S. Foster. 1982. The maintenance of community structure in a central California kelp forest. J. Exp. Mar. Biol. Ecol. 64: 189-201.

Davis, N., G.R. VanBlaricom and P.K. Dayton. 1982. Man-made structures on marine sediments: effects on adjacent benthic communities. Mar. Biol. 70: 295-303.

Dawson, E.Y. and M.S. Foster. 1982. Seashore plants of California. University of California Press, Berkeley. 226 p.

Dayton, P.K. 1985. Ecology of kelp communities. Annu. Rev. Ecol. Syst. 16: 215-245.

Dayton, P.K. and M.J. Tegner. 1984. Catastrophic storms, El Niño, and patch stability in a southern California kelp community. Science 224: 283-285.

Dayton, P.K., V. Currie, T. Gerrodette, B.D. Keller, R.Rosenthal and D. Ventresca. 1984. Patch dynamics and stability of some California kelp communities. Ecol. Monogr. 54: 253-289.

Dean, T.A. 1985. The temporal and spatial distribution of underwater quantum irradiation in a southern California kelp forest. Estuarine Coastal Shelf Sci. 21: 835-844.

Dean, T.A. and L.E. Deysher. 1983. The effects of suspended solids and thermal discharges on kelp, p. 114-135. *In* W. Bascom (ed.), The effects of waste disposal on kelp communities. South. Calif. Coastal Water Res. Proj., Long Beach.

Dean, T.A. and F.R. Jacobsen. 1986. Nutrient-limited growth of juvenile kelp, *Macrocystis pyrifera*, during the 1982-1984 "El Niño" in southern California. Mar. Biol. 90: 597-601.

Dean, T.A., S.C. Schroeter and J.D. Dixon. 1984. Effects of grazing by two species of sea urchins (*Strongylocentrotus franciscanus* and *Lytechinus anamesus*) on recruitment and survival of two species of kelp (*Macrocystis pyrifera* and *Pterygophora californica*). Mar. Biol. 78: 301-313.

Devinny, J.S. and L.A. Volse. 1978. Effects of sediments on the development of *Macrocystis pyrifera* gametophytes. Mar. Biol. 48: 343-348.

Deysher, L.E. and T.A. Dean. 1984. Critical irradiance levels and the interactive effects of quantum irradiance and quantum dose on gametogenesis in the giant kelp, *Macrocystis pyrifera*. J. Phycol. 20: 520-524.

Deysher, L.E. and T.A. Dean. 1986a. *In situ* recruitment of sporophytes of the giant kelp, *Macrocystis pyrifera* (L.) C.A. Agardh: effects of physical factors. J. Exp. Mar. Biol. Ecol. 103: 41-63.

Deysher, L.E. and T.A. Dean. 1986b. Interactive effects of light and temperature on sporophyte production in the giant kelp *Macrocystis pyrifera*. Mar. Biol. 93: 17-20.

Deysher, L.E. and T.A. Norton. 1982. Dispersal and colonization in *Sargassum muticum* (Yendo) Fensholt. J. Exp. Mar. Biol. Ecol. 56: 179-195.

Domning, D.P. 1978. Sirenian evolution in the North Pacific Ocean. University of California Press, Berkelely. 176 p.

Druehl, L.D. 1970. The pattern of Laminariales distribution in the northeast Pacific. Phycologia 9: 237-247.

Duggins, D.O. 1980. Kelp beds and sea otters: an experimental approach. Ecology 61: 447-453.

Duran, L.R. and J.C. Castilla. 1989. Variation and persistence of the middle rocky intertidal community of central Chile, with and without human harvesting. Mar. Biol. 103: 555-562.

Ebeling, A.W. and D.R. Laur. 1985. The influence of plant cover on surfperch abundance at an offshore temperate reef. Environ. Biol. Fishes 12: 169-179.

Ebeling, A.W. and D.R. Laur. 1988. Fish populations in kelp forests without sea otters: effects of severe storm damage and destructive sea urchin grazing, p. 169-191. *In* G.R.VanBlaricom and J.A. Estes (eds.), The community ecology of sea otters. Springer-Verlag, Berlin.

Ebeling, A.W., D.R. Laur and R.J. Rowley. 1985. Severe storm disturbances and reversal of community structure in a southern California kelp forest. Mar. Biol. 84: 287-294.

Ebeling, A.W., W. Werner, F.A. DeWitt and G.M. Cailliet. 1971. Santa Barbara oil spill: short-term analysis of macro-plankton and fish. U.S. Environmental Protection Agency, Water Pollut. Control Res. Ser. 15080EAL02/71. Natl. Tech. Info. Serv., Springfield. VA. 68 p.

Elner, R.W. and R.L. Vadas. 1990. Inference in ecology: the sea urchin phenomenon in the northwestern Atlantic. Am. Natur. 136: 108-125.

Eppley, R.W., A.F. Carlucci, O. Holm-Hansen, D. Kiefer, J.J. McCarthy and P.M. Williams. 1972. Evidence for eutrophication in the sea near southern California sewage outfalls. Calif. Coop. Oceanic Fish. Invest. Rep. 16: 74-83.

Estes, J.A. and C. Harrold. 1988. Sea otters, sea urchins, and kelp beds: some questions of scale, p. 116-150. *In* G.R.VanBlaricom and J.A. Estes (eds.), The community ecology of sea otters. Springer-Verlag, Berlin.

Estes, J.A. and G.R. VanBlaricom. 1985. Sea otters and shell fisheries, p. 187-235. *In* J.R. Beddington, R.J. Beverton, and D.M. Lavigne (eds.), Marine mammals and fisheries. Allen and Unwin, London.

Estes, J.A., R.J. Jameson and A.M. Johnson. 1981. Food selection and some foraging tactics of sea otters, p 606-641. *In* J.A. Chapman and D. Pursley (eds.), The worldwide furbearer conference proceedings, Vol. 1. Worldwide Furbearer Conference, Inc., Frostburg, Maryland.

Fisher, R.A. 1971. The design of experiments. 9th ed. Hafner, New York. 252 p.

Foster, M.S. 1975a. Algal succession in a *Macrocystis pyrifera* forest. Mar. Biol. 32: 313-329.

Foster, M.S. 1975b. Regulation of algal community development in a *Macrocystis pyrifera* forest. Mar. Biol. 32: 331-342.

Foster, M.S. 1982. The regulation of macroalgal associations in kelp forests, p. 185-205. *In* L. Srivastava (ed.), Synthetic and degradative processes in marine macrophytes. Walter deGruyter, Berlin.

Foster, M.S. and D.R. Schiel. 1985. The ecology of giant kelp forests in California: a community profile. U.S. Fish and Wildlife Service. Biol. Rep. 85: 1-152.

Foster, M.S. and D.R. Schiel. 1988. Kelp communities and sea otters: keystone species or just another brick in the wall?, p. 92-115. *In* G. VanBlaricom and J.A. Estes (eds.), The community ecology of sea otters. Springer-Verlag, Berlin.

Foster, M.S., M. Neushul and R. Zingmark. 1971. The Santa Barbara oil spill part 2: initial effects on intertidal and kelp bed organisms. Environ. Pollut.2: 115-134.

Foster, M.S., J.W. Carter and D.R. Schiel. 1983. The ecology of kelp communities, p. 53-69. *In* W. Bascom (ed.), The effects of waste disposal on kelp communities. So. Calif. Coastal Water Res. Proj., Long Beach.

Foster, M.S., T.A. Dean and L.E. Deysher. 1985. Subtidal techniques, p. 199-231. *In* M.M. Littler and D.S. Littler (eds.), Handbook of phycological methods: ecological methods for macroalgae. Cambridge University Press, Cambridge.

Frey, H.W. 1971. California's living marine resources and their utilization. Calif. Dep. Fish Game Rev., Sacramento. 148p.

Gaines, S.D. and J. Roughgarden. 1987. Fish in offshore kelp forests affect recruitment to intertidal barnacle populations. Science 235: 479-480.

Gerard, V.A. 1976. Some aspects of material dynamics and energy flow in a kelp forest in Monterey Bay, California. Ph.D.Thesis, University of California, Santa Cruz. 173 p.

Gerard, V.A. 1982. *In situ* rates of nitrate uptake by giant kelp, *Macrocystis pyrifera* (L.) C. Agardh, tissue differences,environmental effects, and predictions of nitrogen-limited growth. J. Exp. Mar. Biol. Ecol. 62: 211-224.

Grant, J.J., K.C. Wilson, A. Grover and H.A. Togstad. 1982. Early development of Pendleton artificial reef. Mar. Fish. Rev. 44 (6-7): 53-60.

Green, R.H. 1979. Sampling design and statistical methods for environmental biologists. Wiley, New York. 257 p.

Grigg, R.W. and R.S. Kiwala. 1970. Some ecological effects of discharged waste on marine life. Cal. Fish Game 56: 145-155.

Grove, R.S. 1982. Artificial reefs as a resource management option for siting coastal power stations in Southern California. Mar. Fish. Rev. 44 (6-7): 24-27.

Harger, B.W.W. and M. Neushul. 1983. Test-farming of the giant kelp, *Macrocystis*, as a marine biomass producer. J. World Maricult. Soc. 14: 392-403.

Harris, L.G., A.W. Ebeling, D.R. Laur and R.J. Rowley. 1984. Community recovery after storm damage: a case of facilitation in primary succession. Science 224: 1336-1338.

Harrold, C. and S. Lisin. 1989. Radio-tracking rafts of giant kelp: production and regional transport. J. Exp. Mar. Biol.Ecol. 130: 237-251.

Harrold, C. and J.S. Pearse. 1987. The ecological role of echinoderms in kelp forests, p. 137-233. *In* M. Jangoux and J.M. Lawrence (eds.), Echinoderm studies Vol. II, A.A. Balkema, Rotterdam.

Harrold, C. and D.C. Reed. 1985. Food availability, sea urchin grazing, and kelp forest community structure. Ecology 66:1160-1169.

Hines, A.H. and J.S. Pearse. 1982. Abalones, shells, and sea otters: dynamics of prey populations in central California. Ecology 63: 1547-1560.

Holbrook, S.J., R.J. Schmitt and R.F. Ambrose. 1990a. Biogenic habitat structure and characteristics of temperate reef fish assemblages. Aust. J. Ecol. 15: 489-503.

Holbrook, S.J., M.H. Carr, R.J. Schmitt and J.A. Coyer. 1990b. Effect of giant kelp on local abundance of reef fishes: the importance of ontogenetic resource requirements. Bull. Mar. Sci. 47: 104-114.

Hunt, D.E. 1977. Population dynamics of *Tegula* and *Calliostoma* in Carmel Bay, with special reference to kelp harvesting. M.S. Thesis, San Francisco State University, San Francisco. 81 p.

Hurlbert, S.H. 1984. Pseudoreplication and the design of ecological field experiments. Ecol. Monogr. 54: 187-211.

Jackson, G.A. 1977. Nutrients and production of the giant kelp *Macrocystis pyrifera* off southern California. Limnol. Oceanogr. 22: 979-995.

Kastendiek, J. 1982. Competitor-mediated coexistence: interactions among three species of benthic macroalgae. J.Exp. Mar. Biol. Ecol. 62: 201-210.

Kenyon, K.W. 1969. The sea otter in the eastern Pacific Ocean. U.S. Dept. Interior, N. Amer. Fauna 68. 352 p.

Kimura, R.S. and M.S. Foster. 1984. The effects of harvesting *Macrocystis pyrifera* on the algal assemblage in a giant kelp forest. Hydrobiologia 116/117: 425-428.

Kvitek, R. G. and J.S. Oliver. 1988. Sea otter foraging habits and effects on prey populations and communities in soft-bottom environments, p. 22-47. *In* G. R. VanBlaricom and J.A. Estes (eds.), The community ecology of sea otters. Springer-Verlag, Berlin.

Lawrence, J.M. 1975. On the relationships between marine plants and sea urchins. Oceanogr. Mar. Biol. Annu. Rev. 13: 213-286.

Leighton, D.L. 1971. Grazing activities of benthic invertebrates in southern California kelp beds. Nova Hedwigia 32: 421-453.

Leighton, D.L., L.G. Jones and W.J. North. 1966. Ecological relationships between kelp and sea urchins in southern California. Proc. Int. Seaweed Symp. 5: 141-153.

Levin, S.A. 1988. Sea otters and nearshore benthic communities: a theoretical perspective, p. 202-209. *In* G.R. Vanblaricom and J.A. Estes (eds.), The community ecology of sea otters. Springer-Verlag, Berlin.

Lobban, C.S. 1978. Growth of *Macrocystis integrifolia* in Barkley Sound, Vancouver Island, B.C. Can. J. Bot. 56: 2707-2711.

Lowry, L.F. and J.S. Pearse. 1973. Abalones and sea urchins in an area inhabited by sea otters. Mar. Biol. 23: 213-219.

Luning, K. 1980. Critical levels of light and temperature gulating the gametogenesis of three *Laminaria* species (*Phaeophyceae*). J. Phycol. 24: 181-191.

Luning, K. and M. Neushul. 1978. Light and temperature demands for growth and reproduction of laminarian gametophytes in southern and central California. Mar. Biol. 45: 297-309.

Mann, K.H. 1973. Seaweeds: their productivity and strategy for growth. Science 182: 975-981.

Mattison, J.E., J.D. Trent, A.L. Shanks, T.B. Akin and J.S.Pearse. 1977. Movement and feeding activity of red sea urchins (*Strongylocentrotus franciscanus*) adjacent to a kelp forest. Mar. Biol. 39: 25-30.

McPeak, R.H. and D.A. Glantz. 1984. Harvesting California's kelp forests. Oceanus 27: 19-26.

Meistrell, J.C. and D.E. Montagne. 1983. Waste disposal in southern California and its effects on the rocky subtidal habitat, p. 84-102. *In* W. Bascom (ed.), The effects of waste disposal on kelp communities. South. Calif. Coastal Water Res. Proj., Long Beach.

Miller, D.J. and J.J. Geibel, 1973. Summary of blue rockfish and lingcod life histories; a reef ecology study; and giant kelp, *Macrocystis pyrifera*, experiments in Monterey Bay, California. Calif. Dept. Fish Game, Fish Bull. 158: 1-137.

Miller, R.J. 1985. Seaweeds, sea urchins, and lobsters: a reappraisal. Can. J. Fish. Aquat. Sci. 43: 2061-2072.

Nelson, B.V. and R.R. Vance. 1979. Diel foraging patterns of the sea urchin *Centrostephanus coronatus* as a predator avoidance strategy. Mar. Biol. 51: 251-258.

Neushul, M. 1963. Studies of the giant kelp, *Macrocystis*. II. Reproduction. Am. J. Bot. 50: 354-359.

North, W.J. 1963a. Kelp habitat improvement project: Annual report 1962-1963. W.M. Keck Laboratory of Environmental Health Engineering, California Institute of Technology, Pasadena. 123 p.

North, W.J. 1963b. Ecology of the rocky nearshore environment in southern California and possible influences of discharged wastes. Int. J. Air Water Pollut. 7: 721-736.

North, W.J. 1971a. Growth of individual fronds of the mature giant kelp, *Macrocystis*. Nova Hedwigia 32: 123-168.

North, W.J. 1971b. The biology of giant kelp beds (*Macrocystis*) in California. Nova Hedwigia 32: 1-600.

North, W.J. 1974. Annual report of the kelp habitat improvement project, 1 July, 1973 - 30 June, 1974. W.M. Keck Laboratory, California Institute of Technology, Pasadena. 137 p.

North, W.J. 1976. Aquacultural techniques for creating and restoring beds of giant kelp, *Macrocystis* spp. J. Fish. Res. Board. Can. 33: 1015-1023.

North, W.J. 1983. The sea urchin problem, p. 147-162. *In* W. Bascom (ed.), The effects of waste disposal on kelp communities. South. Calif. Coastal Water Res. Proj., Long Beach.

North, W.J. and C.L. Hubbs. 1968. Utilization of kelp bed resources in southern California. Cal. Dept. Fish Game, Fish Bull. 139: 1-264.

North, W.J. and J.S. Pearse. 1970. Sea urchin explosion in southern California coastal waters. Science 167: 209.

North, W.J., M. Neushul and K.A. Clendenning. 1964. Successive biological changes observed in a marine cove exposed to a large spillage of mineral oil, p. 335-354. *In* Proceedings of the symposium on pollution of marine organisms, Prod. Petrol., Monaco.

Norton, T.A. 1978. The factors influencing the distribution of *Saccorhiza polyschides* in the region of Lough Ine. J. Mar. Biol. Ass. U.K. 58: 527-536.

Pearse, J.S. and A.H. Hines. 1979. Expansion of a central California kelp forest following mass mortality of sea urchins. Mar. Biol. 51: 83-91.

Pringle, J.D. 1986. A review of urchin/macro-algal associations with a new synthesis for nearshore, eastern Canadian waters. Monogr. Biol. 4: 191-218.

Quast, J.C. 1971. Observations on the food of the kelp bed fishes. Nova Hedwigia 32: 541-579.

Reed, D.C. 1990. The effects of variable settlement and early competition on patterns of kelp recruitment. Ecology 71:776-787.

Reed, D.C. and M.S. Foster. 1984. The effects of canopy shading on algal recruitment and growth in a giant kelp forest. Ecology 65: 937-948.

Reed, D.C., D.R. Laur and A.W. Ebeling. 1988. Variation in algal dispersal and recruitment: the importance of episodic events. Ecol. Monogr. 58: 321-335.

Riedman, M.S. and J.A. Estes. 1988. A review of the history, distribution, and foraging ecology of sea otters, p. 4-21. *In* G.R. VanBlaricom and J.A. Estes (eds.), The community ecology of sea otters. Springer-Verlag, Berlin.

Rosenthal, R.J., W.D. Clarke and P.K. Dayton. 1974. Ecology and natural history of a stand of giant kelp, *Macrocystis pyrifera*, off Del Mar, California. U.S. Natl. Mar. Fish. Serv. Fish. Bull. 72: 670-684.

Schiel, D.R. 1982. Selective feeding by the echinoid, *Evechinus chloroticus*, and the removal of plants from subtidal algal stands in Northern New Zealand. Oecologia 54: 379-388.

Schiel, D.R. 1988. Algal interactions on shallow subtidal reefs in northern New Zealand: a review. N. Z. J. Mar. Freshwater Res. 22: 481-489.

Schiel, D.R. In Press. The enhancement of paua (abalone) populations in New Zealand. *In* S.A. Shepherd and M.J. Tegner (eds.), Abalone of the world: their biology, fisheries, and culture. Blackwell, Oxford.

Schiel, D.R. and M.S. Foster. 1986. The structure of subtidal algal stands in temperate waters. Oceanogr. Mar. Biol. Annu. Rev. 24: 265-307.

Schiel, D.R. and B.C. Welden. 1987. Responses to predators of cultured and wild red abalone, *Haliotis rufescens*, in laboratory experiments. Aquaculture 60: 173-188.

Seymour, R.J., M.J. Tegner, P.K. Dayton and P.E. Parnell. 1989. Storm wave induced mortality of giant kelp, *Macrocystis pyrifera*, in southern California. Estuarine Coastal Shelf Sci. 28: 277-292.

Simenstad, C.A., J.A. Estes and K.W. Kenyon. 1978. Aleuts, sea otters, and alternate stable-state communities. Science 200:403-411.

Smith, B.M. 1979. The effects of copper on gametophytes of *Macrocystis pyrifera*. M.S. Thesis, California State University, Hayward. 101 p.

Tegner, M.J. 1989. The feasibility of enhancing red sea urchin, *Strongylocentrotus franciscanus*, stocks in California: an analysis of options. Mar. Fish. Rev. 51(2): 1-22.

Tegner, M.J. and R.A. Butler. 1985. The survival and mortality of seeded and native red abalones, *Haliotis rufescens*, on the Palos Verdes peninsula. Cal. Fish Game 71: 150-163.

Tegner, M.J. and R.A. Butler. 1989. Abalone seeding, p. 157-182. *In* K.O. Hahn (ed.), Handbook of culture of abalone and other marine gastropods. CRC Press, Boca Raton, Florida.

Tegner, M.J. and P.K. Dayton. 1977. Sea urchin recruitment patterns and implications of commercial fishing. Science 196: 324-326.

Tegner, M.J. and P.K. Dayton. 1981. Population structure, recruitment and mortality of two sea urchins (*Strongylocentrotus franciscanus* and *S. purpuratus*) in a kelp forest. Mar. Ecol. Prog. Ser. 5: 255-268.

Tegner, M.J. and P.K. Dayton. 1987. El Niño effects on southern California kelp forest communities. Adv. Ecol. Res.17: 243-279.

Tegner, M.J. and L.A. Levin. 1983. Spiny lobsters and sea urchins: analysis of a predator-prey interaction. J. Exp.Mar. Biol. Ecol. 73: 125-150.

Underwood, A.J. 1981. Techniques of analysis of variance in experimental marine biology and ecology. Oceanogr. Mar. Biol. Annu. Rev. 19: 513-605.

Underwood, A.J. and E.J. Denley. 1984. Paradigms, explanations, and generalizations in models for the structure of intertidal communities on rocky shores, p. 151-180. *In* D.R. Strong, D. Simberloff, L.G. Abele and A.B. Thistle (eds.), Ecological communities: conceptual issues and the evidence. Princeton University Press. Princeton.

Vadas, R.L. 1972. Ecological implications of culture studies on *Nereocystis luetkeana.* J. Phycol. 8: 196-203.

Vadas, R.L. 1977. Preferential feeding: an optimization strategy in sea urchins. Ecol. Monogr. 47: 337-371.

VanBlaricom, G.R. and R.J. Jameson. 1982. Lumber spill in central California waters: implications for oil spills and sea otters. Science 215: 1503-1505.

VanBlaricom, G.R. and J.A. Estes (eds.). 1988. The community ecology of sea otters. Springer-Verlag, Berlin. 247 p.

Weaver, A.M. 1977. Aspects of the effects of particulate matter on the ecology of a kelp forest (*Macrocystis pyrifera* (L.) C.A. Agardh) near a small domestic sewer outfall. Ph.D. Dissertation, Stanford University, Stanford. 174 p.

Wharton, W.G. and K.H. Mann. 1981. Relationship between destructive grazing by the sea urchin, *Strongylocentrotus droebachiensis,* and the abundance of American lobster, *Homarus americanus,* on the Atlantic coast of Nova Scotia. Can. J. Fish. Aquat. Sci. 38: 1339-1349.

Wheeler, W.N. 1980. Effect of boundary layer transport on the fixation of carbon by the giant kelp, *Macrocystis pyrifera.* Mar. Biol. 56: 103-110.

Wheeler, P.A. and W.J. North 1981. Nitrogen supply, tissue composition and frond growth rates for *Macrocystis pyrifera.* Mar. Biol. 64: 59-69.

Wilson, K. and H. Togstad. 1983. Storm caused changes in the Palos Verdes kelp forest, p. 301-307. *In* W. Bascom (ed.), The effects of waste disposal on kelp communities. South. Calif. Coastal Water Res. Proj., Long Beach.

Wilson, K. and R. McPeak. 1983. Kelp restoration, p. 199-216. *In* W. Bascom (ed.), The effects of waste disposal on kelp communities. South. Calif. Coastal Water Res. Proj., Long Beach.

Wilson, K., P.L. Haaker and D.A. Hanan. 1978. Kelp restoration in southern California, p. 183-202. *In* R. Krauss (ed.), The marine plant biomass of the Pacific Northwest coast. Oregon State University Press, Corvallis.

Woodhouse, C.D., R.K. Cowen and L.R. Wilcoxon. 1977. A summary of the knowledge of the sea otter, *Enhydra lutris,* in California and an appraisal of the completeness of biological understanding of the species. U.S. Mar. Mammal Comm. Spec. Rep. No. MMC-76/02. U.S. Mar. Mammal Comm., Washington, D.C. 71 p.

Zimmerman, R.C. and J.N. Kremer. 1984. Episodic nutrient supply to a kelp forest ecosystem in southern California. J. Mar. Res. 42: 591-604.

Zimmerman, R.C. and J.N. Kremer. 1986. *In situ* growth and chemical composition of the giant kelp, *Macrocystis pyrifera*: response to temporal changes in ambient nutrient availability. Mar. Ecol. Prog. Ser. 27: 277-285.

CHAPTER 8

Restoring Stream Habitats Affected by Logging Activities

K. V. Koski

National Marine Fisheries Service
Alaska Fisheries Science Center, Auke Bay, Alaska

Abstract

Most of the 5,225,000 km of streams in the United States have been degraded by land-use practices including agriculture, grazing, channelization and logging. Salmonids are important to the nation's economy; restoration of streams is needed to return the carrying capacity of the habitat to a previously existing level. Logging alters a hierarchy of environmental factors (water quality, energy source, physical structure, flow regime and biotic interactions) which can limit salmonid production. A fundamental concept of stream restoration is that removal of such limiting factors will increase production. Early efforts to restore streams failed because of inadequate knowledge of limiting factors, stream dynamics and an over-reliance on hatchery propagation. Increased knowledge of the structure and function of stream ecosystems and of salmonid-habitat relationships has provided the present scientific basis for effective stream restoration.

The stream's carrying capacity to produce salmonids is controlled by the structure and function of the riparian zone. The woody debris function has been most affected by logging and development. Physical structures emulating channel stability and habitat complexity created by woody debris are the focus of most restoration projects. Effective restoration programs must be holistic in scope and include procedures for restoration of the watershed, stream channel and fish resources. Program planning must include an inventory of fish and habitat in the watershed, determination of habitat requirements of the species involved, assessment of land-use activities, analysis of limiting factors and a project evaluation plan. Because restoration projects are costly and may never attain pre-existing conditions, restoration must not be done in lieu of adequate protection. The best alternative is to provide good watershed management and to maintain healthy riparian buffer zones.

Introduction

From all over the land there is a murmur of complaint about the pollution of our creeks and rivers by manufacturing companies, dyeing establishments, saw mills, and the like.... (Mather 1875)

History of Stream Abuse

Deforestation and pollution were recognized as a threat to streams and fish habitat over a century ago by Mather (1875). Van Cleef (1885) appealed to the American Fisheries Society for "habitat restoration" and "protection of riparian habitat" to restore trout habitat destroyed by logging. Unfortunately, settlers and developers disregarded such concerns and left a legacy of degraded streams.

Streams were dammed to power mills and bellows of forges; forests were cleared for crops, fuel and lumber; and runs of fish were overfished. By the late 1700s, Atlantic salmon (*Salmo salar*) showed serious declines because of pollution, soil erosion and loss of habitat. The Exeter River lost striped bass (*Morone saxatilis*) and sturgeon (*Ascipenser* spp.) from overfishing by 1762, and its alewife (*Alosa pseudoharengus*) run from damming by 1790 (Bowen 1970). The great Atlantic salmon runs in the Connecticut River were extinct by the 1820s because of small hydropower dams (Jones 1986). The Michigan grayling (*Thymallus tricolor*) became extinct from unregulated logging of the Great Lakes pine forests (Behnke 1991).

During the late 19th century, many great salmon (*Oncorhynchus* spp.) runs in the West (e.g., Columbia, Sacramento and Rogue rivers) declined (Childerhose and Trim 1979). In 1883, 19 million kg of Columbia River chinook (*O. tshawytscha*) salmon were caught commercially; recent salmon and steelhead (*O. mykiss*) runs in the Columbia River have averaged about 9 million kg (2.5 million fish), mostly of hatchery origin (Behnke 1991).

In 1990, the U.S. Department of Commerce, National Marine Fisheries Service (NMFS) was petitioned to list the Snake River stocks of chinook and sockeye (*O. nerka*) and the lower Columbia River coho (*O. kisutch*) salmon stocks as "endangered" or "threatened" species under the 1973 Endangered Species Act (Behnke 1991). The winter run of chinook salmon in the Sacramento River has already been listed as threatened.

Value of Anadromous Salmonids

Salmon are the Nation's most valuable marine fish species (U.S. Dept. Commer. 1990) and have been a mainstay to the cultural and economic life-styles of Alaska and the Pacific Northwest for nearly two centuries. The average commercial catch (1985-89) of all Pacific salmon in the northeastern Pacific (Figure 1) is about 170 million fish, a value to fishermen of about $773 million (Talley 1990). Alaska accounts for about 73% of the total catch (64% of its total value), Canada for 21%, and Washington, Oregon and California combined for 6%. Habitat degradation is at least partially responsible for the small proportion of the total salmon catch currently shared by Washington, Oregon and California.

Logging Industry

America's timber wealth produced a transient logging and lumbering industry, which began in New England in the 1700s, migrated to the Great Lakes states and Gulf Plain region in the late 1800s, and finally expanded into the Pacific Northwest (Robbins

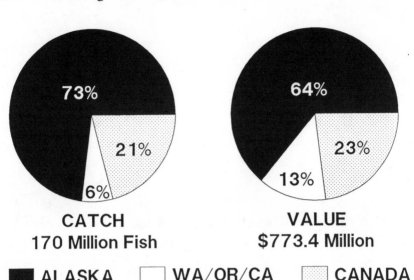

CATCH
170 Million Fish

VALUE
$773.4 Million

■ ALASKA ☐ WA/OR/CA ▦ CANADA

Figure 1. Production of Pacific salmon in the northeastern Pacific Ocean, as measured by average commercial catch and value, 1985 to 1989 (Talley 1990).

1988). Lumbermen existed in an economic environment of over-production and cutthroat competition; in the quest for profits, there was no consideration for soil erosion, selective logging, or the environmental consequences of splash dams and cut-and-run practices (Figure 2). For over 150 years, the forest industry was a prime example of migrating capital, rapid liquidation of resources and boom-and-bust cycles for towns dependent on the forest bounty.

In the early 1800s, the dominant feature of pristine streams in North America was the abundance of logs, rootwads, branches or fragments of trees which had fallen into or over the stream (Sedell et al. 1984). For over 150 years, however, channel clearance for navigation, large-scale clearcutting and splash-damming systematically cleared woody debris from streams (Sedell and Luchessa 1982). This loss of woody debris significantly changed the physical and bi-

Figure 2. Logs in small stream awaiting water from splash dam to float them to the mill. Photograph by USDA Forest Service, from Sedell et al. 1988.

ological conditions of the streams and may have had the greatest single impact on salmonid habitat.

Today, less than 5% of the original old-growth forest remains on this continent (Findley 1990). Alaska, by virtue of its remoteness, harsh climate and rugged terrain, has become the "last frontier" of pristine forests, streams and wild salmon habitat in North America. Large-scale logging began in Southeast Alaska in the mid-1950s, and now threatens to devastate the largest remaining stand of old-growth forest in North America, as was done in New England, the Great Lakes states, the Gulf Plain and the Pacific Northwest.

Magnitude of the Problem

After about 200 years of nearly unregulated logging, grazing, agriculture and development, most of the 5,225,000 km of streams in this nation need restoration. In 1982, the U.S. Environmental Protection Agency and the U.S. Fish and Wildlife Service conducted a National Fisheries Survey to assess the biological conditions of streams in the United States by examining aquatic habitat and fish communities (Judy et al. 1984). Results indicate that 81% of the nation's fish communities had been adversely affected by a variety of factors. Water quality and quantity were adversely affected in 56% and 68% of streams, respectively.

Habitat conditions limited the fish community in at least 49% of streams, predominated by a lack of habitat for adults, juveniles, eggs and alevins, and absent or degraded pools, spawning gravel, overhead cover and riffles. Major causes of habitat problems were excessive siltation, bank erosion and sloughing, natural causes and channelization.

As of 1972, over 320,000 km of stream channel had been modified in America (Little 1973 as cited by Wesche 1985). The U.S. General Accounting Office showed that in some states as much as 90% of federally managed streams were in a degraded condition (Hunter 1991). Of the 800,000 ha of riparian habitats in the eleven Western states, about 85% have been altered and are in need of restoration (James R. Sedell, pers. comm.). Of the 27,000 km of streams in western Oregon, 50-85% are debris-poor because of splash-damming, stream cleaning, agriculture conversion, fire, streamside logging and snagging (McMahon 1989). In 40 years, about 3,700 km of streams have been altered by the clearcutting of over 200,000 ha of forest, including 23,742 ha of riparian habitat on Southeast Alaska's Tongass National Forest (USDA Forest Service 1990).

The rejuvenation of stream restoration has produced a multitude of reports, workshops and training sessions (Canada Department of Fisheries and Oceans 1980; Gore 1985; Hunter 1991; Reeves et al. 1991). The effects of logging on streams and salmonid habitats have been studied intensively (Salo and Cundy 1987; Macdonald et al. 1988; Hicks et al. 1991). The objectives of this paper are to briefly review the (1) stream environment and salmonid habitat requirements; (2) impacts of logging; (3) concepts of stream restoration; (4) procedures for effective restoration; and (5) need for protection. Because the literature on the effects of logging, salmonid habitat requirements and stream restoration is extensive, only a synthesis is presented here.

This paper focuses on anadromous salmonids that inhabit streams flowing into the Pacific Ocean, but is generally applicable to all salmonids. Concepts and procedures for restoring stream habitats affected by logging activities are applicable to many other land-use activities including road building, channelization, livestock grazing, urbanization and mining.

Characteristics of Streams

> A river is a tangible and dramatic force, whether it winds through the spruce and hardwood covered ridges of the Laurentian Shield, the tundra of the north or the mountainous rain forests of the west coast. It is part of a vast mysterious entity that reaches from the gravel of the river bottom to the wind currents thousands of feet above which carry insects to destinations unknown to man. (Russell 1980)

Physical Characteristics

A stream is defined as a natural water course containing flowing water at least part of the year, supporting a community of plants and animals within the channel and the riparian vegetation zone (Helm 1985). The term river is often reserved to denote the main stream or larger branches of a drainage system (Morisawa 1968). Although their geographical and geological environments are diverse, streams have similar organization and characteristics such as sediment and debris transport, width-depth ratios, velocity and discharge, and sinuosity and channel meandering.

Hynes (1970, 1975) describes the stream as a product of the valley through which it flows and as a product of the water that flows along it, more or less adjusted to the pattern of discharge. Because of riparian vegetation, stream channels tend to be rectangular in cross section, although curves are skewed. Channels also tend to have alternating pools and riffles, especially where there is an abundance of gravel and cobbles. Riffles tend to be spaced at more or less regular intervals, five to seven stream-widths apart, particularly in gravel-bed streams. In most streams, boulders, cobbles and gravel tend to congregate on the bars that form the riffles; the bars themselves slope first towards one bank and then towards the other, producing some sinuosity of flow even in straight channels (Hynes 1970). The substrate varies depending on the geology and topography of the watershed, as well as the velocity and specific flow patterns in a given reach of stream.

The total length of streams in the United States was estimated by Leopold et al. (1964) to be about 5,225,000 km. Some 86% of the estimated 2 million streams are small (1st-3rd-order) streams, 12% are medium-sized (4th-6th-order) streams and only 2% constitute large (7th-10th-order) rivers. Strahler (1957) used stream order to describe the channel network, and designated small headwater streams possessing no tributaries as 1st-order streams. Two 1st-order streams unite to form a 2nd-order stream; joining two 2nd-order streams forms a 3rd-order stream, and so on.

Functional Values

Trees are the major ecological link between the riparian zone and the stream ecosystem. Functional perspectives developed by Meehan et al. (1977) and Swanson et al. (1982) define riparian zones as three-dimensional zones of direct interaction with stream ecosystems (Figure 3). Critical functions provided to the stream by riparian zones include source of energy, maintenance of water quality, source of physical structure, and regulation of flow. The struc-

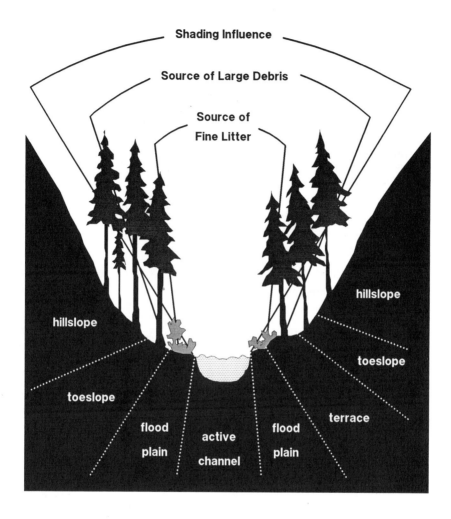

Figure 3. Riparian functions depend on distance and elevation from stream and on type and quality of vegetation. Some of these critical functions include shade and temperature control, litter and detrital sources of energy, transport of soil nutrients, maintenance of water tables, stream bank structure and stabilization, interception of sediment, source of woody debris, and maintenance of fish habitat. Modified from Meehan et al. 1977.

ture and function of a particular stream ecosystem and its quality of fish habitat are largely determined by the characteristics of the associated vegetation, the physical setting of the riparian zone, and the overall condition of the watershed (Gregory et al. 1987).

The forest influence is greatest in small headwater streams where detritus (needle and leaf litter, twigs and branches) forms the major energy base for consumer organisms. This influence diminishes as stream size increases, sufficient light penetrates the canopy, and consumer populations are structured by both particulate detritus and algae (Cummins 1974). The most productive habitats for anadromous salmonids are small, low-gradient streams associated with old-growth coniferous forests, where woody debris from fallen trees is the principal determinant of channel complexity and aquatic biota (Sedell et al. 1988). Although larger streams can provide habitat for reproduction and rearing, they serve mainly as migration corridors; salmonids prefer small, low-gradient streams and medium-sized streams for reproduction and rearing.

Of the many structural factors important to the functioning of stream ecosystems, none appears more important than the physical structure provided by woody debris. Woody debris plays a major role in controlling channel morphology, in routing and storing sediments and organic matter, and in forming complex and diverse habitats (Figure 4).

The physical structure of woody debris creates the complexity of stream channels by affecting stream hydraulics: channels are widened by the lateral deflection of current, depth is increased by obstructions scouring downstream pools and secondary channels are formed by debris accumulations (Keller and Swanson 1979; Hogan 1985).

Salmonids favor pools and backwater areas because the slow current requires little effort for maintaining a feeding station (Bisson et al. 1987). Pools and backwaters with abundant woody debris also provide critical overwintering cover for salmonids. Number and volume of pools are positively correlated with debris pieces in low-gradient streams. A single log (as shown in Figure 5) or several pieces of woody debris anchored in the streambed can form an up-stream-stored sediment deposit that salmonids use for spawning, and a downstream-plunge pool utilized for rearing (Heede 1972, 1985). The importance of woody debris in streams has been well documented (Wallace and Benke 1984; Angermeier and Karr 1984; Sedell et al. 1984).

Stream Classification

A classification system is an important tool in describing a stream's physical and hydrological dimensions, its potential carrying capacity for salmonids, and the characteristics that differentiate it from other streams. Knowledge of a given stream type allows assessment of the potential impacts associated with various land-use management practices and evaluation of possible methods of restoration. Failed restoration efforts can most often be attributed to inadequate knowledge about the character and dynamics of an individual stream.

Stream classification systems developed by Paustian et al. (1984) and Rosgen (1985) categorize stream reaches by geomorphic and hydrologic characteristics of their channels and riparian vegetation. The Channel Type Stream Classification System of Paustian et al. (1984) has been used successfully to evaluate salmonid habitat (Edgington et al. 1987; Murphy et al. 1987), develop stream carrying capacity models (USDA Forest Service 1990), and assess the potential risk of different logging prescriptions (USDA Forest Service 1986). Based on Rosgen's (1985) system, Rosgen and Fittante (1986) derived a guide for determining the suitability of in-stream structures for restoration.

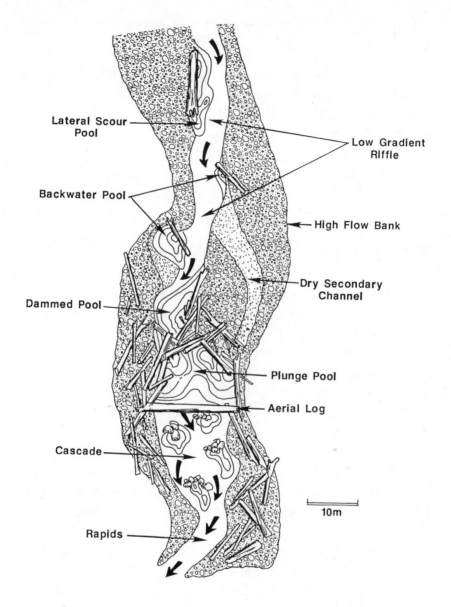

Lateral Scour Pool

Low Gradient Riffle

Backwater Pool

High Flow Bank

Dry Secondary Channel

Dammed Pool

Plunge Pool

Aerial Log

Cascade

10m

Rapids

Figure 4. Diagram of habitat diversity created by woody debris in streams.

Figure 5. Function of a single natural log in storing gravel for spawning and in creating a plunge pool for rearing.

Habitat Requirements of Anadromous Salmonids

The natural homes of the trout are the sluggish waters of the swamp, cold and pure, with their bed of ooze and sheltered bank, or in our upland streams the deep cavities under banks or roots.... Let the home of the trout be regarded as his castle. Entice him from it if you can, but do not invade it. (Van Cleef 1885)

Distribution

Indigenous and transplanted stocks of salmonids (i.e., *Oncorhynchus* spp., *Salmo* spp. and *Salvelinus* spp.) occur in streams throughout the temperate regions of the world. Approximately 50% of the total stream miles in the United States (48 contiguous states) contain salmonids; 11% contain anadromous salmonids (Judy et al. 1984). The distribution of anadromous salmonids in the western United States extends from Monterey Bay, California, to Alaska's north slope (see Meehan and Bjornn 1991 for details), and corresponds to the distribution of the coniferous forests. Twelve important species of salmonids occur in the forested watersheds of western North America, including Alaska. All are dependent on streams for the successful completion of their life cycle. Most are anadromous, but some complete their entire life cycle in fresh water.

Life History Strategies

Salmonids all have the same basic habitat requirements for reproduction and rearing, but differ greatly in behavior and life-history strategies. The most intriguing characteristic of anadromous salmonids is their innate ability to return from the ocean to spawn in streams where they were spawned and reared. This homing characteristic has enabled the development of distinct and separate stocks, each adapted to the particular conditions of its natal stream (Larkin 1974).

All anadromous salmonids utilize the nearshore coastal areas and open ocean during the majority of their growth, but require streams for spawning, egg incubation and juvenile rearing. Natural mortality in streams is high, often exceeding 90% of deposited eggs. Thus, the number and quality of fry and smolts migrating from streams to the ocean control the number of returning adults, although variable marine conditions cause significant variation in survival. Species such as pink (*O. gorbuscha*) and chum (*O. keta*) salmon that have a short freshwater residence time are more likely to have their population limited by reproductive success or insufficient spawning habitat. Species with long freshwater residence times (e.g., coho, chinook, and sockeye salmon and steelhead trout) are less likely to be limited by reproductive success, but usually are limited by diversity of rearing habitat and environmental conditions.

Habitat Requirements

Fish habitat is defined as the aquatic environment and the immediate surrounding terrestrial environment (i.e., riparian zone) that afford the necessary biological and physical support systems required by fish (Helm 1985). The specific environmental factors which are preferred by a species are termed habitat components; a set of specific preferences or components form the habitat for that species. Habitat requirements of anadromous salmonids in streams change with the stage of their life cycle (i.e., embryo, fry, parr, smolt or adult) and the season. Habitat requirements can be divided into the following segments based upon salmonid life history stage: upstream migration of adults (access), spawning and incubation of eggs (reproduction), summer rearing and overwintering of juveniles (rearing), and downstream migration of juveniles and smolts (access). Bjornn and Reiser (1991) have described in detail the specific habitat components, including salmonid preferences and tolerance levels, that are included in the habitat requirements at each segment for the different species.

Environmental Factors

The environment, according to Ryder and Kerr (1989), comprises the total physical, chemical and biological surroundings of an organism, including habitat and other biota. Many environmental factors have influenced the evolution of stream biotas and salmonid habitat preferences. Karr et al. (1986) have grouped these environmental factors into five major categories: energy source, water quality, physical structure, flow regime and biotic interactions.

Energy Source. Stream ecosystems are largely determined by the physical environment, its nutrient sources, and its mix of energy from either autochthonous or allochthonous inputs (Murphy and Meehan 1991). Energy flow or trophic state of a stream is sensitive to organic loading from the terrestrial environment, amount of sunlight, and nutrients.

A fundamental postulate of stream ecosystem theory is that many of the process-orientated attributes, such as production, respiration, energy flow, nutrient cycling and trophic dynamics, change as streams increase in size from headwaters to mouth (Cummins 1974). Thus, a stream system can be considered a continuum, from a strongly heterotrophic headwater regime (1st-3rd-order) dependent on regular input of organic detritus from leaves, needles, twigs, and branches, to a seasonal or annual regime of autotrophy in mid-reaches (4th-6th-order), gradually returning to heterotrophic processes in downstream reaches (7th-12th-order). As a result of physical and biological processes, the particle size of organic material in transport down the continuum becomes progressively smaller.

Man-induced alterations to the type, amount, timing and particle size of organic matter disrupt ecosystem stability by causing changes in the quantity and quality of organic matter available for processing. Depending on the extent of alteration, the response of the biological community can be transmitted throughout the continuum, and invertebrates and fish assemblages may change in response to alterations in energy flow and reduced biotic integrity (Karr et al. 1986).

Water Quality. The quality of water is the most important category of environmental factors affecting the biota of stream ecosystems. This category includes light, temperature, turbidity, dissolved oxygen, nutrients, chemicals, heavy metals, toxic substances, pH and other components, which when altered or added to the stream ecosystem can significantly change the biota. Because water quality is interrelated to all other major categories of environmental factors, an alteration of a single factor in any other category, except perhaps for biotic interactions, could impact water quality and vice versa.

Unlike the physical alteration of habitat, many problems associated with water quality are invisible. Thus, the chemical characteristics of water may not normally be considered within the context of fish habitat, as would water quantity (e.g., minimum flow, stream depth, flow regime), which are readily apparent (Nassichuk 1986).

Water quality can be degraded by nearly all of man's activities that affect the landscape and cause pollution. Point-source pollution from industrial and municipal effluents can be controlled by permits, government regulations and treatment technologies; however, control of nonpoint pollution is much more elusive and typically requires a holistic approach to watershed management and restoration.

Physical Structure. Physical structure plays a major role in determining stream biota. Structure, according to Ryder and Kerr (1989), is the physical, chemical or biological component of the environment which forms a center of organization. Geomorphology and characteristics of the riparian vegetation in the basin are primarily responsible for forming the complexity and diversity of physical structure in the stream. Consequently, logging that reduces physical structure can have significant effects on the complexity and diversity of the habitat. Habitat with a varied physical structure acts as an attractor and a refuge for fishes where they may feed, rest, breed or seek shelter from predators or an inhospitable environment. Therefore, in addition to physical structure such as logs and boulders, habitat is also structured by water masses differing in velocity, depth, temperature, etc.

There are a number of aquatic units or habitat types within streams which have equivalent structure, function and response to disturbance (Helm 1985). Bisson et al. (1982) classified fish habitat in streams into a number of macrohabitat types according to location within the channel, pattern of water flow, and nature of flow-controlling structures. Accordingly, stream macrohabitat consists of three types of riffles, six types of pools, and glides (an intermediary type between pools and riffles). Microhabitat, a further division, consists of that specific combination of habitat components (i.e., velocity, depth, cover, substrate, woody debris) in the locations selected by the fish for specific purposes and/or events (Helm 1985). Microhabitat is the more specific and functional aspects of habitat and cover, whereas, macrohabitat provides the general structure of the habitat essential for various life stages of salmonids.

Flow Regime. The variation in stream flow as a result of seasonal differences in precipitation is a characteristic feature of most biotic habitats in running water (Hynes 1970). Aquatic organisms have evolved to exploit the spatial and temporal changes in flow regimes (Karr and Dudley 1981; Cummins et al. 1984).

Fish, invertebrates and algal species all show specificity for different microhabitats related to velocity, depth and volume. For each salmonid species, at each life history stage, there are microhabitat requirements that are all a function of the flow regime of a particular stream, and these need to be taken into consideration for a successful restoration activity. As Sullivan et al. (1987) have shown, the amount of stream area available for spawning and rearing of salmonids is strongly dependent on the hydraulic characteristics of streams. Hydraulic diversity created by physical structure and by different flow regimes enhances species diversity by providing space for a variety of species and life stages.

Biotic Interactions. Habitat requirements and life-history strategies of salmonids reflect their adaptation to the physical and chemical characteristics of their natal stream and their interaction with other organisms. As a result, several biotic factors, including competition, predation, disease, parasitism and perhaps fitness (e.g., hybridization) are responsible for the abundance and distribution of a species in a stream (Karr et al. 1986). Habitat partitioning in streams is the result of species interaction (Stein et al. 1972; Allee 1982; Reeves 1985). Alteration of habitat factors can elicit changes in biotic interactions which can then upset the natural equilibrium within and between communities. Results could include shifts in species composition and abundance, altered trophic structure, increased predation, or altered primary or secondary production.

Limiting Factors

Distribution and abundance of salmonids in streams are affected by various combinations of environmental factors. Environmental factors can affect the growth and survival of the individual, the vitality of a population, or the condition of the entire stream ecosystem. Fry (1971) classified environmental factors according to their effects on animal metabolism, and demonstrated that an organism responds to the total environment rather than to a single factor. Fry's factors apply to the well-being of the individual and are basically density-independent, such as dissolved oxygen and temperature.

In Blackman's (1905) concept of limiting factors to which Liebig's law of the minimum applies, materials and energy limit or regulate the productivity of the population through density-dependent mechanisms, such as food and space. Bisson (1990) distinguished between factors controlling individual fitness and those regulating populations. He stated that habitat preference and population abundance result from individual choices that are the result of selective pressures to maximize fitness.

Based on these concepts, a limiting factor can pragmatically be defined as any environmental factor which, when altered, significantly changes the production of a desired species. This definition of limiting factors is fundamental to the established relationships between habitat components (i.e., environmental factors) and salmonid production and survival. As such, any one factor in the major categories just discussed could become limiting. Therefore, the concept of habitat restoration is based on the premise that alleviating a limiting factor(s) will increase fish production; consequently, a critical step in the restoration process is analyzing those factors that might limit salmonid production.

Ryder and Kerr (1989), point out that there are fundamental factors in the environment, like dissolved oxygen, water temperature, light or nutrients, that are necessary for fish survival and at times can act as either a controlling or a limiting factor as described by Fry (1971). However, some factors (e.g., dissolved oxygen, temperature or toxic substances) can be at lethal levels and directly affect fish survival—these are the survival determinants referred to by Ryder and Kerr (1989) or lethal factors by Fry (1971). Therefore, Ryder and Kerr believe that to understand what factors are limiting a population, the stream ecosystem first should be examined to ensure that the basic survival determinants meet minimal standards for viable populations to exist. Thus, the physical, structured habitat should not be considered limiting until the more fundamental factors of the environment which affect survival are addressed.

Although the limiting factor concept is useful in identifying limitations to salmonid production, it can oversimplify complex ecological processes (Hall and Baker 1982). As a demonstration of the difficulty in identifying a limiting factor, Everest and Sedell (1984) calculated that as many as 73 factors could potentially limit fish production in a hypothetical stream with three or more salmonid species, each with different age classes and habitat requirements.

Changes in a watershed and stream ecosystem that can occur with logging will most likely affect several factors, one or more of which could easily become limiting during or at some time following the disturbance. A thorough knowledge of the habitat requirements and life history strategies of the species endemic to a particular stream can assist in identifying the factors that limit production and suggest approaches for restoration.

Carrying Capacity

Carrying capacity is the maximum number of salmonids the stream, stream reach, or specific habitat can produce over time. Given suitable water quality and recruitment, the carrying capacity of a stream is primarily a function of discharge and physical structure. Because stream environments are variable, carrying capacity changes annually, fluctuating up to a maximum set by site characteristics (Milner et al. 1985). Habitat components that predict numbers or biomass of stream salmonids have been identified by a number of researchers (e.g., Lewis 1969; Binns and Eiserman 1979; Lanka et al. 1987). These fish/habitat relationships have been used in models to predict the effects of land-use activities on carrying capacity. No single model or set of components, however, can be used universally to predict the carrying capacity of a stream, but precise models can be developed for specific sites if streams or reaches are stratified into homogeneous units (Fausch et al. 1988; Shirvell 1989). To predict consequences of habitat alteration, the model must also include components that can be affected by management. A model to predict coho salmon smolt carrying capacity in pristine stream habitats in Southeast Alaska is being developed (Kessler 1989). Based on this model, carrying capacity or capability coefficients were computed from fish/habitat relationships for stream reaches classified by the Channel Type Stream Classification System (Paustian et al. 1984). The model can be used to make an objective assessment of the potential smolt yield in a given stream, from which various forest management prescriptions, including restoration needs, can be judged (USDA Forest Service 1990).

Our knowledge of all the factors affecting the carrying capacity of the habitat is far from complete; however, the concept is important in understanding habitat relationships, analyzing limiting factors, or assessing potential changes from habitat alterations, including restoration.

Effects of Logging on Anadromous Salmonid Habitat

After the fullest investigation of these and other streams, I have become satisfied that the destruction of the trees bordering on these streams and the changed condition of the banks produced there-by, has resulted in the destruction of the natural harbors or hiding places of the trout, that this is the main cause of depletion, and that until these harbors are restored, it will be useless to hope for any practical benefit from restocking them. (Van Cleef 1885)

The principal effect of logging is to alter the linkage between forest and stream by removing or changing the character of trees within the riparian zone and, as a result, disrupt the many riparian functions that control the complex biotic and abiotic interactions in streams. Logging activities, therefore, can have major effects on habitat and salmonid populations. The effects of logging on streams and salmonids has been debated and studied intensively for nearly half a century (Salo and Cundy 1987; Macdonald et al. 1988; Meehan 1991).

Logging can be either beneficial or detrimental to fish, depending on the extent of changes in habitat, the species or life stage affected, or the time interval involved. Logging impacts have been difficult to quantify because of the natural variability in adult production and because of multiple watershed activities (e.g., hydropower development, mining, livestock grazing and overfishing).

Case-history studies, such as the Hubbard Brook Experimental Forest (Likens et al. 1977), the Alsea Watershed Study (Hall et al. 1987) and Carnation Creek (Hartman et al. 1987), have resulted in a better understanding of the functioning of stream ecosystems and effects of logging. Though knowledge of the impacts of forest harvest on salmonids remains incomplete, scientists have been able to recognize and predict the response of salmonids to habitat alterations caused by logging (Hicks et al. 1991).

Alteration of Habitat

Logging activities potentially alter all of the major environmental factors affecting the productive capacity of habitat and streams (Table 1). Logging of riparian zones and upland areas, construction of roads and management of second growth can cause multiple and long-term cumulative impacts on streams. In addition, methods of logging and rate of deforestation can play a significant role in determining the level of these impacts. The potential influences of logging on the stream, habitat quality and salmonid growth and survival have been summarized by Hicks et al. (1991). The principal changes in the stream environment from logging include: increased solar radiation and temperature; decreased canopy cover, streambed stability, water quality and woody debris; changes in channel morphology; streambank erosion; and altered streamflow and nutrient runoff.

Positive Changes. Changes in the stream can result in transient or long-term changes in habitat quality. Transient changes (e.g.,

Table 1. The influence of timber harvest and possible potential changes on the stream environment, habitat quality, and salmonid growth and survival. From Hicks et al. 1991.

Forest Practice	Stream Environment	Salmonid Habitat Quality	Salmonid Growth and Survival
Timber harvest from streamside areas	Increased incident solar radiation	Increased stream temperature; higher light levels; increased autotrophic production	Reduced growth efficiency; increased susceptibility to disease; increased food production; changes in growth rate and age at smolting
	Decreased supply of large woody debris	Reduced cover; loss of pool habitat; reduced protection from peak flows; reduced storage of gravel and organic matter; loss of hydraulic complexity	Increased vulnerability to predation; lower winter survival; reduced carrying capacity; less spawning gravel; reduced food production; loss of species diversity
	Addition of logging slash (needles, bark, branches)	Short-term increase in dissolved oxygen demand; increased amount of fine particulate organic matter; increased cover	Reduced spawning success; short-term increase in food production; increased survival of juveniles
	Erosion of streambanks	Loss of cover along edge of channel; increased stream width, reduced depth	Increased vulnerability to predation; increased carrying capacity for age-0 fish, but reduced carrying capacity for age-1 and older fish
		Increased fine sediment in spawning gravels and food production areas	Reduced spawning success; reduced food supply

Table 1 continued.

Forest Practice	Stream Environment	Salmonid Habitat Quality	Salmonid Growth and Survival
Timber harvest from hillslopes; forest roads	Altered streamflow regime	Short-term increase in stream-flows during summer	Short-term increase in survival
		Increased severity of some peak flow events	Embryo mortality caused by bedload movement
	Accelerated surface erosion and mass wasting	Increased fine sediment in stream gravels	Reduced spawning success; reduced food abundance; loss of winter hiding space
		Increased supply of coarse sediment	Increased or decreased rearing capacity
		Increased frequency of debris torrents; loss of instream cover in the torrent track; improved cover in some debris jams	Blockage to migrations; reduced survival in the torrent track; improved winter habitat in some torrent deposits
	Increased nutrient runoff	Elevated nutrient levels in streams	Increased food production
	Increased number of road crossings	Physical obstructions in stream channel; input of fine sediment from road surfaces	Restriction of upstream movement; reduced feeding efficiency
Scarification and slash burning (preparing soil for reforestation)	Increased nutrient runoff	Short-term elevation of nutrient levels in streams	Temporary increase in food production
	Inputs of fine inorganic and organic matter	Increased fine sediment in spawning gravels and food production areas; short-term increase in dissolved oxygen demand	Reduced spawning success

stream temperature) last only a matter of years; long-term changes (e.g., woody debris) last decades or even centuries. Growth, survival and abundance can either increase or decrease, depending on the extent of changes in habitat quality and the species and life stage affected. Increased light reaching the stream after canopy removal can stimulate primary and secondary aquatic production. Summer abundance of young-of-the-year salmonids (fry) can be greater in recently logged areas (<15 year old) than in nearby forested areas because of increased food (Koski et al. 1984; Gregory et al. 1987). Fish growth may also be enhanced after logging by warmer stream temperature which accelerates fry emergence and lengthens the growing season (Holtby and Hartman 1982; Scrivener and Andersen 1984; Thedinga et al. 1989).

Negative Changes. Salmonid populations can be adversely affected by logging because of detrimental changes in habitat quality and reduced survival from a number of factors, including inclement weather and predation (Figure 6). Detrimental changes in habitat quality include: excessive sedimentation and increased turbidity; reduced primary production and invertebrate abundance; reduced water quality from increased water temperature; reduced concentration of dissolved oxygen; loss of habitat structure including changes in channel morphology, loss of woody debris, decreased pool volume and increased riffle area; collapsed streambanks and decreased channel stability; and altered streamflow. Effects could include: changes in biotic interactions including increased predation, reduced survival at all life-history stages, reduced growth, altered age structure, decreased abundance or altered distribution.

Depletion of Woody Debris

Prior to large-scale clearcutting of forests, the dominant feature of pristine streams across North America in the early 1800s was the presence of wood from the large quantities of naturally downed

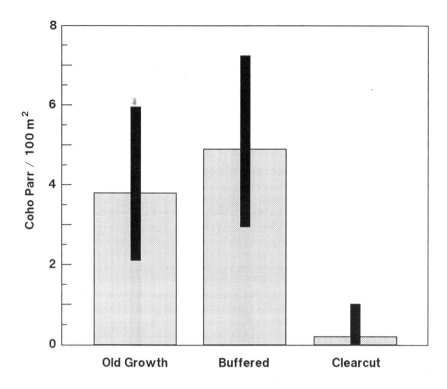

Figure 6. Relationship between coho parr density in winter and different logging treatments of stream reaches. Coho parr density was significantly greater in old-growth reaches as compared to clear-cut reaches, P < 0.01. Dark bars represent 95% confidence limits. Data from Murphy et al. 1986.

trees (Sedell et al. 1984). These streams were systematically cleaned of downed trees, snags, and woody debris for over 150 years (Sedell and Luchessa 1982; Sedell et al. 1984; Triska 1984). During the 1950s-1970s, many fishery managers believed that woody debris in streams prevented fish passage, created log jams, and caused channel scouring during floods (Narver 1971; Hall and Baker 1982). Consequently, stream clearance of debris to benefit fisheries was initiated on a major scale from California to Alaska, and old log jams on many of the major coastal rivers were removed.

Debris removal is still a part of many salmon improvement programs in several Western States, but not on the scale of past decades. The combination of debris removal for fish passage, splash damming, and navigation has left entire drainage systems almost devoid of large wood and has changed the physical and biological conditions of the streams (Sedell and Swanson 1984; Harmon et al. 1986).

Even when woody debris in the stream remains undisturbed during logging, debris quantity declines over time because second-growth forests provide insufficient new debris to replace the woody debris that decays or washes downstream (Grette 1985; Andrus et al. 1988; Murphy and Koski 1989). A model developed by Murphy and Koski (1989) to demonstrate the long-term effects of logging shows that only 26% of the remnant woody debris would remain after 100 years and it would take 250 years for new woody debris from second-growth to reach 85% of the prelogging level (Figure 7). Thus, total amount of woody debris would decline to 30% of the old-growth level 90 years after logging. Because of the positive correlation between the presence of woody debris and high rearing density of coho salmon (Figure 8) (Koski et al. 1984), the potential carrying capacity to produce coho would be reduced accordingly.

In corroborative studies, Grette (1985) and Andrus et al. (1988) found that even-age, second-growth forest does not contribute sufficient woody debris for streams in the Pacific Northwest for at least 50-60 years after logging. In a study of streams 21-140 years after they were affected by logging or wildfire in Oregon, Heimann (1988) found that woody debris from second-growth was not large enough to anchor stable woody debris accumulations in streams until after 120 years. He concluded that an 80-year timber harvest rotation without buffer strips along streams would permanently reduce total debris in the streams to 20% of prelogging levels and would eliminate large pieces.

The loss or reduction in physical structure accompanying the de-

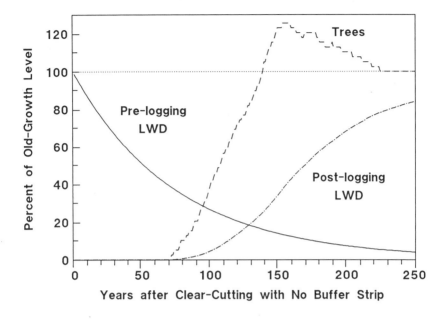

Figure 7. A model of changes in amount of large woody debris (LWD) in small (2nd-3rd-order) streams after clear-cut logging without a buffer strip along the stream banks. The decline in remnant prelogging LWD was calculated from the natural depletion rate as a result of decay, abrasion and export. Second-growth (trees) was based on yield tables for western hemlock-Sitka spruce stands on productive valley bottom sites. Post-logging LWD is the contribution of "new" woody debris to the stream from the second growth (Murphy and Koski 1989).

pletion of woody debris has become a focal point for protection and restoration of riparian habitats because of the direct relationship to salmonids.

Concept of Habitat Restoration

This subject (pollution) is new to us, but is well worth our attention. So far, we have bent our energies to producing fish in

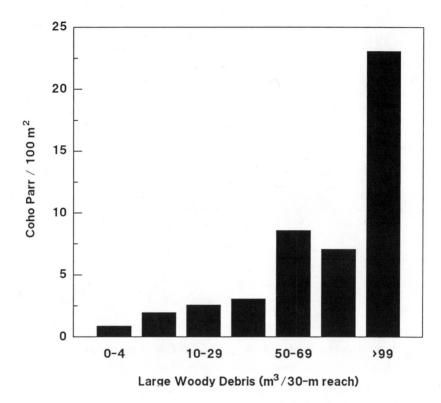

Figure 8 . Relationship between the volume of woody debris in streams and the density of coho parr in winter (Koski et al. 1984).

vast numbers, with but little consideration of the many delicate conditions necessary to their future growth.... (Mather 1875)

Hatchery Propagation

In the early era of fisheries management in North America, hatchery propagation and restocking of fish were perceived as a panacea for restoring fish runs depleted by overfishing, pollution or stream degradation. Thus, in 1871 when the first salmon hatchery was built in the United States (Netboy 1974), the apparent ease of propagating salmonids established hatcheries as the method for restoring streams instead of habitat improvement or regulations. Fish stocking had great public appeal because it was a positive action as opposed to regulations, which were restrictive and created no immediate, visible results (Bowen 1970). Thus followed a proliferation of hatcheries in North America from 1875 to 1925 (Childerhose and Trim 1979). Unfortunately, much of the same philosophy regarding hatcheries has persisted and may ultimately result in the demise of many populations of wild salmonids.

Chronology of Stream Restoration

In the 1930s, concerns of anglers and renewed interest in conservation eventually led to the first major attempts to improve stream habitat. This early chronology of stream habitat restoration has been reviewed by Hunter (1991) and Reeves et al. (1991) and is summarized here. The first large-scale attempt to manage habitat in streams was initiated in 1934 by the U.S. Bureau of Sport Fisheries. Stimulated in part by the availability of a large labor force from the Civilian Conservation Corps, this pulse of activity led to a large number of projects between 1933 and 1937. For example, from 1933 to 1935, a total of 31,084 stream structures were constructed on 406 mountain streams.

The apparent success of the efforts in the Midwest and East was followed by a number of projects in the West. Much of the pioneering work on habitat improvement, however, was done with little knowledge of stream behavior, sediment routing or energy dissipation. Consequently, many projects failed as a result of freshets, particularly on the west coast. Because of structure failure and lack of evaluation, fisheries managers took a pessimistic view of habitat restoration, except for clearing woody debris from streams for fish passage and, as noted earlier, this had detrimental effects. Many evaluations of west coast efforts concluded that failure was more common than success. Nonetheless, rehabilitation and enhancement continued at a significant pace in the Midwest, and several manuals for habitat improvement were produced by state and federal agencies (Davis 1935; USDI Bureau of Land Management 1968). A USDA Forest Service handbook (1952) discussed the many earlier mistakes, the need for careful planning, and the benefits of streamside conservation.

Rejuvenation of habitat restoration in the 1970s was largely a result of studies confirming habitat as a determinant of salmonid abundance (e.g., White and Brynildson 1967; Lewis 1969; Hunt 1971; Binns and Eiserman 1979). In the late 1970s, the revelation of the significance of woody debris in the functioning of stream ecosystems and in the formation of diverse habitats helped focus attention on the importance of stream habitat and its relationship to the terrestrial surroundings.

Natural Recovery

Recovery of streams following chronic spills or chronic stress has demonstrated that streams are able to recover from degraded water quality (Cairns and Dickson 1977; Herricks 1977). Benthic macroinvertebrates can recolonize a stream reach in a short period (i.e., 75-150 days), but the establishment of a stable benthic community may take 300-500 days or longer; this is dependent on stream velocity, depth and substrate in the developing habitat (Gore 1985). Restoration of benthic invertebrates is important because of their role as food for juvenile salmonids.

The recovery of physical features, such as stream channels, is largely controlled by geological factors in the basin that determine sediment transport processes, including sediment size and routing, and by characteristics of the riparian vegetation (Hartman et al. 1987; Sullivan et al. 1987). For example, four case-history studies of the effects of logging on the response and recovery of stream channels indicated that channel recovery ranged from 5 to 60 years and recovery of riparian vegetation ranged from 100 to 200 years (Sullivan et al. 1987). The population of cutthroat trout in a small coastal stream in Oregon had not recovered ten years after logging (Hall et al. 1987). Hartman et al. (1987) concluded after a 15-year study on Carnation Creek that the stream system would not return to the prelogging old-growth state unless it remained undisturbed for many decades.

Because streams have an inherent ability to cleanse themselves, stream restoration can be viewed as a process of "recovery enhancement" (Gore 1985). Restoration enables the stream ecosystem to stabilize at a much faster rate than through the natural physical and biological processes of succession. Habitat development and the rate of recovery are highly variable and dependent on the biota, the site characteristics, and the extent and type of degradation.

Definition of Restoration

The term restoration is often confused with the terms rehabilitation and enhancement. The term to describe the habitat manipulation process must be defined by the desired product or outcome. Thus, restoration infers returning to the original state. The end product of restoration should be an ecosystem resembling predisturbed conditions or conditions similar to undisturbed ecosystems in the surrounding area (Gore 1985). As pointed out by Herricks and Osborne (1985), however, the definition of restoration in streams heavily impacted by man may be either impossible because baseline conditions are not known and undisturbed streams are not available, or inaccurate because the system available for comparison is inappropriate.

Rehabilitation or enhancement is the process of returning a site to a condition which is biologically acceptable to plants and animals. Enhancement refers to improving the current state of the ecosystem without reference to its initial state, whereas rehabilitation may be a pragmatic mix of nondegradation, enhancement, and restoration (Magnuson et al. 1980). Salmonid enhancement usually implies creating more suitable habitat than would naturally occur

(Reeves and Roelofs 1982), although the Canada Department of Fisheries and Oceans (1980) uses the term to mean restoring, protecting, and improving the capability of streams to produce anadromous salmonids.

In this paper, I define restoration as the means of bringing the carrying capacity of the habitat back to a previously existing level, to a level comparable to similar undisturbed sites, or to some predetermined level consistent with multiple use of the stream system.

Goals of Restoration

The most applicable and desirable long-term goal for restoration is to maintain the existing wild stocks of anadromous salmonids and, when possible, preserve genetic variability. Increasing the number of adult fish available to a fishery or the creel could be considered the general working goal of most stream habitat restoration projects. In the case of salmon streams in Southeast Alaska degraded only by logging, returning the stream ecosystem to conditions emulating the functions of a pristine old-growth forest would most likely be the desired outcome. In other parts of the country where there are multiple demands for limited stream resources, restoration to functional pristine conditions may not be feasible, and maximizing salmonid production through restoration and enhancement may suffice as the goal.

Though the original concept of salmonid restoration was based primarily on a hatchery ideology, the current paradigm for stream restoration is founded on stream ecosystem theory, fish/habitat relationships, the concept of limiting factors, and a growing awareness of man's impacts on stream ecosystems. Forfeiting stream habitat protection because of a reliance on hatcheries will hopefully become a thing of the past as restoration and protection become part of management ethics. Restoration should not be regarded as an exemption from stream protection or considered as a solution to mitigation.

Habitat Restoration Procedures

I believe it is possible to restore most of our streams.... The remedy I would suggest is briefly as follows: First—Prohibit the further destruction of either tree or bush upon or near the bank of the stream. Second—Where the soil is wet and suitable, protect the pools by an abundant growth of alders or other bushes. Third—Plant trees on the banks wherever feasible, especially where their roots will protect the surface of the ground, and at the same time permit the washing away of the soil underneath, so that large hollows may be formed as hiding places for the fish. Fourth—In each year, after the spring freshets are over, protect every pool as far as practicable by placing stumps, or trees or bushes in them, so that fishing with nets will be impossible. And also that the trout will be provided with artificial harbors until the natural ones are again restored. (Van Cleef 1885)

Had the wisdom of Van Cleef, Mather and others been followed over the past century, stream restoration would now be more than an art, and the magnitude of stream degradation would be less. Methods used today for restoring physical structure on riparian habitat differ little from the ideas suggested in 1885. However, restoration issues and procedures today are more complex because of cumulative impacts and increased population demands.

To ensure that all major environmental factors affecting the stream ecosystem are addressed, effective restoration requires a holistic approach directed at the entire watershed. An example of such an approach is illustrated by the Klamath Basin Fishery Restoration Program (Iverson 1991). In 1986, Congress authorized a 20-year, $42 million program to rebuild the fish resources of the 15,000-square-mile Klamath River Basin after the dramatic decline in the river's once abundant salmon and steelhead was clearly shown to be a result of nonpoint sources of pollution. A holistic watershed restoration approach was implemented to reverse the decline in fish populations because the piecemeal restoration approach used over the previous 30 years had resulted in nearly 700 individual fishery and stream restoration projects that were not working.

Restoration of Logging-Damaged Streams

Restoration of streams and anadromous fish habitat affected by logging should be approached as a three-step process. First, a program of "watershed restoration" needs to be applied to the watershed to initiate and obtain long-term recovery of the biotic integrity of the entire stream ecosystem. This should include a broad-based water quality control program, control of erosion from uplands and roads, restoration and maintenance of natural flow regimes, revegetation and second-growth management, and management of all watershed and stream uses. The South Fork Salmon River in Idaho is an example of what can be done when active watershed management is applied (Platts and Megahan 1975). Here, the spawning habitat was nearly destroyed by sedimentation from landslides and erosion, caused by a combination of intense rainfall, and logging and road construction on steep, unstable terrain. The U.S. Forest Service declared a moratorium on logging and on road construction in the watershed in 1965. Curtailment of the sediment sources allowed fine sediments in the stream to eventually move through the system with subsequent recovery of spawning habitat. Because of habitat improvement, the moratorium was cautiously lifted in 1978 (Megahan et al. 1980).

Next, the stream channel and instream habitats must be stabilized and the level of fish production increased through a program to restore "habitat carrying capacity." This is an interim "fix" of factors considered to be limiting production until the natural long-term recovery of the channel and watershed has begun. The primary focus would be on physical structure (e.g., stabilization of stream banks, the retainment of spawning gravel, or creation of additional rearing pools).

Finally, the "fish resources" should be managed through the different fisheries to ensure sufficient spawning populations for maximizing the restored carrying capacity of the habitat; and the species should be managed to control predation and/or competition which may have resulted from the degraded conditions.

Review of Restoration Methods

Since the 1930s, a wealth of literature has been generated on stream improvement (Duff et al. 1988). Information on methods and structures is abundant, but relatively few studies have taken a holistic watershed approach. The resurgence of interest in stream restoration has resulted in a multitude of projects, workshops, training programs and reports that have established restoration as a viable management strategy for increasing fish production and other stream attributes (e.g., Hassler 1984; House et al. 1990; Hunter 1991; Reeves et al 1991). Most approaches to habitat restoration have been directed at improving migration, reproduction, rearing, or riparian habitats that were considered to be limiting production. Barriers have been removed and fishways constructed to restore former habitat or access new ones. Culverts have been improved or replaced to restore access to former habitat. Stream gravels have been cleaned, mechanically loosened or trapped to restore spawning habitat. Spawning and egg incubation channels have been constructed to restore or enhance reproductive capability of streams. Complex and diverse structures have been created in stream channels to restore and provide summer and winter rearing habitat. Stream banks have been protected and trees planted to restore riparian habitat. The majority of instream habitat manipulation has involved the use of physical structures to improve the stream and its habitat.

Physical Structures. Most artificial structures attempt to mimic factors which shape and stabilize the channel, store sediment, create pools, dissipate stream energy, increase velocity and provide diverse habitat for either reproduction or rearing (Figure 9).

Although watershed and riparian management are generally the most critical issues, installation of physical structures has received the most attention. Their success or failure, however, is dependent on other environmental factors and the health of the riparian zone and watershed (Klingeman 1984; Anderson 1990; Lisle and Overton 1990). Many structure failures can be related to altered flow regimes, increased sedimentation or debris torrents from land-use activities upstream. Instream structures should only be applied as an integral part of an overall watershed and riparian restoration program (Crispin 1990).

There are a multitude of types of instream structures used in stream restoration projects (Wesche 1985; Payne and Copes 1986; Crispin 1990). In the Pacific Northwest, most structures have been installed in streams of 4th to 5th order with normal peak flows of about 6-60 cm s^{-1}. Some log and boulder structures have functioned well in flows from 100-400 cms (Crispin 1990). Most structures are placed in streams with gradients from 1-3%.

Currently, the emphasis in the Pacific Northwest is to use materials that simulate the natural habitat of the particular stream. For example, if woody debris is the dominant feature, log structures will be used, whereas, if bedrock or boulders dominate, boulder structures will be used. In large streams, boulders are the preferred structure because they conform to the channel and provide good interstitial space for rearing fish. Whole trees with rootwads, logs, or individual rootwads are used for main-channel structures, instream cover, and bank stabilization. Combinations of boulders and woody debris are also used extensively (Crispin 1990).

House et al. (1989) reported on a restoration project utilizing 812 structures on 10 streams ranging from 3-20 m in width. Past logging practices had removed the large structural elements and caused stream channel degradation. Over an 8-year period, they found 86% of the structures were fully functional, and they were successful in restructuring the degraded stream reaches by increas-

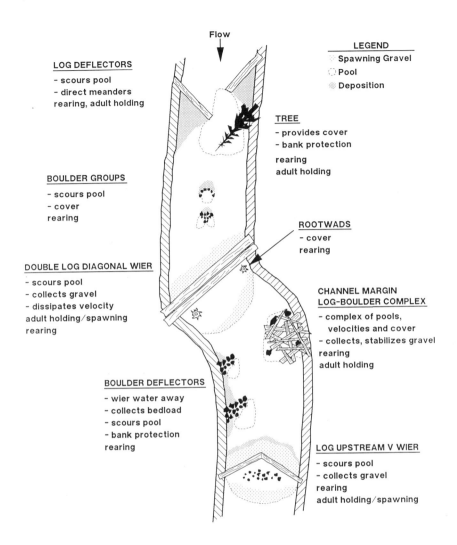

Flow

LEGEND
Spawning Gravel
Pool
Deposition

LOG DEFLECTORS
- scours pool
- direct meanders
rearing, adult holding

TREE
- provides cover
- bank protection
rearing
adult holding

BOULDER GROUPS
- scours pool
- cover
rearing

ROOTWADS
- cover
rearing

DOUBLE LOG DIAGONAL WIER
- scours pool
- collects gravel
- dissipates velocity
adult holding/spawning
rearing

CHANNEL MARGIN
LOG-BOULDER COMPLEX
- complex of pools,
  velocities and cover
- collects, stabilizes gravel
rearing
adult holding

BOULDER DEFLECTORS
- wier water away
- collects bedload
- scours pool
- bank protection
rearing

LOG UPSTREAM V WIER
- scours pool
- collects gravel
rearing
adult holding/spawning

Figure 9. Example of the types of physical structures which are used in streams to restore stream stability, create pools, dissipate stream energy, increase velocity and provide diverse habitat for fish.

381

ing the gravel substrate, instream cover, pool habitat and total usable habitat. The structures also dramatically increased water storage capacity and caused channel aggradation, helping to reestablish riparian vegetation.

Olson and West (1990) evaluated the performance of different types of man-made structures in the Klamath River basin to determine which type most effectively restored salmonid spawning and/or rearing habitat. The most cost-effective structures were log deflectors, boulder deflectors, small boulder weirs, boulder groups with woody cover and free-boulder weirs (i.e., partial spanning).

White and Brynildson (1967), Canada Dept. Fisheries and Oceans (1980), Wesche (1985), Payne and Copes (1986), Hunter (1991) and Reeves et al. (1991) have discussed in detail the role and function of the various types of structures.

Dams—Dams simulate natural debris jams, step-pool formations, or the pool-riffle sequences in sinuous streams. Dams are used primarily to create or improve pools downstream and to collect and hold gravel upstream of the structure.

Deflectors—Deflectors, sometimes referred to as wing dams or jetties, have historically been one of the most commonly used structures to improve fish habitat. Deflectors simulate obstructions which direct the streamflow. They are multipurpose structures that are used to force the stream into a more meandering pattern, create pools, dissipate energy, increase velocity, narrow and deepen streams, protect eroding banks or provide cover.

Cover—Cover consists of overhead and instream components which are provided naturally by tree canopy, overhanging vegetation, debris jams, root wads, woody debris, boulders, substrate interstices, aquatic plants, turbulence and depth. Riffles, glides and pools without cover do not provide effective winter habitat for salmonids (Heifetz et al. 1986). Cover used in conjunction with other structures often provides the diversity and complexity required to make the structures usable as fish habitat.

Stream Banks and Riparian Areas—Restoration of stream banks often involves some form of stream-bank revetment constructed from rock, coniferous trees or brush bundles to deflect the flow away from the bank. Planting shrubs and trees has helped reestablish the root system that stabilizes the bank and provide natural overhangs for fish habitat. Activities that encroach upon the riparian corridor must be moved in order for the bank to recover.

Forest practices regulations can be established to protect riparian vegetation and its habitat functions from disturbance. For example, logging is prohibited within buffer zones 30 m on each side of anadromous fish streams in Alaska to protect riparian and fish habitats (Figure 10) (Tongass Timber Reform Act, Public Law 101-626-November 28, 1990; State of Alaska Forest Practice Act 1990).

Off-Channel Habitat. In many larger stream systems, side channels, natural ponds and beaver ponds provide important spawning and rearing habitat (Cederholm et al. 1988). These areas are usually protected from extreme high flows and often have moderated temperature and stream-flow regimes because of their location away from the main channel and because of their influence on groundwater. Off-channel habitat has been discussed in detail by Cowan (1990). He categorized this habitat into three types: (1) groundwater channels—used primarily for spawning and incubation, (2) overwintering ponds—used as refuge habitat for overwintering coho salmon, and (3) side channel development—used for increasing spawning habitat.

Planning and Evaluation

Everest et al. (1990) recommend that two levels of planning precede habitat manipulation: program planning and project planning. At the program level, managers should consider the following items: coordination of financial resources, selection of sub-basins, selection of species, selection of personnel and coordination of eval-

Figure 10. Buffer zones on streams negate the need for restoration by providing the least costly, most effective, and the best long-term method of maintaining salmonid habitat.

uation programs. Goals, objectives and priorities are established at this level, as are the distribution of funds. The highest priority for restoration should be the most seriously damaged habitats; healthy habitats should be left undisturbed, but protected. The target species to be restored and the proposed methodology must be carefully considered so that one species or stock is not emphasized at the expense of others. Generally, experienced personnel and experts from other disciplines must be consulted to ensure success. Program planning must also include plans for thorough evaluation of the restoration work to determine if objectives are met (Fontaine 1990).

At the project planning level, specific details of proposed habitat restoration activities, including the size of the area being considered and the time allotted for inventory of habitat and fish use, need to be developed. Everest et al. (1990) recommend that a minimum spatial scale should be a sub-basin of not less than 50 km² because of seasonal changes in fish distribution. Ideally, data on fish distribution and habitat availability should be available for an entire basin for a minimum of one year. Such data allow the analysis of limiting factors to isolate problems and develop restoration methodologies.

Project planning must be closely coordinated with program planning and adhere to the following stepwise sequence to assure that projects are meeting program goals [see Everest et al. (1990) for details]:

1. Pre-improvement inventory
2. Analysis of limiting factors
3. Site selection
4. Selection of techniques and materials
5. Selection of implementation procedures
6. Evaluation of projects

Effectiveness of Restoration

Effectiveness of stream restoration has been difficult to assess because of the general lack of project evaluation (Hall 1984; Fontaine 1990). Understanding the causes of success and failure, however, is critical to the development and acceptance of the restoration concept. Quantitative evaluation relies on the response of salmonid populations to changes in the habitat from restoration. Everest et al. (1990) suggest that evaluations should examine three aspects of habitat improvement simultaneously to provide the most meaningful results: (1) changes in habitat quality, (2) changes in biological production, and (3) cost effectiveness.

Evaluation of Habitat Quality. The purpose of this evaluation is to determine if the factors suspected of limiting fish production have been alleviated. Everest et al. (1990) suggest that an effective evaluation should include a "pre-treatment" quantification of all habitat within the sub-basin for at least one year before improvement work begins. Quantification of "post-treatment" habitat after sufficient time for recovery and stabilization provides a comparison to assess changes. Robison (1990) suggests using the paired-stream approach where a similar stream without habitat modifications is compared to the manipulated one.

Evaluation of Biological Production. For anadromous salmonids, the ultimate measure of success is the smolt, because this life stage is the main determinant of adult production (Thedinga et al. 1989). Successful restoration should result in an increase in smolt abundance and an optimum size, age and emigration time characteristic of the particular species or stock. However, changes in the structure of the fish community must also be assessed because all species may not be equally affected by certain habitat modifications. A thorough evaluation of the effectiveness of restoration must demonstrate conclusively that any increase in smolt yield or catchable trout is attributable to the specific habitat improvement and is representative of the entire sub-basin.

Evaluation of Economic Benefits. Probably the most important step in the restoration process is evaluating the cost effectiveness. Such analyses must consider not only planning, implementation and maintenance costs of restoration, but also the benefits derived from the increased fish production and other attributes within the basin over the longevity of the project. Unfortunately, this is rarely done because of the time required to thoroughly evaluate changes in habitat and/or biological production. Economic evaluation will not only reveal important facts to aid planning, implementing, and managing future restoration projects, but it will demonstrate the high costs of restoring habitat as compared to habitat protection. The benefit-cost ratio and present net worth of habitat improvement projects can be assessed following procedures described by Everest and Talhelm (1982) or Everest and Sedell (1984).

Restoring stream habitats is expensive, and costs vary considerably depending upon a number of factors including species, limiting factor(s), land use, size of stream, number of streams in basin, restoration methods, geographic area, type of fisheries and project objectives. A review of projects (Table 2) that used mostly physical structures for stream restoration had costs which averaged about $24,000 per km (excluding the lowest and highest values) and ranged from about $2,000 to $1,214,000 per km of stream (House et al. 1989; Hunter 1991).

The cost of revegetating riparian areas can also be expensive, particularly if the objective is to create a riparian area similar to the predisturbed site. Anderson and Ohmart (1985), working in the desert Southwest, estimated that revegetating a 400-ha riparian plot in the lower Colorado River valley would take 10 years and cost from $8,750 to $11,250 per ha. They concluded that if the costs of revegetation seemed high, then an agency should explore other alternatives (e.g., buffer strips) prior to destroying a reach of valuable riparian habitat, or be prepared to meet the high cost of replacing it.

Restoration projects aimed at improving conditions in major river basins also can be expensive. As discussed earlier, Congress recently authorized $42 million to be spent over a 20-year period to restore the fish resources of the 39,000-km^2 Klamath River basin in northern California and southern Oregon (Iverson 1991).

Though the cost of restoration may seem high, the cost of foregone value as a result of degraded habitat is also high, as demonstrated by the reduced salmon production of the Columbia River.

Table 2. Examples of the costs for restoring streams and fish habitat using primarily physical structures. Costs vary widely depending on limiting factor, land use, stream characteristics, prescription, objectives, etc. Data from House et al. 1989; Hunter 1991.

Name of Project	State	Year	Limiting Factors	Prescription	$ Cost km^{-1}
John Day River	OR	1985	High temp, no holding pools	Stabilize banks, exclude livestock	21,000
Bear Creek	OR	1976	Rapid runoff, poor riparian zone	Altered grazing pattern	52,310
Lawrence Creek	WI	1964	Inadequate hiding/resting cover	Cover/deflector installation	16,280
Spring Creek	MN	1984	Shallow water, heavy siltation	Fencing, silt removal, revegetation	9,320
Cranberry River	WV	1987	Acidic water, no cover	Lime stream	19,890
Confederate Gulch	MN	1986	Lack of spawning gravel	Install log sills to trap gravel	54,700
Little Plover Creek	WI	1973	Lack of undercut banks	Alter bank vegetation	1,885
Rapid Creek	SD	1977	Lack of pools, holding areas	Cover/deflector installation	67,407
West Valley Creek	PA	1978	Wide, shallow water with no cover	Narrow channels	9,470
Little Crabby Creek	PA	1985	NA-proposed development	Reconstruct stream	1,214,000
Nestucca/ Alsea Rivers[a]	OR	1981	Lack of pools, cover	Restructure degraded stream reaches	25,245

[a] Fifteen projects within watersheds.

Economic losses of salmon and steelhead due to operating the Columbia River system predominately for hydropower are $370 million annually; cumulative losses since 1960 have totaled $6.5 billion; and future loss will be $3.7 billion (1980 dollars) per decade (Blumm 1986; Williams and Tuttle, this volume).

A study conducted in Southeast Alaska compared the value of coho salmon habitat formed naturally by woody debris with the value of old-growth timber on Federal land in a 30-m wide buffer zone along both sides of the stream (Koski et al. in prep.). Preliminary analysis of the values discounted at 4% over 100 years indicates that fish habitat at $7.19 per lineal meter of stream was higher than timber at $4.10 per lineal meter of stream.Comparing these values with the $24.00 per lineal meter estimated cost for restoration from Table 2, indicates that the best alternative may be to retain the riparian habitat.

Studies by House et al. (1989) and House and Crispin (1990) concluded that the best economic alternative for protecting and maintaining salmonid productivity of coastal Oregon streams was by maintaining mature coniferous riparian zones. Their restoration work achieved structural, habitat, biological and economic success, but stream rehabilitation and other artificial attempts to increase anadromous salmonid production were more expensive than maintaining healthy, self-sustaining riparian zones capable of supplying a continuous input of mature conifers to the channel.

Needs and Considerations

What we need and must somehow find in this last part of the twentieth century is a land and water ethic—perhaps better, an ethic of land, air, and water. What it will take is an attitude of mind, an understanding that blacktop destroys, dirty water pollutes, bad air destroys, and solid determination to control or prevent abuses even if it seems to cost something in immediate material effects.... (Haig-Brown 1980)

Research and Management Needs

Restoration must be used as a management tool in correcting past land-use mistakes; it must not be considered an excuse or mitigative measure for poor land or stream management. Though restoring physical structure may be the solution in some cases, most often it is done because of a lack of understanding of the problem, or because of its appeal of accomplishment. Needs for stream restoration include (1) regional inventories of the status of stream habitats and their needs for restoration, (2) the development of local/ regional stream classification systems to aid restoration planning, (3) the formulation of riparian habitat protection policies based on habitat structure and function, (4) evaluations of the effectiveness of restoration projects for enhancing stream and habitat recovery, and (5) development of procedures for restoring and managing riparian and upland areas affected by logging.

Conservation Ethic

Unless there is a dramatic change in America's conservation ethic, habitat abuse and neglect will most likely continue, as history has shown. The growing need and use of finite resources by a rapidly growing world population will continue to threaten pristine streams and exacerbate neglected ones. The perspective of what constitutes a healthy stream ecosystem or productive fish habitat must change if we are to begin to meet our obligation to be good stewards of the land and water. Thus, programs to restore fish habitat in streams must not only use effective restoration procedures, but must also implement stream protection policies, and continue to develop environmental education programs. The remaining pristine habitats must be protected and abused habitats restored to a level of productivity that will meet current demands and allow options for the future. Perhaps this volume on habitat restoration will serve as a catalyst to replace the rhetoric of yesterday by action in the 1990s for conserving and restoring one of America's most important natural resources.

Acknowledgments

Comprehensive reviews by Gary Gunstrom, Mike Murphy and Gordon Thayer were greatly appreciated. Slides were generously loaned to me by Dick Aho, Neil Anderson, Jeff Cederholm, John Hamilton, Bill Hauser, Bob House, Bill Meehan, Ted Meyers and Jim Sedell for use in the oral presentation of this paper. Some hard-to-find references were provided by Phil Brna, Kevin Brownlee, Ron Dunlap and Ron Josephson.

I also thank Gordon Thayer and other members of NOAA's Coordinating Committee for the opportunity to participate in the exciting and important symposium which led to this book.

Literature Cited

Allee, B.A. 1982. The role of interspecific competition in the distribution of salmonids in streams, p. 111-122. *In* E.L. Brannon and E.O. Salo (eds.), Proceedings of the salmon and trout migratory behavior symposium. University of Washington, Seattle.

Anderson, B.W. and R.D. Ohmart. 1985. Riparian revegetation as a mitigating process in stream and river restoration, p. 41-79. *In* J.A. Gore (ed.), The restoration of rivers and streams. Butterworth, Boston.

Anderson, J.W. 1990. Design and location of in stream structures. *In* R. House, J. Anderson, P. Boehne and J. Suther (eds.), Stream rehabilitation manual emphasizing project design, construction and evaluation. Sec. 6. Oregon Chapter, Am. Fish. Soc. 20 p.

Andrus, C.W., B.A. Long and H.A. Froehlich. 1988. Woody debris and its contribution to pool formation in a coastal stream 50 years after logging. Can. J. Fish. Aquat. Sci. 45:2080-2086.

Angermeier, P.L. and J.R. Karr. 1984. Relationships between woody debris and fish habitat in a small warm water stream. Trans. Am. Fish. Soc. 113:716-726.

Behnke, R.J. 1991. America's changing fish fauna. Trout Unlimited 32(2):35-38.

Binns, N.A. and F.M. Eiserman. 1979. Quantification of fluvial trout habitat in Wyoming. Trans. Am. Fish. Soc. 108:215-228.

Bisson, P.A. 1990. Importance of identification of limiting factors in an evaluation program. *In* R. House, J. Anderson, P. Boehne and J. Suther (eds.), Stream rehabilitation manual emphasizing project design, construction and evaluation. Sec. 2. Oregon Chapter, Am. Fish. Soc. 22 p.

Bisson, P.A, R.E. Bilby, M.D. Bryant, C.A. Dolloff, G.B. Grette, R.A.-House, M.L. Murphy, K V. Koski and J.R. Sedell. 1987. Large woody debris in forested streams in the Pacific Northwest: past, present, and future, p. 143-190. *In* E.O. Salo and T.W. Cundy (eds.), Streamside management: forestry and fishery interactions, Proceedings of a Symposium held at University of Washington, Feb. 12-14, 1986. College of Forest Resources, University of Washington, Seattle. Contrib. No. 57.

Bisson, P.A., J.L. Nielsen, R.A. Palmason and L.E. Grove. 1982. A system of naming habitat types in small streams, with examples of habitat utilization by salmonids during low streamflow, p. 62-73. *In* N.B. Armantrout (ed.), Acquisition and utilization of aquatic habitat inventory information. Proceedings of an Am. Fish. Soc. Symposium, Oct. 28-30, 1981, Portland, Oregon.

Bjornn, T.C. and D.W. Reiser. 1991. Habitat requirements of salmonids in streams, p. 83-138. *In* W.R. Meehan (ed.), Influences of forest and rangeland management on salmonid fishes and their habitats. Am. Fish. Soc. Spec. Publ. 19.

Blackman, F.F. 1905. Optima and limiting factors. Ann. Bot. 19:282-295.

Blumm, M.C. 1986. Why study Pacific salmon law? Idaho Law Rev. 22:629-656.

Bowen, J.T. 1970. A history of fish culture as related to the development of fishery programs, p. 71-105. *In* N.G. Benson (ed.), A century of fisheries in North America. Am. Fish. Soc., Washington, D.C.

Cairns, J., Jr. and K.L. Dickson. 1977. Recovery of streams from spills. *In* J. Cairns, Jr., K.L. Dickson and E.E. Herricks (eds.), Recovery and restoration of damaged ecosystems. University of Virginia Press, Charlottesville.

Canada Department of Fisheries and Oceans, British Columbia Ministry of the Environment. 1980. Stream enhancement guide. Vancouver. 82 p.

Cederholm, C.J., W.J. Scarlett and N.P. Peterson. 1988. Low-cost enhancement technique for winter habitat of juvenile coho salmon. N. Am. J. Fish. Manage. 8:438-441.

Childerhose, R.J. and M. Trim. 1979. Pacific salmon and steelhead trout. University Washington Press, Seattle. 158 p.

Cowan, L. 1990. Off-channel habitat enhancement. *In* R. House, J. Anderson, P. Boehne and J. Suther (eds.), Stream rehabilitation manual emphasizing project design, construction, and evaluation. Sec. 10. Oregon Chapter, Am. Fish. Soc. 34 p.

Crispin, V. 1990. Main channel structures. *In* R. House, J. Anderson, P. Boehne and J. Suther (eds.), Stream rehabilitation manual emphasizing project design, construction, and evaluation. Sec. 9. Oregon Chapter, Am. Fish. Soc. 30 p.

Cummins, K.W. 1974. Stream ecosystem structure and function. BioScience 24:631-641.

Cummins, K.W., G.W. Minshall, J.R. Sedell, C.E. Cushing and R.C. Petersen. 1984. Stream ecosystem theory. Verh. Int. Ver. Limnol. 22:1818-1827.

Davis, H.S. 1935. Methods for the improvement of streams. Memorandum I-133. U.S. Bureau of Fisheries, Washington D.C. 27 p.

Duff, D.A., N. Banks, E. Sparks, W.E. Stone and R.J. Poehlmann. 1988. Indexed bibliography on stream habitat improvement. 4th rev. Dept. Fisheries and Wildlife, College of Natural Resources, Utah State University, Logan. 121 p.

Edgington, J., M. Alexandersdottir, C. Burns and J. Cariello. 1987. *In* Channel type classification as a method to document anadromous salmon streams. State of Alaska, Dept. Fish Game Info. Leaflet No. 260. 70 p.

Everest, F.H. and J.R. Sedell. 1984. Evaluating effectiveness of stream enhancement projects, p. 246-256. *In* T.J. Hassler (ed.), Proceedings of the Pacific Northwest stream habitat management workshop, held at Humboldt State University, Arcata California, Oct. 10-12, 1984. Western Division, Am. Fish. Soc.

Everest, F.H., J.R. Sedell, G.H. Reeves and M.D. Bryant. 1990. Planning and evaluating habitat projects for anadromous salmonids. *In* R.

House, J. Anderson, P. Boehne and J. Suther (eds.), Stream rehabilitation manual emphasizing project design, construction and evaluation. Oregon Chapter, Am. Fish. Soc. 28 p.

Everest, F.H. and D.R. Talhelm. 1982. Evaluating projects for improving fish and wildlife habitat on National Forests. U.S. Forest Service, Pacific Northwest Forest and Range Experiment Station, General Tech. Rep. PNW-146, Portland Oregon. n.p.

Fausch, K.D., C.L. Hawkes and M.G. Parsons. 1988. Models that predict standing crop of stream fish from habitat variables: 1950-85. U.S. Forest Service, General Tech. Rep. PNW-GTR-213. 52 p.

Findley, R. 1990. Will we save our own? Natl. Geogr. Mag. Sept. 1990:106-136.

Fontaine, B. 1990. Biological evaluation. *In* R. House, J. Anderson, P. Boehne and J. Suther (eds.), Stream rehabilitation manual emphasizing project design, construction, and evaluation. Sec. 15. Oregon Chapter, Am. Fish. Soc. 15 p.

Fry, F.J. 1971. The effect of environmental factors on the physiology of fish, p. 1-98. *In* W.S. Hoar and D.J. Randall (eds.), Fish physiology, Vol. 4, Environmental relations and behavior. Academic Press, New York.

Gore, J.A. 1985. The restoration of rivers and streams. Butterworth Publishers, Boston. 280 p.

Gregory, S.V., G.A. Lamberti, D.C. Erman, K.V. Koski, M.L.Murphy and J.R. Sedell. 1987. Influence of forest practices on aquatic production, p. 233-255. *In* E.O. Salo and T.W. Cundy (eds.), Streamside management: forestry and fishery interactions, Proceedings of a symposium held at University of Washington, Feb. 12-14, 1986. Institute of Forest Resources, University of Washington, Seattle. Contribution No. 57.

Grette, G.B. 1985. The role of large organic debris in juvenile salmonid rearing habitat in small streams. M.S. Thesis, University of Washington, Seattle. 105 p.

Haig-Brown, R. 1980. Bright waters, bright fish, an examination of angling in Canada. Timber Press, Forest Grove, Oregon. 142 p.

Hall, J.D. 1984. Evaluating fish response to artificial stream structures: problems and progress, p. 214-221. *In* T.J. Hassler (ed.), Pacific Northwest stream habitat management workshop proceedings. Western Division, Am. Fish. Soc. Humboldt State University, Arcata California.

Hall, J.D. and C.O. Baker. 1982. Rehabilitating and enhancing stream habitat: 1. Review and evaluation. *In* W.R. Meehan (ed.), Influence of forest and rangeland management on anadromous fish habitat in western North America. USDA Forest Service, Gen. Tech. Rep. PNW-138, Portland Oregon. 29 p.

Hall, J.D., G.W. Brown and R.L. Lantz. 1987. The Alsea Watershed Study—a retrospective, p. 399-416. *In* E.O. Salo and T.W. Cundy (eds.), Streamside management: forestry and fishery interactions, Proceedings of a symposium held at University of Washington, Feb. 12-14, 1986. Institute of Forest Resources, University of Washington, Seattle, Contrib. No. 57.

Harmon, J.E., J.F. Franklin, F.J. Swanson, P. Sollins, S.V. Gregory, J.D. Lattin, N. H. Anderson, S.P. Cline, N.G. Aumen, J.R. Sedell, G.W. Liendaemper, K. Cromack, Jr. and K.W. Cummins. 1986. Ecology of coarse woody debris in temperate ecosystems. Adv. Ecol. Res. 15:133-302.

Hartman, G., J.C. Scrivener, L.B. Holtby and L. Powell. 1987. Some effects of different streamside treatments on physical conditions and fish population processes in Carnation Creek, a coastal rain forest stream in British Columbia, p. 330-372. *In* E.O. Salo and T.W. Cundy (eds.), Streamside management: forestry and fishery interactions, Proceedings of a symposium held at University of Washington, Feb. 12-14, 1986. College of Forest Resources, University of Washington, Seattle, Contrib. No. 57.

Hassler, T.J. (ed.). 1984. Proceedings of the Pacific Northwest stream habitat management workshop, held at Humboldt State University, Arcata California, Oct. 10-12, 1984. Western Division, Am. Fish. Soc. 329 p.

Heede, B.H. 1972. Influences of a forest on the hydraulic geometry of two mountain streams. Water Resour. Bull. 8:523-530.

Heede, B.H. 1985. The evolution of salmonid streams, p. 33-37. *In* F. Richardson and R.H. Hamre (eds.), Proceedings of the symposium Wild Trout III, September 24-25, 1984.

Heifetz, J., M.L. Murphy and K V. Koski. 1986. Effects of logging on winter habitat of juvenile salmonids in Alaskan streams. N. Am. J. Fish. Manage. 6:52-58.

Heimann, D.C. 1988. Recruitment trends and physical characteristics of coarse woody debris in Oregon Coast Range streams. M.S. Thesis. Oregon State University, Corvallis. 121 p.

Helm, W.T. (ed.). 1985. Glossary of stream habitat terms. Am. Fish. Soc., Western Division, Habitat Inventory Committee. 34 p.

Herricks, E.E. 1977. Recovery of streams from chronic pollutional stress—acid mine drainage. In J. Cairns, Jr., K.L. Dickson and E.E. Herricks (eds.), Recovery and restoration of damaged ecosystems. University of Virginia Press, Charlottesville.

Herricks, E.E. and L.L. Osborne. 1985. Water quality restoration and protection in streams and rivers, p. 1-20. In J.A. Gore (ed.), The restoration of rivers and streams. Butterworth, Boston.

Hicks, B.J., J.D. Hall, P.A. Bisson and J.R. Sedell. 1991. Responses of salmonid populations to habitat changes caused by timber harvest, p. 483-517. In W.R. Meehan (ed.), Influence of forest and range management on salmonid fishes and their habitats. Am. Fish. Soc. Spec. Publ. 19.

Hogan, D. 1985. The influence of large organic debris on channel morphology in Queen Charlotte Island streams, p. 263-273. Proceedings of the Western Association of Fish and Wildlife Agencies.

Holtby, L.B. and G.F. Hartman. 1982. The population dynamics of coho salmon Oncorhynchus kisutch in a west coast rain forest stream subjected to logging, p. 308-347. In G. Hartman (ed.), Proceedings of the Carnation Creek workshop, a ten year review. Feb. 24-26, 1982, Malaspina College. Pacific Biological Station, Nanaimo, BC.

House, R., J. Anderson, P. Boehne and J. Suther. 1990. Stream rehabilitation manual emphasizing project design, construction, and evaluation. A training in stream rehabilitation workshop, Oregon Chapter, Am. Fish. Soc., Feb. 6-7, 1990.

House, R. and V. Crispin. 1990. Economic analyses of the value of large woody debris as salmonid habitat in coastal Oregon streams. U.S. Dept. Interior, BLM, T/N OR-7:6512, 11 p.

House, R., V. Crispin and R. Monthey (eds.). 1989. Evaluation of stream rehabilitation projects—Salem district (1981-1988). U.S. Dept. Interior, Bureau Land Manage. Tech. Note OR-6. 50 p.

Hunt, R.L. 1971. Responses of a brook trout population to habitat development in Lawrence Creek. Tech. Bull. 48. Wisconsin Dep. Natural Resources, Madison. 35 p.

Hunter, C.J. 1991. Better trout habitat—a guide to stream restoration and management. Montana Land Reliance, Island Press, Washington D.C. 320 p.

Hynes, H.B.N. 1970. The ecology of running waters. University of Toronto Press. 555 p.

Hynes, H.B.N. 1975. The stream and its valley. Int. Ver. Theor. Angew. Limnol. Verh. 19:1-15.

Iverson, R. 1991. Twenty-year Klamath River program links fish restoration and water quality protection actions, p. 11-12. EPA News-Notes, March, 1991.

Jones, D.A. 1986. Atlantic salmon restoration in the in the Connecticut River, p. 415-426. *In* D. Mills and D. Piggins (eds.), Atlantic salmon: planning for the future. Proceedings of the Third International Atlantic Salmon Symposium, Portland Oregon.

Judy, R.D., Jr., P.N. Seeley, T.M. Murray, S.C. Svirsky, M.R.Whitworth and L.S. Ischinger. 1984. 1982 National Fisheries Survey. Vol. I, Technical Report: Initial findings. U.S. Fish and Wildlife Service, FWS/OBS-84/06. 140 p.

Karr, J.R. and D.R. Dudley. 1981. Ecological perspective on water quality goals. Environ. Manage. 5:55-68.

Karr, J.R., K.D. Fausch, P.L. Angermeier, P.R. Yant and I.J. Schlosser. 1986. Assessing biological integrity in running waters, a method and its rationale. Illinois Natural History Survey. Spec. Publ. 5.

Keller, E.A. and F.J. Swanson. 1979. Effects of large organic material on channel form and fluvial processes. Earth Surface Processes 4:361-380.

Kessler, S.J. 1989. Habitat capability and effects model for Dolly Varden and coho salmon in southeast Alaska. Unpublished draft report, USDA Forest Service, Forest Plan Interdisciplinary Team, Juneau, Alaska.

Klingeman, P.C. 1984. Evaluating hydrologic needs for design of stream habitat modification structures, p. 191-213. *In* T.J. Hassler (ed.), Proceedings of the Pacific Northwest stream habitat management workshop, held at Humboldt State University, Arcata California, Oct. 10-12, 1984. Western Div., Am. Fish. Soc.

Koski, K V., J. Heifetz, S. Johnson, M. Murphy and J. Thedinga.1984. Evaluation of buffer strips for protection of salmonid rearing habitat and implications for enhancement, p. 138-155. *In* T.J. Hassler (ed.), Proceedings of the Pacific Northwest stream habitat management workshop, held at Humboldt State University, Arcata California, Oct. 10-12, 1984. Western Division, Am. Fish. Soc.

Koski, K V., L.E. Queirolo and J.R. Mehrkens. In prep. Value of salmonid habitat in an old-growth forest/stream environment in Southeast Alaska.

Lanka, R.P., W.A. Hubert and T.A. Wesche. 1987. Relations of geomorphology to stream habitat and trout standing stock in small Rocky Mountain streams. Trans. Am. Fish. Soc. 116:21-28.

Larkin, P.A. 1974. Play it again Sam—an essay on salmon enhancement. J. Fish. Res. Board Can. 31:1433-1457.

Leopold, L.B., M.G. Wolman and J.P. Miller. 1964. Fluvial processes in geomorphology. W.H. Freeman, San Francisco. 522 p.

Lewis, S.L. 1969. Physical factors influencing fish populations in pools of a trout stream. Trans. Am. Fish Soc. 98:14-19.

Likens, G.E., F.H. Bormann, R.S. Pierce, J.S. Eaton and N.M. Johnson. 1977. Biogeochemistry of a forested ecosystem. Springer-Verlag, New York. 146 p.

Lisle, T. and K. Overton. 1990. Utilizing channel information to reduce risk in developing habitat restoration projects. *In* R. House, J. Anderson, P. Boehne and J. Suther (eds.), Stream rehabilitation manual emphasizing project design, construction, and evaluation. Sec. 5. Oregon Chapter, Am. Fish. Soc. 18 p.

Macdonald, J.S., G. Miller and R.A. Stewart. 1988. The effects of logging, other forest industries and forest management practices on fish: an initial bibliography. Can. Tech. Rep. Fish. Aquat. Sci. No. 1622. 212 p.

Magnuson, J.J., H.A. Regier, W.J. Christie and W.C. Sonzogni. 1980. To rehabilitate and restore great lakes ecosystems, p. 95-112. *In* J. Cairns, Jr. (ed.), The recovery process in damaged ecosystems. Ann Arbor Science Publishers, Ann Arbor, Michigan.

Mather, F. 1875. Poisoning and obstructing the waters, p. 14-19. Proceedings of the American Fish Cult. Assoc., 4th Annual Meeting.

McMahon, T. 1989. Large woody debris and fish, n.p. *In* Silvicultural management of riparian areas for multiple resources. A COPE Workshop, Dec. 12-13, 1989, Gleneden Beach Oregon. USDA Forest Service, Pacific Northwest Research Station and Oregon State University Forestry College, Corvallis.

Meehan, W.R. (ed.). 1991. Influences of forest and range land management of salmonid fishes and their habitats. Am. Fish. Soc. Spec. Publ. 19. 751 p.

Meehan, W.R. and T.C. Bjornn. 1991. Salmonid distributions and life histories, p. 47-82. *In* W.R. Meehan (ed.), Influences of forest and rangeland management on salmonid fishes and their habitats. Am. Fish. Soc. Spec. Publ. 19.

Meehan, W.R., F.J. Swanson and J.R. Sedell. 1977. Influences of riparian vegetation on aquatic ecosystems with particular references to salmonid fishes and their food supply. *In* R.R. Johnson and D.A. Jones (eds.), Importance, preservation and management of riparian habitat: A symposium. USDA Forest Service Gen Tech. Rep. RM-43:137-145.

Megahan, W., W.S. Platts and B. Kulesza. 1980. River bed improves over time: South Fork Salmon. Symposium on Watershed Management 1:380-395. Am. Soc. Civil Engineers, New York.

Milner, N.J., R.J. Hemsworth and B.E. Jones. 1985. Habitat evaluation as a fisheries management tool. J. Fish. Biol. 27(A):85-108.

Morisawa, M. 1968. Streams, their dynamics and morphology. McGraw-Hill, New York. 175 p.

Murphy, M.L., J. Heifetz, S.W. Johnson, K V. Koski and J.F.Thedinga. 1986. Effects of clear-cut logging with and without buffer strips on juvenile salmonids in Alaskan streams. Can. J. Fish. Aquat. Sci. 43:1521-1533.

Murphy, M.L. and K V. Koski. 1989. Input and depletion of woody debris in Alaska streams and implications for streamside management. N. Am. J. Fish. Manage. 9:427-436.

Murphy, M.L., J.M. Lorenz, J. Heifetz, J.F. Thedinga, K V. Koski and S.W. Johnson. 1987. The relationship between stream classification, fish, and habitat in Southeast Alaska. USDA Forest Service, Wildl. Fish. Habitat Manage., Note 12. 63 p.

Murphy, M.L. and W.R. Meehan. 1991. Stream ecosystems, p. 17-46. *In* W.R. Meehan (ed.), Influences of forest and rangeland management on salmonid fishes and their habitats. Am. Fish. Soc. Spec. Publ. 19.

Narver, D.W. 1971. Effects of logging debris on fish production, p. 100-111. *In* J.T. Krygier and J.D. Hall (eds.), Proceedings of a symposium on forest land uses and the stream environment. Oregon State University, Corvallis Oregon.

Nassichuk, M. 1986. Water quality—The forgotten restoration issue, p. 7-13. *In* J.H. Patterson (ed.), Proceedings of the workshop on habitat improvements, Whistler, BC, May 8-10, 1984.

Netboy, A. 1974. The salmon, their fight for survival. Houghton Mifflin, New York. 613 p.

Olson, A.D. and J.R. West. 1990. Evaluation of in stream fish habitat restoration structures in Klamath River tributaries, 1988/1989. Annual Report for Interagency Agreement 14-16-0001-89508, USDA Forest Service Klamath National Forest, 1312 Fairlane Road, Yreka California. 36 p.

Paustian, S.J., D.A. Marion and D.F. Kelliher. 1984. Stream channel classification using large scale aerial photography for southeast Alaska watershed management, p. 670-677. *In* Renewable resource management: applications of remote sensing, Am. Soc. Photogrammetry, Falls Church, Virginia.

Payne, N.F. and F. Copes. 1986. Wildlife and fisheries habitat improvement handbook. USDA Forest Service; Wildlife and Fisheries Administrative Rept. (unnumbered). n.p.

Platts, W.S. and W. Megahan. 1975. Time trends in channel sediment size composition in salmon and steelhead spawning areas: South Fork Salmon River, Idaho. USDA Forest Service Gen. Rep. Intermountain For. and Range Exp. Stn., Ogden, Utah. 21 p.

Reeves, G.H. 1985. Interactions between the redside shiner *Richardsonius balteatus* and the steelhead trout *Salmo gairdneri* in western Oregon: the influence of water temperature. Can. J. Fish. Aquat. Sci. 44:1603-1613.

Reeves, G.H., J.D. Hall, T.D. Roelofs, T.L. Hickman and C.O. Baker. 1991. Rehabilitating and modifying stream habitats, p. 519-556. *In* W.R. Meehan (ed.), Influences of forest and rangeland management on salmonid fishes and their habitats. Am. Fish. Soc. Spec. Publ. 19.

Reeves, G.H. and T.D. Roelofs. 1982. Rehabilitating and enhancing stream habitat. Part 2: Field applications. *In* W.R. Meehan (ed.), Influence of forest and range management on anadromous fish habitat in western North America. USDA Forest Service Gen. Tech. Rep. PNW-124. Pacific Northwest For. and Range Exp. Stn., Portland Oregon.

Robbins, W.G. 1988. Hard times in paradise, Coos Bay, Oregon, 1850-1986. University of Washington Press, Seattle.

Robison, G.E. 1990. Physical evaluation of fisheries habitat structures. *In* R. House, J. Anderson, P. Boehne and J. Suther (ed.), Stream rehabilitation manual emphasizing project design, construction, and evaluation. Sec. 16. Oregon Chapter, Am. Fish. Soc. 34 p.

Rosgen, D.L. 1985. A stream classification system, p. 91-95. *In* Proceedings in riparian ecosystems and their management: Reconciling conflicting uses, April 16-18, Tucson, Arizona.

Rosgen, D.L. and B.L. Fittante. 1986. Fish habitat structures--a selection guide using stream classification, p. 163-179. *In* J.G. Miller, J.A. Arway and R.F. Carline (eds.), Fifth trout stream habitat improvement workshop, Proceedings of a Workshop, August 12-14, 1986, Lockhaven University

Russell, A. 1980. Introduction, p. 11-13. *In* Bright waters, bright fish, an examination of angling in Canada. Timber Press, Forest Grove Oregon.

Ryder, R.A. and S.R. Kerr. 1989. Environmental priorities: placing habitat in hierarchic perspective, p. 2-12. *In* C.D. Levings, L.B. Holtby and M.A. Henderson (eds.), Proceedings of the national workshop on effects of habitat alteration on salmonid stocks. Can. Spec. Publ. Fish. Aquat. Sci. 105.

Salo, E.O. and T.W. Cundy (eds.). 1987. Streamside management: forestry and fishery interactions. Proceedings of a symposium held at the University of Washington, Feb. 12-14, 1986. Institute of Forest Resources, University of Washington, Seattle. Contrib. no. 57.

Scrivener, J.C. and B.C. Andersen. 1984. Logging impacts and some mechanisms that determine the size of spring and summer populations of coho salmon fry *Oncorhynchus kisutch* in Carnation Creek, British Columbia. Can. J. Fish. Aquat. Sci. 41:1097-1105.

Sedell, J.R., P.A. Bisson, F.J. Swanson and S.V. Gregory. 1988. What we know about large trees that fall into streams and rivers, p. 47-81. *In* C. Maser, R.F. Tarrant, J.M. Trappe and J.F. Franklin (eds.), From the forest to the sea: a story of fallen trees. U.S. Forest Service Gen. Tech. Rep. PNW-GTR-229.

Sedell, J.R. and K.J. Luchessa. 1982. Using the historical record as an aid to salmonid habitat enhancement, p. 210-223. *In* N.B. Armantrout (ed.), Acquisition and utilization of aquatic habitat inventory information: proceedings of a symposium, Oct. 28-30, 1981, Portland Oregon. Am. Fish. Soc.

Sedell, J.R. and F.J. Swanson. 1984. Ecological characteristics of streams in old-growth forests of the Pacific Northwest, p. 9-16. *In* W.R. Meehan, T.R. Merrell, Jr. and T.A. Hanley (eds.), Fish and wildlife relationships in old-growth forests: proceedings of a symposium. Am. Inst. Fish. Res. Biol.

Sedell, J.R., F.J. Swanson and S.W. Gregory. 1984. Evaluating fish response to woody debris, p. 222-245. *In* T.J. Hassler (ed.), Proceedings of the Pacific Northwest stream habitat management workshop, held at Humboldt State University, Arcata, California, Oct. 10-12, 1984. Western Div., Am. Fish. Soc.

Shirvell, C.S. 1989. Habitat models and their predictive capability to infer habitat effects on stock size, p. 173-179. *In* C.D. Levings, L.B. Holtby and M.A. Henderson (eds.), Proceedings of the national workshop on effects of habitat alteration on salmonid stocks. Can. Spec. Publ. Fish. Aquat. Sci. 105.

Stein, R.A., P.E. Reimers and J.D. Hall. 1972. Social interaction between juvenile coho (*Oncorhynchus kisutch*) and fall chinook salmon (*O. tshawytscha*) in Sixes River, Oregon. J. Fish. Res. Board Can. 29:1737-1748.

Strahler, A.N. 1957. Quantitative analysis of watershed geomorphology. Trans. Am. Geophys. Union 38:913-920.

Sullivan, K., T.E. Lisle, C.A. Dolloff, G.E. Grant and L.M.Reid. 1987. Stream channels: the link between forests and fishes, p. 39-97. *In* T. Cundy and E.O. Salo (eds.), Proceedings of a symposium on streamside management-forestry fisheries interactions. February 12-14, 1986. Institute of Forest Resources, University of Washington, Seattle.

Swanson, S.V. Gregory, J.R. Sedell and A.G. Campbell. 1982.Land-water interactions: the riparian zone, p. 267-291. *In* R. L. Edmonds (ed.), Analysis of coniferous forest ecosystems in the western United States. US/IBP Synthesis Series 14. Hutchinson Ross, Stroudsburg, Pennsylvania.

Talley, K. 1990. Stats Pack—everything you ever wanted to know about landings, prices, and predictions for the major Pacific Coast fisheries. Pac. Fishing 11:(4)126.

Thedinga, J.F., M.L. Murphy, J. Heifetz, K V. Koski and S.W. Johnson. 1989. Effects of logging on size and age composition of juvenile coho salmon *Oncorhynchus kisutch* and density in presmolts in southeast Alaska streams. Can. J. Fish. Aquat. Sci. 46:1383-1391.

Triska, F.J. 1984. Role of wood debris in modifying channel geomorphology and riparian areas of a large lowland river under pristine conditions: a historical case study. Verh. Int. Ver. Limnol. 22:1876-1892.

USDA Forest Service. 1952. Fish stream improvement handbook. USDA Forest Service, Washington D.C. 21 p.

USDA Forest Service. 1986. Aquatic habitat management handbook. FSH 2609.24, Alaska Region, June 1986. n.p.

USDA Forest Service. 1990. Analysis of the management situation for their vision of the Tongass Land Management Plan. Vol. 1, Chapter 3, Jan. 31, 1990, 773 p.

U.S. Department of Commerce. 1990. Fisheries of the United States, 1989. Supplemental, 4 p. NOAA, NMFS, Fish. Stats. Div., 1335 East West Highway, Silver Spring, Maryland.

U.S. Department of the Interior, Bureau of Land Management.1968. Stream preservation and improvement. U.S. Bureau of Land Management, Washington D.C. Manual, Sec. 6760. 49 p.

Van Cleef, J.S. 1885. How to restore our trout streams, p. 51-55. Am.- Fish. Soc., 14th Annual Meeting.

Wallace, J.B. and A.C. Benke. 1984. Quantification of wood habitat in subtropical Coastal Plain streams. Can. J. Fish. Aquat. Sci. 41:1643- 1652.

Wesche, T.A. 1985. Stream channel modifications and reclamation structures to enhance fish habitat, p. 103-163. *In* J.A. Gore (ed.), The restoration of rivers and streams. Butterworth, Boston.

White, R.J. and O.M. Brynildson. 1967. Guidelines for management of trout stream habitat in Wisconsin. Wisconsin Dep. Nat. Resour. Tech. Bull. 39.

CHAPTER 9

The Columbia River: Fish Habitat Restoration Following Hydroelectric Dam Construction

John G. Williams

National Marine Fisheries Service, Northwest Fisheries Center
Seattle, Washington

Merritt E. Tuttle

National Marine Fisheries Service, Northwest Region
Portland, Oregon

Abstract

Investments in restoring anadromous fish habitat within the Columbia River system represent one of the most comprehensive and significant habitat restoration efforts known. About 1.5 billion dollars has been spent to date, with the bulk of the funding going to mitigate for the loss of anadromous fish resources caused by construction and operation of dams. Habitat research, especially to solve fish passage problems at dams, is a vital component in wisely guiding each investment. Much remains to be done, however, both in research and in correction of habitat problems. National Marine Fisheries Service intends to protect, restore and enhance the habitat to achieve significant benefits for the economies of the Northwest and the Nation.

Introduction

Background

The Columbia River watershed drains 680,000 km^2; its largest tributary is the Snake River (Figure 1). Historically, spawning territory for anadromous fish included over 12,500 km of riverine habitat and resulted in estimated salmon and steelhead (*Oncorhynchus* spp.) production of between 10 and 16 million fish (NWPPC 1987). These fish have been important to the people of the Columbia River Basin since prehistoric times.

The river system includes sockeye (*O. nerka*), coho (*O. kisutch*), chum (*O. keta*) and chinook salmon (*O. tshawytscha*) and steelhead (*O. mykiss*). Chinook salmon make up the majority of fish and the river system once was the largest producer of them in the world (Netboy 1980). Adult fish return to the river throughout the year; however, winter returns are mostly steelhead. Ages of returning adults range from 2 to 6 years, with the largest fish weighing generally more than 20 kg. Chinook salmon are generally designated as spring, summer or fall depending upon when adults enter the river on their way to the spawning grounds. Spring fish travel the furthest distance (up to 1,300 km) and into the smaller tributaries to spawn in August. Fall fish are mainstem spawners in October to November, while summer fish spawn somewhere between the spring and fall.

There are two major life history patterns for juvenile salmon. Mainstem-spawning fall chinook salmon and some stocks of summer chinook salmon produce juveniles that undergo smoltification and migrate to the ocean as subyearling fish during the summer following spawning. Spring chinook, some stocks of summer chinook, coho and sockeye salmon, and steelhead juveniles rear in freshwater for the whole year following adult spawning. They undergo smoltification and migrate to the ocean as yearlings the next spring (about 1.5 years after the adults spawned).

Figure 1. Map of the Columbia River Basin showing existing fish habitats and dams.

407

Although all species of salmon still exist in the Columbia River, they unfortunately have declined precipitously from historical levels (Netboy 1980). This is particularly true for wild stock fish migrating to and from upper reaches of the river system. Initial declines resulted from overfishing, but losses since have been primarily the result of hydroelectric dam construction (Raymond 1988; Williams 1989). Suitable habitat previously utilized by anadromous salmonids was decreased by approximately 35% when Grand Coulee Dam (Figure 2) on the mid-Columbia River and Brownlee Dam on the Snake River blocked upstream access for adult fish and were too high for effective passage of juveniles out of the reservoirs (Table 1). Consequently, no fish passage facilities were constructed for Chief Joseph and Hells Canyon Dams, which are now the upstream limits to fish passage. Construction of 13 other mainstem dams during the period from 1940 through 1975 caused cumulative habitat degradation by changing free-flowing rivers into slow-moving reservoirs, inundating areas previously used by some fish stocks for spawning, and increasing travel time for migrants.

Figure 2. Grand Coulee Dam on the mid-Columbia River.

Juvenile outmigrants also suffer direct mortalities when passing through the dams. As a result of the construction of huge reservoirs in the upper reaches of the basin, water storage capacity in the Columbia and Snake rivers nearly doubled from 40.2 km³ in 1965 to 75.5 km³ in 1973. Historically, flows ranged from 1,400 to 28,300 m³ s⁻¹ and the magnitude was primarily the result of the timing of glacial and snow melt. However, the increase in storage capacity decreased peak flows during the spring outmigration so that present day flows generally range from 3,500 to 9,900 m³ s⁻¹ (Figure 3).

The habitat changes as a result of dam construction substantially altered the conditions under which the life history traits of the salmon and steelhead evolved. The population decreases in upper Columbia River salmon runs correspond with these cumulative effects (Table 1).

Definition of Habitat

We define habitat as the freshwater migration pathway for anadromous salmonids and limit our discussion to alteration and restoration of this degraded habitat that resulted from dam construction of the dams on the mainstem Columbia and Snake rivers. Habitat related to spawning and rearing in small tributary areas will not be discussed here but is covered by Koski (this volume).

Past Habitat Degradation

Changes in habitat from dam construction affected the upstream adult migration and the downstream juvenile migration. Except for the few large dams that blocked fish passage entirely, all other mainstem dams contained fish ladders because it was recognized that without adult passage, the upriver salmonid runs could not exist. Thus, the greatest efforts for limiting salmon loss after the construction of the earliest dams were directed toward effective adult passage. Studies were conducted to refine early ladder designs

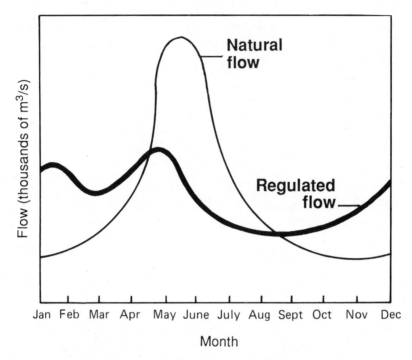

Figure 3. Typical effects of regulated flows on runoff in the Columbia River. After Raymond 1988.

to decrease delay within the ladders. However, delays in migration of up to one week occurred because it was difficult for fish to find the ladder entrances. The ladders and their collection channels have small attraction flows compared with the large flows from turbine discharges and spillways. In addition to delay, some (generally small) percentage of the migration did not pass each successive dam it encountered. Further, the energy reserves expended by adults seeking successful passage routes may have translated into decreased spawning success. With increasing dam construction, remaining adult passage problems were considered less important compared with potential problems with juvenile fish. Initial concerns for juvenile migrants were low because few dams were in place. However, concern for juveniles was heightened with the

Table 1. Trends in adult returns of upper river spring and summer chinook salmon to the Snake and Columbia Rivers compared with hydroelectric dam construction.

Four-Year Average	Number of Adults (1000s)	Dam (Year of Construction, Head in Meters)
pre-1950		Rock Island (1933;12.1)
		Bonneville (1938;14.8)
		Grand Coulee (1941;85.8)
1950-53	299	McNary (1953; 26.1)
1954-57	401	Chief Joseph (1955; 43.8)
		The Dalles (1957; 21.5)
1958-61	316	Brownlee (1958; 68.0)
		Priest Rapids (1959; 21.0)
		Ice Harbor (1961; 30.3)
		Oxbow (1961; 35.5)
		Rocky Reach (1961; 28.2)
1962-65	270	Wanapum (1963; 24.2)
1966-69	266	Hells Canyon (1967; 66.7)
		Wells (1967; 21.8)
		John Day (1968; 31.8)
		Lower Monumental (1969; 30.3)
1970-73	286	Little Goose (1970; 30.3)
1974-77	152	Lower Granite (1975; 30.3)
1978-81	112	
1982-85	98	
1986-89	139	

plans to construct a series of dams on the lower Snake and Columbia rivers. Early research indicated that the pools behind each dam increased travel time by approximately three-fold (dependent upon flow) (Raymond 1979) and caused juvenile mortalities of 8-19% from turbine passage (Bell 1981). The dams also concentrated migrating juveniles, attracting large numbers of predators. Estimates of predation on juvenile subyearling chinook salmon have been as high as 40% for one reservoir (Beamesderfer and Rieman 1988).

The cumulative negative impact to the juvenile migration is considerable. Raymond (1979) estimated that during the low flow year of 1973 (a result of an abnormally low snowfall year in the basin), only 4-5% of the juvenile salmonids passing Little Goose Dam survived to the Dalles Dam.

Further damage to adult and juvenile migration habitats occurred soon after construction of these additional dams. High spring river flows which originally caused little or no problems to migrants when the earliest dams (also lowest head) were in place became lethal because they generally exceeded the powerhouse capacity. The newer dams had higher spillways and the considerable spill created concomitant increases in supersaturated atmospheric gases (particularly nitrogen) in the water. The continuous series of reservoirs increased the levels of supersaturation because of minimal equilibration of gases with the atmosphere between dams. Consequently, migrants were exposed to higher levels of supersaturation for longer times as they migrated through the reservoirs. The result was "gas bubble" disease (Figure 4) which caused mortalities in up to 35% of adult and juvenile fish (Ebel and Raymond 1976).

On some of the major tributaries to the Snake and Columbia rivers, irrigation withdrawals from smaller dams have been particularly devastating to juvenile salmonid migrants. Water diverted to irrigation canals also diverted migrants. The fish eventually died in the irrigated fields or orchards.

Habitat Restoration

The National Marine Fisheries Service (NMFS) conducted much of the initial research toward restoration or rehabilitation of anadromous fish and their mainstem migration habitat in the Columbia and Snake rivers. This work, particularly since the inception of the Pacific Northwest Electric Power Planning and Conservation Act of 1980 (see later), has been fully coordinated through interagency task forces composed of representatives from regional federal and state fisheries agencies, federal regulatory agencies, and public and private power producers. Not all problems are solved, and research continues on many of the issues.

Improvements to adult passage facilities, designs for new facilities, and modifications to discharge patterns from spillways and powerhouses decreased delays of migrants from many days to less than one day under optimum passage conditions. Delays for adult fish also decreased as a consequence of decreased river flows stemming from additional storage capacity within the basin.

To decrease "gas bubble" disease, the base of the spillway at most mainstem dams was modified with deflectors to direct spill horizontally into the tailrace. This decreased nitrogen supersaturation by preventing water from plunging into the tailrace. Since

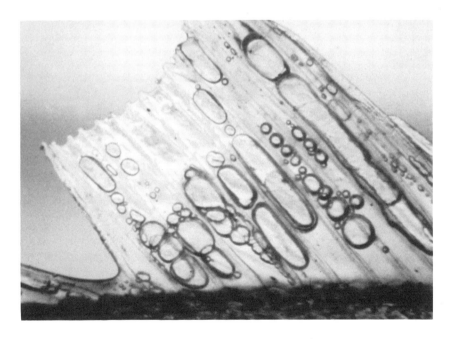

Figure 4. "Gas bubble" disease resulting from exposure to nitrogen supersaturated water.

modification of the spillways in 1976, mortalities due to "gas bubble" disease have been minimal.

Based primarily upon studies that related juvenile travel time to flow (Raymond 1979; Sims and Ossiander 1981), a water budget was developed by the Northwest Power Planning Council as a mitigative measure to compensate for the decrease in natural spring river flows resulting from water storage used for power production later in the year (NWPPC 1987) (Figure 3). The water budget is a block of stored water that can be used between April 15 and June 15 to create an artificial freshet to speed juvenile fish to the ocean. It reduces the firm energy load carrying capacity of the hydropower system throughout the year; nonetheless, it is too small to protect all fish stocks. Additionally, the increased flows provided by the water budget may not substantially decrease travel times for fish that are not physiologically ready to migrate. This is particularly true for hatchery-bred fish. Research to understand juvenile fish migrant behavior in relation to changing river flows, as well as negotiations to improve flows, are ongoing and need to be continued.

Juvenile fish bypass systems have been installed at turbine intakes on many hydroelectric dams to divert fish from turbines into safe passage routes (Figure 5). Turbine-related mortalities in juvenile fish have been decreased by 50-75% where effective bypass systems have been installed. Research and planning continue to develop bypass systems at dams where facilities do not presently exist and to improve the effectiveness of installed systems.

Without nitrogen supersaturation, survival through spillways is generally high (Bell and DeLacy 1972); thus, some spill is now used to increase juvenile migrant survival at dams where turbine bypass systems are inadequate. At Lower Granite, Little Goose and McNary dams, the juvenile bypass systems have been linked with a fish transportation program. Juvenile fish entering the bypass system are loaded into barges or trucks rather than released into the river immediately downstream from the dam. The transported fish

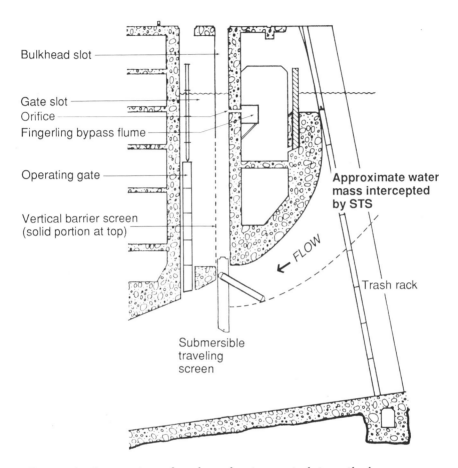

Bulkhead slot

Gate slot
Orifice
Fingerling bypass flume

Operating gate

Vertical barrier screen
(solid portion at top)

**Approximate water
mass intercepted
by STS**

FLOW

Trash rack

Submersible
traveling
screen

*Figure 5. Cross-section of a dam showing typical juvenile bypass system
components.*

are released at a point below Bonneville Dam (the last dam before
the ocean) and, thus, do not face the cumulative impacts of passage
through a number of reservoirs and dams. Survivals of transported
fish for most species are 1.5-2.5 times higher than for fish passing
downstream inriver. To prevent losses of fish to irrigation, screening devices have been installed at the entrance to canals (see below
under the Mitchell Act).

Finally, efforts began in 1990 to remove northern squawfish (*Pty-chocheilus oregonensis*). It is a native predatory fish to the Columbia River system, but its population has increased tremendously because favorable habitat has been created in the reservoirs formed behind the dams. Because the reservoirs slow the river flow and the dams concentrate juvenile migrants, the northern squawfish have become much more effective predators. Gill nets, baited long-lines, and hook and line sampling in conjunction with angler rewards of $3.00 per fish were used to remove northern squawfish in the 1990 test fisheries. It is too early to determine how successful these methods will be. Large electroshocking gear, purse seining and expanded fishing efforts will be tried in 1991. Additional research is also needed to understand predator-prey dynamics in the system, and what effect predator removal efforts will have on the population structure of remaining fish.

Legislation

There is a strong legal basis under federal law for restoring anadromous fish and fish habitat of the Columbia River system. Descriptions of some of the most significant laws follows.

Fish and Wildlife Coordination Act, 1934

This act provides the opportunity for NMFS habitat managers to provide recommendations regarding every federal action that impacts the region's waters. NMFS habitat managers review all public notices for permits involving structures, fills or pollution discharges. Also reviewed are construction actions proposed by federal agencies such as the U.S. Army Corps of Engineers and the Bureau of Reclamation. In addition, all federal licenses affecting water resources are subject to NMFS review and comment. These licenses include the actions of the Federal Energy Regulatory Commission (FERC) and the Nuclear Energy Regulatory Commission.

NMFS also engages in an extensive inspection program of all fishways for anadromous fish at federal and federally licensed hydropower projects. This inspection program is conducted by NMFS fish passage engineering staff, located in Portland, Oregon.

Federal Power Act, 1935

Section 18 of the Federal Power Act gives the Departments of Commerce and Interior the right to "prescribe" fishways for federally licensed hydroelectric projects. Such prescriptions are mandatory.

Under Section 10(j) of the Federal Power Act, NMFS makes recommendations for license terms and conditions for protection and enhancement of anadromous fish resources at federally licensed hydropower projects.

NMFS is also involved, pursuant to the Federal Power Act, in negotiated settlement agreements regarding development of juvenile fish passage facilities at public utility district hydropower plants on the mid-Columbia River.

NMFS was involved, as petitioner, in important litigation (Yakima Indian v. FERC, 746 F.2d 466 [9th cir. 1984]) which helped establish FERC obligations regarding pre-licensing consideration of fishery protection, mitigation issues and FERC obligations under the National Environmental Policy Act.

Mitchell Act, 1938

This act is specific to the Columbia River system. It is the basis for NMFS investment in several habitat restoration efforts. NMFS finances the operation of two different types of fishways, those for adult anadromous fish on their journey upstream and those for juvenile anadromous fish on their journey downstream. Fishways for adult anadromous fish allow the fish to ascend barriers, both natural and man-made, so that the fish may spawn upstream. Fishways for juvenile anadromous fish are also designed to divert fish away from irrigated fields. These fishways consist of screening devices that separate the fish from irrigation water and bypass them back to the river where they continue their seaward journey.

Endangered Species Act, 1973

The purposes of this legislation are to "provide a means whereby the ecosystems upon which endangered species and threatened species depend may be conserved ..." and "to provide a program for the conservation of such endangered species and threatened species ..." Pursuant to this act, and in response to petitions, NMFS is conducting a review of the status of populations of Snake River chinook and sockeye salmon and Lower Columbia River coho salmon. This review will determine if these fish populations should be listed as endangered or threatened species, thus qualifying them for extensive federal protection

The Pacific Northwest Electric Power Planning and Conservation Act, 1980 (NWPA)

This act provides for two regional goals: (1) development of an efficient and reliable energy supply and (2) the restoration of anadromous fish resources damaged by the development of hydroelectric energy supplies. The act established the Northwest Power Planning Council (NWPPC) to oversee power planning and fishery restoration activities. The act also requires that federal agencies give "equitable treatment" to fish and wildlife in all their actions in the Columbia River Basin. Equitable treatment has been interpreted as a substantive standard.

Production of adult salmon and steelhead in the early 1980s was approximately 2.5 million fish. The NWPPC's Fish and Wildlife Program has, as its interim goal, a doubling of the Columbia River anadromous fish run to 5 million fish. After 10 years of implementation under the NWPA, significant improvements in hydropower operations are still needed to meet this interim goal.

PL 100-216 (Swan Falls Legislation), 1987

This legislation called for NMFS, the Department of the Interior, and Idaho Power Company to negotiate an agreement on, and to fund studies regarding, protection, mitigation and enhancement of the fish and wildlife resources of the Snake River Basin. The parties successfully negotiated the agreement, and anadromous fish studies have begun.

These studies involve a search for water to augment Snake River flows to stimulate juvenile anadromous fish migration. They also involve routing and delivery of water in the Snake River according to the hydraulic capacity of the river channels. The studies were completed in late 1991.

The Federal Investment

Between 1951 and 1989, the federal investment in Columbia River anadromous fish resources was about $1.5 billion (Table 2). Without such an investment, the Columbia River salmonid stocks would be confined to a few streams near the ocean that are not blocked by dams.

Future Needs for Anadromous Fish Migration Habitat

The important needs for anadromous fish are as follows:

1. FLOWS: Additional flows to aid seaward migration and, in some cases, upstream migration during spring and summer. As described, NMFS is at the forefront of this effort.

2. BYPASS: Screen and bypass all Columbia River and Snake River hydroelectric turbine bays encountered by juvenile anadromous fish. There is a continued need to improve fish passage survival through research and application of research findings. NMFS is

Table 2. Federal expenditures within the Columbia River Basin (1951-1989).

Columbia River Fishery Development Program (NMFS funded)	$200,000,000
Construction of fish ladders, juvenile passage facilities, research, and some hatcheries (COE funded)	$363,000,000
Lower Snake River Compensation Plan (COE and USFWS funded)	$177,000,000
Fish and Wildlife Activities (Bonneville Power Administration funded)	$195,500,000
Foregone power production due to fish passage obligations	$504,500,000
Total	$1,500,000,000

at the forefront of bypass research and advocates completion of bypass systems at both federal and non-federal dams.

Satisfying These Needs Will Benefit the Northwest and the Nation

Anadromous fish are a significant part of the economy of the Pacific Northwest and are sold throughout the nation and the world. They are an integral part of the quality of life of the Northwest and are one of the reasons people settle and stay here. Under the Pacific Salmon Treaty between the United States and Canada, NMFS has a stewardship obligation for this resource. These fish are also a major part of the settlement of Indian treaties, where significant lands were exchanged for the right to fish in common with non-Indians. These fish have intense public and political support.

Columbia River habitat restoration directly impacts the residents of Alaska, California, Idaho, Oregon, Washington and British Columbia. It is possibly the largest coordinated and sustained habitat restoration effort. NMFS intends to protect, restore and enhance this resource.

Literature Cited

Beamesderfer, R. C. and B. E. Rieman. 1988. Predation by resident fish on juvenile salmonids in a mainstem Columbia River reservoir: Part III. Abundance and distribution of northern squawfish, walleye, and smallmouth bass, p. 211-248. *In* T. P. Poe and B. E. Rieman (eds.), Predation by resident fish on juvenile salmonids in John Day Reservoir, Vol. I. Final report of research to Bonneville Power Administration, Contracts DE-AI79-82BP34796 and 35097. (Available from Bonneville Power Administration, Division of Fish and Wildlife, P.O. Box 3621, Portland, Oregon 97208.)

Bell, M. C. and A. C. DeLacy. 1972. A compendium on the survival of fish passing through spillways and conduits. Final report to U.S. Army Corps of Engineers, Contract DACW57-67-C-0105. 121 p. (Available from North Pacific Division, U.S. Army Corps of Engineers, P.O. Box 2870, Portland, Oregon 97208-2870.)

Bell, M. C. 1981. Updated compendium on the success of passage of small fish through turbines. Final report to U.S. Army Corps of Engineers, Contract DACW57-76-C-0254. 201 p. (Available from North Pacific Division, U.S. Army Corps of Engineers, P.O. Box 2870, Portland, OR 97208-2870.)

Ebel, W. J. and H. L. Raymond. 1976. Effect of atmospheric gas supersaturation on salmon and steelhead trout of the Snake and Columbia Rivers. U.S. Natl. Mar. Fish. Serv., Mar. Fish. Rev. 38:1-14.

Northwest Power Planning Council (NWPPC). 1987. Columbia Basin Fish and Wildlife Program. Portland, Oregon. 246 p.

Netboy, A. 1980. The Columbia River salmon and steelhead trout. University of Washington Press, Seattle, Washington. 180 p.

Raymond, H. L. 1979. Effects of dams and impounds on migrations of juvenile chinook salmon and steelhead from the Snake River, 1966 to 1975. Trans. Amer. Fish. Soc. 108:505-529.

Raymond, H. L. 1988. Effects of hydroelectric development and fisheries enhancement on spring and summer chinook salmon and steelhead in the Columbia River Basin. N. Amer. J. Fish. Manage. 8:1-24.

Sims, C. W. and F. J. Ossiander. 1981. Migrations of juvenile chinook salmon and steelhead trout in the Snake River from 1973 to 1979. Report to U.S. Army Corps of Engineers. 31 p. (Available from Northwest Fisheries Center, 2725 Montlake Blvd. E., Seattle, Washington 98112-2097.)

Williams, J. G. 1989. Snake River spring and summer chinook salmon: can they be saved? Regulated Rivers Research Manage. 4:17-26.

CHAPTER 10

Restoring Wetland Habitats in Urbanized Pacific Northwest Estuaries

Charles A. Simenstad
Fisheries Research Institute, University of Washington
Seattle, Washington

Ronald M. Thom
Battelle Marine Sciences Laboratories
Sequim, Washington

Abstract

Urbanized estuaries probably present the most difficult challenge for restoration of wetlands because of the scope of historical losses, water and air pollution, the potential for disturbance, and the limited opportunities for viable restoration. Our experience with more than ten wetland mitigation and enhancement projects in the Pacific Northwest over the past five years has mandated a more functional, rather than mimicry, approach to the design and evaluation of restored or created wetlands. These projects encompassed diverse wetland habitats, including small-scale enhancement of rip-rap stabilized channel banks with mudflat "terraces," nourishment of beaches depleted of fine-grained sediments, and development of a full-scale brackish marsh-mudflat complex. This functional approach has altered our strategy of evaluation from that based on fish and wildlife use to a quantitative method based upon the attributes of estuarine wetlands that promote the communities of which fish and wildlife are but elements. This protocol also provides a mechanism of positive feedback where quantitative information on the performance of a restored wetland enhances the technology base, adaptively increasing the probability of successful restoration designs in the future.

Introduction

Most of the experience in estuarine habitat restoration and creation in the Pacific Northwest derives from regulatory-mandated mitigation projects. Such compensatory mitigation projects are broadly constrained by the requirements of the permit process, the economics of a process driven predominantly by the developer's cost-benefit ratios, and a history of poor scientific evaluation (Strickland 1986; Kunz et al. 1988). Few, if any, dedicated estuarine habitat restoration projects in the region have gone through a thorough scientific design, implementation and monitoring which places severe constraints upon the status of this "ecotechnology" (Zedler 1986). Therefore, in evaluating the potential for effective restoration strategies, we are forced to draw from the few mitigation projects that involve elements of habitat restoration and are sufficiently documented to allow preliminary conclusions about their success.

Our objectives in this review are to provide an overview of (1) the estuaries of the Pacific Northwest; (2) the habitat structure of these estuaries; (3) the loss of estuarine wetland habitat that has occurred over the past 150 years; (4) the sources of habitat loss and degradation; (5) the consequences to biotic resources that are dependent on estuarine habitats; (6) examples of different approaches to habitat restoration that have been attempted in the region; (7) new approaches to the design and monitoring of the function of restored habitats; and (8) gaps in the ecotechnology of estuarine habitat restoration that need to be addressed by future scientific research and policy analysis.

Estuarine Wetlands in the Pacific Northwest and Their Status

Estuaries in the Pacific Northwest are diverse in distribution, size and type. Within the Columbian Province (Cowardin et al. 1979), there are approximately 26 primary estuaries along the exposed coast (Simenstad 1983). In addition, Puget Sound, Hood Canal and the Strait of Juan de Fuca compose a complex, fjordal inland sea that is made up of a myriad of over 56 "subestuaries"[1] (Simenstad et al. 1982). The drainage basins of these estuaries are similarly diverse, from regional landscapes encompassing the second largest watershed (Columbia River, approximately 660,500 km²) in the United States to short, steep-gradient basins less than 20 km² in area. All of the classes of estuaries—drowned river valleys, fjords, bar-built and tectonic (Prichard 1967; Russell 1967)— are represented. Coastal, hypersaline lagoons do not occur in the region, but one example of a man-made estuary does.

Given the region's diversity of landscape, geological and hydrological settings, the estuaries also differ dramatically in habitat structure, from broad, braided deltaic flats with monotypic stands of emergent marsh or expansive, unvegetated flats, to mainstem channels cutting through bedrock beach terraces. For instance, littoral flats in those estuaries which have been mapped constitute between 8% (Siletz Bay, coastal Oregon) to half or more (Humbolt Bay, California; Tillamook Bay, Oregon; Willapa Bay, Washington) of the total estuarine wetland area (Simenstad and Armstrong, In press). Unlike most East Coast estuaries, expansive areas of emergent marsh are not characteristic of the broad coastal estuaries,

---

[1] A drowned river valley or other type of estuary (e.g., bar-built) that intersects another estuary, such as the fjord estuaries that form the mail basins of Puget Sound.

where they tend to form more "fringing" marshes. Similarly, expanses of contiguous eelgrass (*Zostera marina*) habitat are also uncommon except in a few estuaries (e.g., National Estuarine Research Reserve site at Padilla Bay) and tend to occur as narrow and elongated (in fjords) or irregular patches.

An important aspect of restoration of Pacific Northwest estuarine habitats is that many of the types, characteristics and the processes of estuaries in this region are measurably different and, therefore, appear to be less interpretable, from estuaries on the East Coast of the United States, where habitat restoration may be more advanced. These differences derive principally from distinct geological history, tidal regimes and hydroperiods, all of which influence sediment/soil structure and chemistry, hydric conditions, frequency and duration of tidal inundation, suspended sediment loads and salinity regimes. These variables are, by and large, primary determinants of vegetation and benthos assemblage community structure.

Historic Habitat Loss

Many estuaries in the Pacific Northwest region have suffered significant losses (Table 1), although net loss of estuarine habitat has not approached that of many of the other regions. Although data are not comprehensive, available information indicates that approximately 42% of the tidal wetland habitat in coastal estuaries has been lost or converted (e.g., to non-tidal), and Puget Sound estuaries have lost approximately 71%. Other, shallow sublittoral estuarine habitats have also been altered or destroyed, but quantitative documentation is generally lacking. Thom and Hallum (1990) have estimated that eelgrass habitat in selected areas of Puget Sound that had historical documentation illustrated between −70% and +430% change in total area, but the positive changes were associated uniquely with invasive growth by the exotic species of seagrass, *Z. japonica*. Similarly, Thom and Hallum also attempted to

Table 1. Sources of natural habitat modification, degradation and loss in Pacific Northwest estuaries.

Agriculture	diking only; pastureland
	diking, ditching and some filling; cropland
	freshwater diversion
	pesticide and fertilizer run-off
Logging	deforestation of riparian zone
	log rafting and debris dams
	diversion and other modification of freshwater flow
	increased sedimentation from upland logging practices
Industry	diking and filling
	dredging and channelization
	freshwater diversion
	pollutant input
General Urbanization	shoreline stabilization
	road construction; filling and pollutant run-off
Aquaculture	intensive oyster culture; involving mechanical harvesting and other changes to surface of littoral flats; pesticide application
	salmon culture

quantify kelp habitat changes in Puget Sound and found variable changes between −100% to +200%.

The greatest magnitude of change has generally occurred in the littoral wetland habitats of large river deltas of Puget Sound that have become heavily urbanized, as characterized by the Snohomish River delta (near Everett; −74.4%), Duwamish River delta (Seattle, Washington; −89.9%) and Puyallup River delta (Tacoma; −99.6%). Habitat modification produced in the urbanization of these deltas involved principally dredging and filling activities in shallow subtidal and intertidal channels and flats (both mudflats and emergent marshes). However, the loss of estuarine wetlands, at least to tidal action,[2] in comparatively undeveloped estuaries is not insignificant

[2] A considerable portion of the wetland in undeveloped estuaries involved conversion of wetland by diking into farmed wetlands, which still have hydric soils and often hydric vegetation.

in deltas under intensive agriculture, where diking for "reclaimation" of farmlands has accounted for conversions as high as 33.3% (Skokomish River) to 96.4% (Samish River estuary) in some instances (Table 2), usually of emergent marsh habitat that could be converted directly into pastureland or plowed for crops after dewatering (installation of drainage ditches and tidegates).

The history of habitat loss in the Duwamish River estuary (including Elliott Bay) described by Blomberg et al. (1988) is relatively representative of most temporal sequences in the more urbanized estuaries. The authors documented that, between 1854 and 1908, over half of the tidal swamps were lost, while tidal marshes, littoral flats and shallows, and medium depth water decreased between approximately 5% and 22%. Between 1908 and 1940, medium depth water actually increased to its historical level as dredging was used to fill the remaining tidal swamps and another 60-70% of the historical areas of tidal marshes and littoral flats and shallows. Changes between 1940 and 1985 have involved continued slow decline of tidal marshes and littoral flats and shallows until only 2% of the original habitat areas remains today. While medium depth water declined in area rather slowly and continuously over the 131 years, deepwater habitat increased dramatically between 1908 and 1940 when dredging of navigation channels for deep-draft vessels occurred.

The historical sequence of habitat loss has been more variable in the non-urbanized estuaries. For example, Boulé et al. (1983) illustrated that, while the wetland acreage in the Snohomish River estuary declined precipitously between 1880 and 1940 due to diking and agriculture, the coastal estuaries of Willapa Bay and Grays Harbor did not experience measurable emergent marsh loss until between 1933 and 1974, and 1960 and 1973, respectively.

The sources of estuarine habitat loss in this region are generic to most coastal wetlands throughout north temperate coasts, i.e., the cumulative effects of early colonization with pastoral and agricultural "reclamation" and logging, the expansion of the industrial age around population centers on estuarine waterways, and the exploding urbanization of progressively more shoreline (Table 2). In addition to absolute loss of habitat, the continuous development of shorelines and watersheds continues degradation of the already fragmented remaining estuarine habitats. Disruption of surfacewater and groundwater hydrology, non-point pollutant runoff from developed uplands, and separation of natural corridors to natural upland habitats promote further indirect degradation. Even supposedly benign uses of estuarine habitats, such as intensive oyster culture on littoral flats and salmon pen rearing in enclosed bays, have the potential to degrade estuarine habitats, although this typically occurs over a matrix of comparatively pristine habitats rather than in urbanized environs.

Resource and Ecosystem Responses

When estuarine habitats are lost or degraded, so are the important functions they provide—groundwater recharge and flood desynchronization, sediment retention and other mechanisms of shoreline erosion control, water quality improvement, trophic energy (food web) support, fish and wildlife habitat, recreation, resource harvest, energy sources, education and science, aesthetic appreciation, promotion of biodiversity and the maintenance of microclimate characteristics. Of these functions, fish and wildlife support, and to a lesser extent food web support, are the focus of most concern in restoring estuarine habitats. Evidence for a direct, causal link between estuarine habitat loss and degradation and significant declines of fish and wildlife populations is lacking in most instances because of a multitude of confounding factors. In the Pacific Northwest, there are a number of strong inferences of such a relationship, in which there have been dramatic resource declines or disappearances coincident with major changes in estuarine habitats:

Table 2. Total and relative wetland habitat loss in selected Pacific Northwest estuaries; data complied from Bortleson et al. (1980)[a], Simenstad et al. (1982), Boulé et al. (1983), Thomas (1983), Burg (1984), Boulé and Bierly (1987), Blomberg et al. (1988) and Simenstad (unpubl.).

| Estuary | Area (km²) | | Change | |
	Historical	Modern	Area (km²)	Percent (%)
Coastal				
Chetco River	0.13[b]	0.11	- 0.02	-15.4
Coquille River	1.70[b]	1.50	- 0.20	-11.8
Coos River	16.20[b]	11.10	- 5.10	-31.5
Umpqua River	4.20	1.40	- 2.80	-66.7
Siuslaw River	6.07[b]	5.91	- 0.16	- 2.6
Alsea River	5.30	2.60	- 2.70	-50.9
Yaquina Bay	9.40	3.30	- 6.10	-64.6
Siletz River	1.31[b]	1.30	- 0.01	- 0.6
Nestucca River	3.30	0.90	- 2.40	-72.7
Sand Lake	3.20	2.80	- 0.40	-11.1
Netarts Bay	0.43[b]	0.42	- 0.01	- 2.3
Tillamook Bay	15.50	4.30	-11.10	-72.1
Nehalem River	3.90	1.30	- 2.60	-64.8
Necanicum River	0.23	0.12	- 0.11	-47.4
Columbia River	73.20	24.00	-49.20	-67.2
Grays Harbor	195.00	136.00	-30.30	-30.3
Subtotal	339.07	197.06	-142.01	-41.9
Puget Sound				
Dungeness Spit	0.50	0.50	0.00	0.0
Nooksak River	4.50	4.60	+ 0.10	+ 0.2
Lummi Bay	5.80	0.30	- 5.50	-94.8
Samish Bay	11.00	0.40	-10.60	-96.4
Skagit River	29.00	12.00	-17.00	-58.6
Stillaguamish River	10.00	3.60	- 6.40	-64.0
Snohomish River	39.00	10.00	-29.00	-74.4
Duwamish River	18.00	1.80	-16.20	-89.9
Puyallup River	10.00	0.04	- 9.96	-99.6
Nisqually River	5.70	4.10	- 1.60	-28.1
Skokomish River	2.10	1.40	- 0.70	-33.3
Other areas	3.00	2.50	- 0.50	-16.7
Subtotal	31.60	9.20	- 22.4	-70.8

[a] Bortleson's pre-mapping estimate, based on wetland vegetation distributions, was used rather than the more conservative map differencing estimates.

[b] Minimum; acreage converted to diked marsh not included in this estimate.

• Most species of Pacific salmon (*Oncorhynchus sp.*), probably the single most influential fisheries resource to the region's indigenous people and still a major factor in today's economy and culture, are directly dependent upon estuarine habitats (e.g., Dorcey et al. 1978; Healey 1982; Simenstad et al. 1982; Levings et al. 1986). In general, their populations have been depressed and survival continues to be decreased in the more developed estuaries (see Williams and Tuttle, this volume); coincidentally, many of our urban centers have developed on estuaries which supported some of the most extensive salmon runs in the region. Factors other than estuarine wetland loss have contributed measurably and concurrently to these declines; overexploitation, destruction of freshwater spawning and rearing habitats, river-flow regulation, maladaptive hatchery practices, etc. have all been implicated. But, when compared to relatively undisturbed estuaries, salmon stocks in urbanized estuaries continue to be depressed despite improvement in these other exogenous factors and inter-estuarine conditions, such as water quality.

• Also a culturally, commercially and ecologically important species, Pacific herring (*Clupea harengus pallasi*) spawn extensively, although not exclusively, on subtidal and low intertidal macrophytes such as eelgrass and kelps. Within the distribution of these macrophytic habitats, herring spawning is concentrated at unique sites, the characteristics of which biologists are still at a loss to explain effectively. Although kelp beds are not prominent features of estuaries, eelgrass beds are common features of the euhaline-polyhaline regions of estuaries where mud or sand substrates occur. Loss of many of these habitats has undoubtedly contributed to decreased herring spawning in Puget Sound, especially central and southern Puget Sound, where kelp habitats suitable for spawning are comparatively limited.

- Shorebird (e.g., dunlin, sandpipers, yellowlegs) populations are dependent upon estuarine habitats, especially unvegetated littoral flats, for foraging and roosting during their spring and fall migrations along the Pacific flyway (Simenstad 1983; Simenstad and Armstrong In press). While accurate estimates of trends in migratory shorebird populations are not as available for the Pacific Coast (as has occurred on the East Coast) over the period of estuarine development, most experts agree that the decline and continuing fragmentation of estuarine littoral flat habitats could be contributing to decreased fitness of shorebirds for the physiological demands of their extensive migrations.

Even in the most opportune cases, however, quantitative documentation of the population and population process changes (e.g., reproduction, mortality) that have occurred as a result of estuarine habitat loss and degradation have begun to emerge over only the last 10-15 years.

These fish and wildlife resources are examples of high profile, economically-important taxa. However, there are suites of lesser-appreciated animals, such as small mammals and amphibians, that have undoubtedly suffered considerably from estuarine habitat loss and degradation, but for which little scientific information exists. Similarly, other functions, such as food web support and flood desynchronization, that estuarine habitats like wetlands provide have obviously been affected, though to a thoroughly unknown degree. There is neither scientific nor societal (including economic) justification for developing estuarine habitat restoration strategies that do not include all the major functions and comprehensive, ecosystem-level goals. Unfortunately, our experience to date continues to emphasize fish and wildlife support as the sole functional goal of restoration.

History and Experience of Estuarine Habitat Restoration

In all but a few cases, estuarine habitat restoration in the Pacific Northwest has been attempted through compensatory mitigation; only two of the examples we will cite—the Duwamish River terraces and Jetty Island—originated through non-regulatory activities (Table 3) The strategies that have been applied can be divided into the three broad categories identified by Kusler and Kentula (1989): (1) *restoration* of a site from historic modification, returning it to the original habitat state; (2) *enhancement* in the form of augmentation of a portion or element of a degraded habitat; and (3) *creation* of wetland characteristics where wetland characteristics have not existed formerly, or dramatically different wetland conditions existed. These strategies have been described in a relative order of their perceived (but not documented) probability of success.

Table 3. Categories and examples of estuarine habitat restoration strategies in the Pacific Northwest.

Category	Example	Size (ha)	Date Initiated
Restoration	Salmon River estuary dike breach	21	September 1978
	Elk River estuary dike breach	16	Spring 1987
Enhancement	Lincoln Park beach nourishment	<0.01	Winter 1988-1989
Creation	Lincoln Avenue wetland complex	4	February 1986
	Blaine Marina eelgrass bench design	2	Not constructed
	Grays Harbor brackish slough	2	August 1990
	Jetty Island	8	Winter 1990
Elemental	Duwamish River estuary benches	<0.01	August 1990
	Commencement Bay-Slip 1-gravel beach	<0.01	Winter 1985

Restoration includes modifications to dikes, tidegates and other structures that inhibit normal estuarine hydrology to influence an otherwise unmodified site (e.g., dike-breaching). In the Pacific Northwest, this strategy has also involved more elaborate measures, such as removal of fill on a former wetland, restoration to tidal flow, and planting of emergent or submergent plants that were known to exist there historically. Usually, the historical habitat type is the goal and template of the restoration. In general, the opportunity for success of this restoration is high as long as the primary processes delimiting the habitat type are still effective at that site, e.g., salinity intrusion, sedimentation sources and processes, corridors to other natural estuarine and upland habitats.

Enhancement involves relatively minor restoration or enhancement of substrate, vegetation or some other wetland element that has been lost or degraded due to changes in local physical processes. For example, bulkheading and other development of shorelines typically depreciate sediment sources for maintenance of beach substrates in Puget Sound (Downing 1986); "nourishment" of beaches or other modified shorelines with natural substrates is an example of such a restoration strategy. In this region, enhancement by restoration or creation of a habitat element within the matrix of a developed shoreline has involved extracting an important characteristic of natural estuarine habitats (e.g., low-gradient, unconsolidated sediment "flats") and implanting it within a developed shore (e.g., a stabilized, "rip-rap" bank or piers and pilings).

Creation entails more elaborate restorative reconstruction of natural wetland conditions, including both physical (e.g., topographic, hydrologic) as well as biotic (e.g., vascular plants) elements. This strategy generally involves "implanting" a created wetland into a coastal shoreline where this type of wetland did not exist historically. Examples of wetland creation in the Pacific Northwest would be planting emergent vegetation at a site that was historically unvegetated littoral flat or sculpting tidal channels and sloughs where they had not existed previously. Even such alteration of estuarine wetland habitat composition is still considered a less than advantageous approach because of considerable uncertainty in the outcome. Creation of estuarine habitats from uplands or other regions that did not historically have wetland characteristics has to date been discouraged by natural resource and other agencies regulating wetlands, is intended to be the less preferred strategy, and will not be proposed as such in this synopsis.

The following examples typify approaches to each of these restoration strategies that have been attempted in the Pacific Northwest. They illustrate both the potential and the pitfalls of undertaking what are essentially experiments based on minimal or no empirical relationships between estuarine wetland characteristics and their functions. Where monitoring or other research data that tests their functions are available, we have the means by which to evaluate both the effectiveness of the design as well as the design or strategy modifications that might be employed to correct dysfunctional habitats. It is important to keep in mind during this review that such an adaptive approach to restoration demands at the minimum: (1) explicit goals and objectives; (2) quantitative, systematic baseline and post-construction monitoring of parameters that are an index of habitat function; and (3) the institutional means to tweak the experiment and modify dysfunctional elements in order to learn from the mistakes.

Restoration

Although often cited as the preferred approach to restoration because of the potentially greater predictability for success (Jones and Stokes Assoc. 1988, 1990), breaching of dikes or levees and similar strategies for restoring wetlands have not been exceedingly common nor well monitored. Two examples of this restoration strategy are located on the Salmon River estuary, coastal Oregon, and the Elk River estuary, in Grays Harbor on the Washington coast.

The Salmon River estuary is a small (< 2 km^2), drowned river valley estuary located on the central Oregon coast, immediately south of the major coastal feature of Cascade Head, and is one of Oregon's most pristine estuaries (Wilsey and Ham, Inc. 1974). It is moderately impacted by land uses in the estuary and watershed, which has resulted in increased turbidites and temperatures and reduced freshwater flows (Proctor et al. 1980). In September 1978, a 21-ha segment of diked brackish marsh was reconnected to tidal inundation and flow after having been diked in 1961 for use as pastureland. Based upon initial studies by Mitchell (1981), Frenkel and Morlan (1990) were able to assess vegetation and soil characteristics at the restored estuarine marsh 11 years after the breaching of the dike. In interpreting the natural recolonization of the site, they utilized two 15-ha natural high marsh habitats that occurred on either side of the dike-breach marsh as reference sites (Figure 1). The authors documented that the sequence of vegetation recolonization at the restored marsh did not proceed on a trajectory toward the control site communities, but rapidly developed into a low marsh *Carex lyngbyei* community. This was explained by the lower elevation of the diked marsh, which had subsided 35-40 cm over the 17 years of use as pastureland. Fortunately, Frenkel and Morlan (1990) were able to estimate sedimentation rates on the marsh surface, which they found to average between 5 and 6 cm y^{-1} (range, 3-9 cm y^{-1}) as compared to an average of 4 cm y^{-1} (range 2-9 cm y^{-1}) in the control marsh; sediment accretion in the restored marsh was also found to be measurably higher at lower tidal elevations than at higher elevations. The authors used net primary production (NPP) as the principal index of wetland function and described NPP after a decade of renewed estuarine influence to be 2,300 g m^{-2}y^{-1} in the restored marsh despite the completely different vegetation community structure, as compared to 1,200 g m^{-2}y^{-1} in the reference marshes. However, as is representative of all such restorations in this region, other functions involved in the recolonization and development of

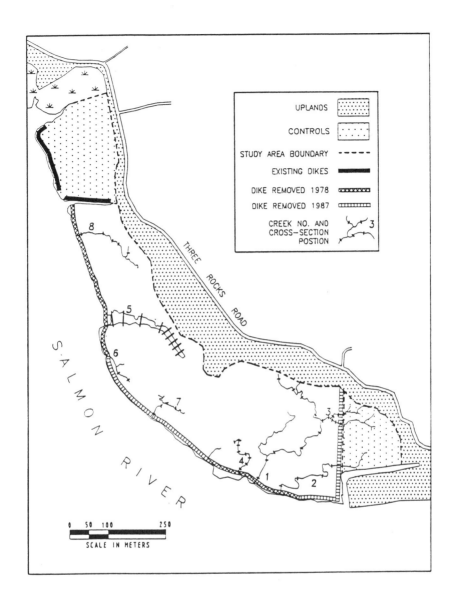

Figure 1. Areal map of dike-breach wetlands in Salmon River estuary; numbers and cross-section positions indicate Frenkel and Morlan (1990) study sites.

the restored marsh system (e.g., fish and wildlife utilization, benthic infauna or epibenthos, nutrient cycling, etc.) were not monitored. Thus, the specific goals and objectives of this restoration, as well as the unidimensional (i.e., NPP) "currency" used to evaluate functional equivalency of the restored habitat, dictated the determination of its success or failure, depending upon the strict adherence to complete restoration of pristine conditions or simple restoration to a functioning natural salt marsh, respectively.

The Elk River is a small estuary in Grays Harbor that drains the southwest corner of the Harbor. The watershed has been and continues to undergo extensive logging activity, but estuarine habitats have been modified only slightly; the primary effect has likely been the construction of a bridge and road along a levee across emergent marsh habitat. The highway formed one side of an approximately 16-ha diked high salt marsh that had been utilized as pastureland for over 50 years until it was breached in June 1987 as part of a wetland mitigation activity. Over the period it was diked, the site was colonized extensively by facultative freshwater wetland plants common in infrequently saturated soils, including the exotic species *Phalaris arundinacea* (reed canarygrass). Like the Salmon River site, the diked habitat in the Elk River estuary had subsided considerably but the precise extent of subsidence had not been measured. Restoration of the site to tidal influence involved excavation of an approximate 10-m gap at one point in the dike. The consequence of the changes in the diked area and the accretion of sediments, and apparent increase in tidal elevation of the salt marsh habitat on the Grays Harbor side of the dike, was an unusual gradient in tidal elevation from low marsh to high marsh to low marsh with distance between the primary estuarine channel and the upland (roadbed) shore. Monitoring of the habitat changes that have resulted has been limited to simply an annual survey of percent coverage of primary emergent wetland plants at five established points across the diked site. These observations indicate a rapid decline in dominance by the predominantly freshwater plant assemblages to recruitment and increased dominance by facultative and obligate estuarine species of wetland plants such as *Salicornia virginica* (pickleweed), *Atriplex patula* (saltweed) and *Carex lyngbyei* (Lyngby's sedge) (Figure 2a). In addition, the combined effect of the narrow dike breach and the dramatic elevation change between the higher, former "foreshore marsh" and the lower, new "back marsh" appears to be responsible for rapid erosion of a tidal channel at the point of the dike breach (Figure 2b). Drainage of the tidal waters from the restored marsh area is slowed because of the limited channel capacity. Many persistent features of the diked-wetland system, such as the existing dike and standing water continue to attract several wildlife species, including waterfowl and river otter (R. Zeigler, Washington Department of Wildlife, pers. comm.). Thus, the dike breach approach in the Elk River estuary is not necessarily progressing along the same trajectory as the Salmon River marsh restoration. The changes that have occurred at the Elk River site during the period of its isolation from the estuary and its location at the shore end of the tidal gradient may be prolonging the processes that will promote return of the site to high salt marsh. Modifications to this restoration strategy that might accentuate this recovery trajectory would be addition of surface sediments to historic elevations, removal of the entire dike (as was eventually done at the Salmon River site), and transplanting of high marsh vegetation. Expanded monitoring of this site, as well as further dike removal and modifications (e.g., channel deepening), were planned for this site in 1991.

Although these examples do not provide any evidence for enhanced fish and wildlife function of breached dike estuarine wetlands, other examples from the region indicate that juvenile salmon accessing similarly restored slough habitats appeared to obtain adequate and representative prey, rear longer and grow larger (Tutty et al. 1983).

(a)

(b)

Figure 2. A formerly-diked emergent marsh at Elk River, Grays Harbor, Washington: (a) approximately three years after breaching of dike; Scirpis *and* Salicornia *form the predominant emergent plants in the background marsh surface; and (b) eroding tidal channel at the mouth of the dike breach. Photographs by C. Simenstad, FRI/UW.*

440

Enhancement

The "nourishment" of the recreational beach at Lincoln Park, on the Seattle shoreline of Puget Sound, was designed originally as a construction project to repair a concrete seawall and control erosion. However, this approach has developed an element of habitat restoration in terms of enhancement of fish and wildlife use of the restored beach. Persistent erosion of beach sediments (e.g., coarse sand and small gravel particles) in the Puget Sound region has become a more common occurrence as the fluvial and morainal deposits in the cliffs that "feed" these beaches are cut off by bulkheading of the shoreline and by modification of longshore transport patterns. In 1983, the U.S. Army Corps of Engineers proposed to augment the eroding beach sediments by artificially "nourishing" the eroding beach by depositing a mixture of sand and gravel from the tidal elevation of +18 ft. MLLW at the top of the existing seawall to an elevation of +3 ft. MLLW along a 1:4 gradient (Figure 3). Because the Corps was concerned about the impact upon natural resources that persisted in the eroded beach, baseline studies of benthic infauna, benthic plants (macroalgae, eelgrass) and epibenthic prey organisms of juvenile salmonids were conducted from 1985 through 1988; these augmented earlier baseline studies that occurred in the mid-1970s. Nourishment of the beach was implemented in 1988 and subsequent biological monitoring and evaluation of the developing beach community continued through 1990. Comparison of pre- and post-nourishment biota was the primary criteria for assessing both impact and functional equivalency. In the case of epibenthic crustaceans, organisms considered to be important prey of juvenile salmon migrating through the area, sampling typically found these prey taxa abundant on the new beach (Hiss et al. 1988; Hiss 1990), but the sampling design did not effectively assess the impact from the loss of the pre-nourishment habitat. Further sampling in 1990 showed that the new beach was not yet colonized by taxa of algae and bivalves that are typically

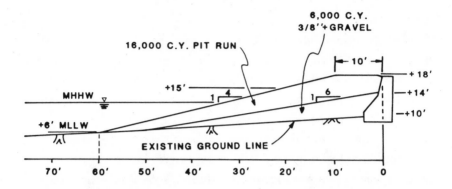

Figure 3. Cross-sectional view of beach nourishment design for enhancement of sand ("pit run") substrate along shoreline of Lincoln Park, Puget Sound, Washington. Figure from U.S. Army Corps of Engineers, Seattle District.

found in similar habitats in Puget Sound (Thom and Hamilton 1990).

One approach to estuarine habitat enhancement involves the incorporation of elements of natural estuarine habitats into the design or refitting of developed shorelines. This elemental approach to the enhancement has evolved out of the concept of wetland attributes as developed in the Estuarine Habitat Assessment Protocol (Simenstad et al. 1991; see below). Attributes are the physical and biological characteristics of wetland habitats which foster fish and wildlife utilization by facilitating reproduction, foraging, refugia (from predation, disturbance, etc.) and physiological adaptation even if they occur within highly modified habitats. The principle behind this approach is that estuarine habitat attributes can be incorporated as elements of modified habitats of urbanized estuaries and might increase fish and wildlife function despite that fact that they were not operating within the matrix of a natural habitat.

One of the first implementations of this strategy was the construction of littoral mudflat "terraces" on rip-rap stabilized shoreline in the Duwamish River estuary. The impetus for this project originated from a unique arrangement between the Port of Seattle and the Muckleshoot Indian Tribe: the Port guaranteed that a percentage (0.25%) of any Port expenditure for habitat mitigation in the Duwamish River estuary would be set aside in a fund that would support non-mitigation fish and wildlife enhancement. One advantage of this fund was that more experimental approaches could be attempted outside the context of the compensatory mitigation arena. In 1988, the Port and the Tribe agreed to test the "terrace" design along the stabilized bank at the Port's Terminal 106 facility in the heart of the highly-industrialized center of the estuary (Figure 4a). The design involved implanting two approximately 48.2-m long × 3.7-m wide benches longitudinally along the 2:1 riprap bank; one to be placed at 0.0 ft MLLW and the other at +2.0 ft MLLW (Figure 4b). The "terrace" was constructed by forming a rip-rap berm, lining the basin with filter fabric, cutting and filling with existing soils and adding an approximately 23-cm thick sand-silt layer over the surface of the constructed flat (Figure 4). To measure the enhancement potential of the "terraces," pre-construction monitoring of epibenthic crustacean prey of juvenile salmon occurred in October-November 1987 and June-July 1988 (Simenstad and Cordell, In revision). Construction of the "terraces" occurred in July-August 1990 (Figure 4d). Post-construction monitoring of epibenthic crustaceans was initiated in March 1991. Visual examination of the completed "terraces" in August 1990 indicated benthic infauna were colonizing and shorebirds had frequented the constructed flats, and that accretion, rather that erosion, of sediments was occurring. However, whether juvenile salmon actually forage on the littoral flats must be verified with more extensive experiments in contrast to basic monitoring.

As with the strategies of estuarine habitat restoration, enhancement strategies such as beach nourishment and incorporation of mudflat characteristics into stabilized banks have the advantage that the habitat being restored historically persisted on the site and, presumably, many of the processes that contributed to the maintenance of the habitat still operate. However, the absence or deterioration of some critical processes, such as sedimentation rates and longshore transport, will continue to plague the long-term stability of the target habitat, and continued enhancement will be often be required in these instances. For instance, it was estimated that nourishment of the Lincoln Park beach must be repeated every 10 years to maintain a beach profile and substrate equivalent to natural beaches (Thom and Hampel 1985).

Creation

Full-scale restoration of highly-modified estuarine habitats is measurably more risky than simple restoration or enhancement of slightly-modified or existing habitats. Even if the expectation is to mimic the original habitat structure and function, the risk of major changes in the system and our lack of understanding about how complex wetland systems are put together or work can limit the potential of success—the latter probably is the greatest limitation to projects of this nature.

Undoubtedly, the most elaborate restoration of an estuarine wetland habitat to date is that described as the Lincoln Avenue wetland (LAW), located in the Puyallup River estuary near Tacoma, Washington (Figure 5a). This restoration project was developed in mitigation for the loss of a 3.9-ha freshwater wetland-upland habitat downstream from Lincoln Avenue (Parcel 5, Figure 5a). Negotiation of the mitigation plan with regulatory agencies resulted in the design of an estuarine wetland-upland complex at the site for a former solid waste landfill behind a dike that ran along the Puyallup River. The resulting design of the LAW was a 3.9-ha habitat complex that included 2.2 ha of brackish habitat and 1.7 ha of upland in a structure that supported fish and wildlife resources in the approximate proportions of: juvenile salmonids, 50%; waterfowl,

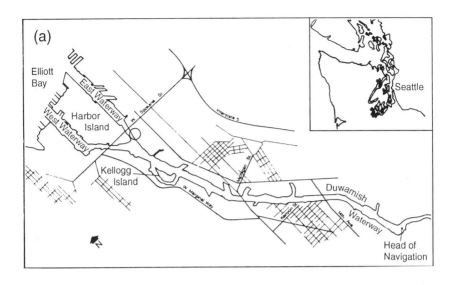

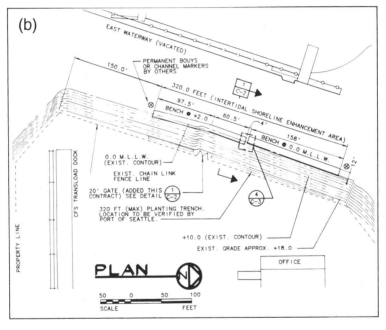

Figure 4. Location (a) and design (b, plan) of littoral flat "terraces" implanted in rip-rap stabilized shorelines at the Port of Seattle Terminal 106 in the Duwamish River estuary, Seattle, Washington. Drawings by George Blomberg, Port of Seattle.

445

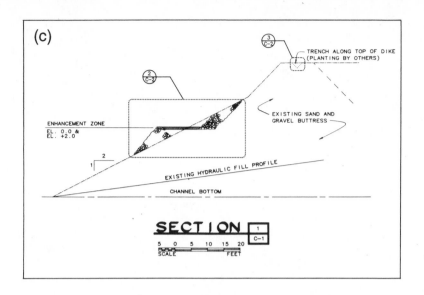

Figure 4. (c) Cross-sectional view of beach nourishment design for enhancement of sand ("pit run") substrate along shoreline of Lincoln Park, Puget Sound, Washington. Figure from U.S. Army Corps of Engineers, Seattle District. (d) View of the created littoral flat "terraces" at terminal 106 in Duwannish River estuary. Photograph by George Blomberg, Port of Seattle.

20%; shorebirds, 10%; raptors, 10%; and small mammals, 10%. Construction of the wetland included building a new dike to surround the system, excavation of 55,000 m³ of solid waste landfill, rerouting of a buried oil pipeline, breaching of the dike to establish tidal inundation, and transplanting of *Carex lyngbyei* sedge plants on the littoral flats. Construction was initiated in July 1985, the dike was breached in February 1986 (Figure 5b), and the sedge plants were planted in March-July 1986 and April-May 1987. Extensive monitoring of sediments, vegetation, infauna, epibenthic crustaceans, fish and birds has occurred annually since 1985 (Figure 5c); in addition, specialized experiments and short-term sampling have been conducted to evaluate emergent insect production and the flux of water constituents, plankton and neuston into and out of the wetland system over a diel period.

The history, design and results of the monitoring have been described in various reports (Thom et al. 1987, 1988b, 1990; Shreffler et al. 1990b), published papers (Thom et al. 1988a; Shreffler et al. 1990a) and student theses (Shreffler 1989). Over the five years since the breach of the dike, these studies indicated: (1) rapidly expanding benthic infauna, epibenthic crustacean, fish and avifauna populations composed of progressively more estuarine species; (2) extensive retreat of the area of planted sedge vegetation, but increased shoot density and net production over the whole brackish habitat of the system; (3) substantial sedimentation (approximately 5 cm y⁻¹ in 1989) on the littoral flats and in the tidal channels; and (4) the wetland system is a sink for some organic matter and a source of inorganic matter. According to the target fish and wildlife resources for which the LAW was designed to support, there continues to be increased utilization, especially relative to the habitat it was intended to replace. In particular, the utilization of the LAW by juvenile salmon appears to be extensive and comparable to the meager data available from natural brackish wetlands in this region (Shreffler 1989; Shreffler et al. 1990a). However, the system is extremely dynamic; basin morphology, sediment characteristics, and vegetation, benthic, epibenthic and fish assemblages are highly variable and changing (Figure 5d). For instance, the area of transplanted and any recruited *Carex* sedge vegetation decreased by 61.5% between 1987 and 1989, but increased slightly (9.1%) between 1989 and 1990, suggesting that some stabilization may be occurring; the total area of constructed channels, however, has fluctuated by +6.8% to −21.1% between years, and shows a net decline of 36.8% over the five years of monitoring (C. Simenstad, unpubl.).

This dynamicism indicates that the LAW persists in an early developmental stage (MacArthur and Wilson 1967) and its equilibrium state, if that is achievable, is difficult to predict. A critical management question is whether the system should be manipulated to maintain the elements of this immature system that support fish and wildlife, or to let it progress toward a more mature system that might not be as utilizable by some species. For instance, continued sedimentation might ultimately preclude juvenile salmon residence for longer than brief tidal cycles, and if expansion of the emergent vegetation across the presently unvegetated littoral flats would preclude shorebird foraging, that would be considered deleterious. Although natural estuarine habitats typically have deep channels and unvegetated flats, the reduced scale and configuration of the LAW and modified physical processes of the surrounding estuary may prevent the system from ever achieving "natural" characteristics despite its present trajectory in that direction.

A more extreme example of attempting to reconstruct estuarine habitats that occurred historically is the Blaine Marine mitigation plan (Thom et al. 1988c). This design was developed as a potential on-site mitigation for removal of a 5.1-ha high (approximately +6.0 ft MLLW) intertidal mudflat that persisted within the confines of an existing marina located in Drayton Harbor, Washington. Drayton Harbor is a small (1,104 ha), shallow estuary located on the southeastern shore of the Strait of Georgia, in which approx-

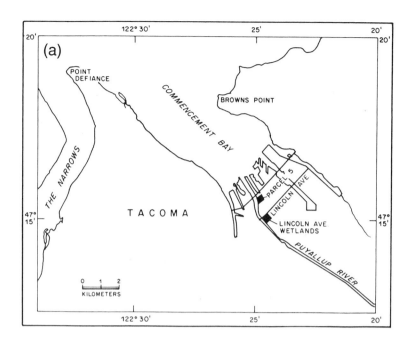

Figure 5. (a) Restored Lincoln Avenue wetland system in the Puyallup River estuary. Location Parcel 5 is the development site that Lincoln Avenue was designed to replace. (b) View of conditions in Puyallup River estuary immediately after dike breach. Photograph from Port of Tacoma.

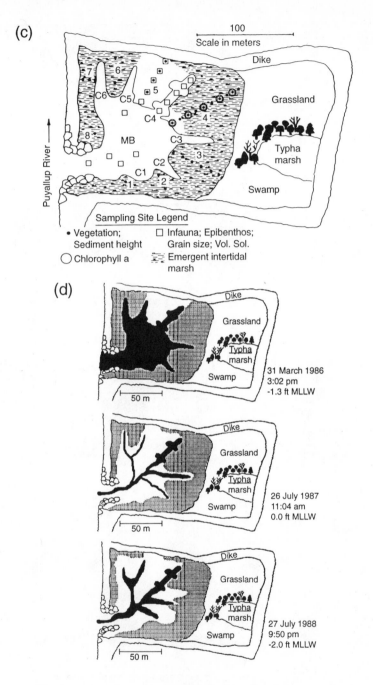

Figure 5. (c) Design and monitoring sites in Lincoln Avenue wetland system. (d) Habitat changes in the first three years of Lincoln Avenue wetland development. Figures from Thom et al. 1988a and Shreffler et al. 1990b.

450

imately one third of the area is dominated by two species of eel-grass (*Z. marina* and *Z. japonica*). The design of the mitigation plan was thus constrained by the available space within the existing marina and the relatively pristine nature of the Drayton Harbor habitats outside the marina. Extensive ecological studies of fish and decapod utilization, prey resources and primary production of emergent marsh, mudflat, slope and eelgrass habitats within and outside the marina in Drayton Harbor in 1988 (Thom et al. 1989) indicated that the mudflat proposed for dredging was providing important wetland functions for certain fish and wildlife (e.g., salmon, flatfish, baitfish [osmerids], and shorebirds). Furthermore, it appeared that some of this function could be unique within the Drayton Harbor system because primary production, fish utilization and prey production on the mudflat tended to increase earlier in the spring within the marina than in either the mudflat or eel-grass habitats outside the marina in Drayton Harbor. Because of the extensively higher production of salmon prey resources (epibenthic harpacticoid copepods) in Drayton Harbor's eelgrass habitat, restoration of eelgrass habitat within the marina was considered as an "on-site" option that might functionally replace the mudflat function for juvenile salmon foraging habitat. Whether this resource would develop in eelgrass established within the marina was entirely speculative; in addition, this approach was not considered to address the loss of mudflat function for other resource animals, such as shorebirds and juvenile flatfish.

A design for eelgrass "benches" was developed that would wrap along the marina breakwater in a 1.2-ha corridor (Figure 6a, b). The low elevation, flat gradient and design to retain standing water inherent in the bench profile was designed to mimic the most suitable physical requirements for eelgrass establishment and growth. The grossly disproportionate replacement of mudflat area by eelgrass area reflected the multiplier of production of salmonid prey resources in eelgrass versus mudflat habitats, but is acknowledgeably inadequate for non-salmonid resources. Given that this strategy has never been implemented in this region, extensive monitoring, pre-development testing of the design, and an extensive contingency plan was considered necessary to maximize functional replacement of the mudflat habitat. This design is currently under consideration by the natural resource agencies.

The creation restoration strategy can approach actual creation in instances where there is marginal evidence that the habitat occurred on the site historically. In these cases, the predictability of habitat function will be considerably higher as long as the scientific basis for the habitat prerequisites are lacking. This can be illustrated by a recent mitigation project in Grays Harbor, coastal Washington State, in which a slough was constructed in a riparian shrub and scrub wetland in the brackish reaches of the Chehalis River portion of that estuary. Such brackish sloughs are naturally distributed throughout the riparian habitat in that reach of the river; thus, the estuarine influence would presumably influence a constructed slough similarly. What is not known is whether surface and groundwater flow, soil conditions, and other characteristics of the natural sloughs are aberrant at the restoration/creation site. This project to create an estuarine slough is the result of mitigation for loss of approximately 0.7 ha of shallow subtidal estuarine channel habitat that will be dredged as a part of the Corps of Engineers Grays Harbor Navigation Improvement Project. In compensation for the loss of juvenile salmon habitat, the Corps constructed a 366-m long, 1.6-ha intertidal-shallow subtidal slough slightly upstream of the site to be dredged. The new slough was designed to include shallow subtidal channel, fringing salt marsh, unvegetated mudflat, channel margin and riparian habitats (Figure 7a); large organic debris (LOD, e.g., waterlogged logs) will be introduced into the slough during initial construction and *Carex lyngbyei* will be transplanted into the constructed wetland to accelerate natural recruitment of these structural and production elements. Much of the

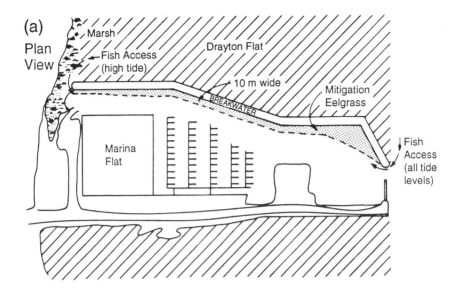

(a)
Plan
View

Marsh

Fish Access
(high tide)

Drayton Flat

10 m wide

BREAKWATER

Mitigation
Eelgrass

Marina
Flat

Fish
Access
(all tide
levels)

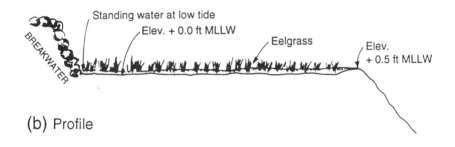

Standing water at low tide
Elev. + 0.0 ft MLLW

BREAKWATER

Eelgrass

Elev.
+ 0.5 ft MLLW

(b) Profile

Figure 6. Design for eelgrass "bench" corridor positioned along breakwater in Blaine Marina, Drayton Harbor, Washington: (a) plan of habitat distribution in marina; and (b) section of eelgrass "bench." From Thom et al. 1988c.

453

design elements were based upon an adjacent natural slough (Anne's Slough) that will form the control site to the mitigation treatment.

The Corps of Engineers is committed to baseline and post-construction monitoring over the next 50 years to ensure that the mitigation is effectively fulfilling its designed objectives and is maintaining its integrity. Utilization of Anne's Slough by juvenile salmon and other fish, the emergent insect and epibenthic crustaceans that constitute primary fish prey resources, avifauna, water quality and vegetation were evaluated in a baseline study late winter-summer 1990; construction occurred in July-August of the same year. While the slough (Figure 7b) was designed with perceived characteristics of estuarine slough habitats in general, the design parameters were not empirically derived from measurements of existing sloughs, and notable differences are evident in some cases. For instance, Anne's Slough completely dewaters during spring low tides, while the created slough was designed to have a considerably deeper channel that retained water during maximum low tides (e.g., −4.0 ft MLLW). While there are valid justifications for this design parameter in terms of fish utilization, and many large sloughs have deeper channels than Anne's Slough, the evaluation of the created slough's functional equivalency to Anne's Slough, the intended control, will be compromised by the differences in these characteristics. The *Carex* transplants and intensive monitoring of both sloughs commenced again in late winter 1991, and supplemental experiments to evaluate juvenile salmon foraging success, growth and residence time in the two sloughs are planned over the next two years.

Another example of estuarine habitat creation in the Pacific Northwest is the Jetty Island wetland and represents the largest wetlands construction project in estuarine waters in the State of Washington. The project consists of construction of a berm on the seaward side of Jetty Island (a former dredge material fill site) to

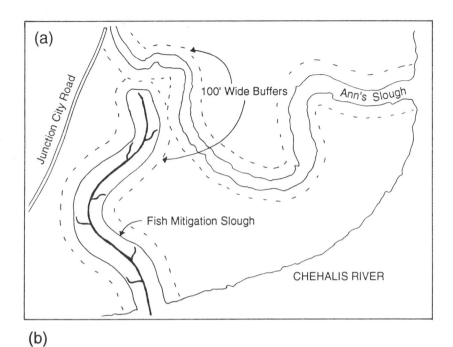

(a)

Junction City Road

100' Wide Buffers

Ann's Slough

Fish Mitigation Slough

CHEHALIS RIVER

(b)

Figure 7. Design (a) and construction view (b) of created estuarine slough habitat in Chehalis River portion of Grays Harbor, Washington. Photograph by C. Simenstad, University of Washington.

455

form an embayment within which a mudflat-marsh habitat was designed to develop (Figure 8a-c; McKenzie et al. 1990). Dredged material from the adjacent Snohomish River navigation channel in the Port of Everett was used to build the berm; the project was specifically intended as a demonstration of the beneficial use of dredged material. The berm was constructed during the winter of 1989-1990 (Figure 8c). To evaluate which marsh plant species would survive and propagate optimally in the created system, transplant experiments were carried out in 1990 (Figure 8b). The experiments showed that three species were appropriate for transplanting and indicated that only the fringing areas of the berm were viable locations for transplants. *Such initial transplant experiments, although often recommended, are rarely carried out.* The experimental phase also allowed time for the dredged material to settle and dewater, and the resultant beach morphology to stabilize. Based on the results of the 1990 experiments, full marsh construction was conducted in 1991. Biological monitoring of the system's performance will be conducted over the next seven years.

Creation may also involve "implants" of habitats or habitat elements that did not occur naturally in specific regions of estuaries, and may achieve some functional success if conditions for maintenance of the habitat are present or the habitat is periodically supplemented as in the example of beach nourishment enhancement. One example of this strategy is the introduction of a coarse gravel beach at Slip 1 in Commencement Bay, Puyallup River estuary, in Puget Sound (Thom et al. 1986). In an effort to partially mitigate for previous losses of shallow water feeding habitat for juvenile salmonids, the Port of Tacoma constructed a 0.08-ha beach at the head of a deepwater slip in Commencement Bay. The beach was constructed between tidal elevations of +6 ft and −6 ft MLLW with large rip-rap boulders as the primary fill material and overlain with a 10-cm thick layer of 2-cm diameter angular gravel. This material, which was dumped in the area with no effort to smooth the gravel surface, resulted in a topo-

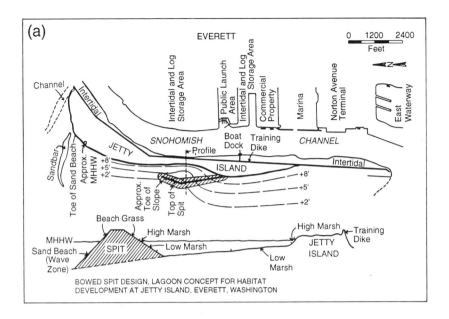

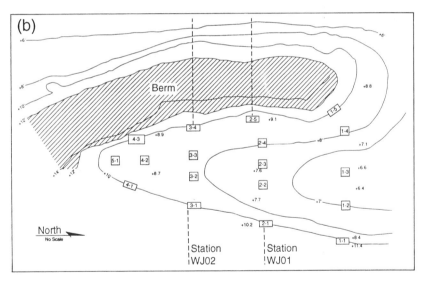

Figure 8. (a) Site plan and (b) profile marsh transplant experimental design for Jetty Island wetland, Snohomish River estuary. Drawings from McKenzie et al. 1990.

457

Figure 8. (c) Aerial view of constructed Jetty Island wetland, Snohomish River estuary, Everett, Washington. Photograph by T. McKenzie, Pentec Environmental.

graphically heterogeneous surface of pits and mounds at a 3.5:1 slope (Figure 9). The final result was an entirely atypical habitat for estuaries such as the Puyallup which, because of an exceptionally high riverine discharge and sediment load, once had (Table 2) only very low gradient, silt-mud to fine sand littoral flats; such high-gradient gravel beaches occurred only in the outer reaches of that estuary. Despite the aberrant nature of the habitat, our preliminary sampling indicated very large populations of epibenthic crustaceans, e.g., harpacticoid copepods such as *Harpacticus uniremis, Tisbe* sp. and *Zaus* spp., that are known to be preferential prey of juvenile salmon; these prey population explosions appeared to be linked to the abundance of epilithic diatoms colonizing the gravel surfaces (Thom et. al. 1986). Thus, implanting of this seemingly aberrant habitat within the matrix of a

Figure 9. Gravel beach designed as foraging habitat for juvenile salmon at Slip 1, Commencement Bay, Puyallup River estuary, in Puget Sound, Washington. Photograph by R. Thom, Battelle Marine Science Laboratory.

highly developed estuary may have promoted and enhanced fish foraging. However, there are very good reasons that such habitats do not occur in the inner regions of estuaries such as the Puyallup, the most probable of which are that: (1) the river is not a source of coarse gravel material and longshore transport does not deposit this material in the inner estuary; (2) the fine suspended sediment material that is introduced into the estuary would rapidly accrete over such material; and (3) sessile organisms such as barnacles and mussels rapidly colonize most hard surfaces that are not prone to heavy sedimentation. All of these mechanisms would predictably preclude epilithic diatom growth on gravel beaches in this region within a few years. Despite this prediction, our monitoring of the Slip 1 mitigation beach had to be curtailed after one year and, to our knowledge, this habitat has not been evaluated subsequently.

Functional Approaches to Restoration "Ecotechnology"

It should be obvious that adequate knowledge of the functions of estuarine habitats is an absolute prerequisite for successful habitat restoration. Unfortunately, this is seldom the case; creation is the typical approach and, even in the case of the seemingly more predictable restoration of constrained wetlands *vis a vis* dike breaches and similar strategies, the data obtained through monitoring tend to identify the outcome of the habitat function rather than the mechanism. A primary example, and one that is the dominant habitat function used in regulating mitigation and developing restoration of estuarine habitats in the Pacific Northwest, is that of fish and wildlife support. The history of monitoring and research illustrates a slow, difficult progression toward acquisition of mechanistic knowledge about estuarine habitat functions for fish and wildlife. The basic, and still prevalent, data are simply presence or absence or density or standing stock of the fish and wildlife species in the habitat. The primary limiting factor in the *occurrence* approach to measuring habitat function is that, except for exclusively resident fauna (e.g., benthic infauna), there is no verification that the habitat is fulfilling a function. The more logical approach is to document utilization *per se*, that is, gather data on the "product" of the function, such as reproduction, feeding, refuge from predation or disturbance, or physiological adaptation. The advantage of the utilization approach is that the "fitness" benefit to the fish and wildlife population would actually be measured or indexed. The disadvantage, however, is that the state of wetland science in this region is so young that we typically do not have enough information to make these measurements in natural wetlands, much less make rigorous comparisons with restored, created or enhanced habitats. Furthermore, this approach is still somewhat deficient in that there is no identification or quantification of the characteristic of the habitat that promoted the utilization.

This dilemma is particularly germane to restoration strategies because we must know *specifically* what characteristics of the habitat promote these functions in order to recreate them. An approach to this mechanistic understanding would be that of identifying and quantifying the attributes of the habitat that directly affect the function. Such functional attributes are particularly critical to restoration because they constitute the *design parameters* for restoring and maintaining restored habitats. Ultimately, however, the requisites of or the processes that create the habitat attributes must be understood because that information is required to sustain the integrity of the attributes.

In evaluating estuarine wetland function for fish and wildlife in the Pacific Northwest, we have developed an attribute-oriented approach to monitoring habitats, this approach strives to gather functional information that can be used to develop increasingly more successful mitigation and restoration projects, i.e., in an adaptive manner. We felt that this unique approach was required because we saw the need for an assessment methodology that was: (1) functional—fish and wildlife monitoring typically does not assess function; (2) standardized—consistency in methods and analyses are absolutely necessary for critical comparison; (3) quantitative—information must be dimensional in order to provide empirical measurements of function; (4) objective—every investigator using the methodology should obtain the same measurement; and (5) adaptive—feedback of information is necessary for evolution of design criteria. The result became known as the *Estuarine Habitat Assessment Protocol* (Simenstad et al.[3]; hereafter called the *Protocol*).

---

[3] The *Protocol* was published in 1991; requests can be sent to Michael Rylko, Puget Sound Program, U.S. Environmental Protection Agency, Region 10, 1200 6th Avenue, Seattle, Washington 98101.

The *Protocol* was developed under the auspices of an ad hoc working group called the Urbanized Estuary Mitigation Working Group (UEMWG) and eventually funded by the Port of Seattle and the U.S. Environmental Protection Agency, Region 10. Design of the *Protocol* involved: (1) categorization of the basic estuarine wetland and associated habitats in the region; (2) identification of fish and wildlife assemblages considered representative of these habitats; (3) hierarchial classification of all the functions that habitats provided to fish and wildlife assemblages; (4) ranking for each function the relative importance of each habitat for each fish and wildlife species in the assemblages; (5) for the functions of reproduction, feeding and foraging, and refuge from predation and for physiological adaptation, identifying the functional attributes of the habitats that promoted that function; and (6) compiling all existing information and data on the attributes as well as their requisites, if known. This "*Protocol* development process" was initially conducted through contact with wetland specialists in the various agencies represented in the UEMWG. At the point of soliciting information and data on the functional attributes, this survey was expanded to approximately 180 wetland specialists in the region and culminated in a three-day workshop to both validate the approach used in developing the *Protocol* and the basic information assembled on the attributes. Based on the results of the workshop and subsequent requests for review comments, the *Protocol* is being revised for publication in hard copy (including appendices with the original survey data and bibliographic support) and as a compiled computer software program. The *Protocol* is being considered as but one of several uniform assessment approaches to be applied to estuarine wetland mitigation projects in this region. It has been proposed as a template for assessment of estuarine restoration initiatives in the region because of its adaptive nature and focus on fish and wildlife resource support. In addition, support is being arranged for field application and verification of the *Protocol's* methods and an update/revision workshop is planned to reevaluate and revise it in several years.

Future Needs and Directions

Based upon our experience in estuarine habitat restoration in the Pacific Northwest, the predictability of successful functional development by specific restoration designs must improve before we can be assured that the loss or degradation of fish and wildlife support, water quality, hydrological and other important functions can be reversed. Predictability could be enhanced by advancing a number of objectives.

Develop Objective Design Criteria

Design criteria for restoration projects should incorporate the best level of understanding about the attributes of the target habitat that promote the desired function. If feasible, designs should incorporate several options or variations of a particular attribute to constitute a legitimate test of the concept, and to provide an adaptive direction toward design modifications. Contingency plans should be developed for each design element in a restoration project such that modifications will evolve logically from the results of monitoring. Design criteria should first derive from the historical context of the original estuarine system, i.e., the natural habitat structure and processes prior to habitat destruction and degradation, and deviate only when it can be shown that subsequent modifications to the estuarine system have been so severe as to obviate historical conditions.

Document Trajectories of Restoration/Succession Sequences

Through rigorous, time-intensive monitoring and research, we need to develop trajectories of attribute values through the developmental sequences ("succession") of restoration and appropriate mitigation projects. Such trajectories must be developed before any single-point measurements (e.g., limited monitoring) assessments can be validly used to interpret the status of restoration habitats. In addition, monitoring of "attribute development" should be conducted in both developed and relatively undeveloped estuaries. This monitoring is necessary in order to determine the more comprehensive effect of system degradation on the ability of a restored habitat to respond given the constraints of degraded water and sediment quality, reduced organism recruitment rates, altered circulation and sedimentation, and proximity to natural habitats.

Implement Comprehensive, Estuary- or Watershed-wide Planning of Restoration

Estuarine habitat restoration, and mitigation for that matter, absolutely cannot be fully effective unless applied within the context of comprehensive restoration planning, e.g., landscape context. Restoration of estuarine habitats is actually a non sequitur if management of other regions of the estuary, uplands or watershed are nonexistent or counterproductive (Josselyn et al. 1989). Comprehensive, system-wide planning should generate scientifically-based information about where to locate restoration projects; their relationship to other habitats; their size thresholds to optimize function; their optimum configuration; and whether it is possible, and how, to accelerate "succession" process by augmenting or subsidizing new projects.

Incorporate Restoration Goals in Regulatory-Mitigation Process

It is probably unreasonable to assume that dedicated restoration can alone accomplish any level of "no net loss" of habitats in the nation's estuaries. To maximize this goal, the same approaches and strategies must be applied in the regulation of mitigation, and specifically compensatory mitigation (e.g., CWA, Section 404). In fact, the mitigation process should be used as additional tests of restoration strategies and to generate alternative strategies.

Support Research to Address Critical Gaps in our Understanding

In addition to the data required to generate successional trajectories and estimates of their predictability, there are several critical gaps that will inhibit the development of restoration ecotechnology unless research to address them is implemented. These include, but are not restricted to:

1. More extensive studies on the basic physical, geochemical and biological processes that structure wetland habitats, in particular sediment accretion and erosion, soil formation, nutrient exchange and natural plant colonization rates and patterns.

2. Application and testing of landscape concepts in restoration.

3. Experiments to determine the optimal genetic mix of plants to maximize transplant survival and propagation.

4. Interdisciplinary studies of restoration designs by engineers, hydrologists and ecologists in order to maximize understanding of interactions among physical and biological processes in wetlands.

5. Testing of functional relationships between wetland fish and wildlife resources and wetland habitat structure.

Induce Restoration Rigor and Ethic

In addition to the institutional mandate for estuarine habitat restoration, the success of any restoration activity will rely considerably upon the rigor of the process and the ethics of the managers and scientists involved in the process. Undoubtedly, one of the most important aspects of promoting rigor in the process is to use scientific peer-review procedures in the evaluation of restoration proposals and encourage publication of the results. Ethics cannot usually be advanced by institutional processes. Instead, there needs to be a movement from within restoration managers and scientists to demand that the final assessment of restoration be measured by the function of the habitat rather than its price tag or some other currency unrelated to its contribution to the estuarine ecosystem and resources of concern.

Expand Restoration Ecotechnology and Research Network National and Internationally

Finally, estuarine restoration should not be pursued as an end to itself, but as a contribution to both the basic science of ecology as well as to the problems and issues of loss and degradation of other ecosystems across the globe. Estuarine restoration initiatives must be considered as one of the integral research units among national and global networks of dedicated research sites. Examples of such National networks include, but are not restricted to: (1) National Estuarine Research Reserve (NERR)—fourteen sites spread across the coastal states and the Great Lakes; (2) the National Science Foundation's (NSF) Long-Term Ecological Research (LTER) and Land-Margin Ecosystem Research (LMER) sites—18 and four sites, respectively, distributed nationally and internationally (i.e., one LTER site is located in Antarctica); (3) EPA Estuaries of Significance; (4) National Parks; (5) the international Man in the Biosphere (MAB) program; and (6) SCOPE.

Summary and Conclusions

Restoration of estuarine habitats in the Pacific Northwest is a young "ecotechnology" employed haphazardly and primarily through the vehicle of regulated compensatory mitigation. Because of the pattern of urban, agricultural and industrial development in the region, most habitat loss and modification has been concentrated in the large estuarine deltas; correspondingly, the vast majority of the need for and attempts at habitat restoration are focused in urbanized estuaries. However, urbanized estuaries probably present the most difficult challenge for restoration of wetlands because of the scope of historical losses, water and air pollution, the potential for disturbance, and the limited opportunities for viable restoration. Evaluation of estuarine habitat mitigation or enhancement projects from the region that apply restoration strategies *per se* indicate that more functional approaches, rather than simple imitation, are required. The projects we have cited encompass four diverse restoration strategies, from incorporation of small-scale *elemental* attributes of littoral mudflats into rip-rap stabilized channel banks using constructed "terraces," *enhancement* of original habitat characteristics by nourishing beaches depleted of fine-grained sediments, to the comprehensive *creation* of the attributes of large-scale brackish marsh-mudflat complexes. This functional approach has altered our strategy of evaluation from that based on fish and wildlife use to a quantitative method based upon the attributes of estuarine wetlands that promote the communities of which fish and wildlife are but elements. This protocol also provides a mechanism of positive feedback where quantitative information on the performance of a restored wetland enhances the technology base, adaptively increasing the probability of successful restoration designs in the future. Ultimately, restoration cannot succeed without taking an estuarine-wide, if not watershed-wide (i.e., landscape), perspective to planning the distribution, composition and character of habitat restoration opportunities. This perspective requires extensive research into both the historical structure and processes once present in the pre-development estuary and the attributes of habitats that promote their various functions. It suggests that to recover some acceptable level of fish and wildlife, hydrologic, sedimentation, water quality, etc., functions of estuarine systems, habitat restoration will have to advance beyond planting salt marsh plants and must embrace upland and upstream management.

Literature Cited

Blomberg, G., C. Simenstad and P. Hickey. 1988. Changes in Duwamish estuary habitat over the past 125 years, p. 437-454. *In* Proceedings First Annual Meeting on Puget Sound Research, Vol. 2. Puget Sound Water Quality Authority, Seattle, Washington.

Bortleson, G. C., M. J. Chrzastowski and A. K. Helgerson. 1980. Historical changes of shoreline and wetland at eleven major deltas in the Puget Sound region, Washington. U.S. Geol. Surv., Denver, Colorado, Atlas HA-617.

Boulé, M. E. and K. F. Bierly. 1987. History of estuarine wetland development and alteration: what have we wrought? Northwest Environ. J. 3: 43-61.

Boulé, M. E., N. Olmsted and T. Miller. 1983. Inventory of wetland resources and evaluation of wetland management in western Washington. Report prepared for Washington State Department of Ecology, Shapiro Assoc., Seattle, Washington. 102 p.

Burg, M. 1984. Habitat change in the Nisqually River delta and estuary since the mid-1980's. M.S. thesis, University of Washington, Seattle. 113 p.

Cowardin, L. M., C. Carter, F. C. Golet and E. T. LaRoe. 1979. Classification of wetlands and deepwater habitats of the United States. U.S. Fish Wildl. Serv., Biol. Serv. Prog., FWS/OBS-79/31. 103 p.

Dorcey, A. H. J., T. G. Northcote and D. V. Ward. 1978. Are the Fraser River marshes essential to salmon? University of British Columbia Westwater Research Center, Vancouver, Tech. Rep. 1. 29 p.

Downing, J. 1986. The coast of Puget Sound: processes and development. Puget Sound Book Series, Washington Sea Grant Program, University of Washington, Seattle. 126 p.

Frenkel, R. E. and J. C. Morlan. 1990. Restoration of the Salmon River salt marshes: retrospect and prospect. Final Report to U. S. Environmental Protection Agency, Department of Geosciences, Oregon State University, Corvallis. 143 p.

Healey, M. C. 1982. Juvenile Pacific salmon in estuaries: the life support system, p. 315-341. *In* V. S. Kennedy (ed.), Estuarine comparisons. Academic Press, New York.

Hiss, J. M. 1990. The effect of beach nourishment on salmonid prey of Lincoln Park beach, Seattle, Washington: post-project conditions. Report to U.S. Army Corps of Engineers, Seattle District, U.S. Fish and Wildlife Service, Fishery Assistance Office, Olympia, Washington. 33 p.

Hiss, J. M., J. L. Schroeder and S. Lind. 1988. The effect of beach nourishment on salmonid prey of Lincoln Park beach, Seattle, Washington: pre-project conditions. Report to U.S. Army Corps of Engineers, Seattle District, U.S. Fish and Wildlife Service, Fisheries Assistance Office, Olympia, Washington. 17 p.

Jones and Stokes Associates, Inc. 1988. Restoration potential of diked estuarine wetlands in Washington and Oregon; Phase I, Inventory of candidate sites. Report submitted to U.S. Environmental Protection Agency, Region 10, 910/988-242. 112 p.

Jones and Stokes Associates, Inc. 1990. Restoration potential of diked estuarine wetlands in Washington and Oregon; Phase II, Inventory of candidate sites in Puget Sound. Report submitted to U.S. Environmental Protection Agency, Region 10.

Josselyn, M., J. Zedler and T. Griswold. 1989. Wetland mitigation along the Pacific coast of the United States, p. 1-35. *In* J. A. Kusler and M. E. Kentula (eds.), Wetland creation and restoration: The status of the science. Vol. I, Regional reviews. U.S. Environmental Protection Agency, Environmental Research Laboratory, Corvallis, Oregon, EPA/600/3-89/038.

Kunz, K., M. Rylko and E. Somers. 1988. An assessment of wetland mitigation practices pursuant to Section 404 permitting activities in Washington State, p. 515-531. *In* Proceedings first annual meeting on Puget Sound research, Vol. 2. Puget Sound Water Quality Authority, Seattle, Washington.

Kusler, J. A. and M. E. Kentula. 1989. Wetland creation and restoration: the status of the science. Environ. Res. Lab., Corvallis, Oregon, EPA/600/3-89/038, Vol. I and II

Levings, C. D., C. D. McAllister and B. D. Cheng. 1986. Differential use of the Campbell River estuary, British Columbia, by wild and hatchery-reared juvenile chinook salmon (*Oncorhynchus tshawytscha*). Can. J. Fish. Aquat. Sci. 43: 1386-1397.

MacArthur, R. H. and E. O. Wilson. 1967. The theory of island biogeography. Princeton University Press, Princeton, New Jersey. 203 p.

McKenzie, T., J. P. Houghton and R. M. Thom. 1990. Analysis of marsh transplant experiment at Jetty Island, Everett, Washington. Draft Report to Port of Everett, Pentec Environmental, Edmonds, Washington.

Mitchell, D. L. 1981. Salt marsh reestablishment following dike breaching in the Salmon River Estuary, Oregon. Ph.D. Dissertation, Oregon State University, Corvallis. 171 p.

Prichard, D. W. 1967. What is an estuary: physical viewpoint, p. 3-5. *In* G. F. Lauff (ed.), Estuaries. American Association for the Advancement of Science, Publication 83, Washington, D.C.

Proctor, C. M., J. C. Garcia, D. V. Galvin, G. B. Lewis, L. C. Loehr and A. M. Massa. 1980. An ecological characterization of the Pacific Northwest coastal region. U.S. Fish and Wildlife Service Biol. Serv. Prog., FWS/OBS-79/14, Vol. I-IV.

Russell, R. J. 1967. Origins of estuaries, p. 93-99. *In* G. F. Lauff (ed.), Estuaries. American Association for the Advancement of Science, Publication 83, Washington, D.C.

Shreffler, D. K. 1989. Temporary residence and foraging by juvenile salmon in a restored estuarine wetland. M.S. thesis, University of Washington, Seattle. 86 p. + append.

Shreffler, D. K., C. A. Simenstad and R. M. Thom. 1990a. Temporary residence by juvenile salmon in a restored estuarine wetland. Can. J. Fish. Aquat. Sci. 47: 2079-2084.

Shreffler, D. K., R. M. Thom, C. A. Simenstad, J. R. Cordell and E. O. Salo. 1990b. The Lincoln Avenue wetland system in the Puyallup River estuary, Washington: Phase III Report, Year three monitoring, January-December 1988. Wetland Ecosystem Team, Fisheries Research Institute, University of Washington, Seattle, Washington, FRI-UW-8916. 54 p.

Simenstad, C. A. 1983. The ecology of estuarine channels of the Pacific Northwest coast: a community profile. U.S. Fish and Wildlife Service Biol. Serv. Prog., FWS/OBS-83/05. 181 p.

Simenstad, C. A. and D. A. Armstrong. In press. The ecology of estuarine littoral flats of the Pacific Northwest Coast: A community profile. U.S. Fish and Wildlife Service Biol. Serv. Prog., FWS/OBS.

Simenstad, C. A. and J. R. Cordell. In revision. Evaluation of intertidal benches in stabilized shorelines for enhancement of juvenile salmonid prey resources: pre-installation assessment of Terminal 106, Duwamish River estuary. Fisheries Research Institute, University of Washington, Seattle, Washington.

Simenstad, C. A., K. L. Fresh and E. O. Salo. 1982. The role of Puget Sound and Washington coastal estuaries in the life history of Pacific salmon: An unappreciated function, p. 343-364. *In* V. S. Kennedy (ed.), Estuarine comparisons. Academic Press, New York. 709 p.

Simenstad, C. A., C. D. Tanner, R. M. Thom and L. Conquest. 1991. Estuarine habitat assessment protocol. Report to U.S. Environmental Protection Agency, Region 10, Fish. Res. Inst., University of Washington, Seattle, Washington, EPA 910/9-91-037. 201 p.

Strickland, R. (ed.). 1986. Wetland functions, rehabilitation, and creation in the Pacific Northwest: the state of our understanding. Proceedings of conference, Port Townsend, Washington, April 30-May 2, 1986. Washington State Department of Ecology, Olympia, Washington, Publ. 86-14. 184 p.

Thom, R. M. and L. Hallum. 1990. Long-term changes in the areal extent of tidal marshes, eelgrass meadows and kelp forests of Puget Sound. Fisheries Research Institute, University of Washington, Seattle, Washington. FRI-UW-9008.

Thom, R. M. and L. Hamilton. 1990. Lincoln Park shoreline erosion control project: post-construction monitoring of eelgrass, infaunal bivalves and macroalgae, 1990. Draft Report to U. S. Army Corps of Engineers, Seattle District, Fisheries Research Institute, University of Washington, Seattle, Washington. FRI-UW-90.

Thom, R. M. and J. J. Hampel. 1985. Lincoln Park benthic resource assessment/enhancement study. Final Report submitted to U.S. Army Corps Engineers, Seattle District. Evans-Hamilton, Inc., Seattle, Washington. 56 p.

Thom, R. M., E. O. Salo, C. A. Simenstad, J. R. Cordell and D. K. Shreffler. 1988a. Construction of a wetland system in the Puyallup River estuary, Washington, p. 156-160. *In* K. M. Mutz and L. C. Lee (tech. coords.), Wetland and riparian ecosystems of the American west. Society of Wetland Scientists, Planning Information Corporation, Boulder, Colorado. 349 p.

Thom, R. M., C. A. Simenstad and J. R. Cordell. 1986. Early successional development of a benthic-epibenthic community at a newly constructed beach in Slip 1, Commencement Bay, Washington: initial observations 1985. Fisheries Research Institute, University of Washington, Seattle, Washington. FRI-UW-8603. 42 p.

Thom, R. M., C. A. Simenstad, J. R. Cordell and E. O. Salo. 1988b. Fisheries mitigation plan for expansion of moorage at Blaine Marina,

Blaine, Washington. Fisheries Research Institute, University of Washington, Seattle, Washington. FRI-UW-8817. 24 p.

Thom, R. M., C. A. Simenstad, J. R. Cordell and E. O. Salo. 1989. Fish and their epibenthic prey in a marina and adjacent mudflats and eelgrass meadow in a small estuarine bay. Fisheries Research Institute, University of Washington, Seattle, Washington, FRI-UW-8901. 27 p.

Thom, R. M., C. A. Simenstad, J. R. Cordell, D. K. Shreffler and L. Hamilton. 1990. The Lincoln Avenue wetland system in the Puyallup River estuary, Washington: Phase IV Report, Year four monitoring, January-December 1989. Wetland Ecosystem Team, Fisheries Research Institute, University of Washington, Seattle, Washington, FRI-UW-9004. 44 p.

Thom, R. M., C. A. Simenstad and E. O. Salo. 1987. The Lincoln Street wetland system in the Puyallup River estuary, Washington: Phase I report; Construction and initial monitoring, July 1985-December-1986. Fisheries Research Institute, University of Washington, Seattle, Washington, FRI-UW-8706. 85 p.

Thom, R. M., C. A. Simenstad, D. K. Shreffler, J. R. Cordell and E. O. Salo. 1988c. The Lincoln Avenue wetland system in the Puyallup River estuary, Washington: Phase II Report, Year two monitoring, January-December 1987. Wetland Ecosystem Team, Fisheries Research Institute, University of Washington, Seattle, Washington, FRI-UW-8812. 80 p.

Thomas, D. W. 1983. Habitat changes in the Columbia River estuary since 1868. Columbia River Estuary Study Taskforce, Astoria, Oregon.

Tutty, B. D., B. A. Raymond and K. Conlin. 1983. Estuarine restoration and salmonid utilization of a previously diked slough in the Englishman River estuary, Vancouver Island, British Columbia. Can. Manuscript Rep. Fish. Aquat. Sci. No. 1689. 51 p.

Wilsey and Ham, Inc. 1974. Estuarine resources of the Oregon coast. Rep. to Ore. Coastal Conservation Development Commission, Portland, Oregon. 233 p.

Zedler, J. B. 1986. Wetland restoration: trials and errors in ecotechnology, p. 11-16. *In* R. Strickland (ed.), Wetland functions, rehabilitation, and creation in the Pacific Northwest: the state of our understanding. Proceedings of Conference, Port Townsend, Washington, April 30-May 2, 1986. Washington State Department of Ecology, Olympia, Washington, Publication 86-14. 184 p.

CHAPTER 11

Restoring and Managing Disused Docks in Inner City Areas

S.J. Hawkins
Port Erin Marine Laboratory, Liverpool University
Isle of Man, United Kingdom

J.R. Allen and G. Russell
Department of Environmental and Evolutionary Biology
Liverpool University, Liverpool, United Kingdom

K.N. White, K. Conlan, K. Hendry
and H.D. Jones
Department of Environmental Biology, Manchester University
Manchester, United Kingdom

Abstract

Port development in the British Isles led to extensive systems of enclosed dock basins along major estuaries. Many have recently fallen into decline or total disuse and are the focus of ambitious redevelopment schemes. The state of water quality in disused docks is broadly appraised on a nationwide basis in the United Kingdom and major problems are identified. These are mainly related to the eutrophic, polluted nature of source waters. Anoxic bottom waters are common in summer when stratification occurs and unsightly dense phytoplankton blooms are also a major problem. Two case studies are examined in detail. Liverpool has docks of high salinity. In one dock, Sandon, an experimental fish farm was run between 1978 and 1983. Dramatic water quality improvements occurred which were attributed to the combination of artificial mixing using an aerator device and dense populations of naturally settled and cultivated mussels acting as a giant biological filter. Anoxic bottom waters were prevented and the water became much clearer. A diverse benthic community dominated by mussels developed and fish proliferated. This work prompted a more detailed study of the effectiveness of mixing and biological filtration by mussels in the nearby South Docks complex which is part of an urban renewal scheme. The effectiveness of artificial mixing combined with use of mussels as a biological filter was confirmed. Additionally, in one enclosed dock, improvements in water quality occurred probably due to the filtering action of large numbers of naturally settling mussels. A diverse mussel-dominated community has also developed in the South Docks.

Preston Docks are a low salinity system at the head of an estuary. Water quality problems are more intractable here with dense algal blooms. Flushing of the dock is one possible solution but it is only partially effective. The benthic community is highly impoverished, and there are no natural candidates to be used as a biological filtering species. A diverse fish community does exist with potential for a recreational fishery.

On the basis of the nationwide survey and the case studies, general solutions to water quality problems are discussed. The strategic solution is water catchment clean-up. Tactical solutions can involve isolation, followed by installation of mixing devices and a biological filter. Chemical methods to strip nutrients and reduce phytoplankton are reviewed, as are more speculative biomanipulative approaches.

Restored disused docks are valuable for water-based recreation, research and education, and in promoting tourism and redevelopment in urban areas. Aquaculture, although tried in the United Kingdom, is less likely to be successful. Restored dock ecosystems are considered to be invaluable in urban conservation; they partially compensate for the disappearance of saline lagoons and can relieve pressure on more fragile natural ecosystems. Priorities for future work are outlined, particularly the need for controlled work on mixing devices in saline systems and further work on biomanipulation, including studies of artificial reefs. In conclusion, restoration of semi-enclosed urban water bodies not only promotes wildlife but aids urban renewal. It is a rare example of development and conservation being complementary activities.

Introduction

From the earliest days of marine transport, ships have operated from sheltered bays and estuaries. Increased trade and larger ship sizes during the 16th and 17th centuries led to the development of more sophisticated harbor facilities involving complexes of quays and warehouses. The next stage in Britain, in the early 18th century, was the development of enclosed or semi-enclosed dock basins usually with entrance lock gates. The building of the world's first purely commercial dock commenced in Liverpool in 1710 (Ritchie-Noakes 1984). This design was subsequently adopted in ports with large tidal ranges which had previously made cargo-handling difficult. Dock basins surrounded by warehouses that formed self-contained units became the standard pattern in the major British ports throughout the 19th and early 20th century. The associated industrialization and urbanization of port areas and their hinterlands have contributed to many United Kingdom estuaries being grossly polluted.

As a result of changes in Britain's industry and commerce in the latter half of the 20th century, many British docks went into decline (McConville 1977). Old docks in the middle and upper reaches of estuaries suffered particularly badly. The larger ships coming into use could not negotiate the estuaries or use dock basins designed for sailing ships and small steamers. Increasing use of containerized cargoes was not compatible with traditional docks in the middle of urban areas or with traditional labor practices. New container terminals with facilities for more rapid turnaround were developed on the seaward edge of existing port conurbations which were able to berth larger ships, and had space for large parking areas with easy access to trunk roads. Switching of cargoes from small coasters to road haulage also diminished demand at many docks. Thus, from the 1970s onwards, large tracts of dockland in many inner city areas became disused and often derelict. This was very pronounced in the United Kingdom, but also occurred elsewhere in the developed world including major east coast ports in North America (see papers in Hoyle et al. 1988 for historical, social and economic background).

British docklands have been the subject of various inner city regeneration projects in the 1980s. These usually involve some combination of shopping complexes, housing, light industry, cultural and recreational provision plus tourism. Dock basins have also been used occasionally for aquaculture and for scientific and technological research. These options are dependent on good water quality. There has also been an increasing appreciation of the important role these aquatic habitats can play in urban nature conservation and recreational fisheries—provided that water quality and ecosystem diversity can be improved (Russell et al. 1983; Cunningham et al. 1984; Conlan et al. 1988; Hendry et al. 1988a, b, c). Unfortunately, many dock basins contain filthy, stagnant, polluted water derived from the industrialized estuaries into which they open (Hendry et al. 1988a, b).

Focussing on the United Kingdom in this paper, we outline the location and number of disused docks. We review the problems associated with industrialized estuaries in general and disused dock basins in particular and present two case studies. Liverpool is an example of an outer estuarine high salinity port in which marked improvements in water quality have been considered possible (Russell et al. 1983; Allen et al. In press). Preston is an inner-estuarine brackish water dock whose water quality problems are more intractable. For each dock complex, we discuss their general ecology and water quality problems, plus possible management measures and their actual or likely success. General solutions are considered from the experience of the case studies and other related work. We discuss the value of restored marine dock complexes in inner city areas, and conclude with proposals for future research work.

The authors wish to emphasize that much of the work is of a

preliminary nature and has been undertaken within the constraints of redevelopment schemes that are underway. Therefore, the approach has not been as rigorously controlled as we would have liked. However, we hope to demonstrate convincingly that diverse aquatic ecosystems can be restored and maintained in the middle of urban areas.

General Problems: A Nationwide Appraisal

Marine and brackish water docks no longer used by commercial shipping occur throughout the United Kingdom (Figure 1). Liverpool and London have the most extensive abandoned dock systems followed by Bristol. Other large disused dock basins occur at Portishead (near Bristol), Cardiff, Penarth, Swansea, Preston, Fleetwood, Barrow, Glasgow, Dundee, Teeside and Hull. Dock areas with much reduced shipping are found at Plymouth, Newport, Barry and on Tyneside. The docks at the end of the Manchester Ship Canal at Salford form the most extensive freshwater system but there is also a large freshwater dock at Gloucester. Most of the docks listed above are subject to some form of redevelopment scheme (Table 1). Docks are invariably located on highly urbanized and industrialized estuaries, rivers or canals. The water quality problems in disused docks generally reflect the condition of their source waters. The more static nature of dock basins can aggravate some existing problems and may prompt new ones.

General Estuarine Problems

The natural characteristics of the estuarine environment present difficulties to the development of a stable, diverse biota. Strong currents make settlement of particles difficult and result in turbid water and unstable substrates. Estuarine organisms must be able to cope with large fluctuations in temperature and salinity, often combined with episodic changes in water flow.

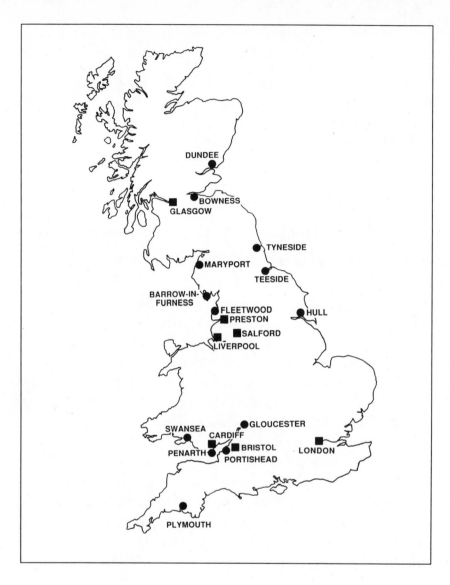

Figure 1. Map of Great Britain showing ports with disused docks. (Key: circle—disused docks; square—site of major redevelopment.)

Table 1. Pollution and development status of some docks in the United Kingdom. Tested metals recorded in bold exceed recommended maximum levels for human consumption.

Location[a]	Dock	Area (hectares)	Coliform Levels[b]	Tested Metals in Fish Flesh.	Present Use[c] (Projected Use)
Newcastle*	Albert Edward	9.20	—	Cd, Cu, Pb, Zn, Hg	Wd
Penarth*	Portway marina	4.80	—	Cd, Cu, Pb, Zn, Hg	Fn, H, la, M, (R), T
Plymouth*	Millbay (total)	17.00	—	Cd, Cu, Pb, Zn, Hg	Bl, Cd, Fn, (H), la, M, (R), T
Liverpool	SandonR	4.25	Low	Cu, Pb, Zn	Recently Infilled
Liverpool†	South	27.2	Low	Cu, Pb, Zn	Bl, H, la, M, R, T, W1, W2
Glasgow*	Princes	5.18	—	Cd, Cu, Pb, Zn, Hg	(H), la, (M), R, T, W2
Preston**	Albert Edward	16.20	Moderate	Cd, Cu, Pb, Zn, Hg	Bl, Fn, (Fe), H, la, M, R, W2
London*	Royal	89.25	—	Cd, Cu, Pb, Zn, Hg	Bl, (Fe), H, la, R, W2
Bristol*	Floating Harbour	20.00	High	—	Bi, Fe, H, la, M, R, T, W2
Barrow*	Cavendish	58.20	—	Cd, Cu, Pb, Zn, Hg	?
Salford*	Quays (total)	11.26	Low	Cd, Cu, Pb, Zn, Hg	Bl, (Fe), H, la, M

[a] References: (*) Hendry et al. 1988a; (†) Conlan 1989; (**) Russell et al. 1983. Bacteriological data from Enticott and Grisdale 1985; Radway et al. 1988; Thames water archives and Alwell 1989.

[b] Coliform levels: Low = max 10,000 100ml^{-1} E.C. limit bathing waters; moderate = max 100,000 100ml^{-1}; high = max above 100,000 100ml^{-1}.

[c] Present use key: Bl—Business/light industry; Wd—Working dock; Fo— Organized recreational fishing; Fn—Unofficial recreational fishing; H—Housing; la—Informal amenity (walking, cinema, gallery etc.); M— Marina; R—retail; T—Tourism; Ur—Unspecified redevelopment; W1—primary contact watersports; W2—secondary contact watersports.

Human influence on estuarine systems may further stress these communities. The discharge of industrial and domestic organic wastes increases biological oxygen demand (BOD) leading to hypoxia or anoxia. Discharge of inorganically (e.g., heavy metals, ammonia) and organically (e.g., oil, organohalines) contaminated effluents and cargo spillages all pollute estuarine waters, sediments and biota. Agricultural and urban runoff into the catchment basin raise the already high natural levels of plant nutrients such as phosphates, nitrates and silicates to hypertrophic conditions. Physical changes to the land-water boundary may also have detrimental effects. For example, reclamation of salt marshes removes a major sink for metals, nutrients and silt from the estuary and from runoff entering from the land. Re-direction of drainage systems speeds flow of water into the river system, strengthening episodic changes and increasing the likelihood of discharge of contaminated stormwaters. The high input of incompletely treated or untreated sewage into most United Kingdom estuaries presents a health hazard from harmful microorganisms.

Dock Water Quality

In addition to polluted source water, cargo spillages, leachates from antifouling paints, residues from graving (repair and paint stripping) activities in dry docks and the dumping of waste material from ships using the docks are sources of contamination, particularly of the sediments. Disused docks are also prone to illegal rubbish tipping. Docks are, therefore, often contaminated with biodegradable organic material, fecal microorganisms and conservative pollutants such as heavy metals, and are eutrophic or hypertrophic.

Docks intended for ocean-going vessels usually consist of one or more relatively deep (7-12 m), rectangular, steep-sided basins. Water surface area varies from less than a hectare (Pomona Dock, Manchester) to 89 ha (Royal Dock, London). The rate of exchange with the adjacent water course may be high, particularly if it is tidal (up to 110% per month at Preston Dock, for example; Conlan 1989). Usually water is impounded (levelling) on spring tides to compensate for leakage and loss during the locking in and out of vessels. Flushing of the dock was not usually a priority in the design of sluices and locks and levelling operations do not always result in significant water mixing. The absence of movement of large vessels further reduces water circulation in disused and redeveloped docks. Dock basins, therefore, have many of the hydrographic characteristics of a lake or lagoon, including the propensity for salinity and/or temperature-induced layering of the water column.

Temperature-induced stratification occurs only in summer in the United Kingdom, due to the cool winters and the greater degree of wind-induced mixing between September and May. Estuarine docks may also be subject to variation in salinity. Basins situated in the upper reaches of poorly-mixed estuaries will be most prone to salinity stratification due to changes in river flow and incursions of high density saline water with the tide (e.g., Conlan 1989). Stratification by whatever means reduces interchange between the surface and bottom layers and has various consequences for water quality.

The presence of large amounts of biodegradable organic material from sewage (and sometimes phytoplankton die-off) results in an appreciable oxygen demand. Water column BODs of over 5 mg L^{-1} are common in United Kingdom docks with maximum values in excess of 15 mg L^{-1} observed at Salford and Preston (Hendry et al. 1988a, b and unpubl.; Conlan 1989). The oxygen requirements of the organically enriched sediments will also exert a considerable demand on the water above (Cross and Summerfelt 1987). Such oxygen demands are not always satisfied in stratified or poorly-mixed basins resulting in long periods of bottom water anoxia. In 6 out of 10 marine, estuarine and freshwater docks, low levels of dissolved oxygen (< 3 mg L^{-1}) or anoxia were observed for periods of up to one month (Hendry et al. 1988a and unpubl.; Conlan 1989; Allen et al. in press).

Anoxia is probably the most serious water quality problem encountered in docks. It detracts from their aesthetic value by promoting anaerobic activity at the sediment and water interface, causing unsightly bubbles of sometimes smelly gases. Nitrification is inhibited, which results in the presence of potentially toxic levels of ammonia and nitrite. The decrease in redox potential increases the release of phosphate (Hallberg et al. 1976; Marsden 1989), a key factor in facilitating algal blooms (see below). Mobilization of some heavy metals from the sediments may also be enhanced (Lu and Chen 1977). Low dissolved oxygen levels will eliminate all but the most tolerant benthic organisms. Such conditions favor colonization by a few opportunistic species, in particular oligochaetes, polychaetes and, in low salinities, chironomid larvae. Often the dock is covered with mats of the bacterium *Beggiatoa*.

Of the 10 United Kingdom docks examined by Hendry et al. (1988a, b, c) 9 contained nitrate and/or orthophosphate at levels (Table 2) likely to result in eutrophic conditions in standing waters (Likens 1975; O.E.C.D. 1982). Similar nutrient concentrations were observed in the adjacent estuaries but algal standing crop and productivity are usually much lower (e.g., in the Ribble at Preston, Conlan 1989; Mersey at Liverpool, Allen et al. in press) due to high turbidity making light limiting and currents flushing plankton seawards (McClusky 1989 for review of British estuaries). In docks such as Liverpool where water exchange is relatively small or intermittent, sedimentation of suspended solids increases water clarity sufficiently to allow phytoplankton growth to reach "bloom" proportions (Russell et al. 1983; Conlan 1989; Allen et al. In press). Data from a "snapshot" study of a further eight United Kingdom docks (Hendry et al. 1988a,b,c) indicate that seven were eutrophic/hypertrophic (mean chlorophyll *a* > 8 μg ml^{-1}: O.E.C.D. 1982).

Table 2. Physical and chemical parameters of disused docks in the United Kingdom. (Maximum-minimum values; nd=below detection limits.)

Dock	Salinity (ppt)	NO_3 (mg L^{-1})	PO_4 (mg L^{-1})	NH_4 (mg L^{-1})	Chl a (µg L^{-1})	Secchi (m)	BOD (mg L^{-1})	Samp. Occ.
Liverpool South	24-28	nd-0.98	0.14-0.81	nd-0.73	<1-79	0.5-7.5	—	12
Sandon	25-31	0.10-1.0	0.05-0.20	—	—	2-10	—	>15
Preston	1-7	0.01-3.1	0.3-3.5	—	4-731	0.6-1.2	2-20	33
Barrow	0.5-2.5	0.36-2.91	0.003-1.4	0.1-0.4	0.1-0.4	0.1-0.4	1-6	2
Bristol	0.6-3.5	3.6-5.2	0.88-1.14	0.5-2.8	17-55	0.5	2-4	11.25-3.44
Glasgow	3.4-19.1	0.48-1.99	0.08-0.50	1.25-3.44	nd-18.5	0.7-0.9	1-15	2
London	5.0-5.4	0.27-1.20	0.76-1.53	0.07-20.0	59-62	0.7-0.9	2-7	2
Newcastle	26-33.5	0.01-0.02	0.02-0.14	—	—	0.9	1-9	1
Penarth	27-29	0.08-0.55	0.05-0.13	—	2.2-35.5	0.8-1.0	5	2
Plymouth	24-35	0.03-0.33	0.01-1.50	0.01-0.10	1.7-3.0	0.5-3.4	<1-4	2

Chlorophyll *a* concentrations of around 50 μg ml^{-1} were observed in the majority of docks we investigated. Such levels impart a marked color to the water and can result in unsightly scums and mats. Nutrient limitation, particularly of nitrogen and phosphorus, is unlikely to be a major check on dock phytoplankton biomass if interchange with the outside estuary or river continues. It may occur, however, in isolated or semi-enclosed systems, particularly if they are of high salinity. Silica has been shown to limit diatom production in the South Docks in Liverpool (Eaton and Russell unpubl. obs.). Mixing of the water column and depth of the photic zone will also influence the type of community present. Preston (brackish) and Salford (freshwater) docks are subject to disturbance caused by flushing and artificial mixing respectively. Both are dominated by cyanobacteria (particularly *Oscillatoria* sp.), a group known to be encouraged by a high frequency of disturbance (Reynolds 1987) coupled with high nutrient levels. The shallower Liverpool docks are subject to dense blooms of the dinoflagellate *Prorocentrum minimum* during June and July (Allen et al. in press). This motile species, as with other dinoflagellates (Holligan and Harbour 1977), may be favored by the relatively stable stratified conditions prevailing at this time. Toxins from algal blooms are known to harm benthos and fish (e.g., Boalch 1979; Widdows et al. 1979; Cross and Southgate 1980) and cause skin irritation and other illnesses in humans (e.g., Tufts 1979; Falconer et al. 1983).

Fecal contamination of urban water courses and estuaries is still prevalent in the United Kingdom with counts of indicator bacteria commonly exceeding European Community (E.C.) standards for bathing waters (Table 1 and Radway et al. 1988, Conlan 1989, Allen et al. in press). Pathogens will enter during water exchange and replenishment, and contact with dock water contaminated by faeces has resulted in morbidity and possibly a fatality (Phillip et al. 1985, 1989). Fecal microorganisms would not be expected to survive indefinitely in open waters (e.g., Faust et al. 1975), and so those introduced into a dock during intermittent discharges will eventually be killed. Fecal coliforms entering a brackish-water dock during locking operations with contaminated estuarine water declined from 8,000 to <1,000 100 ml^{-1} (ca. 10%) within 6 days (Conlan 1989). Fecal bacteria are low in basins subject to limited exchange with contaminated water (Allen et al. in press). However, controversy surrounds the use of indicator organisms in setting water quality standards. A gastroenteritis outbreak occurred following a snorkeling event at Bristol docks although the water conformed to the European Community standard for bathing waters (Evans et al. 1983). Some fecal organisms may remain viable in the sediments for many months (Sawyer et al. 1977) and organic loading, light penetration and salinity will all influence survival (Mitchell and Chamberlin 1975). Therefore, prediction of pathogen behavior and longevity in docks is particularly difficult, although risks will be less in more saline environments.

Disused or redeveloped commercial docks often contain accumulated heavy metal pollution from past shipping activities, supplemented by metals adsorbed to silt entering from outside. Dock sediment metal concentrations are commonly higher than in adjoining riverine and lacustrine sediments (e.g., Bellinger and Benham 1978; James and Gibson 1980; Hendry et al. 1988a). Mobilization from the sediments and uptake from contaminated replenishment water results in elevated heavy metal levels in the food chain (Hendry et al. 1988a; Longsdale 1990). Metal levels in fish are high in some areas and exceed public health guidelines for zinc at London and Newcastle (Table 1).

In summary, the upper estuarine, low salinity docks have the greatest water quality problems due to high nutrient levels and, in consequence, the greatest phytoplankton biomass. Because of organic enrichment, the BOD of most estuarine systems is already high and further increased by the high algal biomass in standing water. As most of these docks stratify, either from thermal or salinity-induced density differences, anoxic bottom waters are common. The BOD can also be sufficiently high to cause low levels of dissolved oxygen even when the water column is unstratified. The most diverse communities of plankton, benthos and fish (Table 3) and motile invertebrates, occur as expected in the more (least stressed) saline systems (Hendry et al. 1988a, b, c). All docks have extremely impoverished sediment communities, the majority of the benthos being found on the dock walls. However, the steep vertical faces of these walls restrict spatial heterogeneity and, hence, the overall diversity of the community. The least diverse communities occur in oligohaline systems such as at Bristol, London and Preston. These do, however, support interesting assemblages of fish with a mixture of estuarine/marine, freshwater and migratory species (Table 3). The diversity of the ecosystem also depends largely on the degree of redevelopment and the quality of management, particularly during the time of cessation of shipping activity.

Case Study 1—Liverpool

When handling its largest number of ships, the port of Liverpool (Figure 2a) possessed over 100 docks, stretching from the river mouth to 10 km upstream. Since the early 1970s, with the decline of the port, many of these docks have become disused or underused (Figure 2b).

Sandon Dock: A Pioneering Project

Historical and Environmental Background. In one such dock, Sandon, an aquaculture trial was set up in the late seventies. Sandon Dock is halfway along the chain of docks on the eastern shore of the River Mersey (Figure 2b). It was used as an experimental fish farm under the guidance of the Freshwater Fisheries Unit, Department of Zoology, University of Liverpool, in a pilot project funded by the Mersey Dock and Harbour Company as part of their diversi-

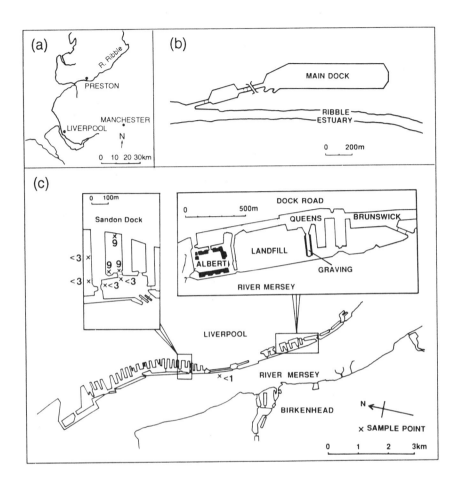

Figure 2. Map showing the northwest of England (a) and the location of the two case studies at (b) Liverpool and (c) Preston. Figures refer to Secchi disc extinction depths (m) in March 1983.

Table 3. Fish species and their relative abundance in 10 U.K. dock basins. Docks are ranked according to salinity (ppt)[a].

Docks	Plym.	Newc.	Livp.	Pen.	Glasg.	Lond.	Prest.	Bar.	Brist.	Salf.
Species					Salinity (ppt)					
	(24.0-34.9)	(25.0-33.5)	(25.0-31.0)	(27.0-29.0)	(0.0-19.1)	(5.0-5.4)	(2.4-5.6)	(0.5-2.5)	(0.6-3.5)	(Freshwater)
Raja clavata — Thornback Ray		+								
Mullus surmuletus — Red Mullet	+									
Crenilabrus melops — Corkwing Wrasse	+									
Acanthocottus bubalis — Sea Scorpion	++		*							
Gobius niger — Black goby	+									
Zoarces viviparus — Viviparous Blenny		+								
Trisopterus minulus — Norway Pout			*							
Gadus morhua — Cod	+		*	+						
Merlangius merlangus — Whiting	++	+	*	++						
Trisopterus luscus — Bib				+						
Pollachius pollachius — Pollock	+				++					
Pollachius virens — Saithe						++				
Ciliata mustela — 5-Bearded Rockling			*							
Gaidropsaris vulgaris — 3-Bearded Rockling			*							
Pleuronectes platessa — Plaice		+	*	+						
Limanda limanda — Dab				*						
Dicentrachus labrax — Bass									*	
Chelon labrosus — Grey Mullet	+								*	
Agonus cataphractus — Pogge		*								

Table 3 continued.

		Plymouth	Newcastle	Liverpool	Penarth	Glasgow	London	Preston	Barrow	Bristol	Salford
Pomatoschistus microps	Common Goby									++	
Platichthys flesus	Flounder	++	++		*	+	++	+++	++	+++	
Sprattus sprattus	Sprat	*			*	+++		‡	‡	+++	
Clupea harengus	Herring	+	+	‡	*	++	‡	‡	‡	+++	
Anguilla anguilla	Eel	+		‡	*	++	‡	‡	‡	++	
Gasterosteus aculeatus	3-Spined Stickleback	*		‡	*		++	+		+	+
Salmo salar	Salmon				*		+	+	+	+	
Salmo trutta	Sea Trout				*		++	+++	‡	+++	
Salmo trutta	Brown Trout				*		+	+	‡	+	
Salmo gairdneri	Rainbow Trout				*					++	
Omerus eperlanus	Smelt					+	++	++		++	
Atherina presbyter	Sand smelt							+	+		
Rutilus rutilus	Roach				*	+	+	+++	‡	+++	‡
Leuciscus leuciscus	Dace				*					*	
Leuciscus cephalus	Chub				*			+		*	
Perca fluviatilis	Perch				*		+			*	
Caprinus carpio	Common Carp				*					*	
Abramis brama	Bream				*					*	
Carassius carassius	Crucian Carp				*					*	
Gobio gobio	Gudgeon				*					*	
Scardinus erythrophthalmus	Rudd				*					*	
Tinca tinca	Tench				*					*	
Phoxinus phoxinus	Minnow				*					*	
Esox lucius	Pike				*					*	

Docks: Plymouth, Newcastle, Liverpool, Penarth, Glasgow, London, Preston, Barrow, Bristol and Salford.
[a] Index of abundance; * no information on abundance; + present (1-4); ++ common (5-19); +++ abundant (>20); ‡ Mean number per multimesh gillnet set.

fication program. When in use as a fish farm, it had a maximum depth of 10-11 m, contained about 425,000 m³ of water, with a maximum change of water level of 1 m due to variations in impoundment in the seaway linking the docks. The lock gates were permanently closed in 1977, but several planks were removed (15 × 45 cm) allowing limited interchange with the outside seaway whose source water is at the seaward end of the docks at Seaforth.

The hydrography of the dock has been described by Russell et al. (1983). Annual variation in salinity was from 26-31 ppt. Temperature varied between 3-18°C with occasional extremes (>20°C in summer 1983, < 0.5° C in winter 1981/1982).

The water quality of the dock was managed by a "Helixor" airlift combined water circulator and aerator (Polcon Environmental Systems). This provided surface oxygenation for cage-reared salmonids, and it destratified the water column, thus preventing the build-up of an anoxic bottom layer (Figure 3 and Russell et al. 1983) Improved water clarity was the most conspicuous feature at Sandon. Secchi disc extinction depths occasionally went as low as 2 m during spring and summer phytoplankton blooms, but usually were between 5 and 7 m. They reached 9-10 m in late winter 1983—much clearer than adjacent docks and the River Mersey (Figure 2b and Figure 4).

The initial aim of the pilot salmonid fish farm in 1978, was abandoned when a "red-tide" of toxic dinoflagellates killed all the stock. This occurred shortly after the Helixor was installed and switched on for the first time, and was possibly due to either a nutrient surge or the resuspension of dinoflagellate cysts from the sediment. Attention switched to culturing bivalves in 1979. Various trials with oyster-rearing generally proved unsuccessful. The main cultivated species was the mussel, *Mytilus edulis.* Mussels were harvested down to 4 m depth from the walls of the dock, where they grew profusely following natural settlement first recorded in summer 1978. Mussels were and still are uncommon in adjacent docks

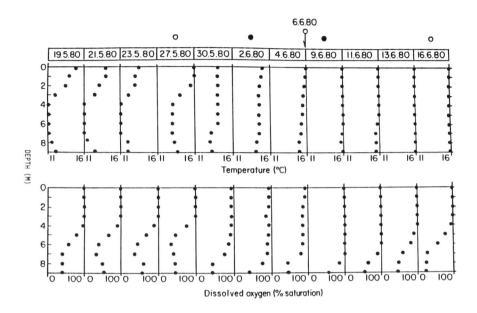

Figure 3. Temperature and oxygen profiles of Sandon Dock, Liverpool (from Russell et al. 1983).

(Figure 4), and their appearance and subsequent survival at Sandon was presumably due to improved water quality consequent on the use of the Helixor. Mussels were also cultured using 4 m long ropes suspended from horizontal ropes tied to the quays and further supported by lines of buoys. Cage culture of trout was successfully restarted in 1982.

Marketable size of mussels (50 mm United Kingdom) was reached within 2 years of rope placement. Growth was very similar to mussels cultured on rafts in Killary Harbour (Rodhouse et al. 1984), compared favorably with cultured populations in Sweden (Loo and Rosenberg 1983), and was faster than most natural intertidal populations in the British Isles (see Seed 1976 for review; Rodhouse et al. 1984). Ropes placed in May 1980 had harvestable mussels for sale during the autumn and winter of 1982. Growth rate during colonization of free space on the walls was similar to

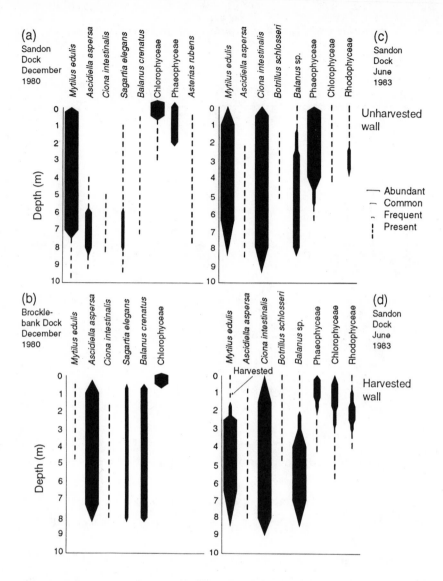

Figure 4. Zonation of major benthos in Sandon Dock in 1980 (a) compared to an adjacent unmanaged dock (Brocklebank, b). Subsequent development of the community at Sandon is shown in 1983 (c, d). Modified from Russell et al. 1983; Cunningham et al. 1984; Hendry et al.1988a.

that on the ropes. It slowed markedly after 2 years, however, presumably because additional cohorts had intensified intraspecific competition.

Heavy metals in the sediment and biota were analyzed at several times over the period of study (Table 4). Levels were higher in animals collected from greater depths on the dock sides. Metals in mussels (*Mytilus edulis*) were monitored over 2 years (Table 4). Zinc levels in the mussels were high but are within acceptable limits for human consumption, as were the other metals. Mussels were also analyzed for coliform bacteria after ultraviolet treatment, but these were absent or insignificant. Levels of heavy metals in oysters, which naturally accumulate zinc even at unpolluted sites, were very high. They could still be sold for human consumption, but clearly there were grounds for concern.

The pilot fish farm ceased operation in October 1983 when the North West Water Authority (N.W.W.A) took up a longstanding option to convert the dock to a sewage sludge-dumping terminal. The dock was dredged and filled in.

Summary of Ecological Studies at Sandon. Phytoplankton were not continuously monitored, but in August 1983 water samples from Sandon and Wellington (adjacent but not destratified) docks were analyzed for phytoplankton and chlorophyll *a* concentration (Liebeschuetz 1985). The mixing effect of the Helixor was shown by the uniform chlorophyll *a* concentration in Sandon. In contrast, Wellington showed high surface values decreasing with depth. The monitoring coincided with a dense "red-tide" of toxic dinoflagellates in the seaway and Wellington dock, but this did not happen in Sandon. A bloom of *Phaeocystis* occurred in May 1983, but disappeared as soon as the mixer was used. The grazing pressure exerted by the mussels, which by this time covered much of the dock walls and several hundred suspended ropes in the dock, was probably a considerable influence in keeping phytoplankton density low

Table 4. Mean levels of zinc, copper and lead (g g dry wt^{-1}) in the water, sediments and selected biota of Sandon Dock (Walmsley 1985). Figures in parenthesis are standard deviation of 3 or more replicate samples. Figures from other sites are included for comparison (Segar et al. 1971; Boyden 1975; Hamlin 1981). [* μg ml^{-1}, n = number of replicates]

	Depth	n	Zinc	Copper	Lead
Water column					
Soluble*	-	2	16.9	6.8	2.8
Total*	-	2	18.7	7.1	3.4
Surface sediment (2 cm)	-	3	759.8 (47.5)	143.9 (6.9	228.1 (15.1)
Mytilus edulis	1	5	144.3 (23.8)	6.0 (0.6)	6.3 (2.3)
	5	5	140.0 (1 2.0)	6.1 (0.9)	3.1 (0.6)
	7	5	1 50.0 (22.8)	7.1 (0.9)	3.9 (0.9)
	10	5	327.1 (36.8)	8.7 (1.7)	7.3 (1.3)
Ascidiella aspersa	1	1	572.8	36.9	19.4
	5	5	458.8 (53.2)	40.0 (2.9)	25.1 (3.3)
	7	5	512.5 (6.5)	40.0 (3.0)	26.7 (0.5)
	10	5	1097.2 (351.6)	67.0 (11.0)	46.4 (3.3)
Ciona intestinalis	5	3	342.1 (74.8)	31.8 (10.8)	144.0 (4.2)
Crassostrea gigas	1	5	6485.1 (1035.0)	494.0 (77.7)	4.7 (0.6)
Mytilus edulis	Irish Sea		9.6	9.1	
From remote sites	Poole Habor		7-11	7-19	
	Brocklebank Dock, Liverpool				

494

(Russell et al. 1983). Other filter-feeding organisms, such as tunicates, probably also contributed.

Zooplankton species diversity was not great, typically 8 or fewer species being found at any time, though the highest number recorded was 13 species. Copepods (particularly the harpactacoid *Microstella norvegica* and *Eurytemora affinis*) and cladocerans (*Evadne spinifera* and *Podon* sp.) were the dominant groups with larvae of mussels and tunicates (*Ascidiella aspersa, Botryllus schlosseri* and *Ciona intestinalis*) occasionally occurring in large numbers. The zooplankton community appeared to decline in abundance with time as the rope-cultured mussel biomass increased (Cunningham et al. 1984). However, the samples taken were not frequent enough before 1981 to be certain that this decline occurred.

The prevention of formation of an anoxic bottom layer by mixing allowed benthos to occur at all depths on the dock walls (see vertical transects in Figure 3). Mussels were not present in the dock in significant numbers in 1977. The most conspicuous and ecologically dominant feature was a dense cover of mussels from a massive initial colonizing cohort in 1978 plus subsequent settlement. The dock wall community became less diverse and mussels became less abundant at depths below 7 or 8 m and the bottom of the walls were almost bare stone with scattered barnacles and tunicates.

During the period of study various successional changes took place in the community associated with the dock walls. Certain species such as prawns (*Palaemon elegans*) and scale worms (*Lepidonotus* sp.) increased in abundance. There were large numbers of tunicates; *Ciona intestinalis* increased with time and *Ascidiella* seemed to become less common. Several species of benthic algae were attached to the dock walls or to mussels. They progressively increased their depth range with time down to 6 m and occurred much deeper in the water than in surrounding docks—presumably due to the greater water clarity. Red algae also became more common with time. In surrounding docks algae are restricted to a narrow <1 m band near the water surface. Associated with the mussels were a diverse array of organisms, including sea anemones, various worms, prawns and crabs. The predatory starfish, *Asterias* also colonized the dock (Russell et al. 1983). The fish farm operators removed these whenever possible. With time, a more patchy community would have been expected to develop, especially if numbers of starfish increased and ate clumps of mussels (Paine 1966). Overall diversity increased from 1980 to 1983.

The sediments on the bottom of the dock had little benthos, except for mussels and tunicates which may have fallen off the walls or other structures. The sediment was virtually anoxic; in parts it was covered by the white, thread-like colonies of the bacterium *Beggiatoa*. Large numbers of shrimps (*Crangon*) and gobies (*Pomatoschistus* spp.) were present on the dock bottom.

Several species of fish were recorded from the dock (Table 3), and some, like whiting and cod, were quite abundant. Cod up to 6 kg were caught in the dock and the entrance to the dock was a popular angling venue. Growth rates of six species were calculated by using otoliths for aging and, despite small sample sizes, were shown to be comparable with natural populations of the same species. Dietary composition of fish were also studied. Gut contents reflected the dock wall and dock bottom communities, and their fullness suggested it was a rich feeding ground. One cod was stomach-tagged with a ultrasonic pinger for several weeks and tracked. This fish did not leave the dock (see Cunningham et al. 1984 for details).

The fish farm was not the highly profitable venture envisaged by the owners, the Mersey Dock and Harbour Company. Although no controlled ecological experimentation was possible due to the need to run the dock as a commercial venture, valuable lessons in dock water quality management and restoration were learned from this project. A diverse community rapidly developed in a dock which had previously contained a rather impoverished one. Improvements in water quality were attributed to the combined action of the mixer and the dense populations of mussels (Russell et al. 1983). The mixer was only used routinely in the summer months, and was generally switched on only when stratification occurred and when bottom oxygen was low. During some hot periods (e.g., summer 1983), and when red tides were common, it was used continuously for several weeks. We presume that declines in phytoplankton observed when the mixer was turned on (Sandon Dock personnel pers. comm.; Liebeschutz 1985), were most probably due to the algae being made more readily available to the mussels, possibly combined with direct effects of the mixing.

The South Docks Liverpool

Background. The knowledge gained at Sandon Dock provided the basis for a new research project which started in 1988 at the nearby South Docks, administered by the Merseyside Development Corporation. It had the aim of exploring further the effects of water mixing and biological filtration on water quality and ecology of the dock system. The development at the South Docks is typical of many such schemes throughout the United Kingdom, incorporating housing, business and recreational facilities. In addition, the striking architecture and attractive waterside location has been exploited as a major tourist attraction.

The South Docks are a chain of interconnected docks of total area 27.2 ha, with double gates onto the River Mersey at each end. Redevelopment began with dredging in 1981, after ten years of total neglect during which almost complete siltation occurred. The docks were dredged to various depths from 3-9.5 m and water from the Mersey was reintroduced in 1985.

Various physicochemical and biological parameters were monitored from June 1988 at three sites in the South Docks (Table 5). The docks all had salinities within the range of 24-28 ppt and temperature varied from 6-24°C. At the start of the research program, water had been returned to some docks for up to 5 years so the initial recovery stages were not recorded. However, improvements in water quality and increases in species diversity were recorded over the sampling period.

Table 5. Summary of physico-chemical and biological measurements in the South Docks and Mersey Estuary, Liverpool (adjacent to the docks). All results based on monthly samples Jan-Dec 1989. Mersey estuary figures from National Rivers Authority data.

		Graving Dock	Albert Dock	Queens Dock	Mersey Estuary
Secchi disc (m) ( ) = on bottom	mean	3.7	3.9	1.8	
	range	1.3-7.5	1.0- (6.5)	0.5- (4)	max. 0.5[a]
Temp. (°C)	range	6.2-21.6	6.3-22.4	6.2-23.4	5-24
Salinity (ppt)	range	24-27	25-28	24-28	
Chlorophyll a (μg L^{-1})	mean	7.6	8.9	18.7	
	range	<1.0-24.9	<1.0-66.4	<1.0-79.1	
Orthophosphate (mg L^{-1})	mean	0.46	0.41	0.40	0.18
	range	0.16-0.81	0.18-0.64	0.14-0.63	<0.05-0.54
Total Inorganic Nitrogen (mg L^{-1})	mean	0.60	0.56	0.52	1.02
	range	0.01-1.18	0.08-1.25	0.07-1.28	0.54-2.0
Total coliforms (100 ml^{-1})	mean		52	128	16,158

[a] Mersey Barrage Company data. Mersey samples taken at high tide.

Water Quality Problems. The water in the South Docks enters directly from an estuary acknowledged to be one of the most polluted in Britain. Water is taken from the Mersey estuary at high tide approximately once every two weeks to replenish losses from evaporation, leakage and locking activities.

Several problems arise from this polluted water source. Large amounts (around 97 tons annually) of contaminated estuarine silt are introduced which will eventually require expensive dredging operations for removal. High levels of fecal indicator bacteria in the replenishment water have prompted concern that pathogenic microorganisms may be present which could infect watersports users. Fecal indicator counts in the docks themselves, however, are generally very low (Altwell 1989). The estuarine water is also a rich source of plant nutrients, stimulating nuisance phytoplankton blooms, which may in turn cause problems of deoxygenation of the water when in decay.

Very low oxygen levels in deeper waters were a common summer problem, often associated with strong thermal stratification. Oxygen depletion has on occasion led to the build up of hydrogen sulfide gas and the release of unpleasant odors when the water is disturbed. In deeper docks, hypolimnetic oxygen can fall to less than 20% saturation for periods of over two months (Figure 5); the effect was to prevent the development of a stable benthic community. Observations made by SCUBA diving in this period showed dead and dying polychaete worms on the mud surface, and flatfish in respiratory distress gathered in the shallower regions of the dock.

Water clarity tends to be poor in spring and summer due to phytoplankton blooms. Dense dinoflagellate populations may turn the water orange-brown. Severe problems with phytoplankton occur most often in the shallow Queens Dock, which is the primary watersports area. Secchi disc extinction depths of less than 1 m are common in this dock in summer months. Dense blooms of the dinoflagellate *Procentrum minimum*, a known toxic species (Okaichi and Imatomi 1979), raises the possibility of fish kills or skin irritation amongst watersports users. No problems have been encountered to date but their occurrence is a future possibility.

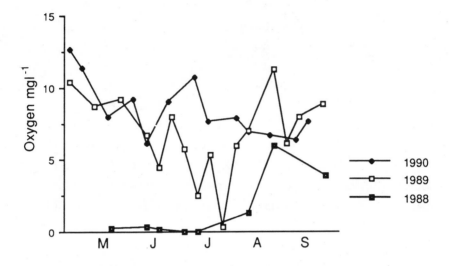

Figure 5. Oxygen concentrations in the Albert Dock, Liverpool showing improvement with time between 1988 and 1990.

Solutions and Water Quality Improvements. Remedial methods investigated have included the use of an artificial aerator and a biological filter. The Graving Dock (Figure 2b) was used as an experimental system and an air lift mixer (Martec Systems Rotamixer) was installed in June 1988. Use of the mixer was not required during winter months, with oxygen concentrations remaining naturally high at all depths, at this time.

At the start of the research project in May 1988, the numbers of mussels throughout the South Docks were very low. A biological filter was introduced to the Graving Dock in the form of a large *Mytilus edulis* population grown in pergolari (nylon) mesh. An initial pilot stock of 600 kg of mussels survived well and in February 1989 the population was increased to 1,450 kg. A natural mussel settlement occurred in September 1988. This settlement was particularly dense in the Albert Dock, thus preventing its intended use as a control dock with a low density of mussels. Consequently, only changes in water quality with time may be considered. The inputs of polluted estuarine water cannot be avoided. However, a policy of replacing water in smaller, more frequent amounts should reduce the penetration of such water into the complex, confining its immediate effects to the first dock of the chain.

After installation of the mixer in the graving dock the oxygen regime improved considerably (Allen et al. in press). Oxygen saturation at depths less than 6 m remained greater than 60% at all times with continuous mixing. The mixer was, however, somewhat underpowered for the size of the dock and low oxygen levels in deeper waters did occur during prolonged periods of hot weather (Allen et al. in press).

Table 6 shows mean summer water clarity over the sampling period and compares this with mussel filtration potential in each dock. A marked improvement in clarity is seen in the two docks with higher filtration ability after natural mussel settlement (the Albert) or additions (Graving Dock). These changes cannot be attributed to zooplankton grazing as total concentrations have declined since May 1988. This observation echoes preliminary observations made in Sandon and strengthening the suspicion that the mussels interact with the zooplankton. The incidence of low oxygen levels in the Albert Dock water has also lessened dramatically over this period (Figure 5). This is possibly due to the increased removal of phytoplankton by filter feeding and consequent reduction in material entering the microbial breakdown pathway.

Development of Benthic and Pelagic Communities. The dock biota has been studied by wall-scrapes, benthic grabs, observations by divers and various fish collecting methods. Table 7 and Table 8 show the changes in dock wall benthos between 1988 and 1990 for the Queens and Graving docks. A marked change in wall benthos in particular has been observed during the sampling period with increases in both biomass and species diversity. Initially

Table 6. Water clarity and filtration rates by mussel populations in the South Docks, Liverpool. Dock volume filtration time = time taken for one dock volume of water to pass through the mussel population in that dock.

	Summer Water Clarity (mean/range Secchi disc extinction depth - m)		Dock Volume Filtration Time (Days)		Mixer Water
	June-Aug. 1988	June-Aug. 1990	Summer 1988	Summer 1990	
Queens Dock	1.2	1.6	negligible	2.3	No
	0.9-1.4	0.9-2.5	filtration		
Albert Dock	1.2	3.9	negligible	0.5	No
	0.8-1.5	2.5-4.5	filtration		
Graving Dock	1.0	5.0	8.1	1.5	Yes
	0.6-1.5	4.0-7.0			

only a very few mussels and bryozoans plus tube dwelling worms (*Polydora*) and amphipods covered the walls until September 1988. Since then the succession has accelerated with addition of species in a similar manner to Sandon to give a mussel-dominated community with associated epifauna of sea squirts and anemones.

Large populations of zooplankton (e.g., *Polydora* larvae, calanoid copepods, *Podon* sp.) present in summer 1988 were dramatically reduced in the following two summers. The zooplankton is now characterized by harpacticoid copepods and barnacle nauplii in low densities.

Fourteen species of fish have been recorded in the South docks to date. These are typical fish of the Mersey estuary, with sprat, whiting, gobies and flatfish being the most common species.

In the Albert Dock in summer 1990, improved bottom oxygen concentrations were followed by the collection of three previously unobserved species in grab samples (*Arenicola marina*, *Tellina tenuis* and an unidentified terebellid polychaete). The development of an infaunal community with filter and deposit feeders will presumably reduce the amount of particulate organic material entering the bacterial pathway and reduce BOD. A bottom-living flora also developed in 1990 in the shallower docks, comprised of *Vaucheria litorea* and loose-lying *Cladophora.*

Lessons from the South Docks and Their Value. Artificial water mixing and biological filtration produced improvements in water clarity and oxygen levels in the Graving Dock. The application of these techniques may be of benefit in similar dock complexes where water quality is of high priority. These two methods are likely to be additive in their effect with increased water mixing reducing refiltration of water and allowing mussels to survive in previously hypoxic waters. Mussel filtration will in turn improve oxygen content of the water as the oxygen demand due to decay of phytoplankton blooms will be reduced. Similar improvements in oxygen saturation and clarity were seen in the Albert dock after a large natural mussel settlement.

The changes observed in benthic communities are attributed to a combination of gradual colonization and succession by species introduced with estuarine water and to the improvements in water quality facilitating the survival of more species.

These observations confirm the work at Sandon Dock. Further research and monitoring are needed, however, over an extended time period, to ensure that the high levels of filter feeders necessary to maintain water quality are sustainable.

The building developments at the South Docks are now well underway. Warehouses have been restored and now house shops, luxury flats, museums, an art gallery and office space. The area is a major tourist attraction having had around 3.5 million visitors in 1989 (more than the Tower of London). The waterspace is used for

Table 7. Changes in major components of dock wall benthos in South Docks: Queens Dock wall communities.[a]

	June 1988				May 1990			
Depth	Surface	1 m	2 m	3 m	Surface	1 m	2 m	3 m
Species								
Bryozoa	A	A	A	A	C	+	+	-
Hydrozoa	-	-	-	-	+	+	+	+
Mytilus edulis	-	-	-	+	C	A	C	P
Botrylus schlosseri	-	-	-	-	+	+	+	-
Molgula manhattensis	-	-	-	-	-	+	-	
Corophium insidiosum	+	+	+	+	+	+	+	+
Gammarus salinus	+	+	+	+	+	+	+	+
Jassa marmorata	-	-	-	-	+	+	+	-
Microdeutopus gryllotalpa	+	+	+	+	+	+	+	+
Balanus improvisus	-	-	-	-	-	+	+	+
Nereis sp.	+	-	-	-	+	+	+	+
Polydora ciliata	+	+	+	+	+	+	+	+
Chlorophyceae	+	-	-	-	C	C	-	-
Fucophyceae	-	-	-	-	C	-	-	-

[a] Key:
Record from: Point quadrants   Wall scrapes (25×25 cm)
               Present-P   Common-C   Abundant-A   Present +; Absent -
               0-5% cover   5-50% cover   50-100% cover

Table 8. Changes in major components of dock wall benthos in South Docks. Graving Dock wall communities.[a]

Depth / Species	June 1988						June 1990					
	Sur-face	1 m	2 m	3 m	4 m	7 m	Sur-face	1 m	2 m	3 m	4 m	7 m
Species Bryozoa	A	A	A	A	C	-	A	A	A	A	P	+
Hydrozoa	-	-	-	-	-	+	+	+	+	-	+	+
Metridium senile	-	-	-	-	-	-	-	-	-	-	-	+
Mytilus edulis	-	-	-	-	-	-	C	C	C	C	P	P
Ascidiella aspersa	-	-	-	-	-	-	-	-	-	P	-	-
Molgula manhattensis	-	-	-	-	-	-	+	+	-	+	P	+
Corophium insidiosum	-	-	-	-	-	-	+	+	+	+	+	+
Gammarus salinus	-	-	-	-	-	-	+	-	-	+	-	-
Jassa marmorata	-	-	-	-	-	-	-	+	+	-	+	+
Microdeutopus gryllotalpa	-	-	-	-	-	-	+	+	+	+	+	+
Balanus improvisus	-	-	-	-	-	-	-	P	P	P	+	-
Nereis sp.	-	-	-	-	-	-	-	-	-	-	-	-
Polydora ciliata	-	+	+	+	+	+	+	+	+	+	+	+
Bangiophyceae	-	-	-	-	-	-	+	+	+	-	+	-
Chlorophyceae	P	-	-	-	-	-	C	C	P	+	+	-
Fucophyceae	-	-	-	-	-	-	P	P	P	-	-	-

[a] Key:

Record from:	Point quadrants			Wall scrapes (25×25 cm)
	Present-P	Common-C	Abundant-A	Present +; Absent -
	0-5% cover	5-50% cover	50-100% cover	

water contact sports such as windsurfing, diving and canoeing and the development of a marina has proved very successful.

Case Study 2—Preston Dock

Hydrography and Ecology

The Albert Edward Dock at Preston has been closed to commercial shipping since 1981. Opened almost a century before, it became uneconomic as a result of its inability to accommodate modern deep-draft cargo vessels and of the decline in trade reflecting the run-down nature of its industrial hinterland. A redevelopment scheme was initiated in 1982 with plans for housing, a shopping mall, recreational use (including fishing, contact water sports and a marina), plus commercial and industrial facilities. These various uses have implications for the required standards of water quality in the dock. The water body must look and smell pleasant, and there should be no risk to human health for people using the water for recreation.

The dock in this case study is situated on the upper Ribble estuary, about 30 km from the sea (Figure 2a, c). The main dock has an area of 16.2 ha and a maximum depth of 8 m. Entrance to this dock is through a small (1.9 ha) holding dock of 4 m maximum depth. The dock is oligohaline with surface salinities varying between 1 and 7 ppt and bottom salinities occasionally reaching 14 ppt. Water lost through seepage and evaporation is replenished from the estuary, generally every two weeks, by opening lock gates on rising spring tides. An average of 33% (maximum 110%) of dock water is exchanged monthly, hence water quality of the dock is largely influenced by the estuary. The water of the upper Ribble is moderately polluted with elevated BOD, nutrients and fecal indicator organisms, and a very high suspended solid load (Conlan 1989). For the majority of the time, however, the dock is isolated from the estuary by lock gates. The system therefore tends to function as a low salinity lagoon (see Dorey et al. 1973; Crawford et al. 1979). The dock is eutrophic, receiving intermittent but high levels of nutrients from the estuary, whose natural nutrients levels are significantly raised by agricultural run-off and sewage discharges. Thus, the stable, shallow, eutrophic conditions provide an ideal environment for the proliferation of phytoplankton (Figure 6).

Diversity of flora and fauna tends to be low, as in most upper estuarine systems, but those species capable of adapting often become abundant. Thus, at Preston, cyanobacteria and the estuarine hydroid *Cordylophora lacustris* dominate the plankton and dock wall benthos, respectively. These form the basis of the biotic community in the dock (Figure 7). Cyanobacteria are poorly grazed by the copepod zooplankton, but are available to rotifers which thrive in the dock.

Further pressure is exerted on the copepod community by the activities of planktivorous fish, comprising both juveniles from marine-estuarine species, such as herring and sprat, and freshwater species (Table 9). Adult fish tend to browse on the fauna associated with *C. lacustris*, mainly chironomids, gammarids and larger crustaceans including mysids and *Crangon vulgaris*. The larger carnivorous fish, sea trout and smelt, are at the top of the dock food web, feeding mainly on clupeids and larger nektonic crustacea. The dock supports virtually no sediment dwelling benthos, probably due to intermittent sediment anoxia and unsuitable substrates for colonization.

Water Quality Problems

The relatively frequent additions of estuarine water generate a number of water quality problems of which the most obvious is unsightly, surface debris largely of estuarine origin.

The influx of higher salinity water (>7 ppt) induces stratification which may persist for over 7 months of the year and over the whole dock for at least two weeks in summer (Conlan 1989). Any temper-

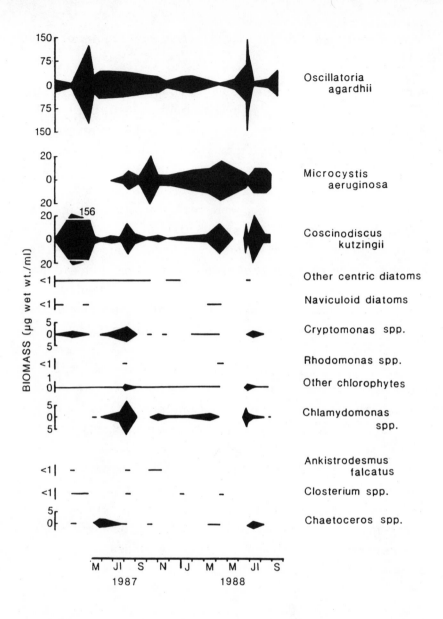

Figure 6. Total biomass of phytoplankton in Preston Dock, east end (Conlan 1989).

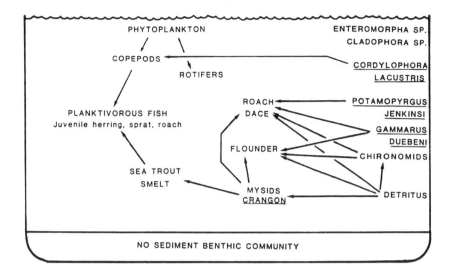

Figure 7. Schematic diagram of Preston Dock ecosystem (Conlan 1989).

ature stratification tends to reinforce salinity differences rather than being the main cause of any pycnocline. Stratification has resulted in long periods of oxygen depletion in bottom waters. In summer 1987, levels of dissolved oxygen of less than 20% saturation occurred for a period of 5 months. These conditions will promote mobilization of nutrients and heavy metals from the sediment. The occurrence of bad smells associated with hydrogen sulfide gas production has not been recorded, but is a potential problem. Clearly the development of a benthic community will be inhibited by low oxygen levels throughout much of the dock.

Very dense populations of the cyanobacteria *Oscillatoria agardhii* and, to a lesser extent, *Microcystis aeruginosa*, develop. Chlorophyll *a* concentrations reach a maximum of 731 µg L^{-1}, which is higher by an order of magnitude than other docks surveyed by Hendry et al. (1988a). The dock always has nutrient levels in excess of those likely to limit algal production. These levels, combined with the stable stratified conditions, often provide an optimum environ-

Table 9. Fish species caught at Preston Docks between May 1987-April 1988 including number of times caught and minimum/maximum numbers, lengths and condition factors for each species (Conlan et al. 1988; Conlan 1989).

	Percent Surveys Caught	Min/ Max Number	Min/Max Length (mm)	Min/ Max Age	Condition Factor (K)
Freshwater Species					
Roach	100	1-51	70-221	1-9	1.0-2.0
Dace	50	2-5	53-193	0-3	1.1-1.3
Chub	17	1-2	233-427	4-9	1.3-1.9
Silver Bream	8	1	108	2	1.7
Brown Trout	17	3	204-310	2-2.sm	1.2-1.4
3-spined stickleback	8	1	46-47	-	1.0
Marine-Estuarine Species					
Herring	100	36-607	40-263	0-5	0.5-1.2
Sprat	50	1-226	46-108	0-1	0.7-1.4
Flounder	100	3-112	78-273	1-4	0.6-1.9
Sand Goby	25	1-3	55-72	0	0.6-1.1
Migratory Species					
Salmon	8	1	567	2.1	1.3
Sea Trout	92	1-22	220-497	1.sm-2.sm	0.9-1.5
Smelt	83	1-10	126-234	1-3	0.7-1.4
Eel	42	1-2	310-527	6-10	0.1-0.2

ment for salinity-tolerant freshwater phytoplankton species, such as *O. agardhii* and *M. aeruginosa*, which are common in poorly mixed oligohaline systems, like Preston. *O. agardhii* is dominant and present in bloom proportions for the majority of the year (Figure 6). During severe blooms, the water becomes very unattractive with the algae turning the water bright green and forming floating mats. In certain cases, this has led to complaints to Preston Borough Council and requests for remedial action (Hoare pers. comm.). Furthermore, the cyanobacteria includes species known to have toxic effects. Fatalities of fish, livestock and birds plus non-fatal illness in humans such as gastroenteritis and skin complaints have been reported (Carmichael 1981; Codd and Bell 1984; Gorha and Carmichael 1988).

Preston Docks fails to comply with the European Community standards for bathing water due to microbial contamination from estuarine inputs (Conlan 1989). Sewage contamination of the dock is not as great as in the estuary, although the dock fails to meet European Community bathing water standards for total coliforms and *Salmonella* in over 20% and 40% of samples, respectively (Conlan 1989). It must be remembered, however, that these are indicator organisms and other pathogens, including viruses and protozoa, may be present in the sewage. As water sports are part of the proposed developments, there is a public health concern about the presence of pathogenic microorganisms.

Heavy metals are not present at concentrations likely to cause great concern in the docks. Siltation from the estuary and contamination from shipping have probably led to the accumulation of metals in the dock sediments, but transfer through the food chain does not appear to be unacceptably high. Levels of heavy metals in flounder, although higher than in uncontaminated estuaries, fall within guideline and/or statutory limits for consumption (Conlan 1989). No action is therefore considered necessary at this time to limit catches and consumption of flounder in the dock or estuary.

Since the other fish species in the dock tend not to be eaten in large quantities, they pose no immediate risk to human health.

Possible Solutions

The improvement of the visual appeal of the water body depends primarily on the removal of surface debris and prevention of water discoloration or scum formation by algal blooms. Surface debris could be controlled with relative ease by using booms across the surface to limit dock entry or by using "gulper boats," as in several docks and harbors (Roberts 1989). In Preston Docks, a boom across the holding dock prevents entry of surface debris into the main dock marina complex.

The physical removal of phytoplankton by draining the dock down, and subsequent refilling on the high tide from the estuary, is probably the simplest and most cost-effective method of algal control. Flushing experiments showed that up to an order of magnitude reduction in phytoplankton could be achieved in the short-term during periods of excessive algal production. At low phytoplankton densities, the population tended to recover quickly to original levels (within a week in the flushing experiment—Conlan 1989). Flushing is, therefore, successful in lowering algal biomass and, if practised on a regular basis, could exercise a control on phytoplankton. Unfortunately, water exchange by introducing high salinity waters conflicts with the need to prevent stratification. Flushing also brings in large quantities of nutrients, debris, silt and pathogenic microorganisms.

Microbial contamination of the dock, resulting from the ingress of polluted estuarine waters, is perhaps the most intractable water quality problem. Total coliforms and *Salmonella* regularly exceed the European Community bathing water directive. The most effective method of reducing microbial contamination in the docks would be isolation from the Ribble estuary, which has been shown in the flushing study to decrease bacterial indicator concentrations quickly.

The only long-term solution is to stop discharge of untreated sewage into the Ribble estuary. At present, the uses to which the dock can be put must therefore be appraised and activities which may increase the chance of human illness curtailed or limited. This approach has recently been implemented at Bristol docks, where a code of health and safety measures has been introduced (Enticott and Grisdale 1985; Phillip et al. 1989). Measures include the banning of certain high risk activities such as swimming, snorkeling and eskimo rolls in canoeing and various rodent control exercises to minimize contamination with leptospiral viruses (cause of Weil's disease).

Installation of artificial mixing devices, as was done at Sandon Dock, would clearly prevent anoxia with its associated problems at the bottom of the dock. Aerobic conditions would encourage a more diverse benthic community and reduce stress on bottom-living fish. The effects of mixing on phytoplankton assemblages are unclear. However, the ability of blue-green algae to regulate movement and hence form floating mats may be impaired in a well-mixed system (Reynolds 1987). Destratification techniques in dock complexes are in their infancy, and no work has been done in low salinity systems such as at Preston. Based on calculations from Radway et al. (1988), it would take more than 3 weeks to turn over the water in the main dock at Preston using one helixor airlift pump, which is capable of entraining water at a rate of 35,000 m^3 day^{-1}. If 5 pumps were installed, then complete turnover of dock water could be achieved in approximately 4-5 days with continuous operation. The necessity for 5 pumps is questionable, as salinity stratification would only pose a problem immediately after dock gate opening and is generally limited to the deeper east end of the dock—probably one or at most two mixers would be sufficient.

When the marina in the docks is fully operational it will necessitate increased utilization of the dock gates and, hence, greater exchange with the estuary, particularly during the summer. At this time of the year, high salinity incursions from the estuary are more prevalent and will cause stratification. The marina will also require the installation of floating pontoons and a wave barrier to reduce wave action in the dock. These may have severe consequences for mixing, as the stratified water column is generally broken down by increased wind-induced wave action. Stratification may well be enhanced under these circumstances, especially if development around the dock leads to further shielding of the water body from wind action. All of these factors need to be taken into account when deciding on the specification of any aeration system.

Value and Uses

At present, the main scientific value of the Preston Dock complex is that it is an artificial saline lagoon—an endangered habitat in Britain. It also has some interesting birds and good fish populations, although the benthos is rather impoverished at present.

The fish assemblage is without doubt its main asset (Table 9). It is an unusual mixture of marine-estuarine water, freshwater and migratory fish species. The size, growth rates and condition of the fish seem to be typical of the estuarine environment (Conlan et al. 1988). From the species mix, abundance and size ranges present, recreational angling should be encouraged. This could be achieved with minimal capital outlay by the dock authorities with costs recouped from rod fee charges or, alternatively, it can be let out to an angling club while maintaining overall control. There is, however, potential for conflict between anglers and other water sports enthusiasts which would need addressing. The most viable alternative is by zoning activities.

Conclusions

There are no easy solutions for improving water quality and ecosystem diversity at Preston Docks. Flushing, which will temporarily decrease phytoplankton biomass, unfortunately introduces debris, silt, nutrients and pathogens and intensifies problems associated with stratification. In the long term, improvement in the catchment basin and estuary will improve the dock. Mixers would help, but the investment would not lead to the improvements brought about at Liverpool. The problems of Preston Docks will remain intractable unless some radical treatments are developed (see further work).

General Solutions for Improvement of Disused Docks

Strategic Approaches

As most water quality problems in docks result from poor water supply, a strategic approach to improve rivers and estuaries must be the long term objective. Large-scale investment in clean technology to reduce the effects of domestic, industrial and agricultural waste will be required along Britain's major rivers. Regular monitoring of water courses, along with effective legislation and real penalties for offenders, are needed to ensure that dumping waste into rivers is not the cheap and easy option. A single authority is required to deal with these aspects of pollution control. In the past in Britain, the downstream effect of river pollution reduced the incentive for some local authorities to tackle offenders. It is hoped that the recent creation of the National Rivers Authority, a nationwide regulatory body with extensive powers and responsibilities, will bring about control of discharges.

One long-term, broad-based attempt at improving the aquatic environment in Britain is the Mersey Basin Campaign. The aims of this scheme are to promote a healthy environment coupled with a thriving economy throughout the Mersey catchment, which includes Greater Manchester and Merseyside in Cheshire and Lancashire. Industrial dereliction and river pollution are major problems in this area, but since the 1950s and 1960s, pollution controls and deindustrialization have reduced inputs and the Mersey Estuary ecosystem is slowly recovering. Occasionally, disasters such as the

Mersey bird kill (Head et al. 1980) or the August 1989 oil spill temporarily halt the overall recovery.

The 25-year campaign to restore the estuary is expected to cost around £4 billion and draws on a mixture of European Community aid, water authority investment, central and local government work, voluntary input and private investment. For inland water quality, the aim is to achieve at least a class 2 (fair water) standard throughout all of the 1,700 km of watercourses by the year 2010. At present, over half of this river length is classified as poor or bad, class 3 or 4 on the Trent Biotic Index (Mersey basin campaign information). The improvement will be achieved by investment in new sewers and sewage treatment works, advice on and regulation of industrial discharges and grants towards improving waste storage and treatment facilities for agriculture.

Joint approaches, such as the Mersey Basin Campaign, are seen by some as the strategic way forward in catchment basin management and as essential for the future health of estuarine and coastal waters.

Tactical Solutions

Isolation of the dock from its polluted water source is an essential first step, and it usually leads to rapid improvements in water quality. Simple settling of sediments from turbid estuarine water allows major improvements in water clarity. Conservative pollutants and plant nutrients bound to this sediment may be permanently removed from the water column, especially if well-oxygenated conditions can be maintained (Lu and Chen 1977; Marsden 1989). A reduction in numbers of contaminating bacteria is soon seen (Conlan 1989), although it is possible that some fecal organisms may persist in the sediments for many months (Sawyer et al. 1977). In addition to the work at Liverpool (Russell et al. 1983; Cunningham et al. 1984; Allen et al. in press) summarized in the previous section, Radway et al. (1988) give an account of the recovery of the freshwater docks at Salford following isolation.

Complete isolation is not always possible: evaporation, leakage and boat movements may cause water losses which must be replaced, although engineering solutions such as blockage of leaks and provision of double dock gates can help to reduce exchange. It should be borne in mind, however, that influx water is the principal source of colonizing flora and fauna, and hence is desirable for ecosystem diversification.

After isolation of the water body, monitoring of water quality and early identification of any problems are both important. Natural improvements may be seen with time as the ecosystem develops, however, manipulation may be required to effect or speed these improvements.

The desirability of high dissolved oxygen concentrations has been discussed previously in this paper. In freshwater, artificial mixing of waterbodies has been used effectively to destratify the water column, improve oxygen concentrations and increase populations of zoobenthos (Fast 1973; Bailey-Watts et al. 1987; Radway et al. 1988). Comparable effects have already been reported in saline docks (Russell et al. 1983; Allen et al. in press).

Hydraulic or pneumatic mixing devices are available (see reviews in Tolland 1977; Pastorok et al. 1981). Hydraulic methods involve the direct transfer of water by pumps and water jets or paddles. Pneumatic methods involve the use of air bubbles emitted in the hypolimnion which, as they rise, entrain water and transfer it to the surface where aeration from the atmosphere can occur. Pneumatic devices tested in lakes and reservoirs have included bubble screens (perforated pipes), diffuser domes or bubble guns. A more refined design is the Polcon "Helixor" which confines the first part of the rise in a tube with a moulded helical insert; this gives the bubble plume a swirling component, which increases oxygen transfer and mixing ability. Helical-type mixers have been used successfully in several dock complexes (Russell et al. 1983; Radway et al. 1988; Allen et al. in press). Pastorok et al. (1981) suggest that, in general, pneumatic devices are cheaper to run and easier to operate. In restored docks, they have the additional advantage of having no visible parts above water and the moving parts requiring mainte- nance are all on land.

Although there is no doubt that artificial aerators can improve the oxygen regime, the direct effects of such artificial destratifica- tion on phytoplankton communities are very unclear. Some studies have reported a decrease in phytoplankton abundance with mixing (e.g., Reynolds et al. 1984; Bailey-Watts et al. 1987) while others report increases (e.g., Knoppert et al. 1970; Fast 1973). Intermit- tent mixing may be the best regime for reducing phytoplankton, by preventing a stable state phytoplankton community from develop- ing (Reynolds et al. 1984). However, this regime reduces the ener- gy efficiency of the mixer and increases monitoring costs. Therefore, the use of continuous low intensity mixing was suggest- ed as a better solution by these authors. Mixing would be expected to favor early season diatoms over later occurring dinoflagellates (high salinity) and blue-greens (low salinity and brackish) but this suggestion is untested in marine shallow water dock complexes.

A recent attempt at harnessing wind power to drive hydraulic mixers has been tried in Liverpool docks (Lake Aid Systems, North Dakota). Unfortunately, these mixers are least efficient when need- ed most (in low wind conditions) and have large moving parts on floating rafts which may be aesthetically and recreationally undesir- able.

Potential beneficial effects on water quality (phytoplankton re- duction, oxygen and clarity improvements) have been reported for many different filter feeding species, e.g., serpulid polychaetes (Davies et al. 1989), clams (Officer et al. 1982), freshwater mussels (Reeders et al. 1989), marine mussels (Russell et al. 1983; Allen et al. in press) and zooplankton (Dorazio et al. 1987).

The use of biological filtration as a management tool can be approached in two ways: the encouragement of natural population increases or the actual addition or culture of large numbers of animals from an outside source. For both these methods, pilot studies in individual docks are required to ensure that water quality is good enough to allow survival of the animals. For natural filtering communities to be exploited, there must be a sufficient larval supply and suitable environmental conditions in the dock. An increase in surface area for settlement may allow increases in benthic filter feeders. The use of shell waste and tire reefs over the soft mud dock bottom for this purpose is being assessed at present in the South Docks, Liverpool.

Increases in zooplankton populations by the removal of planktivorous fish (Riemann et al. 1988; Langerland 1990) or by stocking with piscivorous salmonids (Dorazio et al. 1987) have proved successful in small-scale studies.

The addition of filter feeders to docks is time-consuming and costly, although this has been attempted on a large scale in Liverpool (see case study). The employment of collectors, such as sheets of netting, in areas of high filter feeder settlement and subsequent transfer of these collectors to a dock is one possibility that needs some exploration.

Chemical additions to reduce phytoplankton blooms, although once a common practice in freshwater systems (Landner 1976), have been largely rejected in renovated dock complexes. Chemical treatments may need to be repeated and are, hence, expensive. They may also be harmful to other organisms. Copper sulfate is still the most widely used algicide (Luedritz et al. 1989). Due to its toxicity, the addition of copper sulfate is clearly not compatible with the development of docks for nature conservation.

Several methods have been successfully employed to limit nutrient supplies with the majority either re-routing inputs from anthropogenic sources (Edmondson 1970; Laurent 1971) or stripping nutrients from the water by addition of chemicals, with a subsequent flocculation and sedimentation (Wall 1971; Soltero et al. 1981). These methods tend to be expensive, may need repeating (in the case of chemical additions) and do not always have the desired effect of phytoplankton reductions (e.g., Foy and Fitzsimons 1987). Also regular inputs of nutrient-rich estuary water and consequent resupply of nutrients would negate the use of the above methods.

Costs and Applicability

It is difficult to accurately determine the cost of various remedial treatments, as requirements are very variable; however, some examples from recent work can be given.

At Salford docks, an artificial aeration system consisting of 15 Helixors (Polcon Ltd) was installed in 1989 at a total cost of £260,000. The total electricity requirements per year are 288,000 KW for this system (costing approximately £17,000); however, a continuous mixing policy is operated which is certainly not required at all locations. Generally, power requirements would be lower. The use of mixing devices controlled by microprocessors linked to environmental sensors such as temperature or, with greater difficulty, oxygen probes would be a cost effective way of regulating mixing. The mixer could be switched on when pre-determined environmental conditions became prevalent, such as a marked temperature difference or depleted bottom oxygen.

Costs attached to the use of biological filtration depends very much on the dock ecology and local resources. Netting sheets suitable for the collection, transferral and ongrowth of *Mytilus* spat are available at a cost of 65 pence m^{-2} (Kerrypack Ltd, Bristol, United Kingdom), the amounts required depending on density of mussel settlement and total filtration requirements.

If encouraging natural filter feeding populations is a viable alternative, waste materials suitable as settlement surfaces (car tires, building rubble or shell waste) may be available locally. The use of such materials incurs only labor and transport costs, but their suitability and non-toxicity should be investigated before large-scale introductions are undertaken.

All the management methods described above are applicable to water bodies with a long hydrodynamic residence time. Monitoring of simple parameters such as oxygen, temperature, clarity and chlorophyll *a* over an annual period is required to identify water quality problems. Corrective methods then chosen depend on factors such as severity of problems, water quality requirements and physical characteristics, such as water depth and salinity.

Salinity particularly affects the options available for biological filtration. Greatest difficulty is presented by oligohaline systems as far fewer species are adapted to this environment than either freshwater or marine systems, the lowest diversity being seen in the range 5-7 ppt. Poor species diversity in eutrophic oligohaline waters reduces competition and grazing pressure on algal species and very dense nuisance blooms of single species may develop.

Value of Restored Dock Complexes

Recreation

Restored disused docks clearly have considerable recreational value. They allow urban people to participate in activities which would often involve travelling long distances to more rural coastal locations. Additionally, they enhance informal activities such as walking, jogging, or cycling by providing a pleasant environment and appreciation of wildlife such as birds, fish and other aquatic organisms.

Docks are also suitable for more specialized sports such as fishing, providing steps are taken to prevent negative interactions with wildlife and other users. Fears for public safety and crowded locations have compelled authorities such as the Merseyside Development Corporation, in Liverpool and the London Docklands Development Corporation in London to ban sport-fishing in some places. Large numbers of fish can occur in some docks, including highly valued species such as migratory salmonids. There is also a sufficient variety of them to maintain the anglers' interest, particularly at Barrow and Preston where estuarine and freshwater species coexist. Stocking for recreational fishing is being explored in the freshwater docks at Salford, Manchester, where cyprinids have been introduced. Stocking is less feasible in marine systems, although salmonids could be released in seawater for sports fisheries.

One problem which is difficult to overcome is the presence of pollutants. For example, heavy metals and organohalines in fish flesh could preclude human consumption. The nationwide appraisal suggested that heavy metals are less of a problem than commonly perceived. Contamination can be easily analyzed and decisions made on how to regulate recreational fishing or warn the public of health risks. In many instances, anglers catch fish for sport and release them alive into the water upon capture. As Hendry et al. (1988b) state, the presence of fish and fishermen boosts public and investors confidence in the health of the water body—a very important indirect benefit to any redevelopment in addition to direct enhancement of recreation.

Watersports enthusiasts have long used industrialized rivers, canals and docks. The Oxford to Cambridge boat race on the Thames was run for many years on what was, in effect, London's major sewer—until recent clean-up campaigns restored the upper estuary. In sports such as rowing or dinghy-sailing there is little direct contact with the water and health risks are low. The current boom in marina development, reflecting the increase in coastal cruising in small and medium-sized yachts and motor cruisers, is being partially accommodated in docks no longer used commercially (e.g., Liverpool, Fleetwood, Maryport, Penarth). This development reduces pressure on undeveloped, ecologically sensitive sites in virgin inlets and estuaries.

Activities such as canoeing and windsurfing can involve total immersion in the water and, therefore, the participants are at greater risk from waterborne diseases and toxic or irritating algal blooms. The greatest risks of catching waterborne disease are attached to swimming and sports diving. They are also most dependent on clear, pleasant water. Most dock and redevelopment authorities actively discourage swimming, primarily because of the risks of drowning when swimming in water that is deep and sided by vertical and high quays with limited access points. Disease risks are usually a secondary, but important, consideration.

SCUBA diving is allowed in the South Docks at Liverpool, at the risk of the participants. Sub-aqua clubs using the docks are attracted by the clean water, the relative safety for training of an enclosed water body, and by the marine biota. Equivalent marine communities are more than a two hours drive away. Water quality monitoring of bacterial indicators and selected viruses in both Sandon and the South Docks has suggested that there was negligible risk, and the water was within European Community guidelines—unlike many resort beaches in the north west of England. The magnitude of the risk depends on the degree of exchange with the adjacent polluted Mersey. Since this is minimized in the Albert Dock and biological filtering is high, there should be few free pathogens in the water.

Aquaculture

Sandon Dock, Liverpool, was intended to be a fishfarm. Unfortunately, it was not an economically-viable proposition and operations ceased after the trial period. Water quality and related public health considerations prevent this option in docks on many industrialized estuaries. Marketing problems resulting from public perception of water quality, whatever its status, also reduce the potential sales of products from docks. Work at Sandon Dock showed, however, that shellfish could be produced within United Kingdom public health guidelines for heavy metals. After standard ultraviolet treatment, there were no bacterial problems. These shellfish were sold and were well received in various local restaurants. No studies were made, however, of viruses or organic pollutants which may also present a health risk (but the senior author reports no adverse affects of consumption of mussels, oysters and cod from the dock).

With suitable management, docks can be renovated and managed sufficiently to enable cultivated species such as fish and shellfish to grow well. Whether these should be sold to the general public depends on levels of contaminants and pathogens which can be tested using individuals grown in pilot trials. This particular use of disused docks is unlikely to be pursued further with the current status of most United Kingdom estuaries and associated ports.

Conservation

Port areas and docks in use for commercial shipping have very little wildlife and conservation value as man's activities have considerable environmental impact. Their development also tends to occur on valuable coastal wetlands by reclamation of saltmarshes, and mudflats plus infilling of lagoons. This "hardening" of the coast in port areas markedly changes the range of habitats and their value as nursery grounds for fish and feeding grounds for birds. Once port operations cease, so does disturbance due to repeated impoundment of silty water and stirring of sediments during mooring and spillage during stevedoring. Provided there is still some interchange with the outside estuary or sea, natural colonization can take place by advection of algal propagules and animal larvae, plus the migration of juvenile and adult mobile animals. If the dock has been left open for a long period (e.g., South Docks in Liverpool), recolonization can only occur after dredging and refilling with water. As this process starts from scratch with little or no existing biota, the succession can take some time. In docks such as Sandon where change of use occurred without a phase of neglect, then succession is quicker as many species will already be present.

Water quality management methods, particularly installation of mixers, seem to accelerate succession. More importantly, colonization can occur at all depths. Therefore, a diverse community will develop in a period of between 2 and 5 years. In high salinity docks, a prolific community of algae, sessile filter feeders and associated predators will develop on the dock walls. Colonization of the more inhospitable mud seems to take much longer, but eventually colonization may take place if periods of bottom anoxia are minimized. Docks can also support varied planktonic communities and diverse fish assemblages. These aquatic communities, in turn, prompt the return of water birds, sometimes in considerable numbers.

Certain special features of the docks can promote their wildlife interest. Rare species can be found there. Warm-water effluents can help maintain exotic immigrants introduced by shipping (see Naylor 1965 for review). At Barrow, large populations of waterfowl were attracted to a power station effluent and were much reduced when this ceased operations (Hendry et al. 1988a).

The greatest conservation significance of disused docks is that they resemble saline lagoons. This habitat is under considerable threat in the United Kingdom and other developed countries. In the United Kingdom since the early 1970s, 36 saline lagoons have disappeared, the majority due to infilling as part of coastal developments (La Foley pers. comm.). Docks are by no means "functionally equivalent" to natural lagoons; they have special features such as dockwall communities and they lack typical lagoon fringing habitat.

In essence, restored docks can provide extensive, healthy aquatic ecosytems in the centre of cities. Normally these are very rare. Activities in docks can also reduce human pressure on more pristine coastal environments adjacent to cities.

Education and Research

Disused docks have considerable potential for education at various levels from primary school to university and, perhaps most importantly, for environmental education of the general public. Much of the educational activity is directed towards history (e.g., maritime museums in Exeter, London and Liverpool). However, even at primary school level, awareness of the environment can be heightened by development of teacher packs, good sign-posting and permanent all-weather displays, and exhibitions. In 1989, a small-scale exhibition (afternoons, 5 days a week, for 3 weeks) in a shop unit in the Albert Dock complex attracted over 30,000 people. This simple exhibition involved two small aquaria, a microscope, some touch tanks and a short video presentation. It was expanded to a permanent display for four months in 1990 in liaison with National Museums and Galleries on Merseyside, in part of the Liverpool Maritime Museum. Five larger aquaria, one touch tank, a video microscope, binocular microscopes, drawing areas for children and a video presentation were set up in a small room and manned by full-time demonstrators. What really surprised most visitors was that all the material was collected from the adjacent dock and Mersey Estuary.

Various schemes to build a national aquarium or a major regional aquarium have been explored in Liverpool, but have yet to reach fruition. The aquarium at Boston, Massachusetts, seems an appropriate model for many cities: in addition to providing exhibits about the ecology and environmental problems of the harbor, it provides a base for monitoring and research.

The docks generally could be used for project-based fieldwork at school level and the source of material for natural history, biology and environmental studies. The docks have also been used successfully for practical classes for undergraduates at Liverpool University who investigated phytobenthos and plankton ecology, as well as physico-chemical properties of the water. They provided easy access to a system which could be regularly sampled over a spring bloom. Most of the initial research on disused docks was done by final year undergraduate projects from the Universities of Liverpool and Manchester. They have also been used for short-term research projects for taught Master's degrees. Their use in adult education was made apparent by the response of divers taking evening classes in Marine Biology at Manchester University who enjoyed diving in Sandon Docks and were very surprised by the diversity of marine life.

The polluted nature of many disused docks, particularly high sediment levels of heavy metals, makes them very appropriate for pathway studies on heavy metals and for studies of metal metabolism. Manipulation of these systems using air pumps also provides considerable potential for exploring the effects of redox potential on heavy metal fluxes in marine systems. Fundamental research can also be conducted on the factors affecting phytoplankton and benthic populations and communities in easily manipulated dock basins. Hence, the potential for using disused docks as mesocosms must be emphasized to research funding agencies. Both dock management and research funding agencies need to be made aware of the usefulness of docks for research—especially as an understanding of their ecology may in turn help in the future development of cost-effective management strategies.

Tourism and Redevelopment

Consideration of commercial property values is outside the scope of this paper. However, values of properties in dockside developments in Liverpool, Manchester and London have all risen sharply in recent years. Similarly, demand for location of light industry, commerce and retailing are high in these areas. Unpleasant smells, murky water and unsightly algal blooms would detract from their value—hence, the importance of good quality water.

The attractive waterside location and the appeal of various water-based activities make living and working in these environments pleasant. The architectural merit of restored buildings, and the deliberate development of tourist attractions including shops and restaurants, galleries and museums all attract large numbers of people. The dock ecosystem is therefore an essential ingredient, directly and indirectly, to the commercial success of dockland redevelopment.

Future Work

The future research on disused dock basins should involve two interrelated approaches. Practical engineering research work is required to develop methods for water quality management and habitat diversification. These should be combined with ecological studies to understand and model enclosed eutrophic water bodies to ensure that management is both rational and cost-effective.

There is clearly a need to develop simple mathematical models of enclosed saline ecosystems to aid decision-making. Considerable progress has been made in the Netherlands by groups working on saline lakes/lagoons created by land reclamation and anti-storm surge barriers (e.g., Devkies 1984; Devkies and Hopstaken 1984). Similarly, models of extensive shellfish beds in semi-enclosed bays have been developed by various workers (Incze et al. 1981; Carver and Mallet 1988) and it would be reasonably easy to modify these to encompass docks managed by enhanced filter feeding. Various models for highly enclosed estuaries would also be appropriate starting points (see Knox 1986b for reviews). Before they can be used, however, good input terms for mixing and exchange processes and for filtration rates are required. The relative contribution of advected nutrients and nutrients recycled in situ needs to be known and, especially, the role of anoxia in nutrient fluxes must be understood.

There is a need for much further experimental work on both mixing devices and optimization of mixing regimes in docks. The direct effects of mixing on phytoplankton populations needs further experimental investigation because the results from trials in lakes are so equivocal. The absence of any controlled work done on mixing alone in saline lagoons or docks makes this an urgent requirement. Work in freshwater docks at Salford has shown that the continuous mixing, employed to prevent anoxia, can provide ideal conditions for blue green algae to proliferate (Hendry unpub.). In addition to direct effects on phytoplankton, artificial mixing will increase delivery of cells to the mussels. Also, increasing water flow at first stimulates mussel filter feeding before inhibiting it at higher velocities (Wildish and Miyares 1990). Work is urgently needed to establish an optimal mixing rate which minimizes costs but maximizes destratification and filtering efficiency.

There is a need to establish the optimal mussel stocking density for a particular dock water body; this should maximize filtering capability but minimize inputs of organic matter from pseudofecal and fecal production which can increase biological oxygen demand. Breakdown of biodeposits and excretion will also recycle nutrients (Kautsky and Evans 1987). A population of many fast-growing young mussels would be preferable to fewer old mussels because of their greater filtering ability. It is also likely that bottom culture will present fewer problems than raft or rope culture because a dense infaunal community occurs amongst bottom-living mussels, some of whose members presumably eat biodeposits. Comparisons of the merit of bottom and suspended culture are required, plus development of an easy means of culturing bottom grown mussels. Existing models (see above) would be useful in estimating stocking densities.

The potential of other filtering species needs to be evaluated, particularly of infaunal species such as cockles or clams. Low salinity docks have little filter-feeding biomass. In the Baltic Sea, *Mytilus edulis* has evolved to live in constant low salinities down to 3-4 ppt. These and other genotypes may be suitable for introduction to brackish docks (Preston, London), provided suitable quarantine procedures are followed. In freshwater, the European mytilid *Dreissena polypropha* has been used as a biological filter in the Netherlands (Reeders et al. 1989), and further work is required on its usefulness. Unfortunately, it does seem very sensitive to water quality, feeding rates being much reduced in contaminated water (Coulson 1989). In the United States, various species of bivalves occur in estuaries and saltmarshes (e.g., *Modiolus demissus*) which may be suitable candidates for culture in low salinity basins.

Biomanipulation to increase the variety and abundance of macrophytes will have beneficial affects in increasing habitat diversity and could possibly decrease nutrients. Increased macrophyte cover might also provide refuges for zooplankton as found in freshwater. In high salinity systems, introductions of bladdered forms such as *Ascophyllum* and *Fucus vesiculosus* should be possible. In low salinity waters, rooted macrophytes such as *Ruppia* would be suitable. Trials with introductions of low-salinity tolerant Baltic strains of *Fucus vesiculosus* should be attempted in brackish docks such as Preston. Unfortunately, as eutrophication progresses, most macrophytes succumb to shading primarily by epiphytes and also by increasing biomass of phytoplankton. This certainly happens in freshwater and has been implicated in the decline of fucoids in the Baltic (Kangas et al. 1982; Kautsky et al. 1986).

The clean sharp lines of dockwalls, once essential for ship handling and now beloved of architects, that provide much of the visual appeal of docklands, have low habitat diversity. Curved, gently inclined beaches of large boulders would be better since they would increase surface area for attachment and the number of refuges for fish and large crustacea, and would increase the diversity of cryptic fauna. A solution that would not compromise the architectural merit of the docks would be sub-surface artificial reefs: waste materials such as old culvert piping, builders' rubble generated during redevelopment, power station fly-ash bricks and particularly old car tires would be appropriate and cheap materials to use. Some preliminary trials with tires and scallop shell waste are under way in Liverpool to provide a substrate for mussel settlement on an inhospitable muddy bottom. Certainly the success of artificial reefs in enclosed bays shown by Sheehy and Vik (this volume) suggests this approach is a valid and promising one. In addition to habitat diversification, their introduction should bring about an increase in filter-feeding biomass and, hence, water quality and can be coupled with selective introductions or culture of filter feeders. There is an urgent need for properly designed and controlled experiments on artificial reefs for use in shallow water docks.

To enable this research to be done, some dock basins need to be set aside and used strictly for experiment. It is also desirable that freshwater biologists and limnological engineers become involved in work on saline dock systems—to avoid marine biologists spending a lot of time at reinventing what is already known.

Concluding Remarks

The smaller tidal range (<3 m) of most of the major ports of the United States has resulted in more open port development than in Britain. Waterfront wharves and extensive jetties are the normal pattern rather than enclosed basins. How applicable, then, is our work to the United States? Examination of charts and visits to Washington, New York, Boston (by S.J. Hawkins) and Baltimore (by K. Hendry) suggest that in many port areas there are various semi-enclosed inlets or reaches that could benefit from water quality management. Clearly, management can only occur in areas with limited water exchange. In basins or inlets no longer used for commercial shipping, partial or complete closure can be artificially created by sills, bunds or causeways to isolate the system from outside water of poor quality. Promotion of habitat diversity by artificial reefs, as demonstrated by Sheehy and Vik (this volume), could also be effective in the management of larger bays or estuaries. Some of the management measures discussed elsewhere in this volume could also be applied to new marina developments to overcome water quality problems and promote benthic diversity. Pontoons used for mooring boats make very convenient mussel rope suspension points, to the benefit of water and users alike—provided stocking density does not lead to dense biodeposits causing anoxic bottom water.

Restoration and recovery of disused dock ecosystems occurs through a succession that is largely dependent on a supply of naturally occurring propagules. The process can be speeded by various management techniques and introductions. Reduction of summer anoxia is controlled with relative ease by mixing: this prevents the succession being arrested or reversed from kills each summer. Experience from the Albert Dock suggests that, given time, a diverse ecosystem can develop without mixing, though we suspect that mixing would have accelerated community development. We also suspect that in some dock systems the amount of mixing required will diminish with time to the extent that it would be used rarely or not at all.

Water quality management also hinges largely on preventing dense phytoplankton blooms. Such prevention can be achieved through overgrazing by large numbers of filter feeders such as mussels. Mixing clearly enhances biological filtration. The direct effects of mixing on phytoplankton are unclear. They may be negative if the water is deep and mixing is intermittent. Continuous mixing, although preventing anoxia, may stimulate phytoplankton blooms, particularly in shallow water.

Eutrophic water bodies are strongly "bottom-up," being driven by plentiful nutrients that can result in dense phytoplankton blooms which swamp the trophic level above. In some brackish water docks, diversity of consumers is impoverished and the scope for biological control is limited. Introduction or culture of benthic filter feeders such as mussels reinstates "top-down" regulation of the ecosystem.

Large enclosed waterspaces are rare in urban environments. Although many cities have lakes and many have waterfronts, large saline or freshwater lagoons are much less common. Therefore, it is not surprising that town planners and architects have utilized the unique features of various docklands in inner city redevelopment schemes. Whatever the initial aims of the redevelopment, the variety of uses that disused docks can be put to depends on good quality water. After isolation from the adjoining source, water quality will improve in most instances through natural recovery. The experience gained at Liverpool and Manchester shows that this can be enhanced by judicious management and introductions. The cost of these measures is usually very low, compared to the overall investment in the scheme, and the result can be diverse aquatic ecosystems in the middle of cities.

Dockland developers and their consulting engineers are now accepting the need to incorporate water quality and ecological considerations at the initial stages of planning and design. Sufficient knowledge exists now to identify whether a system will be easily improved or whether it has an intrinsically intractable nature. This early identification of potential problems and scope for amelioration will help define the suitability of the site for various uses. The ecological design ethic espoused by other contributors to this volume must be adopted by decision makers. At present, much of the research and consultancy undertaken on restoring urban aquatic ecosystems is analogous to rescue archaeology: the ecologists are only called after the bulldozers go in. It is accepted without a second thought on land that gardens, parklands and open space should be landscaped and maintained. It is now slowly being realized that aquatic parklands need "gardening" as well—and can be gardened.

Port development and the conversion of marsh, lagoons and foreshore into docks, piers and jetties has decreased the habitat diversity of estuaries and bays. Restoration of disused dock basins and inner harbors can help to recreate old ecosystems or bring into being new ones. More importantly, it provides habitats accessible on a day-to-day basis to large numbers of people for recreation and watersports, fishing and appreciation of nature. This removes pressure on more delicate, vulnerable and pristine natural habitats. Docks can also provide a formidable resource for education and can be very useful for original research.

Restoration of disused dock ecosystems is a prime example of how, with a little care, commercial development and nature conservation can complement each other. Property values and commercial viability of new ventures are directly dependent on attractive, diverse aquatic ecosystems. Restoration of docklands improves the quality of life for urban dwellers. At the same time, aquatic wildlife returns.

To realize fully the potential of restoration of docks and inner harbors, a multi-disciplinary approach is required, involving much further ecological research and the development of environmental engineering expertise. Restoring diversity to polluted and impoverished marine environments is essential to urban renewal in old ports, both in the new and old worlds.

Acknowledgments

We gratefully acknowledge the help of the Mersey Dock and harbor Company and the staff of Sandon Fisheries in allowing us to work at Sandon Dock and for providing much information. We also wish to thank the many students at Liverpool and Manchester

Universities who helped with sampling and data collection—Andrew Walmsley, Julia Liebeschuetz and Paul Naylor deserve special thanks. The Nature Conservancy Council are thanked for their financial support in the early days; without Dr. Mitchell's interest this project would have folded. In recent years, the Merseyside Development Corporation, Preston Development Corporation and Salford Development Corporation have been the source of considerable financial support and have been most cooperative in allowing access to their dock systems. Gary Porteous has provided analytical backup throughout. Ulric Wilson and Elspeth Jack helped type and edit the manuscript. Gordon Thayer's editing and encouragement were much appreciated.

Literature Cited

Allen, J.R., S.J. Hawkins, G.R. Russell and K.N. White. 1991. Eutrophication and urban renewal: problems and perspectives for the management of disused docks. Marine Coastal Eutrophication, International conference, Bologna, Italy, March 1990. Proceedings in press.

Altwell Ltd. 1989. Report of a study conducted to evaluate the microbiological quality of water within the Liverpool South Docks and Wirral Dock complexes. Contract report R/00038-89 to the Merseyside Development Corporation.

Bailey-Watts, A.E., E. J. Wise and A. Kirika. 1987. An experiment on phytoplankton ecology and applied fishery management: effects of artificial aeration on troublesome algal blooms in a small eutrophic loch. Aquacult. Fish Manage. 18(3): 259-275.

Bellinger, E. G. and B. R. Benham. 1978. The levels of metals in dockyard sediments with particular reference to the contributions from ship bottom paints. Environ. Pollut. 15: 71-81.

Boalch, G.T. 1979. The dinoflagellate bloom on the coast of south west England, August-September 1978. J. Mar. Biol. Assoc. United Kingdom. 59: 515-517.

Boyden C.R. 1975. Distribution of some trace metals in Poole harbour, Dorset. Mar. Pollut. Bull. 6: 180-186.

Carmichael, W. W. 1981. Freshwater blue-green algae (cyanobacteria) toxins—a review, p. 1-13. *In* W. W. Carmichael (ed.), The water environment. Algal toxins and health. Plenum Press, New York.

Carver, C.E.A. and A.L. Mallet. 1988. Assessing the carrying capacity of a mussel culture operation: a preliminary study. J. Shellfish Res. 7: 152.

Codd, G. A. and S. G. Bell. 1984. Eutrophication and toxic cyanobacteria in freshwater. Inst. Wat. Pollut. Contr. 84:225-232.

Conlan, K. 1989. Ecology and hydrography of an upper estuarine dock: a case study of Preston Docks. Ph.D. thesis, University of Manchester.

Conlan, K., K. Hendry, K. N. White and S. J. Hawkins. 1988. Disused docks as habitats for estuarine fish: a case study of Preston dock. J. Fish Biol. 33 (Suppl. A): 85-91.

Coulson, J. 1989. An investigation into the feasibility of using the freshwater mussel, *Dreissena polymorpha*, as a biological control for the improvement of water quality at Salford Quays and Preston Docks. M.S. thesis, University of Manchester. 99 p.

Crawford, R. M., A. E. Dorey, C. Little and R. S. K. Barnes. 1979. Ecology of Swanpool, Falmouth. V: Phytoplankton and nutrients. Estuarine Coastal Shelf Sci. 9: 135-160.

Cross, T. F. and T. Southgate. 1980. Mortalities of fauna of rock substrates in south-west Ireland associated with the occurrence of *Gyrodinium aureolum* blooms during autumn 1979. J. Mar. Biol. Assoc. United Kingdom. 60: 1071-1073.

Cross, T.F. and R.C. Summerfelt. 1987. Oxygen demand of lakes: sediment and water column BOD. Lake and reservoir management 3: 109-116. Proc. Sixth Annual Conference and International Symposium of NALMS (1986), Portland, Oregon. NALMS, Washington D.C.

Cunningham, P.N., L.C. Evans, S.J. Hawkins, G.D. Holmes, H.D., Jones and G. Russell. 1984. Investigation of potential of disused docks for urban nature conservation. 60 p. (Report to Nature Conservancy Council, copies available).

Davies, B.R., V. Stuart and M. deVilliers. 1989. The filtration activity of a serpulid polychaete population (*Ficopomatus enigmaticus* (Fauvel)) and its effects on water quality in a coastal marina. Estuarine Coastal Shelf. Sci. 29: 613-620.

Dorazio, R.M., J.A. Bowers and J.T. Lehman. 1987. Food-web manipulations influence grazer control of phytoplankton growth rates in Lake Michigan. J. Plankton Res. 9: 891-899.

Dorey, A. E., C. Little and R. S. K. Barnes. 1973. An ecological study of Swanpool, Falmouth. II. Hydrography and its relation to animal distributions. Estuarine Coastal Mar. Sci. 1: 153-176.

Edmonson, W.T. 1970. Phosphorus, nitrogen and algae in Lake Washington after diversion of sewage. Science 169: 690-1.

Enticott, R.G. and S. K. Grisdale. 1985. Monitoring of water quality in Bristol City Docks—a co-operative approach. Environ. Health 93: 59-61.

Evans, E.J., R. Phillip and R.G. Enticott. 1983. Survey of the health consequences of participating in water-based events in the Bristol City Docks, a report from the Control of Infection Study Unit, Bristol and Weston Health Authority, to the Bristol City Docks Water Quality Study Group. Environmental Epidemiology Report No.14, University of Bristol. 44 p.

Falconer, I. R., A. M. Beresford and M. T. C. Runnegar. 1983. Evidence of liver damage by toxin from a bloom of the blue-green alga *Microcystis aeruginosa*. Med. J. Austr. 1: 511-524.

Fast, A.W. 1973. Effects of artificial destratification on primary production and zoobenthos of El Capitan Reservoir, California. Wat. Resour. Res. 9: 607-623.

Faust, M. A., A. E. Aotaky and M. T. Hargadon. 1975. Effects of physical parameters on the *in situ* survival of *Escherichia coli* MC-6 in an estuarine environment. Appl. Microbiol. 30: 800-806.

Foy, R. H. and A. G. Fitzsimons. 1987. Phosphorus inactivation in a eutrophic lake by the direct addition of ferric aluminium sulphate: changes in phytoplankton populations. Freshwater Biol. 17: 1-13.

Goeham, P.R. and W.W. Carmichael. 1988. Hazards of freshwater blue-green algae (cyanobactera), p. 403-432. *In* C.A. Lembi and J.R. Woaland (eds.), Algae and human affairs. Cambridge University Press. London.

Hallberg, R.O., L.E. Bagander and A.-G. Engvall. 1976. Dynamics of phosphorus, sulphur and nitrogen at the sediment-water interface, p. 295-308. *In* J.O. Nriagu (ed.), Environmental biogeochemistry, Ann Arbor Science, Ann Arbor, Michigan.

Hamlin, J.F. 1981. Trace metal contamination in some dockyard sediments with particular reference to the contribution from ship bottom paints. M.S. thesis, University of Manchester. 115 p.

Head, P.C., B.J. D'Arcy and P.J. Osbaldeston. 1980. The Mersey estuary bird mortality, autumn/winter 1979. North West Water summary report DSS-EST-80-2.

Hendry, K., K. N. White, K. Conlan, H. D. Jones, A. D. Bewsher, G. S. Proudlove, G. Porteus, E. G. Bellinger and S. J. Hawkins. 1988a. Investigation into the ecology and potential for nature conservation of disused docks. Nature Conservancy Council Contract Report (HF3-11-52(3)).

Hendry, K., K. Conlan, K. N. White, G. S. Proudlove, A. Bewsher and S. J. Hawkins. 1988b. Water quality in disused docks: Their potential for recreational and commercial fisheries. Amer. Wat. Resour. Tech. Publ. Ser. TPS-88-1:225-234.

Hendry, K., K. Conlan, K. N. White, A. Bewsher and S. J. Hawkins. 1988c. Disused docks as a habitat for estuarine fish: a nationwide appraisal. J. Fish Biol. 33 (Suppl. A): 239-241.

Holligan, P.M. and D.S. Harbour. 1977. The vertical distribution and succession of phytoplankton in the western English Channel in 1975 and 1976. J. Mar. Biol. Assoc. United Kingdom, 60: 1075-1093.

Hoyle, B.A., D.A. Pinder and M.S. Hussain (eds.). 1988. Revitalising the waterfront. Belhaven Press, London.

Incze, L.S., R.A. Lutz and E. True. 1981. Modelling carrying capacities for bivalve molluscs in open, suspended-culture systems. J. World Maricul. Soc. 12: 143-155.

James, C. J. and R. Gibson. 1980. The distribution of the polychaete *Capitella capitata* (Fabricius) in dock sediments. Estuarine Coastal Mar. Sci. 10: 671-683.

Kangas, P., H. Autio, G. Hållfors, H. Luther, Å. Niemi and H. Salemaa. 1982. A general model of the decline of *Fucus vesiculosus* at Tuárminne, south coast of Finland in 1977-1981. Acta Bot. Fennica 118: 1-27.

Kautsky, N., H. Kautsky, U. Kautsky and M. Waern. 1986. Decreased depth penetration of *Fucus vesiculosus* since 1940's indicates eutrophication of the Baltic Sea. Mar. Ecol. Prog. Ser. 28: 1-8.

Kautsky, N. and S. Evans. 1987. Role of biodeposition by *Mytilus edulis* in the circulation of matter and nutrients in a Baltic coastal ecosystem. Mar. Ecol. Prog. Ser. 38: 201-212.

Knoppert, P.L., J.J. Rook, T.J. Hofker and G. Oskam. 1970. Destratification experiments at Rotterdam. J. Am. Wat. Assoc. 62: 448-554.

Knox, G.A. 1986. Estuarine ecosystems: a systems approach. Vol. 2. CRC Press, Boca Raton, Florida. 248 p.

Landner, L. 1976. Eutrophication of lakes: causes, effects and means of control with emphasis on lake rehabilitation. W.H.O. ICP/CEP200.

Langerland, A. 1990. Biomanipulation in Norway. Hydrobiologica 200: 535-540.

Laurent, C.W. 1971. The condition of lakes and ponds in relation to the carrying out of treatment measures, p. 111-23. *In* Proceedings of the 5th international conference on water pollution research. Pergammon Press, Oxford.

Liebeschuetz, J. 1985. Hydrography and phytoplankton of a reclaimed dock-basin in Liverpool. M.S. thesis, University of Manchester. 105 p.

Likens, G.E. 1975. Primary productivity of inland aquatic ecosystems, p. 314-348. *In* H. Lieth and R.H. Whittaker (eds.), The primary productivity of the biosphere. Springer-Verlag, New York.

Longsdale, K. 1990. The hydrology and ecology of the Albert Dock complex, Liverpool, with particular reference to the fish population. M.S. thesis, University of Manchester. 108 p.

Loo, L.O. and R. Rosenberg. 1983. *Mytilus edulis* culture: growth and production in western Sweden. Aquaculture 35: 137-150.

Lu, J.C.S. and K.Y. Chen. 1977. Migration of trace metals in the interfaces of seawater and polluted surface sediments. Environ. Sci. Technol. 2: 174-182.

Luederitz, V., A. Nicklisch and J.G. Kohl. 1989. Copper as an algicide. Acta. Hydrochim. Hydrobiol. 17(1): 61-73.

Marsden, M.W. 1989. Lake restoration by reducing external phosphorus loading: the influence of sediment phosphorus release. Freshwater Biol. 21: 139-162.

McConville. 1977. The shipping industry in the United Kingdom. International Institute for Labour Studies, Research Series 26.

McClusky, D.S. 1989. The estuarine ecosystem. 2nd ed. Blackie, London.

Mitchell, R. and C. Chamberlin. 1975. Factors influencing the survival of enteric micro-organisms in the sea: an overview, p. 103-114. *In* A.L.H. Gameson (ed.), Discharge of sewage from sea outfalls. Pergammon Press, Oxford.

Naylor, E. 1965. Effects of heated effluents upon marine and estuarine organisms. Adv. Mar. Biol. 3: 63-103.

O.E.C.D. 1982. Eutrophication of waters—monitoring, assessment and control. OECD, Paris.

Officer, C.B., T.J. Smayda and R. Mann. 1982. Benthic filter feeding: A natural eutrophication control. Mar. Ecol. Prog. Ser. 9: 203-210.

Okaichi, T. and Y. Imatomi. 1979. Toxic dinoflagellate blooms, p. 385-388. *In* D.L. Taylor and H.H. Seliger (eds.), Developments in marine biology, Vol. 1. Elsevier, New York.

Paine, R.T. 1966. Food web complexity and species diversity. Am. Nat. 103: 65-75.

Pastorok,R.A., M.W. Lorenzen and T.C. Ginn. 1981. Environmental aspects of artificial aeration of reservoirs. A review of theory, techniques and experiences. Tetra Tech Co. Final Report TC 3400 to Waterways Experiment Station. U.S. Army Corps of Engineers, Vicksburg, Mississippi.

Phillip, R., E. J. Evans, A. O. Hughes, S. K. Grisdale, R. G. Enticott and A. E. Jephcott. 1985. Health risks of snorkel swimming in untreated water. Int. J. Epidemiol. 14: 624-627.

Phillip, R., S. Waitkins, O. Caul, A. Roome, S. McMahon and R. Enticott. 1989. Leptospiral and hepatitis A antibodies amongst windsurfers and waterskiers in Bristol City Docks. Pub. Health 103: 123-129.

Radway, A., L.S. Walker and I. Carradice. 1988. Water quality improvement at Salford Quays. J. IWEM. 2: 523-531.

Reeders, H.H., A. Bij De Vaate and F.J. Slim. 1989. The filtration rate of *Dreissena polymorpha* (bivalvia) in three Dutch lakes with reference to biological water quality management. Freshwater Biol. 22: 133-141.

Reynolds, C. S. 1987. The response of phytoplankton communities to changing lake environments. Schweiz. Z. Hydrol. 489: 220-236.

Reynolds, C. S., S. W. Wiseman and M. J. O. Clarke. 1984. Growth- and loss-rate responses of phytoplankton to intermittent artificial mixing and their potential application to the control of planktonic algal biomass. J. Appl. Ecol. 21: 11-39.

Riemann, B., T.G. Nielsen, S.J. Morsted, P.K. Bjornsen and J. Pock-Steen. 1988. Regulation of phytoplankton biomass in estuarine enclosures. Mar. Ecol. Prog. Ser. 48: 205-215.

Ritchie-Noakes, N. 1984. Liverpool's historic waterfront—The worlds first mercantile dock system. HMSO, London.

Roberts, D.A. 1989. The removal of surface debris from London's river—the way forward. The harbour—an ecological challenge, Hamburg, Sept. 11-15.

Rodhouse, P.G., C.M. Roden, G.M. Burnell, M.P. Hensey, T. McMahon, B. Ottway and T.H. Ryan. 1984. Food resource, gametogenesis and growth of *Mytilus edulis* on the shore and in suspended culture: Killary Harbour, Ireland. J. Mar. Biol. Assoc. United Kingdom. 64: 513-529.

Russell, G., S. J. Hawkins, L. C. Evans, H. D. Jones and G. D. Holmes. 1983. Restoration of a disused dock basin as a habitat for marine benthos and fish. J. Appl. Ecol. 20: 43-58.

Sawyer, T.K., G.S. Visvesvara and B.A. Harke. 1977. Pathogenic amoebas from brackish and ocean sediments, with a description of *Acanthamoeba hatchetti* n.sp. Science 196: 1324-1325.

Seed, R. 1976. Ecology, p. 13-65. *In* B.L. Bayne (ed.), Marine mussels: their ecology and physiology. Cambridge University Press, Cambridge.

Segar, D.A., J.D. Collins and J.P. Riley. 1971. The distribution of the major and some minor elements in marine animals. Part II. Molluscs. J. Mar. Biol. Assoc. United Kingdom. 51: 131-136.

Soltero, R. A., D. G. Nicholas, A. F. Gasperino and M. A. Beckwith. 1981. Lake restoration: Medical Lake, Washington. J. Freshwater Ecol. 1: 155-165.

Tolland, H. G. 1977. Destratification/aeration in reservoirs. Water Research Council Tech. Rep. TR 50.

Tufts, N.R. 1979 Toxic dinoflagellate blooms, p. 403-407. *In* D.L. Taylor and H.H. Seliger (eds.), Developments in marine biology, Vol. 1. Elsevier, New York.

Vries, I.De. 1984. The carbon balance of a saline lake (Lake Grevelingen, The Netherlands). Neth. J. Sea Res. 18:511-528.

Vries, I.De and C.F. Hopstaken. 1984. Nutrient cycling and ecosystem behavior in a salt-water lake. Neth. J. Sea Res. 18:221-245.

Wall, J.P. 1971. Horseshoe lake: Nutrient interaction by chemical precipitation. Inland lake renewal and management demonstration project. University of Wisconsin and Department of Natural Resources, Madison.

Walmsley, A. 1985. A quantitative model describing the distribution of zinc, copper and lead in an estuarine dock. M.S. thesis, University of Manchester. 113 p.

Widdows, J., M.N. Moore, D.M. Lowe and P.N. Salkeld. 1979. Some effects of a dinoflagellate bloom (*Gyrodinium aureolum*) on the mussel *Mytilus edulis*. J. Mar. Biol. Assoc. United Kingdom. 59: 522-524.

Wildish, D.J. and M.P. Miyares. 1990. Filtration rate as a function of flow velocity: preliminary experiments. J. Exp. Mar. Biol. Ecol. 142(3): 213-219.

CHAPTER 12

Developing Prefabricated Reefs: An Ecological and Engineering Approach

Daniel J. Sheehy

ABB Environmental Services Inc.
Wakefield, Massachusetts

Susan F. Vik

Aquabio, Inc.
Annapolis, Maryland

Abstract

Although the traditional goal of restoration has been to turn degraded or disturbed habitats and biological communities to their original or near-original state, this is not always possible. Other innovative alternatives, such as the replacement of damaged or lost habitat with functionally equivalent human-made habitat or the substitution of alternate resources, may be necessary in order to provide the ecological functions needed to maintain important living resources. This paper focuses on how a system-oriented approach can be integrated with ecological design and engineering methods to define and achieve restoration objectives. It describes the application of this approach to develop, evaluate and apply prefabricated modules that are designed to provide critical habitat by functioning as reefs or hardground in coastal areas impacted by extensive disturbance or habitat loss. The results of these studies indicate that designed habitat modules can be used to replace critical habitat functions for selected species or communities and suggest that they are useful tools for coastal restoration projects.

Introduction

Considerable pressure is placed on natural resource trustees to produce measurable results with funding obtained for habitat or natural resource restoration. All too often, however, restoration efforts become mired in indecision and indirection over attempts to answer basic questions:

- What habitat functions have been lost or damaged?
- How can lost habitat or other resources be restored?
- How can restoration priorities be established?
- What technologies should be used for the restoration?
- How can restoration effectiveness be measured?

The difficulties in answering these questions are frequently due in part to the involvement of multiple agencies and interest groups, each with its own agenda and objectives. In addition, difficulties are compounded due to the complex nature of restoration ecology itself, which often seems to be more of a subjective art or craft than a quantifiable science.

When properly applied, however, a number of approaches and methodologies can provide an effective framework for producing measurable restoration results. In this paper, we focus on how a systems analysis approach can be integrated with ecological design and engineering methods to develop a specific technology, prefabricated habitat modules designed to function as human-made reefs, to meet quantifiable restoration objectives.

Restoration Requirements

Restoration planning challenges resource trustees and decision makers to take a positive approach toward compensating for habitat loss and natural resource injury or loss. Restoration is generally the treatment of last choice undertaken after other efforts, if any, to prevent or mitigate adverse effects have failed. Restoration requirements usually originate as the consequence of a series of past decision errors that have caused cumulative impacts or a catastrophic event such as an oil spill. Innovative approaches that can be adaptively managed are needed to restore, replace or acquire the equivalent of damaged habitat or injured natural resources.

The need for estuarine and coastal habitat restoration is due to the adverse effects of chemical contamination and physical habitat alterations on living marine resources dependent on these areas for basic life functions. Chemical alterations include the inputs of nutrients, herbicides, pesticides and toxic compounds that result in eutrophication or contamination, which have direct and indirect negative effects on living marine resources. Physical alterations such as dredge and fill activities, direct impact damage, and changes in circulation and sedimentation also contribute significantly to reducing available habitat for living marine resources, including some of our most important commercial and recreational fish and shellfish stocks.

The discharge of toxics into coastal waters, such as PCB discharges at the New Bedford Harbor Superfund site, have resulted in the contamination of fish and shellfish resources and have forced closure of important fishing grounds, causing significant economic losses. Nutrient loadings from point and non-point sources have also resulted in significant changes in habitat quality by altering community structure, for example, loss of submerged aquatic vegetation or coral mortality and by contributing to hypoxic and anoxic conditions.

Although environmental pollution from dramatic oil spills or widespread contamination of natural resources receive considerable media attention, EPA's Scientific Advisory Board recently ranked physical habitat destruction in the highest category of ecological risk due to its geographic extent and irreversibility (Alm 1990). Extensive areas of coastal marsh, mangrove, seagrass beds and open shallow water habitat have been lost or damaged, particularly in estuarine areas, as a result of dredging and filling. In addition, the cumulative impacts of other types of physical habitat damage, such as coral reef destruction due to vessel groundings or beach nourishment program accidents and sedimentation due to changes in land use, can also result in significant habitat loss. Heavy fishing gear used on "live" bottoms and offshore mining operations are other potential causes of physical habitat damage.

Ecological functions important to fisheries resources that may be lost because of habitat destruction or degradation include shelter, which is essential for many species, food production or concentration, and critical nursery or spawning habitat. Many species or life stages have specific habitat requirements, and when available habitat is degraded or destroyed, the carrying capacity of the ecosystem for these stocks may also be reduced. For example, when the hard bottom substrate necessary for the development of oyster or coral reefs is lost as a result of dredging, filling or erosion-induced siltation, the production of valuable resources may be dramatically reduced. Many fish and shellfish species use marshes, mangroves, and seagrass or kelp beds for shelter and food during critical life stages. When this habitat is damaged or destroyed, competition for any remaining habitat can result in a reduction of the abundance or growth of these resources, or even a shift in populations to other areas that may be less desirable.

Natural reefs, including rocky areas, ledges, hard bottom, coral and oyster beds, provide essential shelter and food for a wide variety of fish and shellfish and clearly affect the abundance and diversity of fish and shellfish populations. Many species are dependent on the availability of this shelter or forage and are generally not resident in areas devoid of bottom structure or other refuge. Recruitment of many fishes appears to be influenced by the amount of shelter (Sale 1968). The complexity of the shelter or substrate also influences competitive interactions (Jones 1988) and, as a result, species composition. Roberts and Ormond (1987) found that the number of holes in coral are the best predictor of fish abundance in Red Sea fringing reefs and Chester et al. (1984) suggested that the roughest bottom is usually the most productive for temperate zone fishes. Similar observations have been made for decapod crustacean species; Abele (1974) found a strong correlation between the number of species and the number of types of substrate available.

The epibenthic communities that colonize reef substrate also provide forage and can actively concentrate planktonic food resources and produce microhabitat suitable for many other food species, particularly crustaceans. Reef epibenthic communities, which are similar to those on disused dock areas (Hendry et al. 1988; Hawkins et al. this volume) or marinas (Davies et al. 1989), also function as biological filters that effectively reduce phytoplankton biomass and potentially limit algal blooms, thereby improving or maintaining water quality, particularly in urban or developed areas.

Reefs also alter water circulation patterns, produce upwelling and cause mixing. These physical changes result in ecological effects. The vortices and eddies generated by a turbulent flow passively concentrate plankton and detritus, thus creating a source of food for both benthic and nektonic species. Planktivorous reef fish species appear to deplete such plankton supplies on a local scale (Bray 1981). The physical stimuli caused by water passing over reefs or shoals has long been known to attract many species, some of which congregate in these areas for spawning. In addition, reefs function as topographic features that alter hydrodynamic characteristics and cause internal waves that may contribute to turbulence production and mixing, particularly in estuarine areas (Kranenburg 1988). In coastal areas, internal waves may cause shoreward transport of larval invertebrates and fish (Shanks 1988).

Systems Approach

The goal of systems analysis in an environmental or fisheries management situation is simply to promote good planning and decision-making in practical but complex situations (Grant 1986). The purpose of applying a systems approach (*sensu* Churchman 1968) to restoration is that it provides a structured framework that forces planners and decision makers to identify precisely what the system that they wish to restore encompasses; to determine overall project objectives; to clarify assumptions; and to define specific functional ecological requirements. Once functional requirements are clearly defined, the ecological design criteria for restoration can be identified and the success with which the criteria are met can be measured.

It is essential that the restoration planning process clearly link the damage or impact assessment, the restoration objectives and the post-implementation performance evaluation. Natural resource damage assessments, impact assessments or baseline studies generally form the basis of information used to support the development of restoration requirements. It is important that this assessment be planned from the start to permit a rational justification for any required restoration, and that restoration requirements be focused, to the extent possible, on replacing the ecological functions of the lost or degraded habitat. Furthermore, to provide a means of accurately assessing whether or not the implemented restoration actually achieved its objectives, the performance criteria developed for post-implementation monitoring and evaluation also must be explicitly stated and agreed to in advance so that plans focus on clear functional performance objectives rather than on vague or poorly defined goals.

In many cases, these three elements—damage assessment, restoration objectives and performance evaluation—have not been linked due to either a lack of adequate data from the natural resource damage or impact assessment, uncertainty inherent in the proposed restoration technology, or a failure to specify clearly the performance criteria that will be applied during the restoration planning process. The lack of continuity in the planning process has often resulted in confusion about the boundaries of the ecosystem that is being restored as well as about the restoration objectives and their eventual translation into functional requirements and performance criteria.

Identifying the boundaries of the system to be restored at the outset of a restoration project is particularly important for two reasons: the tendency to assume that everyone involved defines the system in the same way and the temptation to put the horse before the cart by focusing narrowly on discrete physical attributes or functions to be replaced without considering their interconnection with the entire system. Because of the potential for conflict later in the restoration planning process if agencies or groups have differing ideas about what constitutes the system to be restored, it is necessary to be as explicit as possible at the start; by their nature, ecological systems are open-ended and involved in exchanges with their surroundings, and thus often defy attempts at a general definition (Smith 1988).

Although determining project objectives would seem to be a simple task, this is often not the case, especially when multiple agencies and other groups are involved. Objectives may not be stated clearly, or multiple and possibly conflicting objectives may exist. Attempting to establish objectives before reaching agreement about the boundaries of the system to be restored is a frequent mistake, which results in wasted time and the possibility of conflict and confusion.

Objectives should be stated as precisely as possible, but in broad terms that do not attempt to encompass a specific design solution or approach. A clear statement of the problem should provide the basis for objectives. For example, the problem might be that phytoplankton growth and induced turbidity have significantly reduced the submerged aquatic vegetation (SAV) that provides critical nursery habitat for many fish and shellfish species as well as some natural turbidity control. The cumulative loss of this habitat results in declines in fisheries resources. Objectives might be to restore nursery refuge and turbidity control functions; specific levels of restoration might even be specified as part of the objectives, but in order to allow a range of alternatives to be considered, the means for meeting the objectives should not be specified. Additional objectives or constraints might be related to cost levels and time factors.

Objectives should satisfy the basic requirements posed by the problem and should not conflict with the broader organizational goals of the responsible resource agencies and other groups involved in the planning process. When more than one set of organizational goals are involved, defining objectives may require considerable discussion and compromise. A variety of decision support tools such as decision risk analysis are available to resolve conflicts that arise in restoration planning. Following the systematic and logical process imposed by many of these tools creates an audit trail of the decision making process, which can be advantageous since many environmental decisions may be ultimately scrutinized by the public and possibly the courts (Sheehy and Vik 1988).

Functional requirements should state as explicitly as possible what the restoration measures must do, but should not prescribe the physical design or implementation of these measures. They should provide detailed specifications that answer, at a minimum, these questions:

- What are the restoration measures supposed to do?
- What are the specific performance objectives for the restoration measures?
- How long should the restoration measures last?
- What aesthetic criteria, if any, must be met?

Because the functional requirements provide the basis for both the ecological design for the restoration as well as the measurement criteria for judging the degree of success or failure of the restoration effort, it is essential that these requirements be described as accurately and specifically as possible.

Ecological Design and Engineering for Restoration Applications

Planning for the restoration of estuarine and coastal habitat is an ecological design problem. The process of ecological design is a search for agreement between the past features of the damaged habitat and the set of functional requirements established by the restoration planners and decision makers. Implementation of restoration plans becomes an ecological engineering effort because it involves altering structural aspects of the ecosystem by applying ecological principles to redirect natural self-organizing biological processes. Although the treatment of choice may be to return the disturbed ecosystem to its predisturbance condition (Cairns 1988), this is not always possible. In some cases restoring selected functional attributes or creating an alternative ecosystem may be the only viable options.

To restore, replace or provide equivalent habitat to support living marine resources through ecological engineering, it is necessary to identify processes and functions that are related to the productivity or carrying capacity of the targeted species or communities. The basic approach involved in the application of designed habitat modules such as those shown in Figure 1 is to develop a structural framework that provides refuge, substrate or altered circulation patterns. These changes in the physical structure allow the natural self-organizing capabilities of the ecosystem to respond through colonization, recruitment and immigration in order to promote the development of target populations or communities. In cases where extreme disruption or damage limit or impede natural recovery or response, other restoration actions such as stocking, seeding or transplanting may need to be integrated with the physical structure to provide the desired results.

Many of the basic ecological functions provided by natural reefs or other forms of shelter/refuge such as seagrass or macroalgal beds may be replaced, in part, through the deployment of designed and prefabricated structures. Little doubt exists that habitat structure and complexity can be increased by placing designed reef modules in areas previously devoid of bottom structure. This habitat or substrate complexity increases the number of species (Abele 1974) and may be very important in maintaining community diversity (Smith 1972). Predator-prey interactions are influenced by the availability of prey refuge and both predator and prey energetics are affected by the relative concentration and availability of food. Studies of large scale reef modules indicate that they function much like natural patch reefs (Ino and Sheehy 1978), providing refuge and reducing predation risk by reducing predator efficiency (Levitan and Genovese 1989).

A long-term controversy has been waged over whether human-made or "artificial" reefs simply attract or actually produce fisheries resources. This controversy, however, is based on the wrong question; it is not the origin of the reef material but other factors, such as the species of concern and its life stage, the design, location, size and age of the reef, and the temporal and areal extent of reference, that determine the answer. Properly designed and sited human-made reefs function in primarily the same manner as natural reefs;

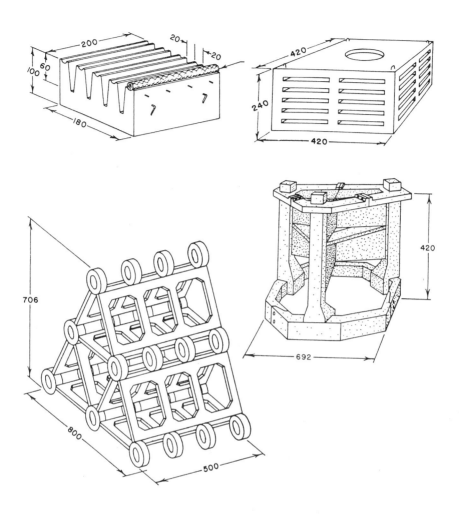

Figure 1. A variety of unit reef habitat modules designed specifically to provide habitat attributes important to target species or communities. Many are generic for groups or guilds of species; however, some are designed for specific species of even life stages. Modules are deployed in a variety of configurations; several different types may be used as part of a reef system. Dimensions in cm.

they both attract and enhance production of fish. Russell (1975) suggests that attraction is actually unlikely, especially for reef fish which have a limited range of movement. Extensive long-term observations on rock and prefabricated lobster shelter reefs in New England indicated that, although initial colonization by lobsters was by immigration, populations were eventually supplemented by recruitment from the plankton (Sheehy 1978).

Many elements of this concentration versus production controversy appear related to the spatial scale of the reefs examined or the restricted temporal scope of the investigation (Sheehy and Vik in press). Long-term observations of natural and human-made temperate zone reefs indicate that, for many species, ecological functions such as refuge, circulation changes and hard substrate that are provided by natural reefs can be largely replaced by human-made structures or rock reefs. The refuge and forage provided by a reef composed of rocks deployed from a barge are functionally the same as that provided by a reef composed of rocks deposited as moraine by a retreating glacier. In both cases, the structural aspects of the reef, particularly the habitat complexity and induced circulation changes, are combined with the natural self-organizing processes that result in the development of a biological community on the reef. The initial epibenthic community development further increases habitat complexity and forage bases to support additional resident and transient species.

Natural reefs are formed from a variety of geological and/or biological processes. Whether formed by geological processes such as moraines, upthrust ridges or submerged coastlines or, by biological processes such as coral or oyster growth, natural reefs are not "optimally" designed for fish or shellfish habitat but instead reflect the nature of their formation. It is possible to apply ecological knowledge about the behavioral ecology of a target species to efficiently manage these natural self-organizing processes in order to produce cost-effective habitat for restoration. Designed modules, prefabricated or not, permit the flexibility to configure reefs to meet specific functional objectives as well as site-specific oceanographic and substrate conditions. Reefs provide artificial substrates that have taxa and processes similar to those exhibited by natural substrates (Schoener 1982).

Some of the shelter and forage base or structural features that are lost when coastal habitat is damaged or destroyed can be replaced through ecological engineering with designed reefs. Ecological engineering is defined as the "environmental manipulation by man using small amounts of supplemental energy to control systems in which the main drives are still coming from natural sources" (Odum 1962). Prefabricated reef or midwater modules can be designed and built to restore some important ecological functions when natural reef, hard bottom or vegetative habitats are degraded or lost or when bottom sediments have become badly contaminated, or are lost as a result of fill activities. Properly designed and sited structures can replace some of the essential functions such as shelter, food and critical reproductive habitat and can also aid in managing water quality and fishing effort.

Advantages of Designed Reefs

All forms of restoration have limitations and constraints. Designed reef modules and related midwater structures, however, have some advantages over traditional American artificial reefs composed of scrap materials as well as other alternative restoration methods such as marsh and seagrass bed creation. The ability to control form and features means that designed reefs can be developed for use in areas where other types of restoration are limited or are extremely expensive because of limited site availability or water quality constraints. Designed reefs are flexible tools that can be used as part of comprehensive restoration plans or as interim measures to maintain critical habitat functions until natural recovery or remediation permits other forms of restoration.

Salt marsh or mangrove creation requires available land, which is difficult to obtain in many urban or port environments due to existing uses and limited areas for creation. Some mitigation activities in California have occurred more than 60 miles from the impact area because of lack of closer land. Designed reefs have no requirement for land acquisition and can be used in areas where sites for marsh creation are limited or nonexistent.

Successful seagrass or SAV (submerged aquatic vegetation) restoration requires adequate water quality (Fonseca this volume). Nutrient and sediment loadings in many areas such as the Chesapeake Bay increase turbidity and eutrophic plankton blooms and result in conditions that can preclude successful SAV growth (Hurley 1990). Designed structures are not as dependent on water quality and can be used in these areas until water quality improves. In fact, the presence of structures that induce improved circulation and hard substrate as well as the fouling organisms using the substrate may actually contribute to the improvement of water quality (Hendry et al. 1988).

Many cases of toxic material release, particularly at some of the coastal Superfund sites, have resulted in benthic habitat contamination. As a result of the loss or damage to the benthic community, the natural filtration in some of these industrial, port and urban areas has been lost or reduced; in other areas, the benthic forage base has been contaminated, resulting in food chain bioaccumulation. Construction of high profile reef or midwater structures with extensive effective surface area can produce substrate for epibenthic communities that may provide substitute food sources and water filtration. Reefs can, in fact, function as large scale biofilters that can help replace some of the original filtration capacity lost as a result of toxic contamination and habitat loss (Sheehy and Vik In press).

Because of water quality or wave exposure conditions, the ability to control the location and elevation of suitable substrate can result in a more diverse and abundant community than would be possible naturally. Chou and Lim (1986) found greater coral diversity and colony numbers on artificial substrate than on nearby natural reef.

The application of designed habitat enhancements for restoration or mitigation should be clearly distinguished from the traditional use in the United States of human-made reefs for recreational fishery enhancement, since these are very different applications and are based on different objectives and materials (Sheehy and Vik 1985a). Artificial reefs for recreational fishing are targeted at sport fish species, with the performance objective being to improve catch per unit effort or access to the fishery and, thereby, benefit industries and services dependent on recreational fishing. Recreational fishing reefs are typically sited to permit easy access by fishermen or for solid waste disposal convenience. For the most part, they were not designed in a formal sense at all but were primarily composed of scrap materials, such as wrecked automobiles, and old tires, or materials of opportunity, such as construction rubble, surplus ships or offshore oil platforms; in fact, solid waste disposal was often at least as strong a motive as fisheries enhancement. Recreational reefs made from scrap or surplus materials have occasionally resulted in problems due to damages caused by improper siting, design, or placement or conflicts between multiple users. These scrap material reefs have limited application for restoration or mitigation because of their lack of design and siting flexibility or reduced life expectancy. This is particularly true in nearshore and estuarine areas where wave, current and substrate conditions impose significant restrictions.

In contrast, designed reefs for restoration are intended to replace habitat and ecosystem functions to support entire biological communities. They are sited based on a more complex set of criteria and should use the best available technology to meet their conservation management objectives (Sheehy and Vik 1985a).

Designed and prefabricated reefs offer several other advantages that make them potentially useful for restoration projects. They are modular fixed cost items that take effective advantage of depth, the third dimension of the water column. Reefs that are properly designed and sited are generally onetime (fixed cost) applications, whereas SAV plantings or marsh creation may require future maintenance or replanting, particularly if water quality conditions are marginal. Prefabricated reefs are modular, can be applied in a phased manner, and do not require a major commitment without some opportunity to evaluate success. Reef performance can be predicted from data from small scale preplacement studies and information on other high profile structures in the immediate area such as docks and bridges. These performance projections can be evaluated in advance of major applications. Since bottom and midwater modules take advantage of depth, they make very effective use of available space which is usually limited because of multiple use conflicts.

Designed Reef Limitations

Designed reefs and midwater structures are not a panacea for all types of restoration and have their own set of constraints. These include both functional and site limitations. Reefs are only applicable where evidence exists that the replacement shelter, forage or circulation benefits match the functions that are lost or damaged. Designed reefs and related structures are best applied as part of comprehensive remediation and restoration plans and should be fully coordinated with fisheries management plans.

Especially when used for out-of-kind applications, designed structures may not replace all functions lost, such as the primary productivity of seagrass systems and the hydrologic functions of marshes. In cases of toxic contamination, the system may need to undergo some remediation or natural recovery before on-site designed structures can be effective.

In addition, these types of structures cannot be rationally applied without having potential impacts on fishery management responsibilities. Augmented areas may require special management zone or sanctuary status to prevent adverse impacts on vulnerable harvested fisheries resources until such time as stocks can recover.

Application of Designed Reef Modules

Elements of this ecological design and engineering approach have been used in a series of projects (Figure 2) employing designed prefabricated habitat modules to produce artificial reefs. A systems approach to design that applied ecological heuristics, engineering failure mode analysis, and value engineering was used to develop, test and evaluate, and apply prefabricated reef technology. Because functional requirements and performance criteria were identified in each case, the degree of success could be measured.

Our efforts to develop and evaluate designed reef technology have been underway for almost twenty years and began with applied studies that addressed some basic questions about whether or not the carrying capacity for American lobster (*Homarus americanus*) could be increased by habitat enhancement using designed modules. Additional investigations of habitat enhancement and extensive abalone (*Haliotus discus discus*) culture as well as a technology transfer program were conducted to examine, select, and transfer applicable elements of East Asian reef technology to the United States. This effort culminated in field operational tests and evaluation comparisons that suggested the potential for using prefabricated reefs for habitat mitigation and restoration. The first major application of this technology for habitat mitigation was completed in Delaware Bay and post-placement performance monitoring is now underway (Sheehy and Vik In press). Prefabricated reefs have recently been proposed to restore coral reef and hardground habitat damaged by beach nourishment activities in Dade County, Florida.

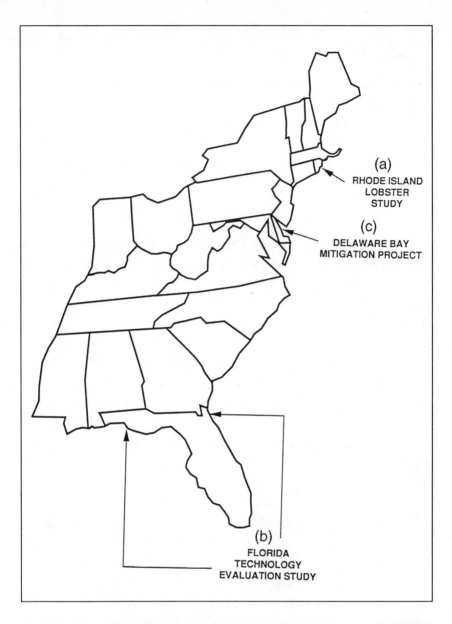

Figure 2. Map illustrating designed prefabricated reef project sites (a) Rhode Island Lobster Study; (b) Florida Technology Evaluation Study; (c) Delaware Bay Mitigation Application.

Proof of Concept: Lobster Shelter Studies

A series of studies were developed at the Marine Experiment Station of the University of Rhode Island to answer basic questions about whether or not the addition of prefabricated shelter units designed specifically for adult lobsters could be used to build reefs that would support lobster abundances similar to those on good natural fishing grounds. The studies also addressed questions related to reef site suitability and lobster behavior ecology as well as reef module design, spacing and orientation in an effort to provide guidance for future applications and to determine if this artificial habitat functioned in the same way as natural reefs.

The American lobster is generally not resident in areas without rock shelter, vegetative cover or substrate that will support burrowing. A number of researchers (Stewart 1970; Briggs and Zawacki 1974) had suggested that in some areas shelter is a limiting factor in the distribution and abundance of nearshore lobsters. Prefabricated shelters made from pumice concrete were used in three habitat manipulation experiments to determine if the addition of human-made shelter to areas formerly devoid of natural shelter would increase the area carrying capacity for lobsters.

To test hypotheses about the influence of available shelter on lobster residency, a preliminary study using reefs composed of single-chamber modules was conducted at several shallow sand bottom sites off Point Judith, Rhode Island. The characteristics and configuration of natural lobster burrows or shelters were used in designing the shelter units, which were prefabricated to reduce unit costs. The addition of prefabricated shelter reefs significantly increased resident lobster populations; lobsters readily occupied the shelters and consumed components of the epibenthic community that developed on the shelters. Observed lobster abundances (Figure 3a) were equal to or greater than those observed on good nearby natural grounds and human-made breakwaters. In addition, results indicated that shelter spacing had a significant effect on oc-

*Figure 3. Lobster Shelter Modules. (a) The original lobster shelter module deployed in Salt Pond and the Harbor of Refuge. Note extensive growth of mussels (*Mytilus edulus*), a primary lobster diet item; (b) modified single chamber module off Block Island.*

cupancy by lobsters and that shelter orientation, with respect to predominant wave and current direction, affected shelter stability and effective life expectancy.

A second study compared reefs composed of the original single chamber shelter and a three-chambered shelter module that afforded approximately the same total available shelter volume. Reefs made from triple-chambered shelters had greater overall percentage of occupancy and supported more abundant lobster populations because of compartmentalization. During this study, all benthic forms of the lobster including post-settlement juveniles were observed on the reefs. These observations suggest that occupancy did not occur through lobster immigration alone but also occurred as a result of planktonic larval settlement.

Although both of the original shelter module designs were relatively effective, some scour-induced subsidence did occur that reduced the life expectancy of the reefs under certain site conditions. To address this problem, laboratory and field studies were conducted (Jones 1974) to develop a more stable design for further testing. The results were also incorporated in a basic simulation program that was used to predict module stability under various combinations of depth, wave, current and substrate conditions. This information was then applied to improve reef module design and siting criteria (Sheehy 1976).

The third phase of this study was a larger-scale controlled test of reefs composed of a new module at six Rhode Island sites that provided a broader range of substrate and oceanographic conditions (Figure 4). Each of these reefs was monitored bimonthly for a year by diver-biologists; the three most stable reefs were monitored more intensely for a second year as part of a tagging program developed to evaluate lobster residence period and movement. During each survey, divers conducted a complete census of each reef, tagged adult lobsters and recorded the shelter location and the size, sex, molt condition, claw number and claw size of each lobster as well as occupancy by other fish and mobile invertebrates. Multidimensional contingency table analysis was used to examine the interaction of variables on the lobster abundance and distribution within the reef.

The results of the third study confirmed earlier observations that the addition of reefs in areas devoid of natural shelter or substrate suitable for burrowing can significantly increase the abundance of lobsters (Figure 3b) and overall community diversity, and helped explain the interaction between variables that influenced lobster occurrence and abundance assessment endpoints. Data and extensive diving and time-lapse camera observations documented that the reefs provided not only shelter but also substantial food from the available epibenthic community that developed on the reef. There

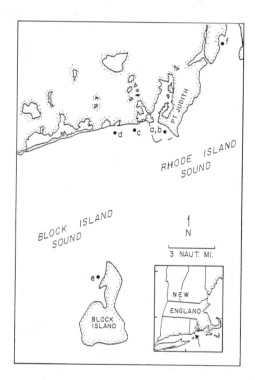

Figure 4. Location of lobster reefs (second study) in Rhode Island waters.

also was an increase in the macrobenthic biomass in the vicinity of the reef. Other species also used the reef for critical life functions such as spawning (*Macrozoarces americanus*) and ovideposition (*Loligo pealii*) (Sheehy 1978).

Despite shelter design improvements and the demonstrated use by lobsters and other species, a range of failure modes that would restrict available sites for future enhancement or would require more significant design changes was observed. Suitable sites for lobster reefs are limited and, for long-term enhancement such as restoration, a careful examination of all relevant site factors is required prior to construction. Reefs must be designed not only to efficiently provide ecological functions, but also to remain effective over the required performance period; otherwise replacement and maintenance must be considered in determining life-cycle costs.

Technology Transfer: East Asian Reef Technology

Based on the results in Rhode Island, work on designed reefs was expanded to include a wider range of target species and objectives. As part of this effort, a new set of functional requirements was defined to improve module stability and life expectancy. In addition to ecological functions, the reefs also have to resist failure that result from:

- Overturning by wave forces.
- Sinking into the substrate.
- Moving from the site as a result of current and wave forces.
- Structural damage.

The Japanese and Taiwanese have invested substantial sums in research and development on reef module design and have extensive laboratory and field test and evaluation data that can be used either to select appropriate sites or to design to site. Rather than reinvent the wheel, we used a technology transfer approach to identify the most promising reef module designs in terms of functional requirements for typical American applications and to try the units here.

The state-of-the-art in reef technology in Japan and the Republic of China is well advanced; a wide range of materials, designs and management alternatives are in use. We conducted systematic field and laboratory studies in both countries during 1977-1978 to evaluate the relative effectiveness and stability of various designs, some of which have been developed for specific species or life stages or are used in conjunction with extensive culture (stocking) programs (Sheehy 1979). Long term retrospective studies (Ino and Sheehy 1978) using catch and effort data indicated that properly designed and managed reefs increase fisheries production, and that integrated habitat and stocking programs offer significant opportunities to improve production of selected species such as the abalone as well as the American lobster.

To help select designs suitable for American applications and conditions, decision risk analysis methods were used to assess potential performance and costs of alternative reef modules. Selection criteria included unit design flexibility, life cycle cost, and proven reliability under site conditions common along Atlantic and Gulf of Mexico coastal areas. A performance database that included the results of numerous detailed studies on reefs was developed to provide a basis for future applications. Significant technical literature was also translated and reviewed (Aquabio 1982) to provide additional application guidance.

Based on the results of these investigation, a unique reef module design made from prefabricated fiberglass reinforced plastic (FRP) components was selected for operational testing in the United States (Figure 5). Comparison studies of the FRP reefs and scrap concrete culvert reefs of a type commonly used in the United States were conducted at sites off the Florida Atlantic and Gulf coasts (Figure 6).

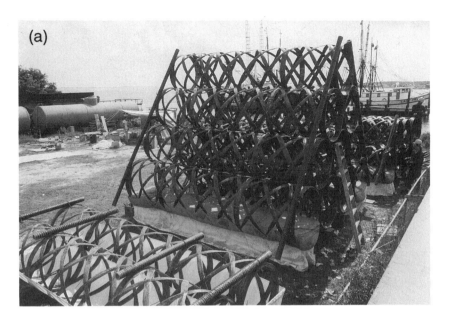

Figure 5. Fiberglass Reinforced Plastic (FRP) prefabricated reefs. (a) High profile module configuration with airbag and ballast concrete in place; (b) lower profile reef module being deployed using air bag system off Panama City, Florida.

567

Reef performance was compared in terms of the functional requirements which included:

- An increase in the diversity and abundance of the fish communities.
- Provision of food sources to support target species of recreational importance.
- Stability on the permit site without substantial degradation in performance.

A number of sampling methods, including visual and cinetransects, time-lapse photography, fouling plates, benthic community

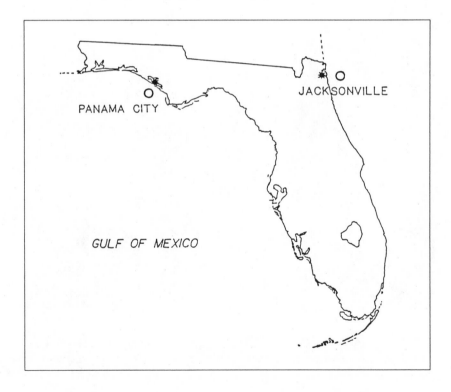

Figure 6. Florida operational test and evaluation sites off Jacksonville and Panama City.

studies, stomach content analysis, and primary production studies were used to evaluate the reef communities that developed (Sheehy et al. 1983). Initial observations indicated dramatic differences between both culvert and FRP reefs when compared with the control sites. When the FRP modules were compared with the culvert reef, however, they clearly supported much more diverse and abundant fish and benthic communities, and provided more complex shelter conditions suitable for juvenile fish and a better forage base to support fish populations. Recent observations, made as part of long-term post-placement studies, revealed that the performance of the culvert reefs in both Panama City and Jacksonville had been significantly reduced due to subsidence. The FRP reef modules, on the other hand, were stable and exhibited no significant performance degradation.

The results of the operational tests and evaluation strongly suggested that stable, complex reef modules could help produce communities similar to natural reefs. The modules demonstrated their potential for habitat mitigation or restoration of lost or damaged hard bottom and reef communities; in addition, they demonstrated their potential to provide alternative forage or shelter when benthic communities or seagrass systems have been damaged (Sheehy and Vik 1985a).

Mitigation Application: Brown Shoal Artificial Reef

Designed reefs have been suggested as a means of habitat mitigation for some time (Sheehy 1979, 1980), and concrete block reefs were successfully used for short-term mitigation of spiny lobster habitat loss during marina construction (Davis 1974). Designed reefs have recently been recommended as mitigation for a number of port and harbor projects along the east coast of the United States and in California.

The Brown Shoal Artificial Reef in Delaware Bay is the first major mitigation application of designed and prefabricated reef technology in the United States. This project was undertaken to provide partial off-site mitigation for a fill project associated with the redevelopment of the Port of Wilmington, Delaware. The project resulted in the loss of 142 acres of shallow open water habitat and, after lengthy interagency discussion and negotiation, a prefabricated reef was accepted as partial mitigation for some of the habitat loss. This type of mitigation requirement is common in many port and harbor areas and in the use of designed reefs for mitigation (Sheehy and Vik 1988); mitigation banking has also been suggested (Sheehy and Vik 1989).

Although the Brown Shoal Artificial Reef project is a mitigation application—specifically, off-site and out-of-kind habitat compensation—the objectives and functional ecological requirements are similar to those needed for many restoration applications, particularly those aimed at providing replacement or equivalent resources. The major difference between compensatory mitigation and replacement restoration is the temporal relationship between the action and the occurrence of the impact or damage. Mitigation is generally applied to deal with ongoing or proposed specific actions whereas restoration addresses the effect of past, cumulative or catastrophic actions.

The objectives for the mitigation reef included:

- Increase local carrying capacities for selected fish and shellfish.
- Be structurally stable and non-leaching.
- Provide effective surface area for epibenthic community development.
- Be accessible to recreational fishermen from multiple launch sites.
- Have an effective life expectancy of at least 30 years.

The initial phase of the effort was a feasibility study to assess the risk associated with the project, to determine logistic requirements and to develop functional requirements for reef design (Sheehy and Vik 1984). This resulted in a preliminary estimate of project costs and identified operational constraints related to technology acquisition and staging areas or placement requirements. The second phase involved the selection of candidate sites and reef module designs suitable for mitigation functional requirements and site conditions in Delaware Bay. A reef site (Figure 7) was selected through a decision analysis approach that incorporated physical, biological, water quality and multiple use conflicts as well as fisheries management objectives. Three candidate reef module designs were also selected, based on both function and use ecological design criteria. All of the recommended candidates were proven designs with extensive effective surface area for an encrusting community, complex shelter, and induced circulation patterns, and adequate bearing surface and stability characteristics (Sheehy and Vik 1985b).

The third phase of the project involved revising and finalizing the module design, reef configuration and deployment plans. The fabrication, placement and configuration plans were developed and accepted; the reef was deployed during the summer of 1989 (Figure 8). Post-placement monitoring studies are currently underway to measure actual reef performance success against assessment criteria and to determine the potential for further reef development. (Sheehy and Vik In press.)

Recommendations

To apply designed or prefabricated reef technology for restoration applications, it is important to improve the planning process for selecting reef designs or sites and to structure restoration programs to take advantage of new information. The following general recommendations are based on the results of past investigations and observations.

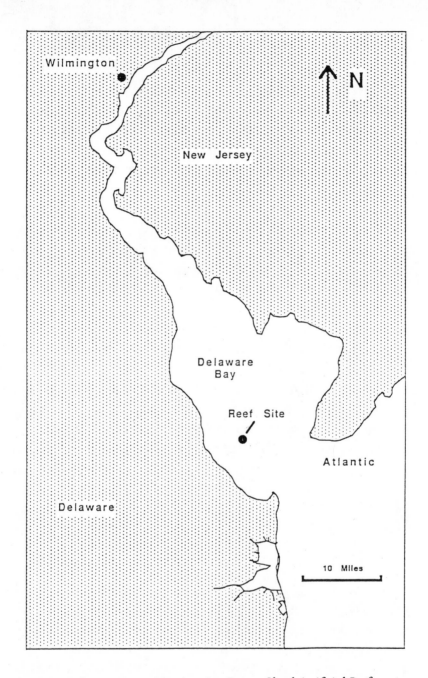

Figure 7. Delaware Bay mitigation site: Brown Shoal Artificial Reef.

Figure 8. Delaware Bay Mitigation Project. (a) Designed prefabricated reef module at staging area; (b) reef modules loaded on barge for transport to the Brown Shoal Artificial Reef.

Apply a Systems Approach

The application of a systems approach can aid planners and decision makers in considering all aspects of the proposed enhancement or restoration. This approach needs a multidisciplinary team and should include at a minimum marine and fisheries biologists, oceanographers, engineers, coastal planners and fisheries managers. Decisions concerning reef sites and design are generally complex, controversial and involve more than one agency. Trade-offs must be made between multiple and related criteria or objectives. Attempts to deal with such complex problems in a piecemeal way often lead to poor management decisions; decision analysis tools can aid in focusing this effort and building a consensus.

Reef Design

The common architectural dictate "form follows function" is relevant to the suggested ecological engineering approach, and reef module design and configuration plans should be based on ecological functions (or attributes) such as shelter, surface area for epibenthic organisms, circulation alteration, and internal wave generation that are targeted for restoration. Functional objectives should be clearly defined and a design approach should be used to restore ecological functions lost as a result of adverse impacts. A clear statement of functional objectives, preferably based on a natural resource damage or impact assessment or on knowledge of the target species or community ecological requirements, helps to ensure that the appropriate functional attributes are used in module design.

Multiple enhancement or restoration objectives frequently must be incorporated into reef designs. It may be necessary to accommodate multiple design criteria for different species, life stages or communities that cannot be achieved with a single module design. In these situations or when site conditions are variable, consideration of a range and mix of individual habitat enhancement units to meet the complex requirements (different species or life stages, variations in site conditions) is both desirable and cost-effective.

When considering alternative designs, all costs associated with the alternatives throughout the project life cycle should be considered in the final restoration plan. A common mistake is to base selection on module acquisition costs without considering site preparation, placement, maintenance, and, if the life expectancy does not exceed the restoration plan specification, replacement costs. Considerable uncertainty is inherent in selecting unproven and untested reef module designs that lack prior performance or engineering stability data. This uncertainty should be weighed when comparing alternative designs.

Site Selection

Site selection should include all appropriate biological, chemical, geological, physical and multiple use factors or considerations. Failure to examine all appropriate factors can result in functional failure, premature structural failure, fisheries management problems or multiple use conflicts. Once the site is selected, the site conditions should be used to refine the module design and configuration plan to reduce the probability of failure. The trade-offs between ecological function and stability or reliability must be considered explicitly and either the reef should be designed to site or other appropriate measures, such as foundation preparation to increase bearing capacity, should be incorporated into the restoration plan. For example, increasing effective surface area to promote a more abundant epibenthic community may also increase the module lift and drag factors that, in turn, may degrade module stability.

Performance Evaluations

Past efforts to evaluate performance of restoration and mitigation actions have often been flawed because of failure to adequately establish baseline conditions, clearly identify performance criteria, and adequately monitor and evaluate performance over a reasonable time frame. To assess performance, it is essential to establish adequate estimates of baseline conditions against which future performance can be compared. This is especially important in estuarine areas where considerable spatial and temporal variability occurs and limited control sites are available.

To improve the performance and predictability of designed reefs, additional retrospective studies and manipulative field experiments are required. Retrospective studies of past reefs urgently need to be expanded; careful analysis of past applications can provide invaluable insight for future projects. Focused manipulative experiments designed to test specific hypotheses are also needed to refine design and siting methods; performance predictions are also needed to establish the scale of required restoration. Finally, the information gained from these retrospective studies and manipulative experiments should be used to develop adaptive management approaches for improving restoration technology.

The use of designed reefs for restoration is a developing technology and initial applications will provide further data needed to improve the state-of-the-art. Although uncertainty is involved in applying new technology, we cannot wait for perfect knowledge. We must learn from the initial mistakes and incorporate this knowledge to improve future applications. Adequate information is now available to make reasonable decisions for restoration applications of designed reefs.

Conclusions

Designed and prefabricated reefs are potential tools for mitigation and restoration, and are particularly effective when used as part of a systems approach that clearly defines objectives, specifications and evaluation criteria. Research to date strongly suggests that properly designed and sited prefabricated reefs function in much the same manner as natural reefs, and can be used to restore or replace shelter and food lost due to habitat contamination or physical alteration. The distinction between natural and human-made reefs may be largely one of material origin, size or age rather than basic biological function. Furthermore, since natural reefs were never "designed" to provide fish or shellfish habitat and their form is largely determined by their geological or biological origin, it is possible to apply ecological design and engineering to improve upon the structure of natural reefs in order to provide specific ecological functions that are targeted for restoration or replacement.

Building designed reefs is an application of ecological engineering: applying small amounts of anthropogenic input (structural change) to redirect a natural self-organizing biological process. A systems approach that applies an ecological design method to plan the restoration of habitat functions can provide a positive framework for restoration planning. These design methods apply biological and ecological knowledge to develop solution-focused strategies to generate satisfactory resolutions to complex, and sometimes initially ill-defined, restoration problems.

Designed prefabricated reefs and related structures are not the remedy for all coastal restoration problems. They should be used as part of integrated plans and are applicable only when the functions that they can provide meet the restoration objectives. Designed reefs may also be useful as interim out-of-kind measures, such as off-site habitat replacement, until water quality, site remediation and/or site availability make other restoration options such as seagrass or marsh restoration viable.

Designed reef modules should be used as tools for redesigning habitat structure to meet specific ecological requirements. Prefabrication methods can improve the cost-effectiveness of this approach, especially with flexible or modular designs that can be tailored to site specific requirements. Although additional studies and applications will aid in refining future approaches, designed reef module technology is now available for operational use and can be applied to meet current restoration requirements.

Literature Cited

Abele, L.G. 1974. Species diversity in decapod crustaceans in marine habitats. Ecology 55:156-161.

Alm, A.L. 1990. Regulatory focus: Reilly promotes new priorities. Environ. Sci. Technol. 24:1624.

Aquabio, Inc. 1982. Japanese artificial reef technology: Translations of selected Japanese literature and an evaluation of potential applications in the U.S., Vol. I. D.J. Aquabio, Inc. Tech. Rep. 604. 380 p.

Bray, R.N. 1981. Influence of water currents and zooplankton densities on daily foraging movements of blacksmith, *Chromis punctipinnes,* a planktivorous reef fish. Fish. Bull. U.S. 78:829-841.

Briggs, P.T. and C.S. Zawachi. 1974. American lobsters at artificial reef sites in New York. N.Y. Fish Game J. 21:73-77.

Cairns, J. Jr. 1988. Increasing diversity by restoring damaged ecosystems, p. 333-343. *In* E.O. Wilson (ed.), Biodiversity. National Academy Press, Washington, D.C.

Chester, A., G.R. Huntsman, P.A. Tester and C.S. Manooch III. 1984. South Atlantic Bight reef fish communities as represented in hook-and-line catches. Bull. Mar. Sci. 34:267-279.

Chou, L.M. and T.M. Lim. 1986. A preliminary study of the coral community on artificial and natural substrates. Malay Nat. J.39:225-229.

Churchman, C.W. 1968. The systems approach. Dell, New York. 243 p.

Davies, B.R., V. Stuart and M. de Villiers. 1989. The filtration activity of a serpulid polychaete population (*Ficopomatus enigmaticus*) and its effect on water quality in a coastal marina. Estuarine Coastal Shelf Sci. 29:613-620.

Davis, G.E. 1974. Notes on the status of spiny lobsters, *Panuluris argus,* at Dry Tortugas, Florida, p. 22-32. *In* D.E. Aska (ed.), Proceedings: research and information needs of the Florida spiny lobster fishery. Florida. Sea Grant Prog., Gainesville, Florida.

Grant, W.E. 1986. Systems analysis and simulation in wildlife and fisheries sciences. Wiley. New York. 338 p.

Hendry, K.H., K. Conlan, K.W. White, G.S. Proudlove, A. Brewsherand and S.J. Hawkins. 1988. Water quality in disused docks: Their potential for recreational and commercial fisheries. Am. Wat. Resourc. Tech. Publ. Ser. TPS-88-1: 225-234.

Hurley, L.A. 1990. Field guide to the submerged aquatic vegetation of the Chesapeake Bay. U.S. Fish and Wildlife Service, Chesapeake Bay Estuary Program, Annapolis, Maryland.

Ino, T. and D.J. Sheehy. 1978. Extensive aquaculture of Abalone (*Haliotus* spp.) in Japan, the Republic of China, and Korea. Prepared by Shimonoseki University for the Japan Society for the Promotion of Science. 224 p.

Jones, A.M. 1974. Analysis and design of an artificial lobster habitat. M.S. Thesis, University of Rhode Island, Kingston.

Jones, G.P. 1988. Experimental evaluation of the effects of habitat structure and competitive interactions on the juveniles of two coral reef fishes. J. Exp. Mar. Biol. Ecol. 123:115-126.

Kranenburg, C. 1988. Long internal waves and turbulence production in estuarine flows. Estuarine Coastal Shelf Sci. 27:15-32.

Levitan, D.R. and S.J. Genovese. 1989. Substratum-dependent predator-prey dynamics: patch reefs as refuges from gastropod predation. J. Exp. Mar. Biol. Ecol. 130:111-118.

Odum, H.T. 1962. Man in the ecosystem. Bull. Conn. Agr. Station, Storrs 652:57-75.

Roberts, C.M. and R.F.G. Ormond. 1987. Habitat complexity and coral reef fish diversity and abundance on Red Sea fringing reefs. Mar. Ecol. Prog. Ser. 41:1-8.

Russell, B.C. 1975. The development and dynamics of a small artificial reef community. Helgol. Wissen. Meeresunter. 27:298-312.

Sale, P.F. 1968. Influence of cover availability on depth preferences in juvenile manini *Acanthurus triostegus sandvicensis.* Copeia 802-807.

Schoener, A. 1982. Artificial substrates in marine environments, p. 1-21. *In* J. Cairns Jr. (ed.), Artificial substrates. Ann Arbor Science, Ann Arbor, Michigan.

Shanks, A.L. 1988. Further support for the hypothesis that internal waves can cause shoreward transport of larval invertebrates and fish. Fish. Bull. U.S. 86: 703-714.

Sheehy, D.J. 1976. Utilization of artificial shelters by the American lobster (*Homarus americanus).* J. Fish. Res. Board Can. 33:1615-1622.

Sheehy, D.J. 1978. An evaluation of artificial reefs constructed from unit concrete shelters designed for the American lobster (*Homarus americanus*). Aquabio, Inc., Belleair Bluffs, Florida. Doc. 78-TN-04. 75 p.

Sheehy, D.J. 1979. Fisheries development: Japan. Water Spectrum 12(1):1-9.

Sheehy, D.J. 1980. Artificial reefs as a means of marine mitigation and habitat improvement in Southern California. Aquabio, Inc. Annapolis, Maryland, Tech. Rep. 80-TT-564. 80 p.

Sheehy, D.J. and S.F. Vik. 1984. Artificial reef feasibility study: proposed mitigation for the development of the Wilmington Harbor South Disposal Area. Aquabio, Inc., Annapolis, Maryland, Tech. Rep. No. 85-TN-627.

Sheehy, D.J. and S.F. Vik. 1985a. Designed reefs for habitat loss compensation, p. 1439-1450. *In* P. Magooneta (ed.), Proceedings of the fourth symposium on coastal and ocean management. American Society of Civil Engineers, New York.

Sheehy, D.J. and S.F. Vik. 1985b. Artificial reef siting and design development plan for Delaware Bay and adjoining coastal waters. Aquabio, Inc., Annapolis, Maryland. Tech. Rep. No. 84-TN-613.

Sheehy, D.J. and S.F. Vik. 1988. Habitat enhancement technology: New methods for resolving port development and fishing conflicts, p. 215-225. *In* Ports and harbors: our link to the water. Proceedings of the 11th international conference of the Coastal Society, Boston, Massachusetts.

Sheehy, D.J. and S.F. Vik. 1989. Extending mitigation banking beyond wetlands, Vol. 2, p.1242-1253. *In* Proceedings of the sixth symposium on coastal and ocean management, New York.

Sheehy, D.J. and S.F. Vik. In press. Artificial reefs: state-of-the-art. Aquabio, Inc., Annapolis, Maryland, Tech. Rep. No. 91-TN-264.

Sheehy, D.J., S.F. Vik and H.H. Mathews. 1983. Evaluation of Japanese designed and American scrap material reefs. Aquabio, Inc. Annapolis, Maryland, Tech. Rep. No. 83-RD-607.

Smith, F.E. 1972. Spacial heterogeneity, stability and diversity in ecosystems. Trans. Conn. Acad. Arts and Sci. 44.

Smith, I.R. 1988. Ecology and design: an introduction. J.Environ. Manage. 26:103-109.

Stewart, L.L. 1970. A contribution to the life history of the lobster, *Homarus americanus* (Milne-Edwards). M.S. Thesis, University of Connecticut, Storrs.

CHAPTER 13

The *Torrey Canyon* Oil Spill: Recovery of Rocky Shore Communities

S.J. Hawkins
Port Erin Marine Laboratory, Liverpool University
Isle of Man, United Kingdom

A.J. Southward
Marine Biological Association
Plymouth, United Kingdom

Abstract

Some background information on usage and the ecology of rocky shores is given before chronic and acute impacts are surveyed. Oil pollution is considered the major acute anthropogenic impact on rocky shores. The Torrey Canyon *oil spill is considered as a case study; 119,000 tons of Kuwait oil were being carried, of which 40,000 tons came ashore, 10,000 tons of dispersant were applied to 14,000 tons of oil which was stranded in Cornwall. Dispersant use had considerable environmental effects, killing most animal life. The main herbivorous species, limpets (*Patella *spp.) were particularly badly affected. Observations on several shores in Cornwall showed their death led to a proliferation of ephemeral green algae giving way to dense growths of* Fucus *which covered the shore from years 2-5, and to the loss of the surviving barnacles. Limpets recruited well under the dense fucoid canopy which eventually* *disappeared leaving the shores barer than usual. At Porthleven the limpet population crashed dramatically after 5 years. Long-term observations on the shore at Porthleven (1967-1990) suggest that stabilization of the shore community to normal levels of spatial and temporal variation definitely takes at least 10 years and probably nearer 15 years. Among the lessons from the* Torrey Canyon *are that dispersant application is unwise, and that time-scales of recovery are long and should be taken into consideration in claims for compensation. Removal is considered a preferable option to dispersal and biological methods hold promise. The need for an experimental component to assess the efficiency of different treatments in the response to the next oil spill is highlighted. "Doing nothing" is considered a sensible option with oil on more exposed shores.*

Introduction

Whether from wrecked ships or in accidents on off-shore or land based installations, oil spills are probably the most dramatic marine pollution incidents. The consequent high mortalities of sea birds and other wildlife and the devastation to littoral and near-shore areas cause grave public concern and have prompted considerable scientific attention.

In this paper, we intend to examine the effects of oil spills on rocky shores. To set the scene, the value and usage of rocky shores are considered followed by a brief account of how rocky shores in the north-east Atlantic are structured. The various impacts on shores are briefly outlined to emphasize the importance of oil spills before focussing on the *Torrey Canyon* incident. Long term investigations on shores in the south-west of England provide an excellent case study to evaluate time-scales and mechanisms of recovery of rocky shores from oil spills. The *Torrey Canyon* spill also provided one of the first major tests of responses and treatments designed to reduce the impact of the spill. Therefore, valuable lessons about how to deal with oil spills can be derived from its study and placed in a wider perspective. Finally suggestions for design of studies of recovery are made.

Background Information on Rocky Shores

Value and Usage

Rocky shores are the most extensive littoral habitat on wave-exposed coasts and a major habitat on "drowned" fjordic coastlines. They have considerable functional links with both inshore and offshore ecosystems (Mann 1982). Rocky shorelines provide spawning and nursery grounds for the eggs and juveniles of a number of fish, including commercial species such as the Pacific herring. Considerable amounts of biomass are exported from these habitats in the form of dislodged or fragmented seaweeds and eelgrass, especially from low down on the shore and in sheltered conditions. Such plant biomass may be transported offshore or can be deposited on sediment shores where it usually forms dense banks along the strandline at the top of the shore. These strandlines support highly productive invertebrate communities exploited in turn by large numbers of birds; birds also feed directly on the shore during low tide. Various mammals forage over the shore. Of particular note are otters which live and feed on the shore as well as the immediate subtidal area. Rocky shores also provide hauling out and breeding grounds for seals and sea lions.

Shores are used in a variety of ways. Since earliest times, man has caught fish and collected shellfish from rocky shores—as large middens at coastal sites testify. Today, in developed societies many delicacies are collected from rocky shores, in some places on a commercial basis or as part of recreational activities. Subsistence-collecting is still a very important source of protein in many less developed societies. Important space-occupying animals (goose barnacles, mussels, oysters), grazers (sea urchins, winkles, limpets) and predators (whelks, crabs, octopus) are all taken. Industrial harvesting of seaweeds occurs in various parts of the world, mainly to produce useful compounds as additives for the food, cosmetics and pharmaceutical industries but sometimes for fertilizer or direct human consumption.

A major use of rocky shores in industrialized societies is for recreation. Fishing and bait collecting are major activities. Many people, especially children, take great delight in exploring the shore in general and rock pools in particular. The appeal of rocky shores is used in environmental and biological education at all levels, from infant school to college and adult education.

Rocky shores have also been used extensively as an ideal model system for ecological research (see Connell 1974). They are two dimensional, easy to sample non-destructively, have sharp environmental gradients, and are inhabited by mainly sessile or sedentary organisms whose taxonomy is generally well known. Of greatest importance is the ease with which they can be experimentally manipulated. Consequently, their study has made a considerable contribution to general ecological theory.

In the future, rocky coastlines may be the sites of renewable energy generation using tidal or wave power. The aquaculture sector may also make increased use of rocky coastlines, where fish farms can take advantage of clean sediment-free water typical of these sites.

Thus, although rocky shores have no great direct economic value, they are an important part of inshore ecosystems. They support a diverse array of plants and animals easily accessible for recreation, education and research. Therefore, rocky shores have considerable conservational value.

Factors Structuring Rocky Shore Communities

Rocky shores have received much attention world-wide. In North America they are particularly well understood because they have been the subject of many elegant field experimental studies (e.g., Paine 1966, 1974; Connell 1970; Dayton 1971; Menge 1976; Lubchenco 1978, 1983; Lubchenco and Menge 1978; Sousa 1979; Paine and Levin 1981; Petraitis 1983, 1987; Cubit 1984; Miller and Carefoot 1990; Chapman 1990) and several comprehensive reviews (e.g., Connell 1972, 1983; Menge and Sutherland 1976, 1987; Paine 1980; Lubchenco and Gaines 1981; Sousa 1984; Menge and Farrell 1989). Much of this work has shown the importance of biological interactions in setting distributions and structuring communities, although some more recent studies have emphasized the importance of recruitment fluctuations (e.g., Gaines and Roughgarden 1985; Roughgarden et al. 1987).

In the North East Atlantic, shores have been studied seriously for a long time (Audouin and Milne-Edwards 1832). There are excellent classical descriptive accounts of the patterns of distribution in relation to the major environmental gradients of tidal height and wave action (see Southward 1958; Lewis 1964 for reviews). Despite being the site of some of the earliest field experimental manipulations (e.g., Hatton 1938; Jones 1946) less work has been done on European shores than elsewhere in recent years. However, the factors structuring the mid shore region (eulittoral, *sensu* Lewis 1964) are reasonably well understood.

In the British Isles, sheltered shores are covered completely by large fucoid seaweeds, arranged in clear zones. Exposed shores are dominated by barnacles, mussels and limpets. Intermediate shores tend to be patchy: with clumps of *Fucus*, interspersed amongst areas of barnacles and bare rock with limpets usually in clumps. The early experimental work done on the Isle of Man (Jones 1948; Lodge 1948; Burrows and Lodge 1950; Southward 1956, 1964) clearly showed that *Fucus* was prevented from completely covering more exposed shores by the grazing activities of limpets. It has been suggested that a dynamic equilibrium exists along the wave exposure gradient which mediates the ability of fucoids on the one hand, or barnacles plus limpets on the other, to dominate a shore

(Southward and Southward 1978; Hawkins and Hartnoll 1983b). In intermediate conditions at the point of balance between fucoids and limpets, shores are patchy with considerable temporal fluctuation. This spatial and temporal variation is partly generated by external recruitment fluctuations but also has an internal cyclic component due to interactions between fucoids, limpets, dogwhelks and barnacles (Connell 1961; Hawkins 1981a, b; 1983; Hawkins and Hartnoll 1982a, b, 1983a, b; Hartnoll and Hawkins 1985).

Impacts on Rocky Shores

Chronic

Rocky shores typically have much water movement with good mixing and dispersal. As a consequence, the influence of effluents tends in most instances to be localized and limited, unless poorly dispersed through discharge into a bay or inlet.

Discharge from sewers can directly smother organisms and increase biological oxygen demand and nutrient levels. Generally, these lead to very localized stimulation of ephemeral green algae at the expense of perennial algae and sessile animals (e.g., Littler and Murray 1975). Similar effects are likely to occur from other types of organic discharge such as pulp waste, industrial effluents and series of small oil spills (Dicks and Hartley 1982). For example, at Milford Haven, United Kingdom, the oil refinery effluent, possibly combined with local effects of small scale application of oil-spill dispersants, has reduced the numbers of the grazing limpet *Patella*, and increased biomass of mainly ephemeral algae (Baker 1976a; Petipiroon and Dicks 1982).

Leachates from tributyltin-based antifouling paints have recently been shown to affect stenoglossan gastropods more severely than other rocky shore biota. At very low levels (nanograms per liter), these induce "imposex" in *Nucella lapillus* and other stenoglossans.

Male sexual characteristics become superimposed on females; they grow a penis and the vas deferens proliferates and blocks the female genital duct causing sterility (Bryan et al. 1986; Gibbs et al. 1990; Spence et al. 1990a). This leads to marked distortion of population structure and decline in numbers, particularly on shores near ports and harbors (Spence et al. 1990b).

Dogwhelks are important predators in shore communities (Connell 1961, 1970; Dayton 1971; Menge 1976). Decreases in their numbers would be expected to lead to greater numbers of barnacles. This may distort the plant-animal balance on rocky shores as greater densities of barnacles will increase the likelihood of algal escapes from grazing since limpets and winkles cannot easily forage amongst dense barnacles (Hawkins 1981a, b; Hawkins and Hartnoll 1982b; Lubchenco 1983; Miller and Carefoot 1990).

Other chronic impacts include battering by logs (from both natural sources and artificially derived from logging; see work by Dayton 1971) which, although unselective, creates new space and tends to remove long-lived species and encourages short-lived opportunists.

Intensive collection of animals for consumption (e.g., Moreno et al. 1984; Hockey and Bosman 1986) and bait (Van Herwerden 1989) also severely disrupts community structure and again tends to favor short-lived species. Trampling during informal recreational activities will also affect shores to varying degrees but little information is available (but see Boalch and Jephson 1978; Beauchamp and Gowing 1982; Ghazanshahi et al. 1982; Bally and Griffiths 1989). Extensive use of shores by field courses probably causes considerable disruption in areas near marine laboratories.

All the above impacts tend to favor shorter-lived ephemeral species, particularly green algae, and thus influence the balance of plants and animals on rocky shores. Productivity may go up, but the community structure will be distorted and species diversity will be lowered.

Acute

Natural catastrophes including movement of shorelines by earthquakes (Haven 1971; Johansen 1971), "red-tides" of toxic dinoflagellates (e.g., Southgate et al. 1984) and climatic extremes such as cold winters (Crisp 1964) and warm events or *El Niños* (Branch 1984; Dayton and Tegner 1984) can all result in widespread mortality and disruption of community structure on rocky shores.

There are few examples of acute anthropogenic impacts on rocky shores. Testing of nuclear weapons can have effects similar to earthquakes (Lebednik 1973). Oil spills are the most frequent and catastrophic type of disaster. Table 1 lists various spills to give an

Table 1. Selected major oil spills (various sources).

Ship or installation*	Date	Oil Spilt (tons)	Amount of Oil Beached (tons)
Persian Gulf*	Jan 91	>1 million	?
Ixtoc 1*	03 Jun 79	500,000	12,000
Amoco Cadiz	16 Mar 78	233,000	80,000
Torrey Canyon	18 Mar 67	100,000	35,000
Khark 5	19 Dec 89	70,000	?
Metula	09 Aug 74	51,000	42,000
Exxon Valdez	24 Mar 89	38,000	4,500
Urquiola	12 May 76	30,000	25,000
Aragon	30 Dec 89	25,000	?

idea of the amounts involved and the proportion which comes ashore. A considerable literature exists on the immediate and longer term effects of oil on shores (e.g., see Baker 1976b; Clark 1982). The toxicity varies with the type of oil and the degree of breakdown at sea before coming ashore. Physical smothering is usually a more important source of mortality than chemical poisoning. Clean-up operations often kill more marine life than the oil itself (Foster et al. 1990), and the problem is explored in the rest of our paper.

The *Torrey Canyon*

Brief History Of Spill

The following brief summary of the spill is based on United Kingdom government publications stemming from inquires into the incident (Great Britain, House of Commons 1968) and from a report by the Marine Biological Association (Smith 1968). The *Torrey Canyon* was registered in Monrovia (Liberia) and belonged to the Barracuda Tanker company (registered office Monrovia, places of business New York and Bermuda, all United States resident shareholders), on 20 year charter to the Standard Oil company of California. The officers and crew were Italian. It was carrying 119,000 tons of Kuwait crude oil for B.P. to the oil refinery at Milford Haven (Figure 1).

At 8:50 a.m. on Saturday, March 18, 1967, the ship stranded on the Pollard Rock, part of the Seven Stones, a notorious danger to shipping, while steaming on auto pilot at a full speed of 16 knots. Wind was northwesterly force 5 and visibility was good. The Seven Stones are clearly shown as a navigational hazard on all charts. In the Admiralty "Channel Pilot," masters are specifically advised not to take the route between the Isles of Scilly and Cornwall. A Light Vessel moored about 4 km north-eastward of the Pollard Rock fired warning rockets as the *Torrey Canyon* approached danger. The Liberian board of enquiry found that the stranding might have been averted at the last minute by prompt and correct over-riding of the autopilot. Her Majesty's Government view was that in addition to the clear negligence on the part of the master (licence re-

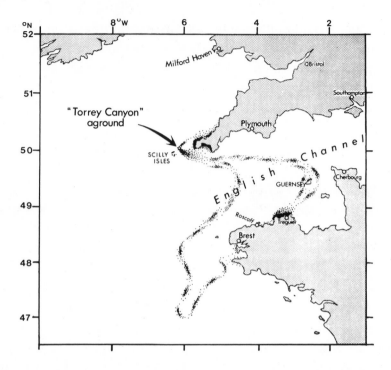

Figure 1. Approximate track of the drifting oil released from the Torrey Canyon *and the coasts where it was stranded. Based on Figure 1 of Southward and Southward (1978).*

voked by the Liberian government), some blame could be apportioned to the owners due to the slack drill on board, especially with respect to operation of the autopilot.

The ship's momentum carried it forward so that bottom damage was suffered over at least half its length. Jettisoning of oil in an attempt to gain buoyancy, and breaching of the forward fuel and cargo tanks resulted in 30,000 tons of oil being lost in the first few hours.

A further 20,000 tons were lost over the next seven days during salvage operations on the vessel. It was decided to attempt to salvage the vessel as the best means of reducing the total amount of pollution. A Dutch team was on board by the afternoon of Saturday, March 18. The highest tides were due on March 26 and 27, and, hence, attempts to refloat the *Torrey Canyon* were delayed until then. Salvage operations were hindered during this week by heavy seas and by an explosion which killed the Dutch chief salvage officer, which stopped all power on ship and the risk of subsequent explosions. A trial pull failed on March 25, and the main attempt on March 26 was delayed by a severe gale. During this attempt, the main cable linking the *Torrey Canyon* to two of the four tugs parted. Shortly afterwards attempts at salvage were abandoned when the ship's back broke. A further 50,000 tons of oil were released. Aerial bombing on three days, March 28-30, burnt the remaining 20,000 tons on board the tanker, which by then was in three parts.

Response to the disaster was rapid. A naval helicopter was over the scene within an hour. Within four hours, ships with dispersant were on their way from Plymouth and began spraying the same evening. A coordinating committee was in place by March 19 under the aegis of the Navy Department.

Scientific advisers were in place by the March 23 with participation by the Ministry of Agriculture, Fisheries and Food, Natural Environmental Research Council, Nature Conservancy. Already a role was being played by the Marine Biological Association (M.B.A.) Laboratory, Plymouth (Natural Environment Research Council funded) which was put on an emergency footing and the whole of its resources devoted to the spill from March 27. Other relevant voluntary bodies such as the Royal Society for Protection of Birds and the Royal Society for Prevention of Cruelty to Animals, also participated in the rescue and treatment of oiled birds.

The 30,000 tons spilt at the time of stranding and the subsequent 20,000 tons released during the week of abortive salvage operations reached the shores of Cornwall on Saturday, March 25 (see Figure 2). By March 27, the coastline from the Lizard to Trevose

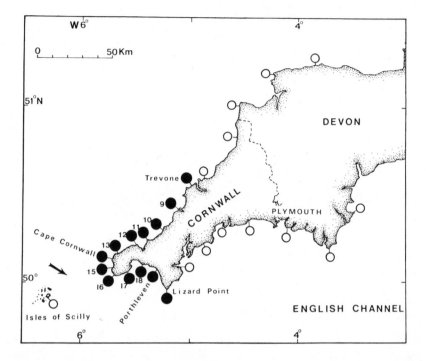

Figure 2. Rocky shore stations monitored by the Marine Biological Association in Devon and Cornwall each year from 1955 to 1987. Open circles, stations not affected by the Torrey Canyon *oil or the clean-up operations. Closed circles, stations affected by oil. The places with numbers are: 9, Newquay; 10, Chapel Point; 11, Godrevy (no direct clean-up operations); 12, St. Ives; 13, Pendeen; 15, Sennen Cove; 16, Porthgwrra; 17, Lamorna Cove and Mousehole (very light oiling); 18, Marazion and Maen Dhu Pt., Perranuthnoe. The arrow marks the Seven Stones reef where the tanker was wrecked*

Head was affected by oil, which had drifted up the north Cornish coast, and then frequently driven ashore by westerly winds. The oil was deposited on very high spring tides (highest for 50 years) and, therefore, was not washed away by subsequent high waters. An area on the north Cornish coast extending from Cape Cornwall to the Camel estuary received most of the oil; the rest of the spill was on the South coast from Mousehole across Mount's Bay to Lizard Point.

The bulk of the oil released when the ship broke up began to approach the Cornish coast, but was swept by a northerly wind on Wednesday, March 29 southwards towards France. Spraying by the Navy at sea continued for 18 days until April 4. Inshore spraying continued for sometime by chartered vessels, then by shore parties until June.

The oil reached Brittany on April 9 and continued to come ashore until the April 28. It was virtually continuous along 90 km of coast and was thicker but higher in water content than in Cornwall, with deposits of up to 30 cm deep. In France, treatment was mainly by removal (often by hand), but considerable steam cleaning and dispersant spraying was carried out in the Cote du Nord region east of Roscoff, at the urgent request of the tourist industry. Over 4,000 tons were removed by April 28. The French Navy also treated much of the oil by sinking it, using sawdust, chalk and other particulates rather than detergents. The last slicks were finally sunk by the French Navy in deep water (> 300 m) in the Bay of Biscay.

Impact of the Torrey Canyon *Oil Spill on Rocky Shores*

The First Ten Years. A thorough account of the immediate consequences of the oil spill and its clean-up has been given by the staff of the Marine Biological Association, Plymouth (Smith 1968). Events over the next two to three years have also been recorded by M.B.A. staff (Bone and Holme 1968; Bryan 1969; Spooner 1970, 1971) and other workers (Nelson-Smith 1968; Stebbings 1970; Crapp 1971a). The first ten years of recovery on selected shores have been described by Southward and Southward (1978). The following brief account is a synthesis of these studies.

What clearly distinguished the *Torrey Canyon* disaster from previous and subsequent oil spills was not the quantity of oil that came

ashore, but the volume and indiscriminate application of dispersants to remove oil from sandy beaches and rocky shores (Southward and Southward 1978). Mortality directly caused by the oil
itself on the shore occurred (O'Sullivan and Richardson 1967) but
was limited. The toxicity of the oil was relatively low to begin with,
and evaporation of the lighter fractions reduced it further during
the time the oil was at sea (Smith 1968; Corner et al. 1968). Thus,
the main ecological consequences of the *Torrey Canyon* oil spill in
Cornwall were due to the intense use of dispersants during spring
1967.

The first generation dispersants available in 1967 proved highly
toxic to marine life. The main product applied consisted of a 12%
non-ionic surfactant plus a 3% stabilizer in a highly aromatic solvent which was a by-product of kerosene refining, containing toxic
components removed when making fuel for domestic use. Unfortunately, at the time of application the toxicity of these mixtures was
not well known, although subsequent studies amply demonstrated
their effects (e.g., Boney 1968; Baker 1971; Crapp 1971b). Laboratory studies showed a 24 hr LC_{50} has also been suggested varying
from 0.5-5 ppm on sublittoral organisms, and 5-100 ppm on intertidal organisms. In the short term, the solvent was the most toxic
ingredient (Corner et al. 1968) but over longer periods all components were very poisonous (Wilson 1968). The most susceptible intertidal animals were limpets of the genus *Patella* (LC_{50} in 24 hours
of 5 ppm). The concentrations that killed marine organisms were
much lower than those needed to disperse the oil, and in most instances were very much lower than concentrations resulting from
field application.

Field observations showed that close to dispersant spraying nearly all animals were killed. Many algae were also damaged or killed.
In the general vicinity of dispersant spraying, herbivorous gastropods (snails), decapod crustacea (crabs and shrimps) and echinoderms (e.g., sea stars and sea urchins) were very badly affected. In

particular there were virtually complete kills of limpets of the genus *Patella*. These are a "keystone" (*sensu* Paine 1966) species in the north east Atlantic, their grazing being responsible for structuring midshore communities on moderately exposed and exposed shores (see Southward 1964; Hawkins and Hartnoll 1983a for reviews). On some rocky shores (e.g., Porthleven), cleaning was so thorough that recolonization started from scratch. Other shores had a few survivors (e.g., Maen Du Point), mainly fucoid algae and more hardy animals such as barnacles and the topshells, *Monodonta lineata.*

All the coastline from the Lizard Point to Padstow was affected to some degree by dispersant application and by drifting patches of dispersant and oil mixtures. Therefore, separation of the effects of oil alone, from dispersants alone, and from oil/dispersant mixture is difficult. Godrevy was the only site of those studied by Southward and Southward (1978) at which dispersants were not directly applied. Their application, not allowed there because this National Trust owned property, was near a breeding ground for Atlantic Grey Seals.

Table 2 summarizes information on recolonization of various shores between 1967-1977. The similarity of the overall pattern allows a generalized account of the course of recolonization on the midshore region (midlittoral of Stephenson and Stephenson 1949, 1972; eulittoral of Lewis 1964). This sequence resembles that found in various smaller-scale clearance and limpet removal experiments undertaken in the 1940s (Jones 1948; Lodge 1948; Southward 1956, 1964) and subsequently (Hawkins 1981a, b; Hawkins and Hartnoll 1983a for review).

Following death of grazing animals, a dense flush of ephemeral green algae (*Enteromorpha, Blidingia, Ulva*) appeared, which lasted up to one year. After six months or so the large brown fucoid algae (mainly *Fucus vesiculosus* and *F. serratus*) began to colonize the shores. The more exposed parts of the reefs were completely covered by the small bladderless form of *F. vesiculosus* (form *linearis* or

Table 2. The time course of recolonization of rocky shores in Cornwall, expressed in years from the date of the *Torrey Canyon* disaster, March 1967 (modified from Southward & Southward 1978).

	Lizard Pt. exposed (Vellan Drang)	Lizard Pt. Sheltered (Polpeor Cove)	Portleven (west of harbor)	Maen Du Pt. Perra-nuthnoe	Sennen Cove exposed (300 m east)	Sennen Cove (near Pier)	Cape Cornwall	Godrevy	Trevone, exposed ("sewer rocks")	Trevone, sheltered (MTL reefs)
Relative exposure to wave action	+++	++	++	++	+++	++	+++	++	+++	+
Amount of oil stranded	+	++	+++	+	++	+++	++	++	++	++
Dispersant treatment	+	+++	+++	++	++	+++	++	0	++	+++
Persistance oil/oil-dispersant mix	<1	<1	<1	<1	<1	<1	<1	<1	<1	<1
Enteromorpha maximum	1	1	1	1	0-1	1	1	none	0-1	1
Maximum *Fucus* cover	2-3	1-3	1-3	2-3	1-3	1-3	1-2	none	2	1-3
Minimum of barnacles	2	2	3	4	3	3	3	<6 mo.	2	2-6
Maximum numbers of *Patella*	?	6	5	5	?	3	3	N.A.	3	5
Fucus vesiculosus starts of decline	4	4	4	4	4	5	3	N.A.	3	4
Fucus vesiculosus all gone	5	6-7	6-7	6	5	6	6	N.A.	5	8
Increase in barnacles	4	6	6	5	4	6	4	1	3	7
Numbers of *Patella* reduced	?	6	8	7	6-7	8	7	N.A.	6	N.A.
Normal richness of species regained	5	9	>10	8-10	9	9	8-9	2	5-6	>9-10

evesiculosus), whilst the bladdered form proliferated on more sheltered areas (Figures 3-5).

Very few animals were present under the dense growths of algae on the shore. Those few limpets which had survived cleared space by grazing amongst *Enteromorpha*, which allowed some barnacle settlement. Generally, any surviving barnacles were overgrown and eventually died. The dense *Fucus* canopy prevented subsequent recruitment by the sweeping action of the fronds and barrier effects (e.g., Hatton 1938; Menge 1976; Hawkins 1983). This was probably reinforced by the dense population of the predatory dogwhelk *Nucella* which built up under the canopy (see Southward and Southward 1978). These whelks eat barnacles (e.g., Connell 1961; Menge 1976) and are known to feed more efficiently under canopies (Menge 1978). In consequence, barnacles reached a minimum on most shores between 1969 and 1971 (see Table 2 and Figure 7 for Porthleven).

Limpets first recolonized during the winter of 1967. Their usual habit is to settle from the plankton in pools and damp places before moving out onto open rock later on in their first year (Lewis and Bowman 1975; Bowman and Lewis 1977). Juvenile limpets prefer *Fucus* clumps (Hawkins and Hartnoll 1983b; Hartnoll and Hawkins 1985). The first settlement and the subsequent larger settlement in winter 1968 survived well and grew quickly in the favorable conditions under the *Fucus* canopy (Figure 5). As the numbers and biomass of limpets increased, subsequent recruitment of *Fucus* was prevented by their grazing activities. As the plants aged, the grazing activity of the limpets around the holdfasts contributed to their rate of loss. Eventually the shores became very bare between 1971 and 1975, with even less algae than the normal isolated small patches typical of pre-1967 (see Figures 3, 5).

Once the dense stands of *Fucus* disappeared the abnormally dense population of limpets had only their usual restricted diet of mainly microalgae to feed upon (see Hawkins et al. 1989 for de-

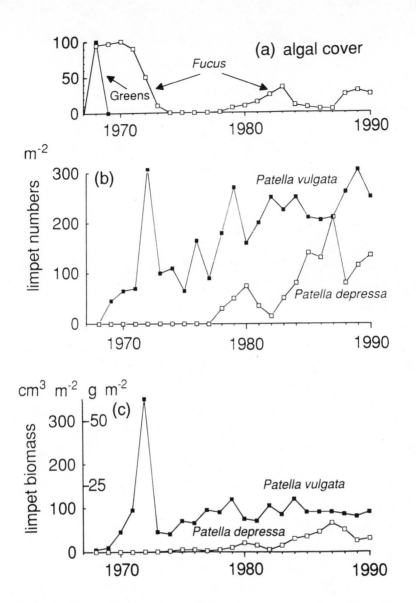

Figure 3. Changes in abundance of algae (a) limpet numbers; (b) and limpet biomass; (c) at Porthleven, Cornwall.

Figure 4. Two examples of the effects of the Torrey Canyon *oil-spill and clean-up showing the widespread nature of the response on wave-exposed shores. (a) Lizard Point, south Cornwall, May 20 1970, at the peak of growth of fucoid algae; (b) the same as (a), May 30 1977 after recolonization by grazers (*Patella *and barnacles* Chthamalus*); (c) Sennen Cove, west Cornwall May 21 1970, showing extent of fucoid cover; (d) the same spot as (c) June 1 1977, after recolonization by grazers and barnacles.*

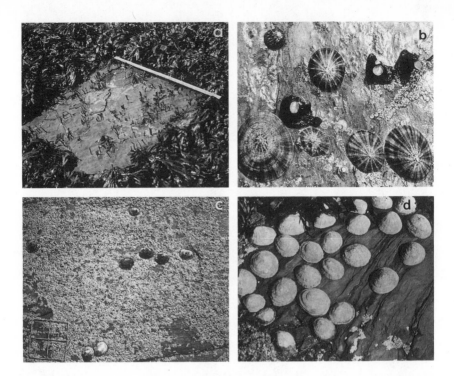

Figure 5. Changes on a moderately exposed shore after the Torrey Canyon *oil-spill and clean-up, Middle reefs Trevone North Cornwall. (a) a 1 m × 1 m quadrat, April 3, 1969, showing settlement of fucoids (20% Fucus vesiculosus and 80% F. serratus). The algae have been cut away to allow counting of fauna; (b) close-up of part of a similar quadrat, June 24, 1971 showing rapid growth of grazers* (Patella)—*largest specimen 35 mm shell length; (c) the same area, April 5 1977 showing results of grazing, and recolonizationn by barnacles* (Chthamalus *and* Semibalanus)*; (d) a cluster of* Patella vulgata, *April 24 1975, the remains of the previous dense population, alongside a pool containing fucoids. The largest shell is 40 mm long; the rings on the shell show rapid growth in the first three years of algal abundance, little thereafter.*

tails of diet). This reduction in available food prompted many of them to forsake their normal homing habit and migrate in lemming-like fronts across the shore (see Figure 6). The pronounced peak shown in 1972 (see Figure 3a) was when this front crossed the area sampled at Porthleven. Similar fronts occurred elsewhere in 1972 and 1973. A few survivors of these fronts could still be seen in the population up to ten years after the incident (see Figure 8). Subsequently, the limpet population became much reduced when the initial colonizing cohorts crashed, but later recruitment led to increased populations. Barnacle numbers increased on all shores once the fucoids declined (Figure 7). The other common limpet in the southwest (*Patella depressa* = *P. intermedia*) was very slow to recover, being very sparse for the first ten years on the mid tide level reefs at Porthleven (Figure 3, see also Southward 1979). Previously it had always been common (10-50% of the total population of limpets) at Porthleven (e.g., Crisp and Southward 1958).

Longer Term Observations, 1978-1990. Since 1978, regular observations have been possible only at Porthleven, one of the worst affected sites. An area of mid-tide flat reefs of about 40 m × 40 m extent has been studied since the spill (see Southward and Southward 1978). Whole-shore photographs have been used to assess seaweed cover. Samples of limpets for assessment of biomass, density and population structure have been taken every spring (March and April). Some destructive sampling has been unavoidable as limpets must be removed for accurate identification to species on this shore where dense growths of barnacles obscure shell characteristics. Limpets have been taken from single 1 × 1 m quadrats until 1980, and from 3-4 0.5 × 0.5 m quadrats from 1980 onwards. Information for the period 1978-1982 has been presented in Hawkins et al. (1983). The data presented in Figures 3-6 from 1983 onwards have not been previously published.

Figure 6. Abundant but starving Patella vulgata *at Porthleven, December 2, 1971, after they had consumed all the algae nearby: (a) looking north-east; (b) looking south-west from the same point, showing the "front" of limpets advancing towards the surviving clumps of* Fucus.

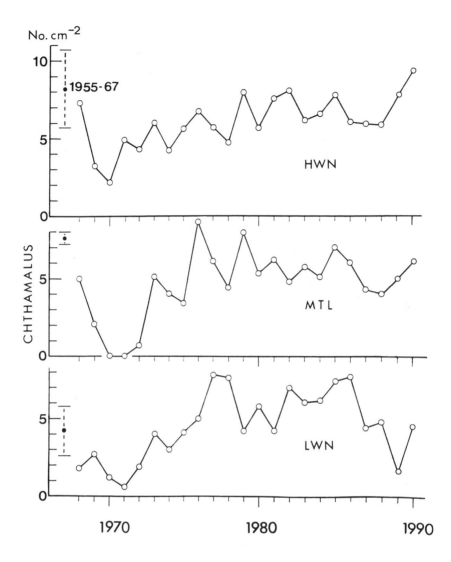

Figure 7. Density of barnacles (Chthamalus *spp.) at three shore levels at Porthleven, Cornwall. HWN high water neaps, MTL mean tide level, LWN low water neaps. The mean and standard deviation of the counts made before the spill (1955-1967) are shown.*

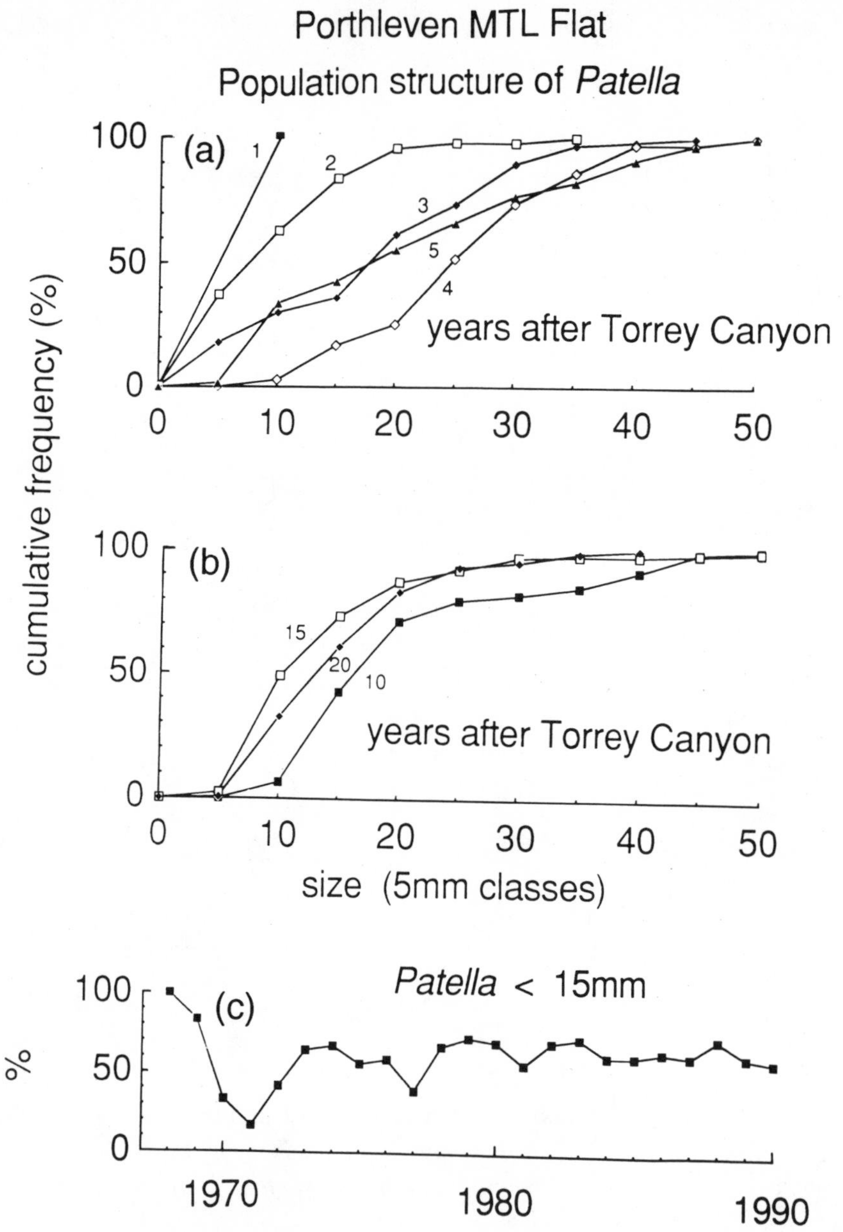

Figure 8. Changes in the population structure of Patella vulgata *with time after the* Torrey Canyon *oil spill (a, b) and the percentage of new recruits in the population at Porthleven, Cornwall (c).*

Following the bare period between 1974 and 1978, the shore went through a phase of increased *Fucus* cover (Figure 3). Although quite dense patches occurred in some places, overall cover never exceeded 40%. This still probably represents, however, a higher abundance than in the pre-spill period (A.J. Southward, unpubl. obs.) and on unaffected shores of comparable wave exposure in the southwest of Britain (Hawkins and Southward, unpubl. obs.). Subsequently, the cover decreased for a few years and then increased again.

The density of *Patella vulgata* increased with some fluctuations from 1975 before dropping in the early 1980s and then rising again. The fluctuations after 1982 probably reflect normal levels of spatial and temporal variation typical of limpet populations (see Lewis and Bowman 1975; Lewis 1976; Hartnoll and Hawkins 1985). Fluctuations of biomass were much smaller and the population was remarkably stable after 1985. *Patella depressa* also increased in numbers from 1978 onwards, dipping slightly in the early 1980s before increasing for several years. Its biomass also increased from 1982 until 1987 before dropping in the late 1980s.

More recent records allow a further appreciation of the changes in the population structure of the dominant species of limpet, *Patella vulgata* (Figure 8; see also Southward and Southward 1978; Southward 1979; Hawkins et al. 1983 for the early part of the recolonization sequence). During the period of dominance by the initial colonizing cohorts, recruitment was low (e.g., 1969-1972) presumably due to severe inter-age class competition. Subsequently, recruitment improved, and with the occasional dip (1977, 1981) the population generally had 60-70% juvenile limpets under the length of 15 mm after 1988. This is around the size at which they take up semi-permanent home scars (Jones 1948).

An overview of the stabilization of the *Patella vulgata* population is given by the reciprocal plot of numbers against biomass (Figure 9). The rapid expansion of the initial colonizing cohorts and

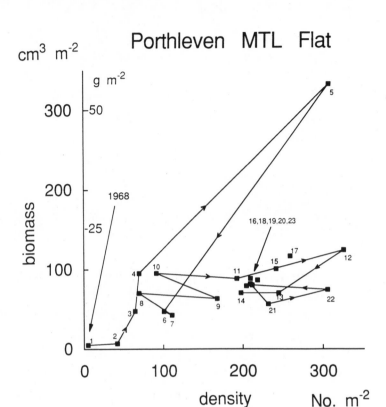

Figure 9. Reciprocal plots of density versus biomass of Patella vulgata *at Porthleven, Cornwall. Biomass is given in terms of total shell volume and dry weight per unit area.*

their subsequent crash is clearly shown. A second surge to a population of many small individuals was seen in years 10-12. Eventually the population stabilizes at around 200-250 individuals m^{-2} with a biomass of 10-20 grams dry weight, reflecting natural temporal and spatial variation. In 1992, 25 years after the disaster, we intend to undertake large scale sampling of the area to give the level of spatial variation. This has been precluded to date by the necessity of avoiding destroying the sampling area.

Barnacles have been monitored since the 1950s as part of a long-term assessment of the effects of climate on the English Channel ecosystem (see Southward 1967, 1980, 1991 for details). Counts have been made at three levels in the eulittoral (see Figure 7). Results for *Chthamalus* have been presented alone since *Semibalanus balanoides* is rare at the entrance of the English Channel (Crisp and Southward 1958; Lewis 1964). The mean and standard deviation for counts in the 12 years prior to the spill give some idea of natural temporal variation. After the initial decline following the spill, the population slowly built up at all shore levels. At high water of neap tides, all counts were within one standard deviation of the pre-spill mean from 1979 onwards although only one value (in 1990) exceeded the mean. At mid tide level, a similar irregular pattern of increase was observed, with exceptionally high counts just after the bare phase in 1976. Subsequently, there was considerable fluctuation and levels were below the pre-spill mean. Low on the shore, there was much greater fluctuation which probably reflected the greater influence of biological interactions, including predation (Connell 1961) and competition with spasmodic settlements of *Mytilus*. The latter may account for the drop observed in the late 1980s. Again, an overshoot in numbers at the end of the bare phase occurred. Low on the shore, numbers returned to natural levels quite rapidly.

Assessing the Time Scale of Recovery. Before recovery can be assessed "normality" must be defined. We consider the normal condition in the eulittoral of exposed shores of Devon and Cornwall to be one of small-scale spatial and temporal fluctuations in the major components of fucoids, barnacles and limpets. Isolated patches of *Fucus* occur, but they are never more than clumps of a few plants and total cover rarely exceeds 20% (e.g., Crisp and Southward 1958; Lewis 1964). The patchiness and fluctuations are partly generated by variation in recruitment and small-scale differences in microhabitat, predation and physical disturbance. There can also be an internal element caused by aggregation of limpets under clumps of *Fucus*, thereby reducing grazing pressure elsewhere, leading to further escapes from grazing (see Southward and Southward 1978; Hawkins and Hartnoll 1983a; Hartnoll and Hawkins 1985 for reviews of work in British Isles). Therefore, we define "recovery" as a return to normal levels of spatial and temporal variation. We have good quantitative information on barnacles for a range of affected and unaffected shores, both before and after the spill to assess this variation. For the other components, we have less frequent records, supplemented photographs and visual memories of visits made to shores in the south west by A.J. and E.C. Southward.

Southward and Southward (1978) suggested that recovery and stabilization of the shore community were occurring through damped oscillations. Subsequent published work (Hawkins et al. 1983) and the results presented here confirm that recovery and stabilization have taken place at Porthleven.

After an initial large increase in the main plant *Fucus*, and a similar but aphasic increase in the key herbivore *Patella*, subsequent fluctuations have been much smaller. *Fucus* cover was clearly abnormal for 11 years, and was perhaps slightly elevated in the early 1980s, before fluctuating at normal levels after 15 years or so. Abundance and population structure of *Patella vulgata* were clearly abnormal for at least 10 years; Figure 8 suggests that at least 13 years is probably a more reasonable estimate of return to normal. The decline in limpet grazing pressure in the 1980s probably allowed *Fucus* cover to increase again to levels higher than normal causing this second smaller amplitude oscillation.

P. depressa has been very slow to recover from the spill, which would be expected with this southern species near its northern limits (Southward and Southward 1978; Southward 1979). At unaffected sites in the west country *P. depressa* also declined in the 1970s and early 1980s (Hawkins and Southward, unpubl. obs.).

This decline was probably broadly related to long-term climatic change with this period being generally colder. Other southern elements of the fauna, such as the barnacles *Chthamalus stellatus* and *Chthamalus montagui*, also became rarer in this period (Southward unpubl. data) and this may have slowed their recolonization as well. Offshore in the late 1970s and early 1980s, northerly fish species became more common and there were changes in the plankton (Southward 1980). *Patella depressa*, along with *Chthamalus* and other elements of the southern fauna, became more common in the mid 1980s following good recruitment during the warm summers of 1983 and 1984. This return to a more southern type of fauna was also seen offshore, with increases in warm temperate fish such as pilchards (Southward et al. 1988).

The second increase in *Fucus,* in the early 1980s, may have been partly or wholly due to causes other than instability induced by the oil spill clean-up. One possible cause could have been the influence of climate. In more northern regions of Europe, shores tend to be dominated by fucoids whereas further south shores are dominated more by limpets and barnacles (see Ballantine 1961; reviewed in Hawkins and Hartnoll 1983a). A temporary switch to colder and damper more northerly conditions could have favored fucoids. Also during this period, the effects of leachates from tributyltin-based anti-fouling paint could have resulted in diminished numbers of dogwhelk predators, thus increasing the probability of escapes of fucoids (see comments in section on impacts). Unfortunately, we do not have good data for dogwhelk numbers throughout the recovery period. The extent of imposex in the population was assessed in 1987, and these data suggested that reproductive output was being slightly interfered with.

On balance, although these factors could have contributed to higher levels of *Fucus* on these shores during this period, we do feel that the second increase was at least in part a genuine component of the damping process induced by abnormalities in the limpet population still apparent 10-12 years after the disaster. The level of *Fucus* found was greater than similar unaffected shores in the south west, and no broad scale increases in *Fucus* were noted in Devon and Cornwall.

In summary, the timescale for recovery seems to be at least 10 years; however, if limpet population structure and barnacle densities are used as criteria, then at least 15 years seems more realistic. Obviously these estimates are in part subjective and intuitive due to limitations in the data. These times are not surprising when the long life spans of the main organisms are considered: *Fucus* 4-5 years (Knight and Parke 1950), *Patella* (up to 20 years, Fischer-Piette 1941, but usually < 10 years), and *Chthamalus* (at least 5 years and possibly 20). If we estimate that the average life span of a limpet is 7-10 years, it is not surprising that population structure takes 15 years to stabilize and that effects are apparent on the trophic level below.

Lessons from the *Torrey Canyon* Oil Spill

It was very quickly learned that large-scale use of dispersants caused acute toxic effects. In the few weeks taken for the oil to cross the English Channel a very different approach was adopted by the French and the Channel Islanders—using manual removal or suction devices in many instance, dispersants were only applied sparingly (Smith 1968; Cabioch 1971). This lesson of not indiscriminately using dispersants was absorbed by those in charge of response to the Santa Barbara blow-out in 1969 (Straughan 1971) and subsequent spills (e.g., *Amoco Cadiz* in 1978). Stimulus was also given to the development of less toxic dispersants. The *Torrey Canyon* also prompted evaluation of operational procedures to minimize damage and maximize effectiveness of dispersants.

The timescale for recovery of shore communities affected by dispersant application has clearly been much longer than thought by many in the early 1970s. The more optimistic viewed dense growths of seaweeds as a sign of recovery rather than being a highly disturbed community, and suggested that return to normal occurred within two years (e.g., Steinhart and Steinhart 1972; Mackin 1973). The more pessimistic still considered that only a few more years were needed for complete return to normal (e.g., Wardley-Smith 1976). Cowell et al. (1972), however, predicted a protracted period for recovery. Southward and Southward (1978) rightly dismissed optimistic forecasts and the "myth" of rapid recovery. At that time they could only assert that some shores heavily treated with dispersants had not returned to normal after 10 years, while many had taken at least 5-8 years. We can now suggest that 15 years or so is a more likely estimate of recovery time on the worst dispersant affected shores. In contrast, recovery at one of the only shores where oil was substantially untreated (Godrevy) was very quick, being almost complete within three years—a quarter of the time for dispersant treated shores. That recovery has subsequently been so protracted strengthens the case against the way in which dispersants were used after the *Torrey Canyon* oil spill.

Another outcome from the work on the *Torrey Canyon* is that a sensible way of assessing ecological damages would seem to be a formula involving a value for acute impacts plus a long term component calculated on the basis of time for recovery: i.e., Compensation (C) = Acute damages (A) + chronic long term damages (L) × years for recovery (Y).

The expectation would be that A would be large and settled immediately. L would be generally smaller and may have to be settled in separate payments with an initial payment based on existing evidence or expert opinion and a subsequent assessment (adjusted for inflation) at suitable intervals until recovery was considered complete. As the legal timescale and the ecological timescale are often similar, getting the initial estimate of recovery time may be less difficult than it seems.

However, to do this "recovery" needs to be defined. On hard substrates, "return of major species to normal levels of spatial and temporal variation in abundance" would seem appropriate. Measures of plant biomass and productivity are less useful. They are only appropriate on naturally macrophyte-dominated shores, for example, in the northeast Atlantic in shelter or low on the shore. On patchy or animal-dominated shores, they are misleading as plant biomass and primary production may actually increase in a severely disrupted community (i.e., in the absence of the next trophic level). Productivity estimates are also difficult to obtain, and the effort would be better spent on obtaining good estimates of population abundance of major species. Usually, only a few major taxa need to be studied. These can be identified from the literature and by experienced shore ecologists. Often this will be less than 10 species or categories (such as ephemeral green algae) and rarely more than 30. These can easily be quantified non-destructively in most instances, with the help of photographic or video methods. Aerial surveys or clifftop pictures of whole shores are also particularly useful.

Wider Perspective

We believe the above estimates of time for recovery to be valid. We also are well aware that they are to some extent based on a subjective assessment. In an ideal world, we would have firmer statistical evidence to separate natural changes from those induced by the spill. As in many spills, long time series from before the incident were not available—except for the barnacles which by accident were part of another study. Comparisons with control areas were hindered by the extent of the spill and compounded by difficulties in comparing communities in the English Channel which is a biogeographic boundary zone with quite rapid change occurring over short distances of < 100 km (see Crisp and Southward 1958). Interest in the ecological effects of the spill waned rapidly after the first few years. Consequently, most of the long-term monitoring we have done has been on a shoe-string budget and largely in spare time from other projects.

The general areas in which oil spills occur are reasonably predictable even if their timing is not. They usually occur in major shipping lanes or in the approaches to ports or at terminals. Despite some recent criticisms, there is still a case for background ecological monitoring of sites at risk by prolonged low frequency monitoring using appropriate methods and survey design (e.g., Cowell 1978; Baxter et al 1985). Surveys of this nature would enable assessments of damage if an acute incident occurred. A by-product would be useful ecological information as long-term studies are all too rare. Once an incident occurs, there is usually time for a snapshot "time zero" survey before oil comes ashore—if resources and skilled man power can be mobilized quickly. An eco-disaster contingency fund, which can be used to pay for rapid deployment of experts, would seem essential in any civilized petroleum-dependent society. There is also clearly the need for provision of long-term funding for studies of recovery. As Foster et al. (1990) convincingly argue, the response to the next major spill should have a properly designed and replicated experimental component. This will test the effectiveness, side-effects and recovery time of communities following different treatments, including all important "do nothing" controls in a range of habitats. An experimental protocol can be worked out in advance on the basis of experience of previous spills for various sections of coast line at risk worldwide.

Our work suggests that after an initial intensive phase of 1-2 years to assess acute impacts, a medium-term but less intensive study should be commissioned for a further 5 years. There should also be provision for prolongation as required. This would involve regular but infrequent sampling at selected sites for at least 10 years and possibly longer. As recovery seems to occur through an oscillatory process, in many instances it is likely that pre-impact levels will be reached and then overshoots or undershoots will be common. Thus, monitoring should be continued after normal levels are first reached.

There is clearly a case for not using detergents on exposed rocky coasts where wind and waves will naturally disperse the oil. What about more sheltered areas with greater sensitivity to oil pollution, or when oils more toxic than Kuwaiti crude come ashore?

Since the *Torrey Canyon* disaster various methods to physically disperse oil or to mechanically or manually collect it have been developed (Table 3). These avoid widespread application of toxic chemicals to the environment. Foster et al. (1990) provide an excellent review of the approaches currently being advocated (e.g., API 1985 a, b; NAS 1989; USCG 1989), and the decisions made in implementing response to the *Exxon Valdez* spill.

Table 3. Methods for cleaning up spilled oil on rocky shores. Modified from Foster et al. 1990 largely based on API 1985b.

Natural Processes
Enhanced Biodegradation
Manual or mechanical cutting of oiled vegetation
Manual scraping of substrate
Manual removal (e.g., shovels)
Mechanical removal (e.g., dredge, backhoe, grader)
Vacuum pumping (suck up oil on water)
Sorbents (e.g., absorbent plastic pom-poms)
Oil skimmers (on water surface at high tide)
Burning
Surface treating (gel/solidify oil)
Flushing and mechanical disturbance (e.g., high pressure nozzles)
Sand or sand/water blasting (sand mixed in water)
Steam cleaning
Chemical dispersants

Physical dispersal, especially if using steam-cleaning, sand-blasting or high temperature or high pressure water flushing will all kill plants and animals. This has essentially the same biological effect as application of toxic dispersants—although superficially it is more environmentally friendly. The new generation of less toxic dispersants have been shown in some cases to accelerate removal of stranded oil and cause little ecological damage (e.g., Crothers 1983; Gilfillan et al. 1983; Cross et al. 1987a, b; Owens et al. 1987). Their application is, however, not recommended by NAS (1989) because of continued concerns about environmental effects and doubts about their efficacy—although API (1985b) consider them still to be a viable if not a preferred method (Foster et al. 1990). Overall dispersal, whether by physical or chemical means, does not seem a good option particularly as the oil or oil and dispersant mixture must go somewhere. It may end up on more ecologically sensitive areas, or be driven into sediments in cryptic habitats under boulders or cobbles and have effects in the water column close inshore. The various collection and removal methods seem a better option (Foster et al. 1990). In most instances, manual methods (whether of oiled plants or of absorbent materials) seem to cause less disturbance than that associated with the trampling and movement of equipment, vehicles and vessels during mechanized operations.

Removal of oiled *Fucus* would be expected to have little long-term impact as generally stands recover rapidly (e.g., Lubchenco and Menge 1978; Lubchenco 1983; Hawkins and Harkin 1985), usually within 2-3 years. Removal of *Ascophyllum* (e.g., on Atlantic coasts) would be more worrying as recovery can take much longer (10 years or more, Keser et al. 1981; Vadas and Wright 1986).

On balance then, it may be preferable to not clean shores because most clean-up methods will increase acute impacts and prolong recovery—even on "sensitive" sheltered shores and the splash zone where oil may dry into long-lasting patches of tar (Foster et al. 1971). We largely concur with this view. However, shores are but one component of the ecosystem and the consequences of not cleaning up for other wildlife (birds, sea otters) must be carefully evaluated. Public concern for "higher" (and "cuddlier") animals applies pressure for their immediate rescue and indirectly for clean-up activity. There is little point, however, in saving vertebrates at great cost if their feeding grounds are disrupted for several years. Cuddly animals and birds feed largely on "creepy crawlies."

Considerable hope has been raised by the possibility of enhancement of natural biodegradation (Floodgate 1972) by adding limiting nutrients to the oil (e.g., N, Gibbs 1975), particularly in oleophilic media (e.g., Atlas and Bartha 1973). Laboratory tests and field trials have been encouraging (e.g., Atlas and Bartha 1973; Cane et al 1981). We await full publication of observations following the extensive use of fertilizers, especially oleophilic preparations, during the *Exxon Valdez* clean-up operation (see Foster et al. 1990). Preliminary results have been equivocal and suggest that the effectiveness of this approach may have been reduced by low temperature (EPA 1989; see also Atlas and Bartha 1972; Atlas 1981). Nevertheless, a treatment accelerating natural degradation processes holds much promise—provided that by-products of bacterial metabolism are not toxic and fertilization does not cause temporary eutrophication of enclosed bays. No doubt these methods and possible addition of recently-developed special strains of oil-eating bacteria will be used in the hotter climate of the Arabian Gulf.

Recovery of shores occurs largely by natural recolonization from algal propagules and larvae from unaffected areas. Spills are rarely of greater extent than larval dispersal of major species such as limpets, barnacles and mussels. Recolonization of animals with direct development (some littorinids, some dogwhelks) without a larval phase would be expected to be much slower. This is the only area where there is scope for active restoration by introducing animals from unaffected areas. To our knowledge, this has not been explored. In instances other than wide-spread dispersant treatment, it would be unnecessary as unaffected pockets usually survive from which recolonization occurs.

Conclusions

During the *Torrey Canyon* incident, virtually all the immediate and longer term damage was done to the shore (Smith 1968; Southward and Southward 1978). Most of the damage could be attributed to large scale application of dispersants during clean-up. The time scale for recovery was much greater than expected on shores treated with dispersant. Recovery of untreated shores was much quicker. Since shores recover by natural recolonization from unaffected sites, the best way of restoring shores is to minimize impact during clean-up.

The *Torrey Canyon* incident was an early example of a major ecological disaster made much worse by inappropriate response. A more considered approach to spills has emerged with subsequent experience. The more sensible procedures which have evolved were implemented during the *Exxon Valdez* spill. But as Foster et al. (1990) point out, during an environmental crisis social pressures to "do something" and the political need to be seen to be doing something, often outweigh ecological considerations. Inappropriate and unnecessary clean-up operations will be authorized which may exacerbate the damage caused by oil. The least expensive and most ecologically sound option for restoring exposed and moderately exposed shores covered by oil is probably to do nothing. On sheltered shores, collecting the oil whilst minimizing disturbance is probably the best approach. Even on these shores, Foster et al. (1990) suggest that cleaning may often be unnecessary.

However, the information on which to make decisions is clearly limited. To ensure that adequate ecological advise can be given in the future, the next oil spill in a technologically advanced country must be treated as an experiment as urged by Foster et al. (1990). We support their proposal of a properly designed survey of sites which have been oiled, non-oiled, and treated in various ways. Such a survey must be done in a range of habitats and replicated properly to overcome spatial variation and allow statistical treatment. Our work has shown the timescale of investigations required for follow-up investigations. Underwood and Peterson (1988) have discussed the sampling criteria that would be desirable in future studies.

An alternative approach would be to carry out more realistic field experiments, but this may not be possible since there are now many countries that forbid application of toxic compounds to natural communities, even in small amounts. The size and number of plots necessary for useful science would be likely to cause an unacceptable environmental impact. Some experiments involving manipulation and toxin application are possible in enclosures or mesocosms, as Bakke et al. (1983) have demonstrated; mature populations, including long-lived *Ascophyllum*, can be transplanted from the wild and given appropriate tide and wave regimes. Combined field and mesocosm sampling of pollutant-induced changes is thus possible (e.g., Bakke 1988), but the costs of the maintenance of enclosures will prevent wide usage. Hence, it might be better to await the next oil spill and then apply the lessons of experimentation and sampling from past spills. Unfortunately, each disaster seems to find the local experts unprepared, and mistakes are repeated.

The *Torrey Canyon* and subsequent accidents have shown that if spills cannot be prevented, then they should be treated as much as possible at sea, before oil is stranded. If oil comes ashore, then the only treatments permitted should be those that have been carefully evaluated beforehand. Too much time and energy is wasted on new and untried formulas for removal or dispersal. On the whole, collection seems preferable to dispersion. Decision makers must be prepared to opt for doing nothing when appropriate—but they then need sound ecological advice to help refute criticism.

Acknowledgments

S. J. Hawkins has been supported in fieldwork for this work by the Natural Environmental Research Council and the Nuffield Foundation. He would like to thank the Marine Biological Association for providing facilities and a base for field work. Special thanks go to Dr. Elspeth Jack for the help and patience during the many years that have involved "holidays" in Devon and Cornwall. We would like to thank Dr. Eve Southward for help in the field and preparation of the data. Sarah Proud deserves special thanks for help in finalizing this manuscript. Dr. Gordon Thayer's encouragement and editorial help was much appreciated.

Literature Cited

American Petroleum Institute (API). 1985a. Oil spill cleanup: options for minimizing adverse ecological impacts. American Petroleum Institute, Washington, D.C.

American Petroleum Institute (API). 1985b. Oil spill response: options for minimizing adverse ecological impacts. American Petroleum Institute, Washington, D.C.

Atlas, R. M. 1981. Microbial degradation of petroleum hydrocarbons: An environmental perspective. Microbiol. Rev. 45: 180-209.

Atlas, R. M. and R. Bartha. 1972. Biodegradation of petroleum in seawater at low temperatures. Can. J. Microbiol. 18: 1851-1855.

Atlas, R. M. and R. Bartha. 1973. Simulated biodegradation of oil using oleophilic fertilizers. Environ. Sci. Technol. 7: 538-541.

Audouin, J. V. and H. Milne-Edwards. 1832. Recherches pour servir a l'histoire naturelle du littoral de la France. 1, Paris.

Baker, J. M. 1971. Comparative toxicities of oils, oil fractions and emulsifiers, p. 78-87. *In* E. B. Cowell (ed.), The ecological effect of oil pollution on littoral communities. Institute of Petroleum, London.

Baker, J. M. 1976a. Ecological changes in Milford Haven during its history as an oil port, p. 55-66. *In* J. M. Baker (ed.), Marine ecology and oil pollution. Applied Science Publishers, London.

Baker, J. M. 1976b. Effect of oil on the marine environment. Regional Marine Oil Pollution Conf. Australia. Pet. Inst. Env. Cons. Exec. Tech. Pap.

Bakke, T. 1988. Physiological energetics of *Littorina littorea* under combined pollutant stress in field and mesocosm studies. Mar. Ecol. Prog. Ser. 46: 123-128.

Bakke, T., T. Bokn, S. E. Fevolden, O. A Frydenberg, S.P. Garner, J. Knuten, D. M. Lowe, E. Lystad, K. Moe, F. Moy, A. Pederson, S. Sporstøl, P. Thome and M. Walday. 1983. Long term effects of oil on marine benthic communities in enclosures. Littoral Rock Community Project, SOI Bergstrand. Progress report No. 3. June 1983.

Ballantine, W. J. 1961. A biologically-defined exposure scale for the comparative description of rocky shores. Fld. Stud. 1(3): 1-17.

Bally, R. and C. L. Griffiths. 1989. Effects of human trampling on a rocky shore. Intern. J. Environ. Stud. 34: 115-125.

Baxter, J. M., A. M. Jones and J. A. Simpson. 1985. Long term changes in some rocky shore communities in Orkney. Proc. R. Soc. Edinburgh. B. 75: 47-63.

Beauchamp, K. A. and M. M. Gowing. 1982. A quantitative assessment of human trampling effects on a rocky intertidal community. Mar. Environ. Res. 7: 279-293.

Boalch, G. T. and N. A. Jephson. 1978. A re-examination of the seaweeds on Colman's Traverse at Wembury. Proceedings of the 8th international seaweed symposium. Bangor, 1974.

Bone, Q. and N. A. Holme. 1968. Lessons from the *Torrey Canyon*. New. Sci. 39: 492-493.

Boney, A. D. 1968. Experiments with some detergents and certain intertidal algae. Fld. Stud. 2 (Suppl.): 55-71.

Bowman, R. S. and J.R. Lewis. 1977. Annual fluctuations in the recruitment of *Patella vulgata* L. J. Mar. Biol. Assoc. U.K. 57: 793-815.

Branch, G. M. 1984. Changes in populations of intertidal and shallow water communities in South Africa during the 1982-83 temperature anomaly. S. Afr. J. Sci. 80: 61-65.

Bryan, G. W. 1969. The effects of oil-spill removers ("detergents") on the gastropod *Nucella lapillus* on a rocky shore and in the laboratory. J. Mar. Biol. Assoc. U.K. 49: 1067-1092.

Bryan, G. W., P. E. Gibbs, L.G. Humerstone and G. R. Burt. 1986. The decline of the gastropod *Nucella lapillus* around south-west England: evidence for the effect of tributyltin from antifouling paints. J. Mar. Biol. Assoc. U.K. 66: 611-640.

Burrows, E. M. and S. M. Lodge. 1950. Note on the inter-relationships of *Patella, Balanus* and *Fucus* on a semi-exposed coast. Rep. Mar. Biol. Stn. Port Erin 62: 30-34.

Caboich, L. 1971. The fight against pollution by oil on the coasts of Brittany, p. 245-249. *In* P. Hebble (ed.), Water pollution by oil. Institute of Petroleum, London.

Cane, P. A., G. D. Floodgate and P.A. Williams. 1981. Biodegradation of beached oil, p. 373-381. *In* T. A. Oxley and S. Barry (eds.), Biodeterialtion 5. John Wiley and Sons, London.

Chapman, A. R. O. 1990. Effects of grazing, canopy cover and substratum type on the abundances of common species of seaweeds inhabiting littoral fringe tide pools. Botanica Mar. 33: 319-326.

Clark, R. B. 1982. The long-term effect of oil pollution on marine populations, communities and ecosystems. Royal Society, London.

Connell, J. H. 1961. Effect of competition, predation by *Thais lapillus*, and other factors on natural populations of the barnacle *Balanus balanoides*. Ecol. Monogr. 31: 61-104.

Connell, J. H. 1970. A predator-prey system in the marine intertidal region. *I. Balanus glandula* and several predatory species of *Thais*. Ecol. Monogr. 40: 49-78.

Connell, J. H. 1972. Community interactions on marine rocky intertidal shores. Annu. Rev. Ecol. Syst. 3: 169-192.

Connell, J. H. 1974. Field experiments in marine ecology, p. 21-54. *In* R. Mariscal (ed.), Experimental marine biology. Academic Press, New York.

Connell, J. H. 1983. On the prevalence and relative importance of interspecific competition: evidence from field experiments. Am. Nat. 111: 1119-1144.

Corner, A. D. S., A. J. Southward and E. C. Southward. 1968. Toxicity of oil-spill removers ('detergents') to marine life: an assessment using the intertidal barnacle *Elminius modestus.* J. Mar. Biol. Assoc. U.K. 48: 29-47.

Cowell, E. B. 1978. Ecological monitoring as a management tool in industry. Ocean Management 4: 273-285.

Cowell, E. B., J. M. Baker and G. B. Crapp. 1972. The biological effects of oil pollution and oil-cleaning materials on littoral communities, including saltmarshes, p. 359-364. *In* M. Ruivo (ed.), Marine pollution and sealife. Fishing News (Books), Ltd., London.

Crapp, G. B. 1971a. Monitoring the rocky shore, p. 102-113. *In* E. B. Cowell (ed.), The ecological effects of oil pollution on littoral communities. Institute of Petroleum, London.

Crapp, G. B. 1971b. Field experiments with oil and emulsifiers, p. 114-128. *In* E. C. Cowell (ed.), The ecological effects of oil pollution on littoral communities. Institute of Petroleum, London.

Crisp, D. J. 1964. The effects of the severe winter of 1962-63 on marine life in Britain. J. Anim. Ecol. 33: 165-210.

Crisp, D. J. and A. J. Southward. 1958. The distribution of intertidal organisms along the coasts of the English Channel. J. Mar. Biol. Assoc. U.K. 37: 157-208.

Cross, W. E., C. M. Martin and D. H. Tomson. 1987a. Effects of experimental releases of oil and dispersed oil on Arctic nearshore macrobenthos. II. Epibenthos. Arctic 40: 201-210.

Cross, W. E., R. T. Wilce and M. J. Fabijan. 1987b. Effects of experimental releases of oil and dispersed oil on Arctic nearshore macrobenthos. III. Macroalgae. Arctic 40: 211-219.

Crothers, J. H. 1983. Field experiments on the effects of crude oil and dispersants on the common animals and plants of rocky sea shores. Mar. Environ. Res. 8: 215-239.

Cubit, J. D. 1984. Herbivory and the seasonal abundance of algae on a high intertidal rock shore. Ecology 63: 1905-1917.

Dayton, P. K. 1971. Competition, disturbance and community organization: the provision and subsequent utilization of space in a rocky intertidal community. Ecol. Monogr 41: 351-389.

Dayton, P. K. and M. J. Tegner. 1984. Catastrophic storms, El Niño and patch stability in a southern Californian kelp community. Science. 224: 283-285.

Dicks, B. M. and J. P. Hartley. 1982. The effects of repeated small oil spills and chronic discharges. Phil. Trans. R. Soc. Lond. B. 297: 285-307.

E. P. A. 1989. Alaska oil spill bioremediation project workshop summary. Prepared for the office of research and development. U. S. Environmental Protection Agency. November 7-9, 1989.

Fischer-Piette, E. 1941. Croissance, taille maxima, et longévité possible de quelques animaux intercotidaux en fonction du milieu. Annls. Inst. Océanogr. Monaco 21: 1-28.

Floodgate, G.D. 1972. Microbial degradation of oil. Mar. Poll. Bull. 3: 41-43.

Foster, M. S., M. Neushul and R. Zingmark. 1971. The Santa Barbara oil spill part 2: Initial effects on intertidal and kelp bed organisms. Environ. Bull. 2: 115-134.

Foster, M. S., J. A. Tarpley and S. L. Dearn. 1990. To clean or not to clean: the rationale, methods, and consequences of removing oil from temperate shores. Northwest Env. J. 6: 105-120.

Gaines, S. D. and J. Roughgarden. 1985. Larval settlement rate: a leading determinant of structure in an ecological community of the intertidal zone. Proc. Natl. Acad. Sci. 82: 3707-3711.

Ghazanshahi, J., T. D. Huchel and J. S. Devinny. 1982. Alteration of southern California rocky shore ecosystems by public recreational use. J. Environ. Manage. 16: 379-394.

Gibbs, C. F. 1975. Quantitative studies on marine biodegradation of oil: I. Nutrient limitation at 14 °C. Proc. R. Soc. Lond. B. 188: 61-82.

Gibbs, P. E., G. W. Bryan, P. L. Pascoe and G. R. Burt. 1990. Reproductive abnormalities in female *Ocenebra erinacea* (Gastropoda) resulting from TBT induced imposex. J. Mar. Biol. Assoc. U.K. 70: 639-656.

Gilfillan, E. S., S. A. Hanson, D. Vallas, R. Gerber, D. S. Page, J. Foster, J. Hotham and S. D. Pratt. 1983. Effect of spills of dispersed and non-dispersed oil on intertidal infaunal community structure, p. 457-463. *In* Proceedings of the 1983 oil spill conference. American Petroleum Institute, Washington, D. C.

Great Britain House of Commons. 1968. Report from The Select Committee on Science and Technology, Session 1967-68. Coastal pollution. Her Majesty's Stationary Office, London.

Hartnoll, R. G. and S. J. Hawkins. 1985. Patchiness and fluctuations on moderately exposed rocky shores. Ophelia 24: 53-63.

Hatton, H. 1938. Essais de bionomie explicative sur quelques espéces intercotidales d'algues et d'animaux. Annls. Inst. Océanogr. Monaco 17: 241-348.

Haven, S. B. 1971. Effects of land-level changes on intertidal invertebrates, with discussion of earthquake ecological succession, p. 82-126. *In* The great Alaskan earthquake of 1964: Biology, Publ. No. 1604, Committee on the Great Alaskan Earthquake of 1964, National Research Council. National Academy Science, Washington, D.C.

Hawkins, S. J. 1981a. The influence of season and barnacles on algal colonization of *Patella vulgata* L. exclusion areas. J. Mar. Biol. Ass. U.K. 61: 1-15.

Hawkins, S. J. 1981b. The influence of *Patella* grazing on the fucoid/barnacle mosaic on moderately exposed rocky shores. Kieler Meeres. 5: 537-544.

Hawkins, S. J. 1983. Interaction of *Patella* and macroalgae with settling *Semibalanus balanoides* (L.) J. Exp. Mar. Biol. Ecol. 71: 55-72.

Hawkins, S. J. and E. Harkin. 1983. Experimental canopy removal in algal dominated communities low on the shore and in the shallow subtidal. Bot. Mar. 28: 223-230.

Hawkins, S. J. and R. G. Hartnoll. 1982a. Settlement patterns of *Semibalanus balanoides* (L.) in the Isle of Man (1977-1981). J. Exp. Mar. Biol. Ecol. 62: 271-283.

Hawkins, S. J. and R. G. Hartnoll. 1982b. The influence of barnacle cover on the growth and behavior of *Patella vulgata* on a vertical pier. J. Mar. Biol. Assoc. U.K. 62: 855-867.

Hawkins, S. J. and R. G. Hartnoll. 1983a. Grazing of intertidal algae by marine invertebrates. Oceanogr. Mar. Biol. Ann. Rev. 21: 195-282.

Hawkins, S. J. and R. G. Hartnoll. 1983b. Changes in a rocky shore community: an evaluation of monitoring. Mar. Environ. Res. 9: 131-181.

Hawkins, S. J., A. J. Southward and R. L.Barrett. 1983. Population structure of *Patella vulgata* during succession on rocky shores in South-West England. Acta Ocean. Special No.: 103-107.

Hawkins, S. J., D. C. Watson, A. S. Hill, S. Hutchinson, S. Harding, M. A. Kyriakides and T. A. Norton. 1989. A comparison of feeding mechanisms in microphagous herbivorous gastropods in relation to resource partitioning. J. Moll. Stud. 55: 151-165.

Hockey, P. A. R. and A. L. Bosman. 1986. Man as an intertidal predator in Transkei: disturbance, community convergence and management of a natural food resource. Oikos 46: 3-14.

Johansen, M. W. 1971. Effects of elevation changes on benthic algae in Prince William Sound, p. 35-68. *In* The great Alaska earthquake of 1964. The Committee on the Great Alaska Earthquake of 1964, National Research Council, National Academy Science, Washington, D.C.

Jones, N. S. 1946. Browsing of *Patella.* Nature, Lond. 158: 557.

Jones, N. S. 1948. Observations and experiments on the biology of *Patella vulgata* at Port St. Mary, Isle of Man. Proc. Trans. Liverpool. Biol. Soc. 56: 60-77.

Keser, M., R. L. Vadas and B. R. Larson. 1981. Regrowth of *Ascophyllum nodosum* and *Fucus vesiculosus* under various harvesting regimes in Maine U.S.A. Bot. Mar. 24: 29-38.

Knight, M. and M. W. Park. 1950. A biological study of *Fucus vesiculosus* and *F. serratus* L. J. Mar. Biol. Assoc. U.K. 29: 439-514.

Lebednik, P. A. 1973. Ecological effects of intertidal uplifting from nuclear testing. Mar. Biol. 20: 197-207.

Lewis, J. R. 1964. The ecology of rocky shores. English Universities Press, London.

Lewis, J. R. 1976. Long-term ecological surveillance: practical realities in the rocky littoral. Oceanogr. Mar. Biol. Ann. Rev. 14: 371-390.

Lewis, J. R. and R. S. Bowman. 1975. Local habitat induced variations in population dynamics of *Patella vulgata* L. J. Exp. Mar. Biol. Ecol. 17: 165-203.

Littler, M. M. and S. N. Murray. 1975. Impact of sewage on the distribution, abundance and community structure of rocky intertidal macro-organisms. Mar. Biol. 30: 277-291.

Lodge, S. M. 1948. Algal growth in the absence of *Patella* on an experimental strip of foreshore, Port St.Mary, Isle of Man. Liverpool. Biol. Soc. 56: 78-83.

Lubchenco, J. 1978. Plant species diversity in a marine intertidal community: importance of herbivore food preference and algal competitive abilities. Am. Nat. 112: 23-39.

Lubchenco, J. 1983. *Littorina* and *Fucus*: Effects of herbivores, substratum heterogeneity, and plant escapes during succession. Ecology. 64: 1116-1123.

Lubchenco, J. and S. D. Gaines. 1981. A unified approach to marine plant herbivore interactions. I. Populations and communities. Ann. Rev. Ecol. Syst. 12: 405-437.

Lubchenco, J. and B. A. Menge. 1978. Community development and persistence in a low rocky intertidal zone. Ecol. Monogr. 48: 67-94.

Mackin, J. G. 1973. A review of significant papers on effects of oil spills and oil field brine discharges on marine biotic communities. Texas A. & M. Res. Found. Proj. 737: 171p.

Mann, K. M. 1982. Ecology of coastal waters. University of California Press, Berkeley.

Menge, B. A. 1976. Organization of the New England rocky intertidal community: role of predation, competition and temporal heterogeneity. Am. Nat. 110: 351-369.

Menge, B. A. 1978. Predation intensity in a rocky intertidal community: effect of an algal canopy, wave action and desiccation on predator feeding rates. Oecologia (Berlin) 34: 17-35.

Menge, B. A. and T. M. Farrell. 1989. Community structure and interaction webs in shallow marine hard-bottom communities: tests of an environmental stress model. Adv. Ecol. Res. 19: 189-262.

Menge, B. A. and J. P. Sutherland. 1976. Species diversity gradients: synthesis of the roles of predation, competition and temporal heterogeneity. Am. Nat. 110: 351-369.

Menge, B. A. and J. P. Sutherland. 1987. Community regulation: variation in disturbance, competition and predation in relation to environmental stress and recruitment. Am. Nat. 130: 730-757.

Miller, K. M. and T. A. Carefoot. 1990. The role of spatial and size refuges in the interactions between juvenile barnacles and grazing limpets. J. Exp. Mar. Biol. Ecol. 134: 157-174.

Moreno, C. A., J. P. Sutherland and H. F. Jara. 1984. Man as a predator in the intertidal zone of southern Chile. Oikos 42: 155-160.

National Academy of Sciences (NAS). 1989. Using oil spill dispersants on the sea. National Academy Press, Washington, D.C.

Nelson-Smith, A. 1968. Biological consequences of oil pollution and shore cleansing. Fld. Stud. 2 (suppl.): 73-80.

O'Sullivan, A. J. and A. J. Richardson. 1967. The *Torrey Canyon* disaster and intertidal marine life. Nature, Lond. 214: 448, 541.

Owens, E. H., W. Robson and C. R. Foget. 1987. A field evaluation of selected beach-cleaning techniques. Arctic 40: 244-257.

Paine, R. T. 1966. Food web diversity and species diversity. Am. Nat. 100: 65-75.

Paine, R. T. 1974. Intertidal community structure: experimental studies on the relationship between a dominant competitor and its principle predator. Oecologia (Berlin). 15: 93-120.

Paine, R. T. 1980. Food webs linkage, interaction strength and community infrastructure. J. Anim. Ecol. 49: 667-685.

Paine, R. T. and S. A. Levin. 1981. Intertidal landscapes: disturbance and the dynamics of pattern. Ecol. Monogr. 51: 145-178.

Petpiroon, S. and B. Dicks. 1982. Environmental effects (1969 to 1981) of a refinery effluent discharged into Littlewich Bay, Milford Haven. Fld. Stud. 5: 623-641.

Petraitis, P. S. 1983. Grazing patterns of the periwinkle and their effect on sessile intertidal organisms. Ecology. 64: 522-533.

Petraitis, P. S. 1987. Factor affecting rocky intertidal shores of New England: herbivory and predation in sheltered bays. J. Exp. Mar. Biol. Ecol. 109: 117-136.

Roughgarden, J., S. D. Gaines and S. W. Pacala. 1987. Supply side ecology: the role of physical transport processes, p. 491-518. *In* J. H. R. Gee and P. S. Giller (eds.), Organisation of communities: past and present. Blackwell Scientific Publications, Oxford.

Smith, J. E. 1968. *Torrey Canyon* pollution and marine life. Cambridge University Press, London.

Sousa, W. P. 1979. Experimental investigations of disturbance and ecological succession in a rocky intertidal algal community. Ecol. Monogr. 49: 227-254.

Sousa, W. P. 1984. The role of disturbance in natural communities. Annu. Rev. Ecol. Syst. 15: 353-391.

Southgate, T, K. Wilson, T. F. Cross and A. A. Myers. 1984. Recolonisation of a rocky shore in South-West Ireland following a toxic bloom of the dinoflagellate *Gyroclinuim aureotum*. J. Mar. Biol. Assoc. U.K. 64: 485-492.

Southward, A. J. 1956. The population balance between limpets and sea weeds on wave-beaten rocky shores. Rep. Mar. Biol. Stn. Port Erin. 68: 20-29.

Southward, A. J. 1958. The zonation of plants and animals on rocky sea shores. Biol. Rev. 33: 137-177.

Southward, A. J. 1964. Limpet grazing and the control of vegetation on rocky shores, p. 165-273. *In* D. J. Crisp (ed.), Grazing in terrestrial and marine environments. Blackwell Scientific Publications, Oxford.

Southward, A. J. 1967. Recent changes in abundance of intertidal barnacles in south-west England: a possible effect of climatic deterioration. J. Mar. Biol. Assoc. U.K. 47: 81-97.

Southward, A. J. 1979. Cyclic fluctuations in population density during eleven years recolonization of rocky shores in west Cornwall following the *Torrey Canyon* oil spill in 1967, p. 85-93. *In* E. Naylor and R. G. Hartnoll (eds.), Cyclic phenomena in marine plants and animals. Pergamon Press, New York.

Southward, A. J. 1980. The Western English Channel—an inconsistent ecosystem. Nature, Lond. 285: 361-366.

Southward, A. J. 1991. Forty years changes in species composition and population density of intertidal barnacles on a rocky shore near Plymouth. J. Mar. Biol. Assoc. U.K. 71: in press.

Southward, A. J., G. T. Boalch and L. Maddock. 1988. Fluctuations in the herring and pilchard fisheries of Devon and Cornwall linked to change in climate since the 16th century. J. Mar. Biol. Assoc. U.K. 68: 423-446.

Southward, A. J. and E. C. Southward. 1978. Recolonisation of rocky shores in Cornwall after use of toxic dispersants to clean-up the *Torrey Canyon* spill. J. Fish Res. Board Can. 35: 682-706.

Spence, S. K., S. J. Hawkins and R. S. Santos. 1990a. The mollusc *Thais haematosta*—an exhibitor of imposex and potential biological indicator for TBT pollution. Mar. Ecol. 11: 147-156.

Spence, S. K., G. W. Bryan, P. E. Gibbs, D. Masters, L. Morris and S. J. Hawkins. 1990b. Effects of TBT contamination on *Nucella* populations. Funct. Ecol. 4: 425-432.

Spooner, M. F. 1970. Some ecological effects of marine oil pollution, p. 313-316. *In* Proceedings of a joint conference on prevention and control of oil spills. New York, Dec. 15-17, 1969. American Petroleum Institute, London.

Spooner, M. F. 1971. Effects of oil and emulsifiers on marine life, p. 375-376. *In* P. Hepple (ed.), Water pollution by oil. Institute of Petroleum, London.

Stebbings, R. E. 1970. Recovery of a salt marsh in Brittany sixteen months after heavy pollution by oil. Environ. Poll. 1: 163-167.

Steinhart, C. and J. Steinhart. 1972. Blow out. A case study of the Santa Barbara oil spill. Duxbury Press, North Scituate, Mass.

Stephenson, T. A. and A. Stephenson. 1949. The universal features of zonation between tide marks on rocky coasts. J. Ecol. 38: 289-305.

Stephenson, T. A. and A. Stephenson. 1972. Life between tide marks on rocky shores. W. H. Freeman Company, San Francisco, California.

Straughan, D. 1971. Biological and oceanographical survey of the Santa Barbara Channel oil spill 1969-1970. Vol. 1. Allan Hancock Foundation.

Underwood, A. J. and C. H. Peterson. 1988. Towards an ecological frame work for investigating pollution. Mar. Ecol. Prog. Ser. 46: 227-234.

United States Coast Guard (USCG). 1989. *Exxon Valdez* oil spill field shoreline treatment manual, Mimeo. manual, unpaginated. U.S. Coast Guard, Valdez, Alaska.

Vadas, R. L. and W. A. Wright. 1986. Recruitment, growth and management of *Ascophyllum nodosum*. Actas II Congr. Algas. Mar. Chil. 2: 101-113.

Van Herwerden, l. 1989. Collection of mussel worms *Pseudonereis variegata* for bait—a legislative anachronism. S. Afr. J. Mar. Sci. 8: 363-366.

Wardley-Smith, J. 1976. Oil pollution of the sea—the world wide scene. Tech. Pap. Regional Marine Oil Pollut. Conf. Australia.

Wilson, D. P. 1968. Long term effects of low concentrations of an oil spill remover ('detergent'): tests with larvae of *Sabellaria spinulosa*. J. Mar. Biol. Assoc. U.K. 48: 183-186.

What Is the United States Government Doing to Restore Habitats?

Panel Participants

William Fox
Panel Chair
Assistant Administrator
National Marine Fisheries Service
National Oceanic and Atmospheric
Administration

Nancy Foster
Panel Moderator
Director, Office of Protected Resources
National Marine Fisheries Service
National Oceanic and Atmospheric
Administration

John Knauss
Under Secretary of Commerce for Oceans
and Atmosphere and
Administrator, National Oceanic and
Atmospheric Administration

Michael Deland
Chair
Council on Environmental Quality
Executive Office of the President

Charles Ehler
Director, Office of Oceanography
and Marine Assessment
National Ocean Service
National Oceanic and Atmospheric
Administration

Jimmy Bates
Chief, Policy and Planning Division
Directorate for Civil Works
U.S. Army Corps of Engineers

Daniel Esty
Special Assistant to the Administrator
U.S. Environmental Protection Agency

Michael Brennan
Executive Assistant to the Director,
U.S. Fish and Wildlife Service

JOHN KNAUSS: The National Oceanic and Atmospheric Administration has the legislative responsibility to serve as trustee to marine resources, and has taken a strong role in a groundbreaking program to recover damages for the cost of environmental damage.

We know a lot more about how to legislate against environmental damage than we do about restoration. Environmental restoration is a kind of ecological engineering. Though we may be like "All the King's horses and all the King's men [who couldn't] put Humpty Dumpty back together again," we can, if we're willing, do a very good job of patching up the marine environment. However, environmental restoration is indeed ecological engineering and there is a cost benefit that must be applied as in any engineering effort.

At what point is the cost of restoring the environment more expensive than the worth of trying to restore it completely or as completely as possible? The Administration and the American public recognize we no longer can accept the loss of either species or habitats, hoping that the natural processes will themselves compensate for human actions. These will be key international issues, not only in this next decade but in the next century.

MICHAEL DELAND: The Council of Environmental Quality or CEQ was formed in 1970 by the National Environmental Policy Act. One of its first actions was to recommend the formation of EPA and NOAA, both of which were created later that year.

I represent an Administration which has steadfastly insisted that a healthy, vibrant economy can and must co-exist with a clean, safe environment—that they indeed are different sides of the same coin. Our policies with respect to forestry, wetlands and the coasts, for example, indicate a growing recognition of the in-place values of functioning ecosystems. President Bush is committed to the vision of Teddy Roosevelt, that we must pass on for the benefit of future generations enhanced natural resources.

In the area of coastal policy, the Administration has taken several important actions. The President proposed and signed into law comprehensive oil spill legislation, including provisions to create a fund for habitat restoration, which is one of the reasons why so many of you are gathered here today. The President has placed vast sections of the coastline off limits to oil and gas exploration and created a permanent marine sanctuary in Monterey Bay. He has essentially imposed a decade-long moratorium on these areas. Exploration and leasing of these lands cannot occur unless comprehensive environmental studies demonstrate that the impacts are both well understood and acceptable. In my judgment, that is a balanced decision.

The Administration also has stepped up law enforcement activities aimed at water polluters. In one coordinated action, for example, we charged 26 facilities in the Chesapeake Bay watershed with violations of the Clean Water Act. In another, we secured consent agreements with numerous jurisdictions to phase out by 1992 the ocean dumping of sewage sludge or else be forced to pay substantial penalties.

Finally, the Administration has embraced a goal of no net loss of wetlands. That goal is unprecedented and though it will be difficult to implement, it will continue to be the framework for all subsequent policy in this area. Thus, we have undertaken a broad public process to determine how best to achieve no net loss. But also, we have requested and obtained more funds, 32 percent in 1990, for research, including an acceleration of the national wetlands inventory. We have made large wetlands acquisitions around the country, such as the tract of 106,000 new acres in the Everglades.

Still, despite the new breeze blowing in environmental affairs, our national record on coastal issues is far from perfect. In June 1991, we presented CEQ's 20th annual report on environmental quality to the President, to his cabinet and economists. During the last 20 years of heightened environmental awareness, we found that the nation's progress against coastal degradation has been slower and more painful than our progress in other areas.

By the turn of the century, nearly 120 million Americans will live in coastal areas, an increase of approximately 40 million people since 1960. Already population density in these areas is four times the national average. Over one-half of the wetland habitats which existed when the European settlers arrived are gone—destroyed by developers, drained by farmers. The remaining wetlands are disappearing all too rapidly.

Among the other problems are excessive harvesting of marine life, non-point source pollution, oil spills, sewage and plastics. If we are to make greater progress in the next 20 years, we must begin now with a fresh approach. If we are to improve the balance sheet and achieve, for example, no net loss of wetlands, or achieve a net gain in the status of the marine environment, we must preserve existing ecosystems to the fullest extent. Moreover, we must add significantly to the restoration column of the ledger. Make no mistake, restoration science is destined to become an essential partner in the nation's campaign for a cleaner, healthier environment.

We need to know the status and trends in science and begin applying its findings and techniques to the policy challenges that confront us. This is especially true for marine habitat restoration, since the questions in this area seem to far outnumber the answers. To begin with, how can we determine the damages to an ecosystem when we do not fully understand the complex variables involved? Many have made significant contributions to the state of knowledge of marine environments, but we are challenged still by what remains unknown.

Second, how can we assess the economic costs of damage or the economic benefits of restoration or of preservation? This question of economics is often a sore spot in environmental affairs, but let's be as open-minded as we ask the economist to be. Insights of economics can be a valuable complement to scientific information when making public policy. If we can provide better information and analysis to the economist, then our national policies can be both efficient and effective.

A third question is how to best coordinate the research effort and inform the policymakers. There are many experts: federal, state, local governments, universities, private sector, citizen groups, international organizations. A mechanism is needed to marshal all that expertise in the service of better policy, though such coordination and long-term support of federal environmental research is the exception rather than the rule.

By its very nature, the ocean environment requires an approach that is integrated, interdisciplinary and long-term. There are several ways this might be addressed in the future. One example is the proposal for a national environmental research institute. A second was suggested by the Carnegie Commission in April 1990, which recommended a greater integration of science and environmental expertise in the White House policy structure. Proposed legislation to elevate EPA to the President's cabinet has also stimulated discussion. As the President's in-house environmental advisor, I participate in these discussions and welcome the views of all those in restoration and related efforts.

Finally, what policies can we advance in the face of considerable uncertainty? Scientific uncertainty is a perennial question in matters of the environment. Americans have never considered research a substitute for action. Parenthetically, I reject the careless argument that equates research with inaction. The dichotomy is a false one. Research and action must proceed in tandem. Our current knowledge of ocean degradation requires serious, meaningful precautionary actions, since restoration of marine habitats may be infinitely more difficult than restoration of smaller bodies of water.

Even as we work aggressively to improve our ability to restore damaged marine environments, we must remain true to our innate common sense—an ounce of prevention is worth a pound of cure. Indeed, medicine might be a good analogy: we know how to fix a lacerated finger, though we're still cautious when we handle knives. We cannot play God. Restoration is not creation. We humans are part of nature, and yet nature has done a remarkable job without our help for millions and millions of years.

In the 1990s, we need to focus increasing effort on preventing pollution in the first place rather than on spending millions, billions, to clean it up afterwards. Pollution prevention makes sense for the environment, but increasingly it makes sense for the regulated community as well. Many leaders in business and government now realize that pollution prevention not only results in healthier people, healthier ecosystems; it also results in healthier balance sheets and permanently so. This is yet another example of what we mean by the convergence of economic and environmental interests.

WILLIAM FOX: Before addressing what the National Marine Fisheries Service and Sea Grant are doing in restoration, two fundamental questions need to be answered.

First, why restore marine-estuarine habitats? Near-shore ocean and estuarine fish and wildlife habitats continue to be lost in substantial quantity and function. A loss of this habitat is probably the greatest threat to the long-term productivity of our fishery resources. With very few of our fisheries strictly dependent upon an oceanic life history, most have important relationships to our coastal and estuarine areas. Our challenge is to assist materially in turning back the threat to these areas.

Second, what do we mean by marine restoration? For many, the thought of restoration primarily involves wetlands, such as coastal marshes and lowland hardwood forests. But marine restoration includes two broad areas: biogeography and restoration techniques. In the area of biogeography, we should consider restoration opportunities across the range of marine estuarine, freshwater and anadromous habitats. Restoration techniques include revegetation of wetlands, restoration of tidal wetland areas by breaching of dikes, restoration of anadromous fish zones through removal of stream obstructions and establishment of forested buffer zones.

Also included in restoration are increased shrimp production through water structure regulation to improve access, habitat establishment of oyster through bed construction, increased angler harvests by construction of artificial reef and kelp bed restoration, restoration of coral reefs and the restoration of coastal urban habitats.

The National Marine Fisheries Service is involved in a number of important areas of habitat restoration. (1) the new Department of the Army/NOAA program to restore and create fisheries habitat; (2) restoration research with in-house resources as part of NOAA's Coastal Ocean Program; (3) technical support within NOAA's Superfund efforts; (4) the Sea Grant program's support of restoration research; (5) restoration in response to Pacific Northwest hydroelectric development.

Taking the first of these, the U.S. Army Corps of Engineers and National Marine Fisheries Service together have a program to restore and create fishery habitats. This is an expanded national program, based on favorable results of a pilot study in four states. The pilot work was carried out from 1986 through 1988 by the Corps, NMFS, state agencies and cooperating parties within the Corps' operations and maintenance projects.

The study was a multi-agency process of site selection, planning, design, construction and monitoring of selected measures. It involved 12 NMFS offices, 15 Corps divisions, waterway experiment stations and 44 state agencies and other parties. Of 69 potential sites, six were selected in California, Maryland, North Carolina and Texas. These include the vegetation of dredge material sites in North Carolina and Texas; construction in Mission Bay, California, where we used broken riprap disposal techniques; construction of oyster beds in Somerset County, Maryland; construction of submerged vegetation bed, Twitch Cove, Maryland; and the reclamation of 1,400 acres of agricultural land, Prospect Island, in the Sacramento River delta, California. Habitat restoration work has been completed on five of the sites, and the sixth is nearing completion. Scientific evaluation of the sites in North Carolina, Texas and Maryland is underway.

NMFS and the Corps are now developing a new national plan for which start-up funding has been allotted for FY 1991. Our primary focal point for estuarine research is the Southeast Fisheries

Center. Our laboratories at Beaufort and Galveston conduct a multidisciplinary, generic research program on estuarine and coastal habitats. The relative value of nursery areas is studied, including marshes, seagrasses, mangroves and non-vegetative habitats and fish species. These are the only federal research laboratories studying seagrass systems.

Seagrass research has been continuous for a decade, and over 50 publications have been produced. This research has been funded from numerous sources at state and federal levels. Examples of some ongoing seagrass research activities include collection and study of data on the success of transplants, in areas of California, Texas, North Carolina, New Jersey, Florida and Connecticut, and development of cost-effective transplanting techniques. Cost benefit evaluations are being done in Florida to compare planting techniques. Other areas of study include the role of seagrass bed sediment microflora, and the effects of dominant North American seagrass species on currents, waves and sediment stabilization.

We have a cooperative seagrass study in Hope Sound, Florida, with the U.S. Fish and Wildlife Service and Florida Department of Natural Resources. This study is determining the relationship between seagrass distribution, abundance and composition with light and hydrological factors and regulatory measures (involving primarily boating traffic), to protect seagrasses that are major habitat for trout, redfish and endangered Florida manatee. A cooperative study in Tampa Bay with the Florida Department of Natural Resources and the University of South Florida is examining if artificially propagated seagrass minerals provide habitat functions similar to natural minerals.

Substantial participation in the National Marine Fisheries Service/Corps of Engineers restoration pilot study comes out of our Southeast Fisheries Center. Similar research projects at the Beaufort and Galveston Laboratories have created salt marsh habitats for fishery organisms.

Restoration research within NOAA's new Coastal Ocean Program began in fiscal year 1990. The Estuarine Habitat Program is a cooperative program of National Marine Fisheries Service, National Environmental Satellite Data and Information Service, known as NESDIS, and the Office of Oceanic and Atmospheric Research within NOAA.

One major area of research interest is habitat enhancement, restoration and construction. The objective is to (1) determine how coastal and estuarine habitats function to support living marine resources, including research on methods for habitat restoration; (2) to locate and measure critical habitats and the rate at which they are being changed or lost, using satellite, aerial, photographic and surface-level surveys; and (3) to synthesize information for use by managers in protecting, conserving and restoring critical habitats.

The National Marine Fisheries Service has cooperative programs with the University of South Carolina, University of North Carolina, North Carolina State University, Texas A&M, and San Diego State. No other national effort addresses these fundamental questions in a comprehensive, integrated program. Examples of ongoing work include developing a manual for assessing restored wetlands and comparing them with natural wetlands at the San Diego State University; accelerating and evaluating the development of restored and created seagrass beds at the Universities of South Florida and North Carolina; developing ecosystem functions in restored and constructed wetlands, at North Carolina State, San Diego State and University of North Carolina; and an ecosystem comparison of transplanted and native salt marshes, and the chronology and development of habitat for fishery species at Texas A&M.

NOAA is the trustee of living marine resources, under the Comprehensive Environmental Response Compensation and Liability Act of 1990 (CERCLA), also known as Superfund. In a team effort with others in NOAA, National Marine Fisheries Service regional centers provide technical support within NOAA's Superfund program by participating in preliminary natural resource surveys; determining the extent of resource injury assessments; developing technical recommendations for NOAA positions; and designing plans for restoration.

NMFS is developing cost-effective restoration alternatives for the coastal waters off Palos Verdes Peninsula in Southern California. The ongoing case involves the releasing of DDT and PCBs by Montrose Chemical Corporation. Damage is also being assessed in New Bedford Harbor at the head of Buzzards Bay, after five corporations, including AVX, released PCBs and heavy metals for over 40 years.

Through grants to universities in Delaware and Florida, Sea Grant researchers are also active in restoration, by applying genetic engineering technologies to develop lower cost and improved plant species for wetlands. In South Carolina, Sea Grant studies on the effects of diking wetlands on estuarine flora and fauna were instrumental in the decision to protect historic wetlands. Concerned Sea Grant researchers in Louisiana have been examining the means to remedy the severe coastal wetland losses there, especially regarding plant stress and wetland configuration. California Sea Grant has been the leader in developing techniques for marsh restoration by improving measurement of functional equivalency between differing habitats.

Despite these many activities, we still know so little. Operating on a case-by-case basis, we are assuming that restoration techniques will increase fish and wildlife productivity. But success is unpredictable, and benefits are difficult to quantify. While marine habitat restoration has been around for quite some time, it has not been demonstrated scientifically to provide the functional values of natural ecosystems. Growth from propagation, for example, does not necessarily mean an area has developed functional relations to those natural areas it has been designed to emulate. We are in for a long period of research and development to enhance the existing technologies and evaluate how well they are working.

We badly need demonstrated successes. We must place high priority not only on long-term research, but on demonstrating that restoration and enhancement can occur with present technology and by promoting cost benefit information. Armed with such information, Congress and the public will be able to see the potential of restoration, which should translate into support for public policy, programs, further technology development and conservation of aquatic habitats.

CHARLES EHLER: The Office of Oceanography and Marine Assessment (in 1992, name changed to Office of Ocean Resources Conservation and Assessment) manages many programs related to habitat restoration, among them coastal management, the estuarine research reserve that provides funds to states to acquire important coastal habitats, and the marine sanctuaries program.

All of these programs are important to the nation because of the opportunities they present to prevent habitat degradation in the first place and, in the long run, reduce the need for habitat restoration activities. We never want to lose sight that prevention is, in fact, the long-term answer to problems of restoration.

We manage the national status and trends program which, since 1984, has monitored toxic chemicals in bottom-feeding fish, in shellfish and in sediments at almost 300 sites around the coastal United States. A major report on the study's results to date describes the state of coastal environmental quality.

Through another important effort, our strategic assessment program, we have organized and published important characteristics of coastal areas of the entire United States, including detailed information on the biogeography of over 400 species of fish, shellfish, birds and marine mammals.

Through the Office of Oceanography and Marine Assessment, we are regularly producing atlas products, including new electronic versions. We have prepared a major digital data base using information contained in the maps of the Fish and Wildlife Service's national wetlands inventory, and together we are publishing a major report on coastal wetlands of the United States. All of this information is indirectly important to habitat restoration.

A major activity of my office is damage assessment and restoration planning, especially related to abandoned hazardous waste for Superfund sites. After 10 years, federal efforts to implement the natural resource provisions of the Superfund Act are now gaining momentum, as a result both of the *Exxon Valdez* and other major oil spills in 1989 and the Administration's pro-active environmental policies.

The Under Secretary for Oceans and Atmosphere, Dr. John Knauss, acts on behalf of the Secretary of Commerce as a federal trustee for natural resources in coastal areas. He is responsible for determining the extent and cost of injuries to natural resources from various short and long-term events, and for seeking compensation from responsible parties to restore the injured resources. NOAA began carrying out these responsibilities as a federal natural resource trustee in 1980 when it referred a claim to the Department of Justice for natural resource damages resulting from PCB contamination in New Bedford Harbor in Massachusetts.

Starting in 1984, our hazardous materials response program initiated an evaluation of nearly 500 coastal hazardous waste sites, all on EPA's national priorities list, with the conclusion that approximately half of those sites could affect NOAA trust resources. The annual summary of these site reports prepared over each of the last five years has provided critical information for identifying candidate sites for natural resource damage claims and for restoration alternatives. More recently, following the *Exxon Valdez* oil spill,

NOAA established a damage assessment and restoration center, professionally staffed with scientists, economists and lawyers.

In spring 1990, we announced two major new damage claims, one for injuries to NOAA trust resources from DDT and PCBs in marine sediments off Southern California's Palos Verdes Peninsula and the other for contaminated-related injuries to the natural resources of Elliott Bay located near Seattle in Puget Sound. Additional claims in other coastal areas are being considered. Restoration of natural resources injured by discharges of oil or releases of hazardous substances is one of the mandated goals of the Superfund Act. Achieving that goal through complex litigation is a long and arduous task. The New Bedford case, which was filed in 1983, is close to a settlement that will lead to restoration of the harbor and its closed shellfishing areas. Restoration options currently being reviewed for New Bedford harbor, assuming resources become available, include artificial reef construction for lobster and recreational finfish, and marsh creation and remediation.

Fortunately, our trust resources have not yet been contaminated in many threatened sites. In these instances, NOAA, through our coastal resource coordinators who are all located in EPA regional offices and our hazardous materials response program, works with EPA to develop cleanup plans that minimize future risks to those resources. We have worked closely with EPA to ensure that ecological issues related to marine and estuarine resources become an integral component of Superfund cleanup activities. We have been involved actively in actions at over 250 sites, providing NOAA recommendations on work plans, evaluating remedial investigations and feasibility studies, and interpreting data at the request of EPA regional offices.

Biological and chemical monitoring programs have been established at many Superfund sites to establish cleanup targets, to ensure that cleanup actions actually protect natural resources and to measure the effectiveness of habitat restoration projects. In some instances, NOAA, working with other federal and state trustees, has been able to negotiate restorative measures without referring claims against potentially responsible parties. For example, at a Superfund site in Texas, responsible parties agreed to create new wetlands to compensate the public for contamination of existing wetlands. At the Munisport landfill site in North Miami, Florida, NOAA, working together with other trustee agencies and environmental groups, was able to encourage a plan that will greatly reduce contaminant discharges threatening mangroves in nearby Biscayne Bay. Trustee actions at the Army Creek landfill in New Castle, Delaware, modified a cleanup plan not only to reduce wetlands destruction, but to create wetlands and mitigate the adverse effects of capping that landfill. The remedy includes fisheries enhancement and wetlands rehabilitation.

Although the federal government has moved slowly to implement the natural resource provisions of the Superfund Act, its efforts to do so have been gathering momentum and will provide the means to repair much of the environmental harm caused by careless waste disposal practices in the past.

DANIEL ESTY: The quality of our marine environment today is of tremendous interest. It can no longer be accepted as a dumping ground for wastes. There is no free lunch, and a lot of what we're doing these days is sending out overdue bills on free lunches from the past 40 years. Nor is it going to be acceptable in the future to consider our marine environment as a buffer to absorb accidents. Increased public attention to oil spills is one indication of a change in the atmosphere. The marine environment truly is a treasure to be guarded, protected and, where necessary, restored. It will require a great deal of commitment and work.

EPA is contributing in three relevant areas. First, wetlands; second, our National Estuary Program; and third, our special work in response to the *Exxon Valdez* oil spill. In the mid-1980s, EPA launched an important program of restoring wetland habitat and gathering scientific information. In particular, EPA initiated a wetlands research program in 1985. Literature was reviewed to establish the current baseline of knowledge on creating and restoring wetlands. A second area of focus has been analysis and evaluation of wetlands mitigation data, required under the Clean Water Act. In particular, Section 404 of the act requires analysis of the environmental impacts of dredge and fill activities. We are developing methods for comparing the functions of created and restored wetlands with natural wetlands. Finally, we're comparing field data on naturally occurring and artificially created and restored wetlands, trying to get out and actually test what is being done.

In 1990 we issued the first document to come out of our research effort, *Wetlands Creation and Restoration: The Status of a Science*. It's an attempt to look back, figure out where we are, and establish the baseline from which we can move forward. A second major product, a handbook on mitigation, has recently been published. Our efforts in these areas are probably far too modest. We are spending some $300,000 a year on this wetlands mitigation project, far too little to really make the kind of difference we are going to need in the future.

We also have been participating in the Wetlands Forum, which brings together parties from a range of public and private sector interests to try to develop an appropriate response to wetlands destruction and restoration. We are acting on a number of the Forum's recommendations. Important among these is the suggestion that we implement stronger mitigation requirements within our regulatory programs so that, taken as a whole, conversions of wetlands are fully offset by wetlands, restoration creation. We are also trying to ensure that where government agencies are responsible for the destruction of wetlands, there is full compensation. Finally, we're working to establish a public/private/national initiative that is going to seek out opportunities to restore wetlands. A particular focus should be on government-owned lands and areas that have been altered by construction of government facilities.

In addition to the general activities of wetlands protection and restoration, EPA is involved in our National Estuary Program. By act of Congress, EPA has designated 17 estuaries around the country—from Casco Bay, Maine, to San Francisco Bay to Galveston Bay on the Gulf—as national estuaries and has set up special programs for trying to preserve, protect and restore these watersheds. Whenever a national estuary is established we implement action plan demonstration projects that are designed to undertake cleanup of the estuary in priority areas. In San Francisco, we are addressing the runoff that is polluting San Francisco Bay. We've identified five areas of concern: (1) floatable contaminants, such as litter; (2) hydrocarbons such as oil from streets and parking lots; (3) toxic contaminants such as pesticides or spilled household and commercial compounds; (4) inert settleable solids such as construction site sand and silt; and (5) nutrients, such as fertilizers, that can lead to elevated oxygen demand in the estuary.

We are establishing and analyzing the use of artificial wetlands to purify and restore the Bay. Artificial wetlands have been used for some time as a way to safely trap and assimilate pollutants from treated municipal wastewater before they enter downstream rivers. We are now exploring whether artificial wetlands might be used to restore and protect the bays and rivers of our country, particularly from urban storm water. The site we've selected in San Francisco is an ecologically stable, 55-acre marsh that was built about 10 years ago for research on storm water flows from urban areas. We have a series of programs to test whether this artificial wetland can serve a restoration function.

We are conducting surveys to establish the marsh's physical, chemical and biological baseline conditions. We're installing and calibrating basic flow, monitoring and sampling equipment. Full-scale demonstrations of the wetland system will be conducted over an 18-month period covering two complete wet weather seasons. We're also analyzing the necessary water column, sediment and plant tissue data to determine the wetland's ability to trap pollutants. And we are estimating the major costs of constructing and operating artificial wetlands at other locations, using this as a pilot. A report on the results will include the feasibility of using wetlands for storm water treatment and the effects on estuaries and bays. The final report will discuss the characteristics of the marsh, the quantities and qualities of storm water and dry weather flows passing through it, inlet and outlet concentrations of selected indicator pollutants, indicators of wetland conditions such as habitat and, finally, water quality in the estuary before and after the effort. This is one example of the 17 national estuary programs that we are operating, a good many of which contain elements aimed at restoration of the marine environment.

A third area of EPA involvement is the effort to clean up the Gulf of Alaska and Prince William Sound following the *Exxon Valdez* oil spill. The sad story is familiar: 11 million gallons of oil, the largest amount ever spilled in United States waters, dumped into one of the most pristine environments this country still has. There were terrible effects on the ecology and the environment, with a variety of natural resources damaged, including fish, birds, mammals, intertidal plants and animals. At the time, it was clear both to President Bush and Governor Cowper in Alaska that the tragedy was a mess and that dramatic efforts above and beyond anything we had done before would be required to address this huge-scale problem.

President Bush, in particular, asked EPA to coordinate efforts among the federal government agencies, to restore Prince William Sound and the Gulf of Alaska to the extent possible and as quickly as possible. Since then, we have worked very closely with NOAA, with the Fish and Wildlife Service and Department of Agriculture and with the other participants under Superfund. The damage assessment process has been successful and has produced a great deal of information, much of which will be put into a public depository. We also have put forward a number of restoration efforts through a planning group, co-chaired by EPA and the state of Alaska, working out of Anchorage. We focused a great deal on public participation in the process, to ensure that the people of Alaska have a chance to say how our response should be handled. We have begun to identify feasible restoration options. We have taken public comment and looked at overall approaches and have adopted an ecosystem-wide approach, really trying to take a broad brush perspective. We've begun to determine the pace of natural recovery and are beginning to determine where direct restoration might be appropriate. We are also beginning to look at the costs and the options of restoration efforts.

Symposiums were held in April 1990, in Alaska and a good deal of public information was provided. Ideas were offered in three broad areas: one, restoration approaches and philosophies; two, the planning process; and three, specific restoration activities. These include planting rye grass and reestablishing it in coastal areas that have been affected by the oil spill and cleanup efforts, and establishing the feasibility of putting back into the system some of the critical intertidal fauna, such as grazers, as well as limpets and predators, such as starfish.

We are also identifying sites in need of protection from coastal erosions and, through a restoration pilot project, examining the practicality of reestablishing supertidal beach rye grass as a way of dealing with this problem. We are working to identify upland habitats used by ducks and other birds and animals in the area. In addition, we are also examining the land status, uses and management plans of areas near the spill in preparation for possible acquisition of resources equivalent to those that were damaged.

The challenge in Alaska is enormous, and the restoration ambitions there are great. The goal the President set for us is an enormous one—trying to restore the area to pristine shape after such a devastating amount of oil was spilled into the water. The growing focus on the environment has raised the public's expectations enormously, and that makes our efforts all the more important. We really do need to create a science of restoration, to be used as a tool in our environmental protection efforts.

JIMMY BATES: In keeping with the President's goals, the U.S. Army Corps of Engineers is focusing on restoration of environmental values. The current Corps budget guidance and program philosophy makes restoration of fish and wildlife habitats a priority status, giving it equal standing with our traditional purposes of commercial navigation and flood control. With appropriate sponsors and cost sharing, we may now pursue environmental restoration in new and ongoing studies.

Restoration, as we have defined it, can include both mitigation and some level of enhancement. It refers to those measures we would undertake to return the existing but degraded productivity of fish and wildlife habitats to their modern historic levels—this means the natural, normal levels of fish and wildlife productivity that have occurred in recent times. For example, we would not set, nor pursue, unrealistic and unpractical goals such as restoring the fish and wildlife habitat of the Chesapeake Bay to the conditions that Captain John Smith found and recorded. We would, and in fact are, cooperating with other federal and state agencies, investigating ways to return the Bay's environmental richness. We also are factoring the goals of the Bay program into the planning and design of Corps water resource projects in that area, including identifying opportunities for the beneficial use of clean, dredged material removed from federal navigation channels.

The Corps has developed considerable environmental engineering expertise in the area of coastal fish and wildlife habitat restoration and has been putting this expertise to work in our ongoing programs. Between 1973 and 1978, the Corps conducted the dredge material research program at our Waterways Experiment Station in Vicksburg, Mississippi. Much information was gained on the effects of dredging and disposal activities in coastal areas. This effort was the beginning of the Corps' extensive knowledge on the feasibility of creating and restoring wetlands, shellfish and other coastal fishery habitats using dredge materials.

Thousands of acres of coastal fish and wildlife habitat have been created or restored using clean dredge material. The Corps has established an extensive technology transfer program within the agency itself that could also be utilized in other resource agencies. This program includes numerous publications, videos, workshops and a combination of classroom and hands-on field training conducted by our Waterways Experiment Station.

One of the greatest challenges is establishing ways to reach consensus among resource agencies on how best to determine and measure the monetary and non-monetary benefits and trade-offs associated with the use of dredge material to create and restore fish and wildlife habitat. For example, when and where is a created or restored vegetative wetland equal to or more valuable than the bay bottom that is to be converted? How can and should we address these considerations so that timely, informed decisions can be made? Until these questions are answered, it will be very difficult for us to effectively pursue coastal habitat creation and wetland habitat initiatives, including opportunities that are authorized under Section 150 of the Water Resource Development Act of 1976 and Section 135 of the Act of 1986.

Dr. Fox discussed the 1985 Department of Army/NOAA fish habitat restoration pilot study. While the amount of fisheries habitat that was restored or created under that pilot study was small, it clearly demonstrated that these two agencies could work well together to identify habitat improvement opportunities, select sites, and plan and execute habitat restoration projects with, in most cases, no significant additional costs to associated Corps projects. This latter is a major consideration for determining the Corps' ability to participate in a habitat restoration or creation project without special approval or additional authority. There is broad, high-level support for establishing an ongoing national program between the two agencies.

Finally, we are studying, in cooperation with most of the agencies represented here, ways to restore wetlands and associated habitats on the coast of Louisiana. This is a major habitat restoration investigation, and, hopefully, it will result in significant benefits to the coastal resources in that state and the Gulf of Mexico. The Corps has had a wetlands research program for a number of years, conducted by the Waterways Experiment Station. This was considerably expanded in fiscal year 1991 to examine and develop state-of-the-art techniques in constructing wetlands, including coastal wetlands. The program will investigate how wetlands are developed and maintained, and what their functions and values are. This new information will benefit all of our agencies' efforts to manage and restore the coastal wetlands. As demonstrated in the joint pilot study, we believe this program will contribute significantly to restoring the amount and quality of fishery habitats in our coastal areas, and will improve the engineering and economic efficiency of our civil works/water resources project.

The Corps is participating in the National Estuary Program studies being conducted around the country and is a major participant in the Gulf of Mexico Program. These studies, under the administration of EPA, will hopefully result in the restoration of much of the presently degraded marine and coastal fish and wildlife habitats in our estuaries. The Corps is using its environmental engineering expertise and extensive knowledge of coastal resources to assist the development of comprehensive conservation management plans to attain these restoration objectives.

The Department of Army, Department of Interior, NOAA and EPA are working together to develop a proposal for a joint federal budget initiative to protect and restore the nation's coastal resources. While these may be tight times to propose such an initiative, the spirit of cooperation and willingness of agencies to work together for a common end—restoration of our coastal resources and habitats—is extremely encouraging. The Corps is also working very closely with the Marine Board of the National Research Council and exploring the roles of coastal engineering, the engineering profession and the government in protecting, restoring, creating and enhancing marine habitats. The result of this jointly funded research should be valuable in guiding future coastal marine-oriented environmental engineering.

We are well aware of the budget realities. There will always be more habitat restoration and creation needs and opportunities than we will be able to pursue at any given time. Therefore, we must move forward together to establish a timely, technically sound process to assess, evaluate and prioritize those needs. Such a process would greatly aid our decision-makers in making the hard choices required to prepare and support budget items, particularly our new environmental initiatives with little, if any, historic precedence to guide them. Every agency and organization represented has special skills, experience, expertise and authorities that contribute toward restoring the nation's marine environment. By working together, we can and will succeed.

MICHAEL BRENNAN: Within the Fish and Wildlife Service, it is becoming increasingly clear that we cannot dither around about what environmental issues are or are not significant. Our agency—and the country—is realizing that habitat and habitat loss is the critical environmental issue, with respect to living resources.

The Fish and Wildlife Service historically has had close relationships with individual species-specific activities. The Service is turning its attention increasingly to those species' common denominator habitats which they need to survive. Nowhere is this more urgent than the marine and coastal environment. Approximately 50 percent of the population has come to live in coastal areas in the United States. Our coastal population currently grows at over four times the national average, 80 percent of that increase occurring between 1950 and 1984. Over one-third of the population lives within the narrow band of counties with coastlines. Over half of the population growth projected in the next 20 years is expected to occur in those coastal counties. This growth constitutes an unprecedented assault on marine and coastal living resources. Along with land-based sources of pollution, we note increasing habitat loss and alteration, nutrification, and steady declines in living resources. Species and habitats—the miner's canary of the coastal environment—are declining throughout the coastal zone. Natural filters for pollution increasingly are being destroyed through pollutants themselves or through development impacts.

The government simply is not doing enough with respect to marine and coastal habitats. While there has been an increasing focus, we need to continue the upswing. We need to become increasingly involved in habitat restoration and perhaps, more importantly, prevent the need for habitat restoration activities. The Service's involvement in coastal resource protection management is fairly broad based. We have trust responsibilities for migratory birds, anadromous fish, certain Great Lakes fishes, threatened and endangered species, some marine mammals and service lands, many of which share that common denominator of a coastal or marine environment. We have over 200 field stations divided up among our ecological services field offices, Fish and Wildlife assistant offices, our refuges, hatcheries and research installations.

The Service's mission is to conserve, protect and enhance fish and wildlife and their habitats for the continuing benefit of people, with an increasing focus on conservation of habitat. In fulfilling that mission, mitigation of adverse impacts, caused by development activities, is often necessary. Our habitat restoration and enhancement activities generally fall under three categories: direct habitat management actions taken on Service refuges; technical assistance to other federal, state and local government agencies and private landowners; and research. In our refuge program, the Service is a major coastal landowner and land manager. We have over 160 coastal national wildlife refuges for fish and wildlife resource protection and management, providing sanctuaries and a superb research laboratory for habitat needs. Our management activities on the refuges include wetland restoration, enhancement, shoreline stabilization and erosion control, water quality improvement and contaminant cleanup work.

In cooperation with other agencies, our managers work to identify and conduct habitat enhancement projects, such as the use of clean dredge materials to create nesting habitat for birds, wetland restoration and stabilizing coastal waterways adjacent to refuge properties. The Service provides biological expertise to other federal agencies, states, industry and the public concerning conservation and improvement of fish and wildlife habitat in connection with land and water development activities.

We recommend that these agencies provide technical guidance to complete restoration and mitigation measures under Sections 10 and 404 of the Clean Water Act, administered principally by the Corps of Engineers with EPA. They also have obligations under the Fish and Wildlife Coordination Act, as well as hydropower planning and relicensing.

Since 1978, we have seen an increasing focus within the Service on forging cooperative agreements directly with private landowners. In wetlands areas in particular, it is critical to involve the private sector in habitat conservation and recovery efforts. Over 60 percent of wetlands throughout the United States are privately owned. Unless you get the landowners to recognize the value of their lands and work with you, you are not going to get very far. So we are increasing our efforts to enter into partnership agreements with the private sector. In fiscal years 1987 through 1989, we entered into more than 1,700 such agreements with landowners, covering some 35,000 acres of wetland habitat. While still a very small proportion of wetlands, it is significant that these agreements were triggered solely by the desire of both parties to work together to try and recover the habitat for living resources.

We're involved in several private land stewardship programs at the local level, such as Partners for Waterfowl Tomorrow in the Lower Mississippi Valley. This is typical of a number of programs that involve landowner agreements, education and outreach activities by the Service, development of wintering habitat for waterfowl and wetland enhancement and management. The Central Valley joint venture in California is a partnership with private organizations and federal, state and local resource agencies. The goal is to restore approximately 120,000 acres of wetlands and enhance natural resource values of an additional 300,000 acres. We are looking at another coalition project for Back Bay restoration in Virginia, including watershed management planning, restoration of wetlands and establishment of filter strips.

We administer the North American waterfowl management plan, under the North American Wetlands Conservation Act. Under that Act, state and conservation organizations and private matching funds are used to protect, restore and manage habitat, including coastal areas critical to migratory birds and fishery resources. Two coastal joint venture projects approved for funding include restoration of about 1,200 acres of marsh at the Delta National Wildlife Refuge in coastal Louisiana, and acquisition and restoration of coastal wetlands as part of a project in southeast South Carolina.

The Service has extensive research program activities related to coastal habitat enhancement and restoration. Approximately 40 percent of our research activities relate to coastal and estuarine resources and problems implemented through our National Wetlands Research Center in Slidell, Louisiana. The Center's research focuses on Louisiana coastal marsh restoration, further development of the wetlands geographic information system, waterfowl research and contaminant impact assessment. The Service also conducts extensive research on coastal fishery resources, including Pacific salmon, striped bass, Atlantic salmon, red drum, sturgeon, shad and herring. We have recently dedicated the Northeast Anadromous Fish Laboratory in Turners Falls, Massachusetts and conduct extensive research on Great Lakes fishes through our National Fisheries Research Center in Ann Arbor.

That's a broad overview of Service activities. In addition, we have two exciting new initiatives: natural resource damage assessment and the coastal estuaries initiative. The Service, acting as a trustee, has had a slow start getting involved with Comprehensive Environmental Response Compensation and Liability Act of 1990 (CERCLA). EPA has been out front, leading the first half of the Superfund cleanup to meet primarily human health obligations. The second half of that program, which from the standpoint of wildlife management is equally important, is restoration of damaged natural resources.

In the past, natural resource damages involvement has been thrust upon us as opposed to a planned activity. This year, the Service successfully proposed as a budget initiative a comprehensive natural resource damages assessment program. This program will result in a significant upswing in the level of Service involvement in the Superfund process with an eye towards restoration, recovery, or acquisition of the equivalent of the damaged resources. It is absolutely critical that natural resource trustees be very actively involved with the parties doing the initial cleanup, to ensure that through the Remedial Investigation/Feasibility Study (RI/FS) process, remedies be chosen that at the very least do not further harm affected natural resources in the area. It may indeed be possible, by factoring natural resource concerns in the front end of the project, to yield a more cost effective cleanup overall, both from the standpoint of public health and of natural resource impact.

We are also involved with the agencies' representative panel in the development of the coastal estuaries initiative. These agencies, which are the primary federal trustees for living resources in the coastal environment, need to show the public and the states that a coordinated program can be extremely effective. By demonstrating the ability to work together, we can encourage a similar governmental alliance at the state level. We think it is absolutely essential that this type of pro-active program be developed to ensure comprehensive coastal living resource protection and to provide a national capability to respond quickly to coastal and marine estuary concerns.

Question-and-Answer Session

AUDIENCE MEMBER: It's admirable that two of the four agencies represented—the Corps and National Marine Fisheries Service—have a formal agreement and joint study, but why were the other two agencies left out? It's one thing to have cooperative efforts, but a formal agreement ensures that your activities are coordinated.

WILLIAM FOX: I agree it would be beneficial to work out additional agreements, or maybe an umbrella agreement.

JIMMY BATES: That's the reason, in establishing the wetlands research program, that we're calling on all the agencies together, so that we avoid duplication.

DANIEL ESTY: There is a tremendous amount of cooperation, if not in the formal sense, between EPA and the other agencies. It's quite remarkable if you remember the bad old days when EPA and the Corps would go to battle every other week over a 404 case.

There is a whole new approach to dealing with wetlands protection and it's quite significant.

AUDIENCE MEMBER: No one would be willing to say that a created wetland has the same functional values as a natural wetland. How does this affect mitigation in the 404 program? Would mitigation no longer be a viable option to receiving a permit from the Corps? Or would you have to change the ratio to 4 to 1 or 5 to 1 for unavoidable losses?

ESTY: It's quite clear, at least at this point, that artificial wetlands don't have the same capacity and same environmental effect as natural wetlands, but improved efforts may make them closer in the future. In addition, we may be able to compensate with volume for a lack of equality and quality.

AUDIENCE MEMBER: I've been working with the agreement to restore habitats with the Corps. I wanted to respond to the first question. Early on in designing the pilot study, it was recognized that to be successful it must include other state and federal agencies. Guidance to the field from day one emphasized working with those agencies. The Fish and Wildlife Service and EPA were brought into the process of site identification and, in some cases, involved in subsequent work.

AUDIENCE MEMBER: We're currently operating under a grant from the Charleston Commissioners of Public Works to assess the effectiveness of mitigation efforts relating to lifeline construction across some saltwater in Charleston Harbor. One of the main reasons we received this grant was because EPA Region 4 was willing to consider this as part of the Commissioners of Public Works mitigation activities. Is there any concerted federal initiative to encourage similar sorts of cooperation in a financial way?

ESTY: EPA is looking for creative ways of establishing partnerships with municipalities, with counties and with states. As resources become tighter, the need to leverage federal funds and to work with others who have resources will only increase.

CHARLES EHLER: Many of NOAA's actions involve some creative financial arrangements, with responsible parties paying not only for some cleanup and damage assessment, but all the way through restoration alternatives.

BATES: The Water Resource Development Act of 1986 instituted new cost sharing requirements in all Corps of Engineer water resource development projects, with the exception of inland navigation. All of our feasibility studies for developing a project have to be shared by a local sponsor. Protecting and restoring our environment requires not only cost sharing, but creative thinking and initiatives.

MICHAEL BRENNAN: Speaking for the U.S. Fish and Wildlife Service, our offices, through the 404 affirmative process, have extensive talks with local project proponents and sponsors, encouraging creativity and minimizing habitat impacts. The Service increasingly relies on private sector partnerships or on getting a local landowner actively involved in habitat restoration. That's a key component to our strategy in working toward the goal of no net loss.

AUDIENCE MEMBER: We have to work much more closely with fishery and waterfowl managers and others who are involved in issues facing natural resource managers. How can we possibly succeed in restoring resources and maintaining natural resource stocks—fisheries stocks—if we don't work together?

FOX: Fishery managers have to take into account the structure, function and quality of the habitat that was producing the recruitment of fish populations. By the same token, we must stress the importance of these fishery resources to the country, to the economy and the dependence of them upon the near-shore environment. We need to improve the connection of these elements.

ESTY: In addition to fishery managers, there are wide categories of other people who need to be brought into the dialogue over how to protect water quality and restore marine habitats. You could begin with transportation planners, those who are building roads and highways. You would also want to include water users, water system maintainers, farmers and a variety of other groups. Agencies increasingly will be working together to have across-the-board participation in decision-making.

BRENNAN: It is absolutely critical that populations affected by the project accept the goals. There must be a broadly-shared, public commitment to pay the environmental costs as we go along. The greatest challenge may be in putting together an integrated approach to making habitat conservation part of the cost of doing business.

BATES: As we develop our feasibility reports, we need to do outreach and incorporate all of these views in the plan.

EHLER: We need to involve not only fisheries managers, but those concerned about managing coastal environments. Many states have an institutional arrangement, called the coastal management program.

AUDIENCE MEMBER: I have two questions. The damage assessment program has gotten off to a slow start, due to lack of money and lack of integration of the cleanup process under Superfund. On the oil spill side, Congress has come up with money for doing damage assessments, but we don't have such a fix for hazardous waste sites. Would it be appropriate to open up Superfund to seed money, to get the assessment process rolling? This would create a threat against the responsible parties, that if they don't pay for the assessment, the government will go forward with it. Second, why hasn't a cleanup program at EPA been integrated with the damage assessment process as envisioned by Congress in 1986?

ESTY: Superfund is an enormously complicated law put through with high expectations. Funding for damage assessments is one area where we have some distance to go. The Superfund reauthorization debate that begins next year would be a good opportunity to see some changes made in restructuring the Act to address your concerns.

EHLER: We had a total absence of resources available for damage assessment until last year when NOAA received $2.5 million from Congress. Before that, the few efforts that we had were funded out of our base program. The long-term prospect on financing will come from settlements, such as New Bedford, which would be used to repay money spent on initial damage assessments and to initiate new assessments. The other is to provide incentives to responsible parties for them to pay up front, rather than 10 years after litigation, for damage assessment and restoration. On the integration side, we're continually learning that there is a close relationship between cleanup, both in terms of oil spills and Superfund sites, and damage assessment activities. One of our responses is assigning NOAA scientists, who we call coastal resource coordinators, to work with the Coast Guard on cleanup activities and to locate them in EPA regional offices.

BRENNAN: When Superfund was first enacted in 1980, it was available for natural resource damage assessments and for damages themselves. But Congress amended the Act in 1986 to preclude those purposes. The need to come up with resources for damage assessment is quite clear. The Service and trustee agencies certainly have limited resources and have "robbed Peter to pay Paul." Trustee agencies will need to be more involved in damage assessments as Superfund costs mount. Private sector responsible parties are going to be unwilling to settle with the United States unless the natural resource trustee of the site has agreed not to sue. Government agencies and the private sector will want greater participation by the trustees in the Superfund process—and the key to that is more resources. We're learning more about Superfund and the need for a coordinated cleanup process. Early impediments to close cooperation among EPA and other trustee agencies are being ironed out.

AUDIENCE MEMBER: Over the past year and a half, we have been working very closely with EPA and negotiating with responsible parties on cleanup sites. Natural resource damage assessment and restoration is achieving an equal footing with the cleanup process. We are working with EPA on a memorandum of understanding to standardize how we approach negotiations with responsible parties.

AUDIENCE MEMBER: I'm responsible for looking at the relationship between wetlands and waste disposal. In relation to Superfund where there is no culpable party, there is no source of funding for the natural resource trustees to do damage assessments or restoration. In preparation for reauthorization of Comprehensive Environmental Response Compensation and Liability Act of 1990 (CERCLA), is there any debate over establishing a general fund for ecological restoration?

FOX: Nobody is aware of it, but that is an excellent idea.

AUDIENCE MEMBER: My first point is on the oil spill in Alaska. There was a credibility gap between what the public thought or what we might have led them to think that we could do. Second, if the President were to turn to the five agencies and ask, "How are we doing?" could you hazard a guess on what our net loss is for wetlands in 1990?

BRENNAN: Your first point is a caveat in terms of recognizing what we do and don't know. Mr. Deland made the point that we cannot rely on the mitigation and recovery of damaged resources alone. Society is far more effective at having an impact on resources than restoring them. As resource managers, we should not overstate our restoration abilities when we consider how we preserve living resources. With respect to wetland loss, our research indicates historically that we are losing up to 500,000 acres a year. That continues to be the trend, both through man's activities and through natural processes.

AUDIENCE MEMBER: How do your agencies feel about mitigation banking, that is, the development or enhancement of habitat in anticipation of future impacts? In Oregon, almost everything from the Cascades to the coast is a wetland. There are a lot of problems for those folks who want to do things on small, isolated areas. There is an interesting idea of doing mitigation and monitoring its effect prior to development.

ESTY: It's a concept that is part of the mix that needs to be explored as we define what no net loss of wetlands means in practice.

BRENNAN: From the Service's perspective, mitigation banking needs to be carefully considered. We have to be honest in considering how much we know about wetland values and functions. Is an acre-for-acre replacement or acre-for-acre drawdown of a mitigation bank an appropriate way to implement the 404 program? There are some key unanswered questions.

AUDIENCE MEMBER: I heard little mention of an aggressive approach to public education, with the exception of very limited funds in EPA, estuary programs and a bit in the National Marine Fisheries Service. In these times of strict budgets, public education is the first to get cut and the last to be restored. I plead with you that you include in your initiatives some funds for public education. A lot of folks, including the Marine Advisory Service of NOAA Sea Grant and the Cooperative Extension Service of Agriculture, should be brought in to help bring under control the habitat degradation. Because if we don't turn it around, we're just pushing noodles up the stream.

What Have We Learned, Where Are We Headed?

Panel Participants

Thomas Campbell

Panel Chair
General Counsel
Office of the General Counsel
National Oceanic and Atmospheric
Administration

Grayson Cecil

Special Counsel for Natural Resources
Office of the General Counsel
National Oceanic and Atmospheric
Administration

Victoria Kincke

General Counsel
Heritage Environmental Services, Inc.

Roger McManus

President
Center for Marine Conservation

V. Kerry Smith

Distinguished Professor of Economics
North Carolina State University

William Conner

Damage Assessment Center
Office of Oceanography
and Marine Assessment
National Ocean Service
National Oceanic and Atmospheric
Administration

THOMAS CAMPBELL: It is important to first clarify terms such as cleanup, restoration and mitigation. Cleanup differs from restoration as washing out a wound differs from performing first aid on that wound. If you cut yourself, the first job is to remove the insult. If you are in a dirty environment, you make sure all of the foreign material is out of the cut. Many wounds will heal themselves without any further proactive steps, though in some instances, there is a need for stitches or antibiotics. Essentially, cleanup is the washing of the wound and, by analogy, restoration is taking the proactive steps to help nature in the healing and restorative process.

It is also useful to distinguish between restoration and mitigation. At times there are trade-offs, and we try to mitigate the damage of losing a wetland, though we are focusing on restoration, not mitigation. When you have lost wetland as a result of past pollution, can it be restored?

Significant opportunities present themselves to us as a result of the Clean Water Act, CERCLA, the Fish and Wildlife Coordination Act, and the Marine Sanctuaries Research and Protection Act. Under these statutes, there will be significant funds made available to the scientific community. During the last year, for example, we have had the Shell Oil settlement for $11 million; the AVX settlement in New Bedford Harbor, with $66 million for cleanup and restoration. As a result of our Sandpater Shelf case, the Los Angeles County Sanitation District will make $12 million available for damage assessment, restoration planning and implementation. Following the Megabord spill, we were able to negotiate $250,000 with the owners of the vessel to do baseline damage assessment work. The list goes on: Apex, in Galveston, $350,000; Simpson Champion, $1 million; Army Creek, $800,000; Bay Way, $600,000; GM Foundries, $630,000. In the first year of real focus on this effort, the federal government has taken in more than $45 million.

The *Exxon Valdez* oil spill has meant a $50 million expenditure for assessing environmental damage. Ultimately, hundreds of millions of dollars will be available for restoration there. In addition, NOAA has been given responsibility for writing the damage assessment regulations under the new Oil Spill Liability Act, and we will be fronting significant support into that effort. Such efforts furnish the seed money for scientific research to go forward. Combined with the current political climate, it is possible for significant progress to be made in this area over the next five years.

WILLIAM CONNER: Most of my experience has been on damage assessment rather than restoration. We have looked at a number of different habitats: salt marshes, submerged grass beds, coral reefs, mangroves, kelp forests, rivers, isolated urban wetlands, disused docks and artificial reefs. From my perspective, there are four main themes.

To begin with, we are working in a developing science. If you look at the state of our environment 150 years ago and consider that we are now on a continuum of time that extends into the future, you start to understand where we are in this science. It was not until the 1960s that the public started to develop a heightened concern about our environment. In the '70s, we began to pass laws to control pollution and to understand and quantify its significance. In the '80s, we are trying to mitigate and even reverse the effects of human activity, all of which have resulted from population growth and the industrial revolution. The message is that we should have patience with this science.

The second theme deals with functional replacement. There is a lot of concern about whether a restored or created habitat is the functional equivalent of the natural habitat. While this concern is valid, if you can develop a habitat that only functions at 40 percent, that still is a benefit to the environment.

This argument relates to my third theme. With regard to mitigation versus restoration, we need to have different thresholds of acceptability. In the case of mitigation, you determine whether to undertake an action when you have control over whether the environmental effects of that action actually occur. In the case of restoration, the insult has already occurred and we need to try and recover the lost resources. While we tend to mix these two concepts together, we need to separate them in our decision making.

Finally, we have a great potential for generating funds to undertake restoration to replace injured environmental and natural resources. To be as efficient and effective as possible, we need to be clear about our goals and objectives. We have to understand what we are trying to accomplish before we can choose the techniques that we want to employ.

Let us choose the best techniques available, while also trying out new techniques. Let us do research and development. Let us include in our restoration planning provisions for the detailed studies that we need to finish the recipes, for example, for marshes, so that we can do the best job possible. In our restoration planning, let us also include provisions for long-term monitoring so that we can determine whether objectives are being met and whether the restoration activities have been successful. And let us make sure that whatever we learn we publish and disseminate to the rest of the scientific community, so that the lessons may be used by others.

V. KERRY SMITH: Is there a role for economics in the emerging science of restoration? To determine that, we need to answer a few questions. What do we mean by restoration? Is it simply functional replacement or is it more than that? Are we talking about replacing a few members of a species? Replacing a functional species? How long do we keep that up? What does it cost, and what is it worth?

The law itself provides the mandate for the economics, as well as the resources indirectly and directly for the research and restoration that is going to take place. CERCLA, Superfund Amendments and Reauthorization Act (SARA) and a recent Court of Appeals decision are laws that suggest the damages associated with releasing oil or hazardous waste into the environment should be measured by the cost of restoring the natural resources, unless those costs are grossly disproportionate to the economic value of the affected resources. The economic value is to be measured including both use and non-use values.

While I cannot say what grossly disproportionate means, I do know that the connection between the costs of restoration and the value of the resources is an important one. It suggests that the level of restoration—over what time frame, and whether we actually address this question of functional replacement or try to define it in some other way—will not only affect the science of what we do, but also the resources that we have available. In short, partial restoration affects the dollar figure that we attach to the cost of restoration. Measuring the value of non-marketed resources is an emerging science in its own right, but there is a great deal of art involved in it as well.

Economists can observe what people do in markets and try to infer their motives, their values and how those values relate to what they purchase. A vast amount of detective work is required to uncover those values. An important element in the detective work is looking at choices that people actually make to select and acquire indirectly the services of resources. We buy a house near the beach. We go to a pleasant beach-front resort for vacations. We travel to scenic resources on weekends. Indirectly we have incurred costs and paid a price for those resources. But those costs do not reflect the full dimensions of what people value. To get that information, we have to move from the comfortable domain that most economists are familiar with to what is called "contingent valuation," or the use of surveys to ask people how and why they value resources.

Scientists are not afraid to get to know the species that they want to study. Economists, on the whole, feel differently. In some sense they feel that getting to know the species they want to study will in some way taint them. But there are a large number of economists actively developing the science of listening to people and coding their responses to very careful, scientifically designed questions. This can help inform the process not only of damage assessment but also of restoration.

Restoration, at least in terms of the law, is intended to restore a functioning resource to its baseline conditions, prior to whatever caused the injuries. An economist might argue that the way to deal with restoration is to go back to that old harping "marginal benefits equals marginal costs." You restore the services of the resource until the point where the marginal benefits equal the marginal costs of restorative activity and that is where you stop.

But the law suggests that you must restore the resource to the baseline conditions. Not that everyone is intended to have equal access to its use in any way they wish. We are to maintain the resource at a position corresponding to its baseline condition and yielding the same values. What that suggests to an economist is that we have to measure all of those values. We cannot rely exclusively on the values that correspond most closely to private values and commodities, but must measure the non-use values that people realize just because they know that there is a clean environment. Just because they know there is a functioning interaction between that environment and the whole set of marine species. That is not easy, but it is just as important to the task of restoration as it is to understanding the interaction between the species in any functioning marine environment.

Can we do it? We can, but only with lots of help, not just from economists, but from the science community as well. We have the task of trying to explain the scientific dimensions to people in a way that they can understand, appreciate and respond to. So it is important that in coding these questions, when asking people why they would value restoration or maintenance of an environment, we have an authentic representation of the science and that it be translated in a reasonable and understandable way.

How do we go from what is essentially a series of promises to a set of deliverables? There is not a systematic program of research that brings together the natural scientists and the economists in developing the kind of information that will have an impact on restoration and how that affects the awards that are available under damage assessment.

Furthermore, there is not an institutional memory of what we have learned about this emerging science in both economics and the natural sciences. We have drawn down the stock of knowledge that exists from indirect sources about as far as we can. We do not always have access to the new knowledge that is available in emerging cases. They cannot be submitted to peer review and we need to change that. Moreover, we do not have a set of guidelines developed about what we do know in the non-market valuation area.

There are a set of very general pronouncements that do not answer the hard questions. Finally, we need to understand that there is a mix of public and private participation that is possible in restoration. There are many opportunities for using private incentives in the process of restoration and not relying exclusively on public intervention. This might well be the most efficient course.

ROGER MCMANUS: I would like to address the question of where we go from here as a political problem. What has been lost has become increasingly clear. It has been cited, for example, that over 90 percent of California coastal wetlands has been lost. For those in the environmental community, that raises a question about balance in our values. A balance in environmental policy is an issue—or a plea or a promotion—that we hear a lot about in making difficult policy decisions. The President himself has invoked the notion that we need to look to balance in solving environmental problems. We can easily identify a balance with respect to marine near-shore and coastal ecosystems. We have given up almost all of the resources in many cases, and so balance would argue that we should not lose any more. This leads to a second theme that is been echoed by policymakers, scientists and attorneys: what we are talking about is a way to make up for what we have lost, not a mitigation excuse for going ahead and destroying what we have left. When proposals for new marinas are found to cause problems for coastal wetlands, the repeated response is, "I'll go build one somewhere else." But this does not always work out as easily as it sounds.

At the Center for Marine Conservation we have been conducting mangrove research in Florida over the last couple of years and have concluded that it is easy to go into the mangrove and plant plants. But what these plants do and whether they fulfill the roles of the ecosystems that have been lost are open questions. So it is important to keep repeating like a mantra: Restoration ecology is a way for making up what we have lost. It is not an excuse for destroying what we have left.

NOAA has certainly begun to take extensive leadership to recoup the damages that we have suffered as a society from destruction and pollution of near-shore and coastal marine ecoystems. This effort suggests that we as a society value these ecosystems and the products and services that they provide.

Economists have trouble valuing these things for obvious reasons. We are not dealing in the open marketplace, and it is difficult to establish values and determine how we should both penalize people who destroy resources and how much we should pay to restore them. But in any case, we do value them. This is not just swampland or idle marshland that has no value to society. The more those values are recognized, the more society will be able to make decisions about what amount of effort should be given to them.

VICTORIA KINCKE: There are people in industry today who are very much aware of restoration and the efforts of NOAA and other agencies. Others, though, are wondering what restoration has to do with natural resources and habitats. The good news is that restoration does work. It can have absolutely outstanding effects if it is planned and implemented well and if it is closely monitored. The difficulty comes in trying to define the preexisting condition of an environment or habitat. Baselines can be very difficult to determine. Obviously if you have a particular environmental incident in an area that has been well studied, establishing a baseline may not be a problem. However, many of us have to deal with situations where baselines have never been established.

In addition, many in industry are concerned that we do not provide compensation for problems that we did not cause. If a spill occurs, for example, you will find that industry willingly recognizes its legal responsibility to provide compensation suitable to that incident. But we are opposed to providing compensation and restoration activities for incidents that happen as a result of urbanization, for incidents prior to ours, and for seasonal fluctuations. Dr. Stephen Hawkins (this volume) refers to baselines in his discussion of the *Torrey Canyon* incident.

Another problem is obtaining the right mix of experts. When you have a Superfund site or a spill, determining who the most appropriate experts might be for a particular problem may not be easy. The timing and integration of scientific and economic experts may always be a problem in any type of restoration activity.

Another issue, and one that Dr. Hawkins also discusses, is cleanup versus further damage versus restoration. When an incident occurs, you may be requested to undertake some form of cleanup, a cleanup that for political, for public, for societal reasons experts argue will further damage the environment. A good example was the use of detergents in the *Torrey Canyon* spill. As policymakers and as a society we have to resolve what trade-offs we are going to accept in terms of making an area look attractive versus doing further environmental damage.

Lastly, in any situation we will always have very different views. In addition to federal legislation, there are state statutes and perspectives on almost any environmental incident. There is going to have to be a blending and mix of federal versus state versus industry versus lawyer versus economist and that is going to raise problems along the way.

Where does industry go from here? We need to deal with education. There are many people in industry who do not have the slightest idea what restoration is and what its objectives are. In any natural resource damage assessment or restoration process, we are going to have the economists. We are going to have people talking about contingent evaluation studies and baseline, about use values and non-use values. These are all terms that many people do not understand.

Lawyers, scientists and economists need to prepare people, before an incident happens, by letting them know that restoration is here to stay and these are the tools you need to deal with it. Industry can then undertake some preplanning, at least with regard to spills and, to some extent, with Superfund. There could be an evaluation of what type of natural resources exist in their manufacturing area. Many industry representatives are spending time in terms of emergency preparedness and crisis management, and we ought to put restoration and natural resource damage assessment up on the same plane.

We have a need for coordination and cooperation between agencies and between agencies and the regulated community. This will avoid spending months talking and arguing, before any of the funds become available to get the resources back on the road to recovery. To the extent that people can come to a table and arrive at a consensus in a tolerant, moderate atmosphere, that is going to speed the process along. Ultimately, we will wind up with a result that industry can live with, and more important, a result that benefits the resources and meets the objectives of society.

Question-and-Answer Session

THOMAS CAMPBELL: I am going to put my policy hat on and ask, is this effort worthwhile? I have been told that the damage assessment process is too costly, too complex and that restoration in many cases simply is not possible. Furthermore, the economics that relate to damage assessment are embryonic. Some people argue that there are no values to these non-market resources, while others say it is impossible to put a specific dollar value on them. Is this damage assessment and restoration worth doing?

ROGER MCMANUS: From a conservation organization response, I see it as absolutely essential. These are valuable resources.

They are difficult to value in terms of assessing the cost and the amount that we should either penalize polluters or the damaging parties or have society fund. But we will better understand how valuable these resources are if we go through this process and try to evaluate and restore them.

It has been suggested that it takes $2 million to restore 10 or 15 acres of wetlands. That is one measure of what we lose when we lose wetlands. That is something that the public and policymakers can start to focus on and give them some basis for making a decision about this issue. In the case of the *Exxon Valdez* spill, at one point Interior regulations suggested that sea otters were worth something like $13.95 because that was the market value of their pelts. But Exxon, for each otter saved, spent some tens of thousands of dollars, so from an economist's perspective perhaps each sea otter is worth $30,000. I do not know which figure is right, but certainly by going through the exercise, we are going to come closer to realizing what the true values of these resources are to us. We will take better care to ensure that the resources are protected and to take reasonable and adequate measures to restore them once they have been damaged.

CAMPBELL: From an economics perspective, can we actually value these resources?

V. KERRY SMITH: I would like to turn the question around. Like it or not, if we do nothing we also place a value on them and that is zero. Any decision we make places a value on the resources. If we make a decision to restore a wetland, it costs us something. The value to society must have been at least as much as that cost. The question is, can we do better? Can we bring the values of the resources and the efforts that we put to try to restore them more in line with the values that people place on the resources, and the answer is a resounding yes. We do that all the time.

There are two or three billion dollars a year spent by marketing research firms to value new products, before those products come on the line. We can do at least as well in valuing non-marketed resources. The economics are in better than an embryonic state. It needs to be expanded, but clearly we can do better than doing nothing. We also can do substantially better both from the perspective of plaintiffs and defendants because the cost of restoration may be grossly disproportionate in relationship to the real value of many natural resources. We need to know that, just as much as we need to know when the cost of restoration is appropriate.

A lot of discussion has centered around the idea of grossly disproportionate. A court decision in Ohio suggested a range of three to one: one dollar's worth of benefit for three dollars' worth of actual restoration cost. Will we ever get to the point in restoration science when we will be able to really restore a system at a three-to-one ratio? We have seen some success stories in terms of seagrasses and mangroves and dock systems, where we have been able to do restoration. Will we ever get to the point where we can economically restore marine systems?

WILLIAM CONNER: Maybe we are putting the cart before the horse. Another part of this question is whether we can really quantify injuries within a three-fold factor of accuracy. If we cannot, the magnitude of error becomes even larger when determining an appropriate restoration and economic valuation. But we can certainly try to quantify injuries and develop reasonable restoration plans, the costs of which would not be grossly disproportionate. It is an integrative process where you have to quantify on both fronts simultaneously as you close in on your objective.

AUDIENCE MEMBER: When NOAA performs its damage assessments, does it make its calculations on the basis of the loss of individuals or does it make its calculations on the cost of restoration?

Do the monies that are assessed go into the general treasury or are they dedicated to a restoration fund?

CAMPBELL: Essentially, the measure of damages is the cost of restoration and a number of economic factors are used to determining whether or not restoration costs are grossly disproportionate to the benefit. There is also the issue of lost use during the period of time between the injury and the time that the actual restoration took place and there is some question whether that is a double recovery, but we do consider that. The major measure of damage, though, is the cost of the restoration process.

In response to the second question, all the money that is recovered goes into fisheries and restoration projects intended to enhance the marine environment. It does not go back into the general treasury.

AUDIENCE MEMBER: I would like to comment on the New Bedford case which has come up several times, and the figure of $66 million that is being bandied about for resource restoration. The actual figure for the cleanup might be close to $600 million. From the viewpoint of the commercial fishermen in the area, they have suffered for about 15 years from the loss of the resource. They are concerned that a lot of the money is being eaten up by past legal costs. At a recent meeting, they were suggesting blacktopping docks and things like that with some of the money rather than actually using it on fishery enhancement.

It is important that when money is appropriated for resource restoration, it be allocated to help the people who are most affected by the pollution. Just by spending this money, the problem still might not be going away.

CAMPBELL: I would have you ask your readers how their fisheries are being managed, from conception to harvest. Many fisheries managers are focusing on dividing up an ever diminishing harvest instead of looking at enhancing and restoring the habitat that will increase the harvest by creating a place for tomorrow's fish to grow.

SMITH: The intention of the measures at least of natural resource damage is not to replace values of private losses incurred by fishermen or any other group. The intention is to value the services of what are really natural assets. It happens that fishermen are users. There are also recreationists and a whole set of users whose values count. The restoration plan presumably has to address all of those parties' concerns, not just any one group.

AUDIENCE MEMBER: Sometimes we fall into the trap of preaching to the choir. Too often, once the choir goes home there is no more music and we do not get anywhere. We are all in a sense custodians of the resources for future generations. We need very soon to devote some attention to public information. There is a lot of public concern but we need to go beyond that. Concern simply is not a substitute for understanding or commitment, and we need that for three reasons.

One is that it is an insurance policy against doing something for the sake of doing something, that is, engaging in meaningless solutions that spend money but really do not deal with the problem. The second reason is that public commitment is essential if we are going to have the financial support that is needed, whether the money comes from the private or the public sector. And finally, public support is the essential underpinning of a conservation ethic and that is really the way we avoid the need for restoration in the first place.

CAMPBELL: Everyone would agree with the final point, but that is an excellent question. What do we do to advance the science of restoration and the economic science of valuing resources?

AUDIENCE MEMBER: What you can do is support science's experimental restoration. There are so many opportunities where you can begin to develop new techniques of restoring habitats without waiting for a disaster or without having to pick a site that is in the path of the next oil spill. There are many ideas that restoration scientists have that are worthy of support and I am disappointed that the panel discussion has been limited to damage assessment and recovery from, in particular, one type of damage, namely, oil spills. I am also disappointed that Ms. Kincke could go from that single example and conclude that restoration works. Restoration does not work when it means the full replacement of ecosystem processes from a whole range of damages that have occurred in the past.

It is like your wounded finger. If the wound is minor and you put a Band-Aid on it, maybe you do not have a scar. But if it is a major damage and you have gone to the point of endangering species with extinction, you are going to end up with a nasty gash or maybe lose the end of your finger. It heals, you can do a little something, but you do not have full functional capacity. So I would urge you to direct efforts in anticipation of improving restoration by supporting experimental science that can give you the answers you will need when that disaster occurs.

CAMPBELL: That is clearly one of the answers. There are going to have to be additional funds expended on doing long-term research on restoration and that will have to come from different sources: the public sector, like NOAA and other agencies; the public interest groups, like the Center for Marine Conservation; states and localities; and the goodwill of industry. The thrust of this discussion is related to the tremendous amount of uncollected data that we, as stewards of the marine resources, have an obligation to go out and collect. In the process, we can further the science of restoration and actually further endeavors in other areas.

AUDIENCE MEMBER: The total cost of getting me to this meeting is greater than the research budget of the work that I did on the *Torrey Canyon* oil spill. There are lots of people here who are professionals who act as middlemen to divert taxpayers' money from its intended use. If you end up constructing a whole series of further bureaucratic structures and extensive peer group review, each set of reviewing probably will cost the biologists or the scientists as much time as actually going out in the field and doing something. You have really got to try to reduce the bureaucracy and make sure something actually gets done.

CONNER: I would like to go back to an earlier point. First, this is not just about oil spills. We are talking about restoration that is undertaken as a result of natural resource damage assessment actions that address not only oil spills, but also releases of toxic substances and degraded areas that result. It also includes injury to sanctuaries under the Marine Sanctuary Research and Protection Act.

So it is a much broader question. And, in fact, the Southern California damage assessment case is one that our office is undertaking and initiating some restoration planning. The reason we cut it this way for the meeting was to look at the potentials and problems of restoring different types of habitats as tools in our toolbox to go out and do the work. You are talking about public money and we are not intending to run primarily on public money. We are intending to recover damages from the responsible parties that will be applied toward implementing the restoration itself. Some prototype work will be needed to test different options before they are undertaken on a larger scale. Long-term monitoring of the project will also be required.

SMITH: I am not the bureaucracy, but I would like to respond a little differently. In the context of restoration, it is not clear to me that we want to go on with business as usual from a scientific research perspective. That is the wrong agenda; it will take too long. We may need to look at specific case studies and find out how we have to change the way we do the science. That is not to suggest that we should delay the activities, but the professions are constituent professions that may not define the interesting problems in a way that actually responds to the needs of restoration.

With respect to the costs of this conference, implicitly most of the people are not paid to be here. They obviously placed a value on the importance of listening. That value is substantially in excess of the cost of whatever the government put out, so it passes the benefit cost test to me by just looking at the value of people that have been here. We all have a commitment to this. How we efficiently and effectively get on with the task is an important and interesting set of problems. But I would argue that we may want to divert the traditional science-based agenda to one based on the nature of the problems.

CAMPBELL: The purpose of this symposium has been to inventory the body of restoration science and to get a group together to advance the work that is being done.

We are trying to create a process where none has existed, and we are trying to push forward a science on the edge of a vast frontier. We are trying to push economics to answer questions that have been difficult to answer and we are trying to press lawyers to craft solutions to our legal problems around the economics and around the science. It is a difficult problem and we are grappling with it.

AUDIENCE MEMBER: Indeed it is a difficult and complex problem. One of the earlier recommendations was that we should think very seriously about creating a restoration discipline in one or more of the universities or academic institutions. Perhaps NOAA could lead this through its Sea Grant institutions or some other organizations.

I was a little uncomfortable with the inference that 40 percent restoration was acceptable. That might send a signal to people unconcerned about full restoration that the government has a policy of accepting something less than even half.

CONNER: Our objective is to achieve full restoration. If that cannot be accomplished one way, then we will use other means to try and achieve some replacement of the resources or some other type of restoration for the public. The message is not that the government is satisfied with partial restoration.

AUDIENCE MEMBER: This symposium has been purposely structured away from CERCLA, the Clean Water Act, the legislative framework that we work under, so we have talked about pure restoration without putting it in those contexts. Inevitably, the panel chaired by a lawyer has brought us back to reality, which is a litigative framework established by Congress. Somewhere down the path maybe we can get away from the litigative framework, which I do not think is the most efficient way to achieve restoration, and begin to work together to correct the ills of the past without pointing fingers and assigning guilt. I think that would be a very useful goal to move toward.

MCMANUS: If we start to value these resources and take efforts to restore them, we will encourage others to take the initiative of protecting them. There are a number of cases, for example, where industries have begun to examine, from the environmental perspective, how they should behave not just because government is telling them what to do. There is a company now with extensive landholdings that is looking at the concept that it should have no net loss of biological diversity on its lands. As government and society start to value these resources, we will see those values reflected in other ways.

VICTORIA KINCKE: Under the Clean Water Act and CERCLA, you are going to have a litigated situation and you are likely to spend a long time and a lot of money before you actually see any restoration work being done. But I think if you get the right mix of people, the right mix of ideology, an overwhelming and a recurring message of cooperation and tolerance, you can work through it even under the current system.

CAMPBELL: I honestly believe that if you do end up in litigation then somewhere the process has failed.

AUDIENCE MEMBER: I have two comments. First, it is critically important that we publish our failures as well as our successes or we are doomed to repeat them. This requires changing a fundamental policy in our effort to reward success in science. Maybe we need a new journal on failed attempts.

The second has to do with money. Senator Nunn in a recent address actually described environmental damage as a threat to our national security. There are Armed Services committees discussing monies on the order of $200 million to be spent for basic research and technological development in the area of environmental damage, to actually predict climate and global change and the whole range of environmental studies. In the future, we may be able to turn to the armed forces for some of the funding that we need to do the assessments.

AUDIENCE MEMBER: Restoration is for now and for the future. Unfortunately, under the guidelines and OMB regulations, the future is not worth very much— $8^7/8$ percent. Also, legally how can you get us out of this really difficult situation that economists find themselves in?

SMITH: The $8^7/8$ or 10 percent figure is huge relative to what the real rate of interest is and should be the discounting factor that is used; still there is a broader issue here. We need to think very carefully about the trade-offs that we as a society make between produced capital, reproducible capital, machines and the like, versus natural capital. That is really what is at issue. There is a well established literature in economics that describes questions where we irreversibly transform a natural resource. We may want to depart in a significant way from traditional economic efficiency mandates, when there is an irreversible transformation to a natural resource and its ability to provide sustained services versus when there is not an irreversible transformation. That is where the emerging restoration science we are all trying to develop connects the decisions about existing damage to resources and future actions that might lead to injury.

I am not as concerned about using some discount rate when we are talking about resources that could be injured and restored. But when the injury leads to a permanent irreversible change in the resource, I would not want to adhere to a strict discounting principle. The current rules do not allow for that and, of course, one of the reasons is the very practical problem of what is an irreversible transformation and which resources are unique and important. The committee documentation that underlies the development of CERCLA showed some attention to these considerations, but because we were not far enough along perhaps in the science of restoration, there were not clear-cut answers. We ought to put that on the agenda for monitoring and managing existing resources that could be affected by future activities that are in themselves valuable. We are not going to shut down all our economic activity to sustain an ever-growing population in the name of preserving the natural environment. We have to identify what is important and put it into perspective. I think the science of restoration is going to help us in that.

CAMPBELL: As a lawyer, there is very little you can do. You simply implement the law that exists.

AUDIENCE MEMBER: This is a wonderfully homogeneous audience, so when you asked was this all worth it, the responses tended to be very self-critical. Before you start bashing yourselves over which body of law, regulation or travel agent you are going to use to consume funds, consider the fact that from my perspective this has been a very enlightening and invigorating exposure to a new tool.

I came here out of curiosity to find out if there was anything behind this concept of habitat restoration because I dwell principally among those who pour concrete and drive pilings. When you consider the value of your work, your comments seemed to imply that the alternative was no action. From my experience, though, the alternative would be more likely bad action, poorly thought out action.

I can only add to the previous admonitions that you print, publish, disseminate failures as well as successes. Yesterday when I saw the slide showing the 35-year-old detonation holes off the Australian shoreline where the seagrasses had never grown back, that was the single, most convincing piece of evidence I have ever seen that says, do not muck around with seagrasses. There probably is among that uncollected, unpublished data and among even the failures some really wonderful ammunition to counterbalance the kinds of actions that are taken. For example, by that retirement community in Florida that put up a wall instead of doing something else to diminish erosion along the coast.

CAMPBELL: There is a reason why very little work was done the first eight years of CERCLA's existence. A great deal of work was done on the cleanup side, very little work was done on the restoration side, and the reason was not only lack of funding. The problem was the questions are incredibly complex and difficult to answer. That is why we brought this group together to begin to grapple with the answers of what we should do with regard to restoration.

AUDIENCE MEMBER: What if you cannot get a complete restoration? For litigation purposes if we cannot get it all back on a one-on-one basis, maybe we need twice as much that is half as good. It is the best we have been able to do. There are dozens of cases around the country where we have accomplished what I call "small" assessments, where we have not gone to court. We have not spent millions of dollars or hundreds of thousands of dollars, but we have negotiated settlements with industry to get some very cost effective restorations done.

Fish and Wildlife did a lot of the initial work on the Shell Oil case, for example. With EPA and NOAA, we have been working on biological work groups where we get together, review Superfund sites, even on sites when there is no responsible party, and we are still looking at what kinds of environmental questions ought to be addressed.

KINCKE: When you are in an industrial or transportation setting, no matter what type of precautions you take, ultimately there will be some type of incident. Some can be anticipated, perhaps not in terms of scale and exact scope, but you can make an assessment in terms of what types of resources are likely to be harmed, what types of equipment would be necessary to have on-site to at least minimize the damages. In addition, we often overlook people and personnel. We need to have some contingency planning in terms of the types of people we are going to bring out to curtail the incident. I also think we ought to have a list of on-site scientists and economists that can hit the ground running and be prepared to assist us in a natural resource damage assessment.

AUDIENCE MEMBER: We have lost 90 percent of our habitat in California and we only have five percent of our forests left. We have lost 70 percent in Washington, and so on. My point is protection really is first. To restore a wetland habitat lost by logging, for example, would cost probably five times what it would for the value of the timber next to the stream. It does not make sense to restore a habitat that is going to cost so much more than what is actually there. I would like to see some of the funds that you are taking in perhaps on litigation be directed toward habitat protection. We have so little left and that is where the effort should go first.

CAMPBELL: I agree that priority has to be habitat protection. The tools that we are discussing are in the restoration arena. There are other things that clearly need to be done outside this context for habitat protection. We will conclude with remarks from each panelist.

CONNER: I am excited about being involved in the natural resource damage assessment area. We are on the brink of a very productive time. There is a real potential for recovering funds to support restoration activities from responsible parties. That sets before us a challenge and a responsibility. We are all going to have to work together to meet those.

SMITH: Joint definitions of problems are very important. Science and economics both have a part to play in the new science of restoration. It may well be that some of the problems raised with respect to the destruction of habitat will at least be partly investigated by the restorative activities and concern to have responsible parties pay for the restoration. In the mining sector, for example, all reserves now are being evaluated not just in terms of the private costs associated with developing resources, but the private and social costs of the kinds of concerns we are talking about. The same thing is happening in the energy sector in the selection of technologies. The private and social costs are being incorporated in what firms are expected to do. In effect, we have set up a set of incentives and the challenge for restoration science is to make sure that we get the technical features of those incentives compatible with the real values people place on the natural resources.

CAMPBELL: The better we articulate the quantitative and qualitative values of the resources we are seeking to protect and restore, the more supportive the public and the policymakers will be of these efforts.

KINCKE: This symposium illustrates a shift in societal thinking. In earlier times, the goal of industry was to provide jobs and products and there was a willingness on the part of society to accept whatever losses might be sustained in habitats and natural resources because of the economic benefits.

Society will no longer tolerate destruction and damage to natural resources and habitats just for the sake of having jobs and products. Industry realizes this shift in expectations and will be taking it head on. Hopefully, in five or ten years industry will not be seen as combative or adversarial, but rather will be seen as part of a team working to restore what resources have been damaged or destroyed.

CAMPBELL: From the beginning of this process I have welcomed a vigorous and vital dialogue. It is vital that this process of easing into a new way of looking at our natural resources happens in an orderly fashion. The fact is we have a number of things at stake here. The ability to actually restore and reclaim damaged portions of our environment is now a very exciting possibility. Societal values are changing. Preservation is seen as increasingly important, but it can only occur by valuing our natural resources. Unfortunately, if we are unable to restore these resources to a certain extent we lose the opportunity to value them. The burden rests upon the scientific community to come up with viable, meaningful, useful ways of actually doing damage assessment and restoration work.

Index

A

abalone 279, 283, 559
absorbent 617
abundance 25, 80, 147, 287, 361, 546
Acanthaster 163
accretion 24, 117, 443
Acropora 163
Acrostichum 226
adaptations 226, 231
   morphological 224
   physiological 224
   reproductive 224
adventitious roots 126
aerator 474, 490
aerial photographs 94
aerial roots 227, 228
aesthetic 158, 429, 551
aesthetic value 279, 482
aggregation 299, 610
agriculture 120, 240, 343, 347, 428, 480
ahermatypes 143
Alaria 282
alewife 344
alfalfa 38
algae 8, 9, 13, 119, 228, 251, 352, 360, 520
algal biomass 485
algal blooms 474, 482, 547
algicide 519
allochthonous 225, 358
Alosa pseudoharengus 344
alteration 116, 358, 365, 409, 545
American lobster 559
ammonia 38, 71, 480, 482
Amoco Cadiz 57
Amphibolis 112, 114, 118
amphipods 66, 123, 502
anadromous 345, 405, 406
anaerobic 168
Anas acuta 14
anchor damage 164, 172

anoxia 12, 24, 473, 480, 481, 513
anoxic sediments 229
anthropogenic 141, 577, 583, 590
antifouling paints 480
application 543, 584, 551, 559
aquaculture 240, 474, 476, 586
aquatic production
   primary 368
   secondary 368
Arabian Gulf 618
arid environments 225
artificial 514
   plantings 246
   reefs 474, 531
   seagrass 132
Ascipenser 344
Ascophyllum 617, 620
asexual reproduction 80
assessment criteria 41, 68
Assiminea californica 11
Atlantic Grey Seals 597
Atlantic salmon 344
atolls 146
augmentation 433
Australia 85
autochthonous 358
autotrophy 358
Avicennia germinans 226
avifauna 447

B

back reef 147
backfilled access canals 93
bacteria 14, 164, 241, 250, 482
bank erosion 348
bank reefs 144
bare-root plants 64, 92
barnacles 311, 495, 502, 583
barracudas 193
barrier reefs 146
barriers 379, 417
baseline data 115, 441, 548
basins 229, 477
Batis maritima 226

beach nourishment 546, 559
beaver ponds 383
Beggiatoa 482
behavior 356, 485
Belding's Savannah sparrow 14, 18
benefit-cost ratio 387
benefits 405
benthic 282, 547, 563
    algae 54, 163, 495
    autotrophs 228
    community 116, 374, 474
    infauna 438, 441
    organisms 482
    plants 441
biodegradable 480
biodiversity 8, 21, 158, 160, 429
biological
    control 189
    filters 547
    interactions 587
    manipulation 324, 530
    oxygen demand 480, 588
    processes 227, 551
biomagnification 84
biomass 88, 223, 229, 284, 585
bioremediation 44
biota 477, 480
biotic interactions 343, 361
bioturbation 83, 102
bird 54, 227, 287
bivalves 8, 11, 441, 490
black band disease 164
black corals 182
black-lipped pearl oyster 192
blade density 88
blowouts 126
blue crabs 33, 67
blue rockfish 300
boat
    basins 56
    harbors 162
    traffic 297
    wakes 250
boulders 350
boundaries 549

brackish 12, 436, 447, 452, 467
breaching 24
breakwaters 160, 185, 296
breeding 157, 158, 161, 585
brown algae 282
brown shrimp 33, 67
bryozoans 293, 502
buffer zones 190, 344, 384
buffering capacity 231
bulkheading 242, 434
butterfly fishes 163
buttonwood 226

C

cadmium 18
calcium carbonate 157
California halibut 15
California least tern 43
calving ground 161
canals 415, 477
canopy 11, 12, 288, 352, 583
canopy density 39
Capitella 11
carbon dioxide 159
carbonate sediments 98
carcinogenic 83
Caribbean queen conch 192
carnivores 14, 29
carnivorous 507
carrying capacity 202, 343, 353, 551
catastrophes 23, 545, 590
catchment basin 480, 515
Catoptrophorus semipalmatus 14
cattails 22
cellulose 121
cementation 147, 182
Cerithidea californica 11
Chaetodon 163
channelization 239, 241, 343, 348, 350
charcoal 227
chemical contamination 545
chemical properties 53, 131
chironomids 482, 507

chlorinated hydrocarbons 44
chronic 84, 103, 583, 588
circulation 118, 196, 481, 545, 547, 556
cladocerans 495
clams 147, 160
clamshell dredge 195
classification 229, 353, 390, 462
clearcutting 346
climate 1, 225, 347
climatic
    processes 228
    stresses 224
climax 80, 83, 86, 118, 121
clones 23
coalescence 99, 124, 135
coastal
    areas 356, 543
    birds 9
    development, 141
    fish habitat 1, 15
    habitats 1
    marine habitats 112
    marsh 546
    morphology 117
    power plants 310
    subtropical 224
    tropical 224
    wetlands 8
    zone 133
cobbles 350
cockles 160, 530
coconut crab 161
cod 496
Coleomegilla fuscilabris 12
coliforms 485, 493
colonization 237, 428, 482, 491, 503, 552
Columbia River chinook 345
commercial 160, 283, 345
    dock 475
    fish 56, 545
    species 56, 585
community 424, 436, 480, 544, 551

development 532
    processes 280
    structure 67, 231, 426, 545, 589
compensation 243, 310, 613
    depth 84
    off-site 570
    out-of-kind 570
compensatory mitigation 424, 433, 443, 466
competition 83, 237, 361, 546
competitive interactions 22
compliance 80, 82, 102, 264
composition 88, 547
conchs 160
concrete 561
configuration 571
coniferous forests 352, 356
Conocarpus erectus 226
conservation 104, 390, 418, 474
constraints 424, 571
constructed wetlands 23
construction 69, 165, 296, 405
    airfield 158
    artificial reefs 184
    bridge 438
    materials 158, 160
    port 158
    road 438
consumption 589
container terminals 475
contaminated effluents 480
contaminated sediments 71
contamination 17, 545
continental coasts 157
contingency plans 98
controls 327, 550, 615
conversion 240, 348, 428
copepods 502
copper 19, 296
coral reefs 141, 227, 546
    bleaching 164
    creation 174
    human impacts on 164
    mortality 545

restoration 174
coral transplantation 178
Coralline algae 147
cordgrass 7
Cordylanthus maritimus 28, 32
corridors 40
cosmetics 160
cost 87, 211, 321, 516, 520, 550
   effectiveness 386, 554
   responsibility 211
coverage rate 80, 82, 103
Cowardin system 81
crabs 8, 83, 157, 160, 279, 496,
   585, 596
creation 23, 68, 86, 118, 122, 243,
   301, 433, 452
Cricotopus 12
criteria 441, 612
critical elements 100
crown-of-thorns starfish 163
crustaceans 115, 157, 228, 441,
   507
cryptic fauna 531
CSIRO 124
cultural practices 53, 69, 160, 345
culture 80, 192, 559, 566
culverts 61, 196, 379
cumulative impacts 365
current velocity 96, 117, 144, 477
cushion star 163
cyanobacteria 484
cyanophyte 188

D

damage 55, 111, 223, 238, 412
dams 382, 344, 405
DDT 296
decapod 451, 547, 596
decision-making 548
decline 123, 141, 344, 428, 431
decomposition
   processes 55
   rates 14
deflectors 382, 413
defoliation 238

deforestation 293, 344
degradation 1, 281, 345, 424, 546
dehiscence 129
dehydration 259
density 66, 202, 607
denudation 21
deposit feeders 29
deposition 24, 118, 147, 228
desiccation 92
design 253, 281, 409, 423, 424,
   475, 533, 543
   criteria 3, 461, 548
   success 424
destruction 55, 111, 118, 121,
   431, 546
   aquaculture 111
   propellers 111
   fishing practice 165, 169
detritus 8, 14, 54, 82, 227, 547
developers 88, 134, 344, 533
development 1, 53, 111, 224, 343,
   423, 474, 552
diagnostic parameters 88
diatom 250, 484
Dictyosphaeria cavernosa 194
die-back 117
diesel fuel spill 296
dike-breaching 434
diked wetlands 24
dike 61, 68, 428
dinoflagellate 484, 498
dipteran 11, 66
discharge 120, 545, 588
disease 64, 164, 188, 241, 361
dispersal 40, 231, 280, 311
dispersant 583, 588, 595
disposal sites 195
dissolved
   nutrients 159
   organic carbon 227
   oxygen 159, 359
distribution 144, 291, 356, 361,
   425
disturbance 83, 230, 235, 423
   regime 234

sites 226
disused docks 476
diversity 84, 112, 144, 147, 227, 486, 546, 552, 557, 589
dock 547, 558
  ecosystems 474
  walls 486
dock basins 473
  enclosed 475
  semi-enclosed 475
dogwhelks 588
dominance 438, 607
donor sites 90, 111, 126, 253
drainage 69, 425, 439
  system 349, 480
dredge 53, 84, 493, 545
  and-fill 164
  disposal 56, 242
  spoils 27
dredging 16, 21, 24, 111, 166, 240, 242, 296, 427
drift algae 293
drift material 286
drilling 53, 57
dry docks 480
dugong 157, 161
dune 21
dunlin 432
dynamite 170

E

echinoderms 147, 596
echinoids 293
ecological 533, 543
  function 102, 158
  recovery 173
  research 586
  resilience 79
  significance 116
  studies 493
  value 279
ecology 476, 583
  functional 223
  structural 223
economic 87, 116, 345

benefits 386
  losses 545
  value 279, 285, 586
economy 343, 405, 431
ecosystem 8,343, 418, 549, 586
  carrying capacity 546
  diversity 476
  functioning 13
  inshore 585
  management 130
  offshore 585
  processes 224
  services 243
ecotechnology 32, 44, 424
ecotypes 61
edges 33
education 158, 228, 390, 429, 474
eelgrass 92, 431, 585
effectiveness 385, 474, 544
effluent 135, 359, 588
El Niño 279, 590
elasticity 232
elevation 24, 53, 69, 248, 556
emergent marsh 425
enclosures 620
encrustation 292, 571
endangered 296, 345
  species 7, 9, 245, 418
endemic species 125
endosperm 127
energy 429
  base 352
  flow 88, 358
  source, 343
  supply 418
enforcement 91
engineering 136, 256, 543, 551
Enhalus 81, 105
enhancement 309, 375, 417, 433, 557
enrichment 296
Enteromorpha 599
environmental 520, 548, 586
  conditions 280
  factors 343

laws 191
pollution 546
studies 526
tolerance ranges 96
epibenthic 29, 438, 556
algae 13
communities 547
community 11
crustaceans 116
epifauna 502
epilithic diatoms 458
epiphytic algae 54, 118
equilibration 412
equilibrium 361
equivalent 551, 570
erosion 24, 82, 223, 295
control 68, 118, 429
estuarine 545
fisheries 227
habitat restoration 424
habitats 1
estuary 112, 473, 474, 475
euryhaline 97
eutrophic 473, 480, 482
eutrophication 55, 87, 111, 164,
168, 197, 545
evaluation 263, 344, 383, 423
evaporation 498, 596
evapotranspiration 225
excavating 256
excavation 23, 98, 438
exotic 525
species 23, 24, 438
invasions 19
weeds 18
experimental
approaches 61
design 134, 281, 327
plots 105
sites 122
explosives 170
Exxon Valdez 57, 618

F
fabrication 571

facultative 439
faecal contamination 480, 484
fan corals 160
farmlands 120
fauna
abundance 88
composition 80
density 26
feasibility 68
feeder
detritivores 118
filter 118
feeding 116
fertilization 71, 284
fertilizer 32, 80, 91, 98, 129, 168,
261, 586
fetch 70, 249
fiddler crab 11, 56
fill 53, 84, 240
filling 16, 21, 68, 166, 443
filter 241, 474, 495
filters 111
filtration 87, 556
finfish 54
fireworm 189
fish 8, 279, 423, 429, 432, 474,
486, 585
commercial 115
communities 348, 474
corrals 160
farm 473, 486, 586
habitat 116, 344
ladders 409
recreational 115
resources 57, 115, 344, 431,
546
fishermen 345, 557
education 190
vocational retraining 190
fishing 279, 416, 545, 546, 555,
586
fishponds 160
fishtraps 160
fishways 379, 417
fjordic coastlines 585

flatfish 499
floating mats 513
flood desynchronization 429
flooding 168, 223
Florida 93
flow regime 343, 360
flowering 231
  factor 96
  plant 80
  shoots 95
flushing 17, 225, 474, 481, 512
fluvial inputs 225
foliar nitrogen 32
food 157, 158, 160, 279
food chain 23, 68, 116, 485, 511
food web 8, 54, 82, 118, 285, 429
forage 33, 443, 547, 554
foraging 432, 442, 459
  habitat 54
  sites 33
foraminiferans 147
forest 344
  physiognomy 229
  structure 223
fouling organisms 556
freshwater 224, 225, 356, 406, 486
  discharges 144
  inflow 22, 97
  kills 168
fringe
  environments 226
  patch 83
  reef 146
fronds 285, 292
Fucus 587, 597, 599
fuel 227, 344
function 23, 343
functional 423
  approaches 460
  attributes 2
  characteristics 229
  ecological 548
  equivalency 7, 68, 255, 438, 525
  recovery 88

requirements 549
restoration 80
roles 82, 88
values 2, 84, 114, 157, 227, 350
funding 100, 112
fungal diseases 252
fungi 14
fungicide 64

G

gametophyte 284
gammarids 507
gastropods 11, 286, 588
generative shoots 126
genetic diversity 23, 40, 80, 89
genetic variability 376, 465
genotypes 23, 86
Geographic Information System 35
geographic range 80, 112
geomorphic processes 224, 228
germination 63, 95, 127, 238
ghost shrimp 8, 11
giant clam 160
giant kelp 281
Gillichthys mirabilis 11
global warming 162
goals 87, 104, 125, 376, 418
gobies 8, 11, 67, 496
goose barnacles 585
gradient 438, 586
grading 33, 70
grain size 70
grass shrimp 33, 67
gravel-bed streams 350
grayling 344
grazing 279, 293, 343, 347
green algae 147, 250, 583
green snail 160, 192
greenhouse effect 119, 172
ground truth 115
grow-out
  timetable 103
growth 63, 224, 361, 491, 546, 583
  annual 82

hormones 129
media 63
perennial 82
rate 13, 25, 79, 80, 125
season 82
guideline 281, 319, 511

H

habitat 112, 243, 279, 343, 423, 585
  abundance 144
  beach 157
  components 357
  creation 1, 196
  damaged 543
  degradation 408
  human-made 543
  land 157
  loss 1, 8, 239, 543
  managers 416
  restoration 344
  structure 424
  type 434
Haliaspis spartina 12
halibut 318
Halimione portulacoides 58, 61
Haliotus discus discus 559
Halodule 81, 86, 114
Halophila 81, 105, 112, 114
  decipiens 92
  ovalis 113, 118, 127
halophytes 8
hammock communities 229
harbors 475, 589
harpacticoid copepods 29, 458
hatchery 345, 372, 431
Hawaiian monk seal 161
hazardous substances 2, 3
headwater streams 350
health hazard 480
heavy metals 19, 45, 84, 359, 480
Hemigrapsus oregonensis 11
herbicides 545
herbivores 163, 280
herbivorous

gastropods 596
insects 12
sea urchins 168
species 583
herring eggs 96
heterogeneity 33
heterotrophic 358
Heterozostera 114, 118, 132
Hibiscus tiliaceus 226
high island reefs 146
high marsh 7, 436
highway construction 53, 57
history 111
holdfast 306, 599
holistic 379
Homarus americanus 559
hormone 136
human
  activity 69
  consumption 522, 586
  destruction 119
  disturbances 20
  induced impacts 131
  intervention 111, 313
humpback whale 161
hunting 279, 295
hurricanes 160
hybridization 136, 361
hydraulic 194, 352, 517
hydrodynamic 40, 131, 547
hydroelectric dam 408
hydrologic 2, 24, 425, 481
  alterations 16
  functions 53
  processes 228
hydrology 8, 16, 68, 130, 225, 429
hydroperiods 229, 426
hydropower dams 344
hypersaline 13, 27, 224, 425
hypertrophic 480
hypolimnion 517
hypoxia 480
hysteresis 232

I

identification 462
immigration 552
impact 4, 21, 55, 118, 119, 173, 243, 288, 347, 441, 545, 556, 584, 588, 589, 595
  adverse 68
  assessment 548
  human 293
  mitigation 176
implementation 24, 424, 551
impoundment 240, 481
Incertella 12
indigenous 356
industrial effluents 588
industry 345
infauna 29, 67
inflorescences 128
inlet 588
inlet closure 21
inorganic 480
  nitrogen 292
  nutrients 225
insect 9, 29, 54, 447
installations
  land based 584
  off-shore 584
internal waves 547
interstitial water 70
intertidal 112
  beds 83
  channels 427
  flats 8
  marshes 7, 8, 54
  zone 54
inundation 8
inventory 205, 390
  fish 344
  habitat 344
invertebrate 8, 14, 29, 116, 282, 359, 585
irradiance 130, 292
irrigation 19, 412
islands 157
isolation 163, 516

isopod 11

J

Japanese mussel 28
jetties 382
Juncus roemerianus 61
juvenile 116, 585

K

kelp 282, 300, 431
  beds 546
  biological factors 280
  communities 280
  forests 279
  growth 280
  physical factors 280
  reproduction 280
  survival 280

L

lagoon pinnacles 146
lagoon slopes 146
Laguncularia racemosa 226
Laminaria farlowii 282
land building 225
land management 200
land reclamation 528
land use 239, 343, 353, 546
landforms 225, 227
landscape 359, 464
  aesthetics 223
  diversity 243
  productivity 223
large-scale planting 111
lava flows 162
leachates 480
leaf
  density 119
  detrital 118
  growth 82
  production 82
  sheaths 121
legislation 515
life history 95, 280, 356
light 118, 286, 352, 482

attenuating 112
climate 130
compensation levels 130
intensity 159
limits 162
light-footed clapper rail 9, 12
limestone reefs 112
limiting 460
factor 61, 343, 361, 561
nutrients 618
Limosa fedoa 14
limpets 583, 585
litigation 417
litterfall 227
littoral 425, 427, 584
lobster 157, 160, 279, 283
logging 165, 343, 344, 429, 438, 589
Loligo pealii 565
long-billed curlew 14
longevity 284
longjaw mudsucker 11
loss 344, 405, 423, 424, 498
low marsh 7, 436
lumber 227, 344
lumbering industry 345
Lyngby's sedge 439

M

macroalgae 441, 552
macrobenthic biomass 565
Macrocystis pyrifera 279, 281, 282
macroinvertebrates 66
macrophytes 54, 431
Macrozoarces americanus 565
magnesium carbonate 157
maintenance 72, 243, 264, 444, 558
mammal 9, 54, 228, 432, 585
management 74, 132, 344, 376, 476, 519, 532, 548
manatees 157
mangrove 191, 224, 546
adaptability 223
associates 226

growth rates 223
habitats 223
management 223
range 225
recovery 223
restoration 223
swamps 157
tolerance 223
manipulation 176, 280, 383, 576, 586
mapping 115
marbled godwit 14
mariculture 57, 160, 185
marina 56, 239, 242, 448
marine 12, 279, 477
algae 157
fish 345
lizards 177
pollution incidents 584
reptiles 177
sanctuaries 169
transport 475
vertebrates 157
market values 83
mechanical planters 125, 135
medicines 158
meristematic 126
metabolism 159, 361, 618
metals 241, 480, 485
methodology 91, 280, 544
microbes 11, 14, 25
microclimate 429
microhabitat 360, 547
microheterotrophs 82
micro-topography 229
migration 352, 357, 432, 486
military construction 166
military defense 57
mineral extraction 158, 349
mitigation 1, 80, 82, 89, 91, 130, 243, 376, 424, 570
application 570
banking 570
plan 444
models 363

molluscs 147, 228
monitoring 3, 40, 74, 79, 91, 130, 198, 260, 303, 424, 435, 441, 515, 615
monk seals 157
mooring systems 111, 174
Morone saxatilis 344
mortality 83, 224, 289, 357, 432, 590, 596
mudflat 8, 250, 423, 427
Musculista senhousia 28
muskrat 72
mussels 160, 490, 585
mutagenic 83
mysid shrimp 66, 297, 507
Mytilus 609

N

national security 56
natural
   causes 348
   colonization 311
   functions 255
   mortality 83
   recovery 176, 280
   reefs 546
   regeneration 223
   resource trustees 544
   resources 441
   responses 111
   seeding 69, 95
   variation 280
navigation 239, 242, 346
navigation channels 56, 242, 428
nektonic 507, 547
nematodes 29
Nereocystis 282
net production 447
nets 170
neuston 447
nitrate 480, 482
nitrification 482
nitrifying 250
nitrite 482
nitrogen 13, 70, 98, 412

fixation 26, 37
fixing bacteria 11
no net loss 104
non-compliance 103
non-point
   pollutant 359, 429
   source 84, 545
nonvascular algae 80
northern squawfish 416
nourishment 434, 441
Nucella lapillus 588
nuclear weapons testing 165
Numenius phaeopus 14
nurseries 27, 111, 115, 116, 158, 227, 253, 546, 550, 585
nutrient 7, 8, 13, 112, 127, 129, 223, 224, 225, 279, 286, 291, 358, 474, 480, 490, 588
   additions 32, 129
   concentration 131, 159
   content 25
   cycling 82, 88, 358, 438
   dynamics 25
   limitation 484
   loadings 545
   recycling 26
   supply 97
   transport 292

O

objectives 104, 243, 384, 424, 435, 543, 544, 557
octopus 309, 585
offshore
   mining operations 546
   petroleum 57
   sand mining 160
oil 2, 83, 480
   exploration 165
   pipeline 447
   pollution 173
   spill 19, 53, 165, 308, 545
   wells 17
old-growth forest 347

oligochaetes 29, 482
Oncorhynchus 345, 357, 406
opportunistic species 482
optimum light levels 130
Orchestia traskiana 11
organic 23, 480, 588
  carbon 14
  discharge 480
  loading 358
  material 286
  matter 7, 25, 35, 118, 447
  nutrient 159
  sediments 25, 97
organohaline 480
orthophosphate 482
ostracods 29
otoliths 496
overfishing 164, 169, 344, 408, 430
overharvesting 1
overuse 164, 171
overwintering 352
oxygen
  demand 481
  organic carbon 159
  production 130
oyster 160, 490, 554, 585

P

Pacific
  coast 8
  cordgrass 10
  flyway 432
  herring 431, 585
  salmon 431
Paralichthys californicus 15
parasitism 361
parrot fish 194
particle size 358
particulate detritus 352
passive rehabilitation 176
patch dynamics 288
patch reefs 552
Patella depressa 607
Patella vulgata 607

pathogenic 252, 484, 499, 511
PCB 545
peer review 100, 134, 281, 326, 465
Penaeus esculentus 117
percent survival 88
performance 82, 424, 548, 557
performance bond 100, 264
Pericoma 11
permit 69, 91, 240, 359, 424
persistence 79, 102, 264
petroleum 44, 57, 83
pH 359
pharmaceuticals 160
phenological plasticity 86, 234
phosphate 480, 482
phosphorus 13, 19, 70
photic zone 484
photosynthesis 85, 117, 127, 130, 165, 296
photosynthetic active radiation 119
Phyllospadix 81, 105
physical
  characteristics 349
  conditions 346
  properties 131
  structure 343
physiography 225
physiological
  adaptation 442
  demands 432
physiology 136
phytoplankton 115, 228, 473, 474
  biomass 547
  blooms 119
  die-off 481
pickleweed 10, 439
piers 296
pilchards 611
pilot project 3, 209, 249, 486
pinfish 67
pintail 14
pioneer species 230
pipeline 53, 57, 93, 98

Pismo clam 301
planktivorous 547
  fish 507
  reef species 193
plankton 447, 482, 547, 554
planning 344, 383, 533, 548
plant
  control 40
  growth 32
  growth rates 7
  hopper 12
  nutrients 480
  propagules 61
  species 69
planting 80, 223, 258
  costs 92
  guidelines 40
  stock 92
  success 223
  techniques. 260
  times 82
  units 65, 126
plants 9, 25, 112, 349
  emergent 434
  submergent 434
plugs 61, 65, 125
pneumatophores 227
point sources 359, 545
poisons 170
policy 80, 243, 390, 425, 501
political considerations 321
pollination 28
pollution 1, 112, 123, 141, 242,
    344, 593
  air 423
  water 423, 476
polychaetes 11, 29, 482
population 280, 429, 474, 546,
    552
  growth rates 119
  structure 416, 589
pore-water nitrogen 37
porosity 25
port development 242, 473
ports 239, 475, 556, 589

Posidonia 112, 114, 117, 129, 131
potentially responsible parties 5
prawns 495
precipitation 360
preconstruction 68, 256
predation 238, 361, 411
predator-prey dynamics 416, 552
predator 163, 279, 308, 411, 585,
    589
predict 485, 547, 563, 615
prefabricated habitat modules 544,
    552
pre-project plan 93
preservation 243
prevention 141, 209, 495
prey 116, 439
primary production 115, 227, 436,
    451
priorities 55, 205, 384, 474, 544
pristine 346, 436
processes 288, 426, 432
production 343, 358, 406, 418,
    566
productivity 88, 115, 279, 285,
    551, 589
progradation 117
progress reports 328
project 544
  goals 68, 79, 87, 100
Prokelisia dolus 12
prop root 260
propagation 40, 61
propagule 41, 118, 124, 135, 231,
    225, 259
prop-scarring 102
protection 79, 184, 185, 264, 308,
    344, 349, 417, 418
protocol 40, 133, 192, 326, 424,
    615
protozoa 511
Pterygophora californica 282
Ptychocheilus oregonensis 416
public concern 584, 618
public notices 416
Puccinellia maritima 58, 61

pulp waste 588

Q

quantitative 102, 610
quays 475

R

rainfall 13, 168, 225
raised reefs 158
Rallus longirostris levipes 9, 12
raptors 15, 447
rare species 7, 158, 160, 525
rate of return 111
rays 83
rearing 352, 431
reclamation 240, 428, 429
recolonization 66, 84, 162, 173,
    223, 374, 436, 597
recommendations 4, 571
reconstruction 434
recovery 111, 119, 123, 141, 374,
    439, 498, 516, 532, 552,
    584, 595, 609
recovery rates 162, 174, 280
recreation 158, 429, 474
recreational 283, 441, 497
    activities 585
    fishermen 190
    fishery 474, 476, 545, 557
recruitment 237, 292, 439, 547,
    552, 554, 587, 588
red algae 282, 495
redevelopment 477
redox potential 482
red-tides 590
reed canarygrass 438
reef 147
    corals 146
    development 144
    fishes 157
    module 552, 566
    natural limitations 161
    structure 147
    surface 147
    transplantation 178

refuge 40, 442, 530, 547
regenerate 121, 233
regulations 169, 243, 359
rehabilitation 55, 187, 243
    techniques 174
reintroduction 66
reliability 575
remedial 4, 500, 511, 520
replacement 243, 570
replanting 102, 558
reproduction 352, 432, 442
reproductive 555
    function 166
    propagules 126
reptiles 228
requirements 357, 424, 546
research 3, 111, 124, 141, 228,
    280, 405, 435, 474, 476
    directions 104
    funds 45
    needs 80
reserves 16, 310, 325
reservoirs 408
residence time 55, 547
resilience 4, 23, 87, 89, 230, 231
resistance 231
respiration 130, 166, 358
responses 55, 118, 230, 288, 298,
    429, 552, 584
restoration 1, 23, 55, 68, 82, 86,
    91, 111, 122, 125, 174, 227,
    239, 243, 280, 281, 301,
    343, 371, 409, 412, 423,
    433, 532, 543
    methods 353
    methodologies 5
    plans 3, 206
    sites 25
    strategies 53, 141, 424
restore 301, 405, 474, 476
resuspension 96, 118
retention 95
revegetation 92, 378
revetments 186
rhizome 39, 55, 80, 82, 111, 112,

126, 130
Rhizophora mangle 226
rhizophytic algae 80
ribbed mussels 56
riparian 19, 343, 344, 383
rip-rap 423, 434
river 225, 349, 477
  deltas 427
  system 406
riverine 229
  environments 226, 406
roads 17, 349
rock reefs 554
rocky
  areas 546
  intertidal 81
  intertidal species 83
  shores 583, 584
root 39, 55, 80, 82, 112
rotifers 507
rototill 38
runoff 119, 225, 296, 480
Ruppia 81, 92, 105, 113

S

Salicornia bigelovii 34
Salicornia virginica 10
salinities 81, 159
salinity 13, 24, 53, 63, 69, 123,
    434, 473, 480, 481
  intrusions 224
  regimes 8
  tolerance 10
Salmo salar 344
salmon 343, 406, 409, 439, 490
Salmonella 511
salt evaporation ponds 17
salt marsh 7, 224
  bird's beak 28
  creation 89, 556
  periwinkles 54
salt water crocodile 161
saltweed 439
saltwort 226
salvage operations 592

sampling design 327, 441
sanctuaries 171, 190
sand 147
  beaches 157, 279, 596
  blasting 617
  dollars 147, 318
  dune 16
  flats 8
sandpipers 432
saturated soils 438
scale insect 12, 28
scale worms 495
scientific 343
  evaluation 424
  methods 134
  research 158, 425
scouring 121, 352
screening devices 415, 417
SCUBA diving 101
sea 224, 591
  anemones 318, 496
  bass 294
  sea birds 584
  cucumbers 147
  sea level 230
  level rise 54, 56
  management 308
  otters 283
  pens 318
  sea snakes 177
  sea stars 318, 596
  sea trout 507
  squirts 502
  turtles 157, 160, 161
  urchins 147, 279, 283, 298,
      585, 596
  whips 160
seagrass 80, 191, 228, 282, 546
  exotic species 426
seagrass beds 111, 116, 120, 157,
    227, 546
  biology 125
  fauna 83
  meadows 111, 120, 157
  restoration 79

transplants 125
sealions 585
seals 287, 585
seawall 441
seaweeds 80, 157, 160, 585, 587
    industrial harvesting 585
secondary
    determining factors 130
    production 29
    productivity 33, 227
    succession 226
sedentary organisms 586
sediment 17, 82, 112, 224, 480,
        481, 585
    accumulation 24
    compactness 26
    microflora 82
    movement 40
    nitrogen 37
    removal 54
    retention 429
    stability 82, 130
    stabilizers 111, 117
sedimentation 22, 53, 69, 164,
        165, 279, 296, 434, 436,
        545, 546
seed 63, 92, 126
    bank 23, 118, 129
    collection 129
    production 61
    set 95
seeding 61, 280, 306
seedling 61, 127, 225, 239, 250,
        258
self-shading 117
Semibalanus balanoides 609
sensitivity 235
services 56, 87, 227
sessile 459, 586
sewage 18, 480
    discharge 182, 296
    effluents 242
    nutrients 168
    outfall 183
    sludge 120, 493

spills 13
sewer 279, 588
sexual reproduction 80
shallow water 22, 546
shellfish 545, 585
shelter 116, 158, 225, 249, 475
ship groundings 164, 172, 187
shipwrecks 184
shoalgrass 97
shoot density 86, 447
shorebird 14, 432
    nesting 96
shoreline 157, 429, 585
shrimp 157, 496, 596
silica 480, 484
silicates 480
siliceous sediments 98
silt 250, 480
siltation 348, 497
silvicultural 246
site 111, 279
    evaluation 70
    preparation 130, 257
    selection 53, 68, 80, 91, 130,
        223
slough 439
sludge 296
snails 147, 596
snow geese 72
socio-economic 160
sockeye 406
soft bottom communities 280
soil 8, 32, 250, 443
    chemical properties 69
    erosion 344
    physical 53
    physical properties 69
    runoff 168
    saline 224
    tests 71
    waterlogged 224
sole 318
solid waste dumps 240
spacing 125, 260
Spartina 8, 61

Spartina alterniflora 29, 60, 61
Spartina foliosa 7, 10, 12
spatial 40, 282, 287, 486
spawning 348, 406, 431, 546, 585
spearfishing 189
species 112, 223, 383, 408, 544
    autotrophic 159
    complements 29
    composition 23, 25, 61, 66
    diversity 498
    endangered 161
    heterotrophic 159
    interaction 361
    threatened 161
Sphaeroma 252
spiders 11, 54
spills 173, 374
spiny lobsters 192
splash dams 346
sponges 158
spores 284
sportfishing 160, 522
sprigs 61, 126
stability 4, 563, 575
stabilization 68, 228, 584
stagnant 12
stagnation 24
starfish 163
statistical adequacy 101
steelhead 345, 406
Stellar's sea cow 294
stem densities 39
stipitate kelps 282
stocking 192, 406, 522, 566
stony corals 182
storms 102, 118, 157, 160
stormwater treatment 40
strategies 178, 230, 433
stratification 473, 481, 497
stream 343, 344
    abuse 344
    channel 344
    dynamics 343
    flow 8, 13
    logging 348

stress 374
    response 164
    tolerance 83
stressors 223
striped bass 344
structure 23, 224, 280, 281, 343, 54
    characteristics 68
    factors 352
sturgeon 344
subestuaries 425
sublethal
    affects 84
    temperatures 171
submerged aquatic vegetation 54, 545, 550
submergence 69
subsidence 24, 53, 230, 438, 563
subsistence fishing 160
substrate 225, 250, 280, 282, 290, 477, 552
subtidal 427, 585
    communities 279
    marine 282
    muddy-sand 112
    sand 112
subtropical 144
success 125, 246, 281, 316, 433, 438, 476, 548, 551, 559
    evaluation 266
    predictability 435
succession 237, 464, 503
sulfate reduction 26
sulfides 25
sulphate 250
sunlight 358
    availability 72
surgeon fish 194
survival 100, 123, 224, 258, 357
suspended particulate material 40, 118, 426
suspension feeders 29
sustainable use 268
symbiotic algae 164
Syringodium 81, 114

Syringodium isoetifolium 118
system 124, 548

T

table reefs 146
tagging 191
target species 211, 384
techniques 91, 312
technological research 476
technology 2, 79, 86, 134, 244,
    515, 544, 544, 559
tectonic forces 158
temperate 144, 279
temperature 92, 123, 130, 224,
    279, 291, 359, 436, 480,
    481
temporal
    fluctuation 588
    variation 584
terraces 443
terrestrial
    corridors 28
    environment 357
    sediment 225
    transplants 125
test plantings 32
Tetragnatha laboriosa 12
Thalassia 81, 86, 92, 105
Thalassodendron pachyrhizum 112,
    114
thallus 290
thermal 164, 171, 172
threat 55, 344
Thymallus tricolor 344
tidal
    creeks 22
    currents 96
    elevations 223
    exchange 61
    flooding 69
    flows 22
    flushing 8, 21, 24
    height
    gradients 587
    marshes 53, 54

prism 17
range 53, 69, 96, 475
regimes 426
restriction 53
tide 223, 585, 593
tiger prawns 117
timing of construction 91, 95, 125
tires 184
tissue culture 136
tolerance 224, 231, 357
topography 24, 33, 311
Torrey Canyon 57, 583, 584
tourism 57, 158, 160, 474, 476,
    497, 527
toxic 168, 296, 482, 556
    compounds 545
    dinoflagellates 590
    materials 18, 25, 55
    materials spills 57
    substances 241, 279, 359
toxicity 590, 596
trade-offs 574
trampling 20, 21
transplant 25, 27, 61, 64, 80, 118,
    126, 132, 260, 280, 306,
    315, 456, 620
transport 225
treatment 516, 545
trees 346
tributary 406
tributyltin 588
tritons 160
trophic 612
    dynamics 358
    energy 429
    groupings 29
    linkages 26
    state 358
    structure 67
tropical seagrass 112
trout 344, 491
tsunamis 160
tubeworms 318
tunicates 495
turban snails 295

turbid water 477
turbidity 84, 195, 228, 279, 296, 359, 436
turbulence 547
turnover times 168
turtlegrass 92
Typha domingensis 22
typhoons 160

U

Uca crenulata 11
unvegetated 116, 432
upland eroded soils 56, 165
uprooting 225
upwelling 547
urban 556
    areas 25
    development 168
    nature conservation 476
    renewal 474
urbanization 7, 349, 423, 427, 475
utilization 438, 447

V

value 55, 87, 228, 476, 521, 559, 585
variation 287, 463
vascular plant 13, 80, 434
vegetative
    cover 89
    fragment recruitment 95
    growth 231
    material 123
    propagules 125
    reproduction 126
    shoots 92
vessel groundings 546
viability 255, 267
viruses 511
volcanic island complexes 146
vulnerability 235

W

waste 515
    material 480

products 159
storage 516
water 17
water
    circulation 159
    clarity 282
    column 118
    course 349
    diversion 241
    filtration 118
    movement 588
    quality 2, 53, 68, 130, 228, 280, 343, 348, 359, 429, 473, 476, 547, 555
    functions 159
    temperature 159
watershed 344, 350, 406, 425
wave 96, 157, 160, 563
    action 118, 587
    climate 53, 69
    damping 82
    energy 96, 249
    exposed coasts 585
    exposure 556
    regime 250
    surge 292
weirs 382
wetland 2, 7, 55, 227
    abundance 81
    creation 53
    habitats 2
    losses 53
    restoration 23, 24
    sediments 25
whelks 585
white band disease 164
Widgeongrass 97
wildlife 418, 423, 429, 432, 474
    activities 69
    habitat 240
    mortalities 584
    predation 53, 71
    traffic 53
willet 14
winkles 585

woody debris 343, 346, 368
worms 9, 496
wrecked ships 584

Y

yearlings 406
yellow shore crab 11

Z

zinc 493
zonation 69, 237
zooplankton 495
Zostera 81, 112, 114
Zostera capricorni 123
Zostera japonica 92, 105
Zostera marina 81, 92, 117
Zostera mucronata 113